Subcellular Biochemistry

Volume 22
Membrane Biogenesis

SUBCELLULAR BIOCHEMISTRY

SERIES EDITOR

J. R. HARRIS, Institute of Zoology, University of Mainz, Mainz, Germany

ASSISTANT EDITORS

H. J. HILDERSON, University of Antwerp, Antwerp, Belgium
D. A. WALL, SmithKline Beecham Pharmaceuticals, King of Prussia, Pennsylvania, U.S.A.

Recent Volumes in This Series:

A Continuation Order Plan is available for this series. A continuation order will bring delivery of each new volume immediately upon publication. Volumes are billed only upon actual shipment. For further information please contact the publisher.

Subcellular Biochemistry

Volume 22
Membrane Biogenesis

Edited by

A. H. Maddy
University of Edinburgh
Edinburgh, Scotland, United Kingdom

and

J. R. Harris
Institute of Zoology
University of Mainz
Mainz, Germany

SPRINGER SCIENCE+BUSINESS MEDIA, LLC

The Library of Congress cataloged the first volume of this title as follows:

Sub-cellular biochemistry.
  London, New York, Plenum Press.
    v. illus. 23 cm. quarterly.
  Began with Sept. 1971 issue. Cf. New serial titles.
    1. Cytochemistry—Periodicals. 2. Cell organelles—Periodicals.
QH611.S84 574.8′76   73-643479

ISBN 978-0-306-44554-5   ISBN 978-1-4615-2401-4 (eBook)
DOI 10.1007/978-1-4615-2401-4

This series is a continuation of the journal *Sub-Cellular Biochemistry*,
Volumes 1 to 4 of which were published quarterly from 1972 to 1975

©1994 Springer Science+Business Media New York
Originally published by Plenum Press, New York in 1994

Contributors

Hélène Bénédetti   Laboratoire d'Ingéniérie et de Dynamique des Systèmes Membranaires, F-13402 Marseille Cedex 20, France

L. S. Cox   CRC Cell Transformation Research Group, Department of Biochemistry, University of Dundee, Dundee DD1 4HN, Scotland

Elizabeth M. Ellis   Biomedical Research Centre, University of Dundee, Ninewells Hospital and Medical School, Dundee DD1 9SY, Scotland

Hudson H. Freeze   Cancer Center, University of California at San Diego, and La Jolla Cancer Research Foundation, La Jolla California 92093

Minoru Fukuda   La Jolla Cancer Research Foundation, Cancer Research Center, La Jolla, California 92037

Stephen D. Fuller   Biological Structures and Biocomputing Programme, European Molecular Biology Laboratory, D6900, Heidelberg, Germany

Vincent Géli   Laboratoire d'Ingéniérie et de Dynamique des Systèmes Membranaires, F-13402 Marseille Cedex 20, France

Rob J. M. Hendriks   Biological Structures and Biocomputing Programme, European Molecular Biology Laboratory, D6900, Heidelberg, Germany

C. J. Hutchison   Department of Biological Sciences, University of Dundee, Dundee DD1 4HN, Scotland

Juan MacFarlane   Institut für Physikalische Biochemie der Universität München, D-80366 München, Germany

Guy P. Mannaerts   Afdeling Farmacologie, Katholieke Universiteit Leuven, Campus Gasthuisberg, B-3000 Leuven, Belgium

Matthias Müller   Institut für Physikalische Biochemie der Universität München, D-80366 München, Germany

Graeme A. Reid   Institute of Cell and Molecular Biology, University of Edinburgh, Edinburgh EH9 3JR, Scotland

Colin Robinson   Department of Biological Sciences, University of Warwick, Coventry CV4 7AL, England

Ajit Varki   Cancer Center, University of California at San Diego, and La Jolla Cancer Research Foundation, La Jolla, California 92093.

Paul P. Van Veldhoven   Afdeling Farmacologie, Katholieke Universiteit Leuven, Campus Gasthuisberg, B-3000 Leuven, Belgium

Gunnar von Heijne   Department of Molecular Biology, Karolinska Institute, Center for Structural Biochemistry, NOVUM, S-141 57 Huddinge, Sweden

Simone Wattiaux-De Coninck   Laboratoire de Chimie Physiologique, Facultés Universitaires Notre-Dame de la Paix, B-5000 Namur, Belgium

Robert Wattiaux   Laboratoire de Chimie Physiologique, Facultés Universitaires Notre-Dame de la Paix, B-5000 Namur, Belgium

Preface

Scientific reviews are now of two complementary types: short, very up-to-date articles, as are found in the *Trends* series, and the more traditional longer reviews, which are more comprehensive but take longer to publish. The *Subcellular Biochemistry* series belongs to the latter category where a number of reviews on a broad topic are collected together in the one volume. It has been the aim of this volume to summarize the present state of knowledge of membrane assembly. It is appreciated that some relevant topics have not been included, and an editor's selection is restricted by the many calls on potential authors who are unable to meet all requests made to them. The absence of a discussion of the roles of lipids is, however, a reflection of the fact that a recent volume in this series, edited by Dr. H. J. Hilderson, has been devoted to this subject (Vol. 16, *Intracellular Transfer of Lipid Molecules*), and readers are recommended to this source.

A sound knowledge of membrane structure was a prerequisite for any investigation of membrane assembly. The elucidation of structure immediately presented the problem of how proteins synthesized in the aqueous environment of the cytoplasm could be inserted into the hydrophobic environment of the lipid bilayer. This problem was apparently resolved with the discovery of co-translational insertion into the endoplasmic reticulum membrane, but soon re-presented itself when completely translated mitochondrial proteins were found in the cytoplasm. This apparent paradox was resolved with the discovery of chaperone proteins.

The second major question of membrane assembly to be addressed was how the diversity of membranes within a cell was generated and maintained, and how could potential membrane constituents be directed to their appropriate destinations. Although complete answers to these questions are not yet available, many of the protein targeting signals have been identified.

The present volume opens with three chapters on subjects of general relevance to most membranes: the nature of signal sequences (von Heijne), the mechanisms for the insertion of proteins into membranes (Géli and Bénédetti), and the glycosylation of membranes (Varki and Freeze)—a timely account of a subject that is likely to explode in the near future with the recent technical advances in glycobiology. There follows a series of chapters directed toward membranes of different organelles of eukaryotes, the endoplasmic/Golgi system, which plays such a central role in membrane dynamics (Hendriks and Fuller), mitochondria (Ellis and Reid), chloroplasts (Robinson), lysosomes (Fukuda), and peroxisomes (Van Veldhoven and Mannaerts). The chapter on peroxisomes includes a more general review of this somewhat neglected organelle. Finally, Cox and Hutchison provide a comprehensive review of the nuclear envelope, where assembly takes on a new dimension with disassembly and reassembly at each mitosis. Prokaryotes are considered in the chapter on bacterial membranes by Müller and MacFarlane. The volume closes with a chapter on ischemic effects on the plasma membrane (Wattiaux-De Coninck and Wattiaux) where the authors discuss the ability of membranes, once assembled, to withstand the stresses of continued existence.

A. H. Maddy
J. R. Harris

Edinburgh, Scotland
Mainz, Germany

Contents

Chapter 2
Insertion of Proteins into Membranes: A Survey
Vincent Géli and Hélène Bénédetti

Chapter 3
**The Major Glycosylation Pathways of Mammalian Membranes:
A Summary**
Ajit Varki and Hudson H. Freeze

Chapter 4

Compartments of the Early Secretory Pathway

Rob J. M. Hendriks and Stephen D. Fuller

Chapter 5
Assembly of Mitochondrial Membranes
Elizabeth M. Ellis and Graeme A. Reid

Chapter 6
The Assembly of Chloroplast Membranes
Colin Robinson

Chapter 7
Biogenesis of the Lysosomal Membrane
Minoru Fukuda

Chapter 10
Membrane Assembly in Bacteria
Matthias Müller and Juan MacFarlane

Chapter 11
**Ischemic Effects on the Structure and Function
of the Plasma Membrane**
Simone Wattiaux-De Coninck and Robert Wattiaux

Chapter 1

Signals for Protein Targeting into and across Membranes

Gunnar von Heijne

1. INTRODUCTION

How can a newly synthesized protein navigate its way through the maze of organelles found inside a cell to reach its intended destination? This basic question has fascinated workers in cell and molecular biology for a long time; yet, we still do not have all the final answers. Some principles are clear, however. The targeting information is encoded within the nascent protein itself in the form of distinct "signals," either in the form of stretches of amino acids or as surface patches made from discontinuous parts of the protein. These signals are recognized by receptors, which guide the protein to import machineries located in the appropriate organelle. The biochemistry of the various import pathways is covered in other chapters in this book; this chapter presents an overview of the sorting signals themselves, their sequences and structures, and possible modes of function.

2. SIGNALS IN THE SECRETORY PATHWAY

The great majority of secretory proteins produced by eukaryotic cells are made on ribosomes bound to the cytoplasmic surface of the endoplasmic reticu-

Gunnar von Heijne   Department of Molecular Biology, Karolinska Institute, Center for Structural Biochemistry, NOVUM, S-141 57 Huddinge, Sweden.
Subcellular Biochemistry, Volume 22: Membrane Biogenesis, edited by A. H. Maddy and J. R. Harris. Plenum Press, New York, 1994.

lum (ER) membrane (see Muesch *et al.*, 1990, for a review of an alternative pathway). The growing polypeptide is co-translationally translocated into the lumen of the ER, where folding of the chain is catalyzed by ER-resident proteins such as BiP, prolyl isomerase, and protein disulfide isomerase (Gething and Sambrook, 1992). Once properly folded (Braakman *et al.*, 1992), the protein is transported through the Golgi and post-Golgi compartments to the plasma membrane and is finally secreted from the cell.

To control the passage through the secretory pathway, a number of signals are used. The default is secretion from the cell; in this case, an N-terminal signal peptide is required to target the protein to the ER, and no additional signals are needed. Proteins designed to be retained in one or other of the compartments along the pathway are provided with specific retention signals in addition to the signal peptide. Finally, some membrane proteins cycle continuously between the plasma membrane and endosomal compartments, and thus need endocytosis signals.

2.1. Signal Peptides

Signal peptides are found both in prokaryotic and eukaryotic proteins and seem to have been largely conserved throughout evolution. A functional three-partite design has been defined both through statistical sequence analyses and molecular genetic studies: a positively charged N-terminal end (n-region), a central, hydrophobic stretch (h-region), and a more polar C-terminal segment (c-region) (Gierasch, 1989; von Heijne, 1990). Targeting information is contained in the n- and h-regions, whereas the c-region serves as the recognition site for a signal peptidase that removes the signal peptide once membrane translocation has been initiated.

2.1.1. The n-Region

In bacterial signal peptides, at least one positively charged lysine or arginine is always found in the n-region. Not surprisingly, when these charged residues are removed or replaced with negatively charged ones, the mutant signal peptide functions less efficiently (Lehnhardt *et al.*, 1988; Puziss *et al.*, 1992). It is not clear exactly why this is so, though *in vitro* it has been shown that positively charged residues in the n-region enhance the affinity of the SecA protein for the signal peptide (Akita *et al.*, 1990). Certain mutations in the *secY* gene can suppress the secretion defect caused by an acidic n-region (Puziss, *et al.*, 1992).

Eukaryotic signal peptides also tend to have a positively charged n-region, but this is not an absolute rule (von Heijne, 1984) and quite a few signal peptides with acidic n-regions are known. The role of the n-region has not been extensively studied in eukaryotic systems, although n-region mutations that affect secretion have been described (Green *et al.*, 1989).

2.1.2. The h-Region

To function well, the h-region must be of a certain minimal length and hydrophobicity. Recent studies on a range of h-region mutants (Hoyt and Gierasch, 1991) suggest that the average hydrophobicity in this region must be ≥ 2.4 on the Kyte-Doolittle scale (Kyte and Doolittle, 1982). Homopolymeric h-regions composed only of leucine, isoleucine, or phenylalanine residues function well, whereas valine, alanine, and tryptophan h-regions work less well or not at all (Chou and Kendall, 1990; Rusch and Kendall, 1992). The minimum length of a functional h-region seems to be ~ 7 residues (von Heijne, 1985; Yamamoto and Kikuchi, 1989; Chou and Kendall, 1990; Hikita and Mizushima, 1992). If, on the other hand, the h-region becomes too long the c-region may be moved out of reach of the signal peptidase enzyme, and the protein becomes permanently anchored to the membrane (Chou and Kendall, 1990). .

It appears that the h-region may interact both with proteins and lipids at different stages during the initiation of translocation. In bacteria, the signal peptide can bind to the SecA protein (Akita *et al.*, 1990; Schatz and Beckwith, 1990; Wickner *et al.*, 1991), and in eukaryotic cells it interacts with the 54 kDa subunit of the signal recognition particle (SRP) (High and Dobberstein, 1991; Lutcke *et al.*, 1992). There is strong circumstantial evidence, mainly from biophysical studies of synthetic signal peptides, that signal peptides also interact directly with the lipid bilayer (Demel *et al.*, 1990; Hoyt and Gierasch, 1991).

2.1.3. The c-Region

The c-region is not necessary for targeting to the secretory pathway; rather, its function is to provide a cleavage site for the signal peptidase enzyme. The cleavage site is defined by the "$(-3, -1)$-rule" (Perlman and Halvorson, 1983; von Heijne, 1983; von Heijne 1986d), which states that only small, uncharged residues are allowed in positions -1 and -3 relative to the cleavage site. The critical importance of these two positions has now been amply documented (Folz *et al.*, 1988; Fikes *et al.*, 1990; Shen *et al.*, 1991). A helix-breaking residue often provides a point of demarcation between the h- and c-regions (von Heijne, 1985) and appears, at least in some signal peptides, to help present the cleavage site in the correct conformation (Yamamoto *et al.*, 1989; Shen *et al.*, 1991).

Prolines are nearly completely absent from position $+1$, and it has recently been found not only that Pro_{+1} prevents signal peptide cleavage in *E. coli*, but also that expression of proteins with this mutation competitively inhibits the signal peptidase I (Lep) enzyme (Barkocy-Gallagher and Bassford, 1992;

Nilsson and von Heijne, 1992). Presumably, the Pro_{+1} does not influence the binding of the signal peptide to the signal peptidase but only prevents cleavage, leaving the signal peptide stuck on the enzyme.

Signal peptidases do not appear to belong to any of the standard groups of proteases (the serine, cysteine, aspartic acid, and metallopeptidase), and may constitute a new protease family with a possible mechanistic similarity to the β-lactamases (Black et al., 1992; Sung and Dalbey, 1992).

2.1.4. The "Charge-Block Effect"

Although not part of the signal peptide proper, a region encompassing the first 10–20 residues of the mature protein is also critical for the initiation of membrane translocation, at least in E. coli (Andersson and von Heijne, 1991). This region normally contains few positively charged amino acids (von Heijne, 1986c), and the introduction of only one or two extra lysines or arginines can dramatically affect secretion (Li et al., 1988; Yamane and Mizushima, 1988; Boyd and Beckwith, 1989; Laws and Dalbey, 1989; Summers et al., 1989; Zhu and Dalbey, 1989; Boyd and Beckwith, 1990; MacIntyre et al., 1990). A similar blocking effect has also been demonstrated in eukaryotic secretory proteins, although only with much higher numbers of charged residues (Kohara et al., 1991; Johansson et al., 1992). One implication of these observations is that many eukaryotic proteins may be difficult to export from bacteria simply because the N-terminal region of the mature chain carries too many positively charged amino acids.

2.2. ER-Retention Signals

A number of proteins are specifically retained in the ER through a retrieval mechanism where molecules that escape toward the Golgi are recognized by a recycling receptor and delivered back to the ER (see Chapter 4). Two distinct retention signals have been discovered: a C-terminal tetrapeptide sequence, Lys-Asp-Glu-Leu (KDEL, or HDEL in yeast) normally found on soluble proteins in the ER lumen (Pelham, 1990), and a cytoplasmically exposed C-terminal tail with two or more apparently critical lysines and possibly some additional important characteristics found on proteins that span the ER membrane (Gabathuler and Kvist, 1990; Jackson et al., 1990). Recently, a yeast membrane protein with a lumenally exposed C-terminus was found to end with a typical HDEL-signal (Sweet and Pelham, 1992), suggesting that membrane proteins with the appropriate topology can also be retrieved through the KDEL (or HDEL) receptor system.

2.3. Golgi-Retention Signals

Golgi-retention signals have been localized to transmembrane domains, including short cytoplasmic and lumenally exposed flanking regions, in at least three different proteins (Hurtley, 1992). The coronavirus E1 protein has three transmembrane segments, but only the N-terminal segment can confer Golgi retention onto non-Golgi fusion partners (Swift and Machamer, 1991). Some point-mutations in this transmembrane segment abolish retention. Retention signals in two single-spanning Golgi proteins, the α-2,6-sialyltransferase (Munro, 1991; Colley et al., 1992) and β-14-galactosyltransferase (Nilsson et al., 1991; Teasdale et al., 1992), have also been found in the transmembrane and polar flanking segments. Interestingly, in the case of α-2,6-sialyltransferase, it has been shown that the hydrophobic segment can be replaced by 17 leucines with no adverse effects on retention; when the hydrophobic segment is lengthened to 23 leucines, retention is no longer observed (Munro, 1991), suggesting that there are cytoplasmic and lumenal determinants that must be kept at a fixed distance apart for efficient retention.

2.4. Lysosomal-Targeting Signals

Soluble lysosomal proteins are diverted from the main secretory pathway in the trans-Golgi network. This is accomplished by the recognition of mannose-6-phosphate residues on the protein. The critical step in lysosomal protein biogenesis is the action of the N-acetylglucosamine-1-phosphotransferase enzyme that is responsible for the first step in the conversion of mannose residues to mannose-6-phosphate (Kornfeld and Mellman, 1989). This enzyme recognizes a "patch" on the surface of the protein rather than a linear string of amino acids (Baranski et al., 1990; Baranski et al., 1991), although it is not known precisely what features of the patch are critical (see Chapter 7).

2.5. Endocytosis Signals

Many plasma membrane proteins are endocytosed in response to a signal located in their cytoplasmic domains. This signal has been shown to contain a critical tyrosine residue (Peters et al., 1990) located in a short segment that by NMR (nuclear magnetic resonance) analysis (Bansal and Gierasch, 1991; Eberle et al., 1991) appears to form a reverse turn. Some lysosomal membrane proteins are first directed to the plasma membrane, then endocytosed in response to a tyrosine signal, and finally delivered to the lysosome (Peters et al., 1990), although other pathways for lysosomal targeting of membrane proteins also exist (Vega et al., 1991).

3. SIGNALS FOR MITOCHONDRIAL PROTEIN IMPORT

The signals and import machinery responsible for the import of proteins into mitochondria have been extensively studied (Glick and Schatz, 1991; Pfanner et *al.*, 1992; see Chapter 5). Four different intramitochondrial locations can be defined: the outer and inner membranes, the intermembrane space, and the matrix; a matching set of targeting signals has evolved to ensure proper routing of the imported protein.

3.1. Matrix-Targeting Peptides

The most well-characterized mitochondrial import signal is the matrix-targeting variety. As in the case of secretory signal peptides, there is no conserved "consensus" sequence; rather, the basic feature is the presence of a positively charged amphiphilic α-helix (Roise *et al.*, 1986; von Heijne, 1986b). The importance of the amphiphilic helix has been demonstrated both by extensive mutational analysis (Bedwell *et al.*, 1989; Lemire *et al.*, 1989) and directly through structure determination by NMR (Endo *et al.*, 1989). It is thought that translocation of the positively charged presequence is driven by the electrical component of the proton motive force across the inner membrane (Martin *et al.*, 1991).

After import, the matrix-targeting signal is cleaved either once or twice by matrix proteases. The first cleavage is catalyzed by a two-subunit enzyme (Pollock *et al.*, 1988; Yang *et al.*, 1988). Although not strictly conserved, the cleavage-site is usually defined by an arginine in position -2, or by an Arg-X-Tyr motif (with cleavage taking place immediately after the tyrosine) (Gavel and von Heijne, 1990a). About a third of all imported proteins are cleaved a second time by a distinct protease (Kalousek *et al.*, 1988), which removes an additional eight (or, in some cases, nine) N-terminal amino acids (Hendrick *et al.*, 1989; Gavel and von Heijne, 1990a). The octa-peptide can only be removed if it has a hydrophobic N-terminal amino acid (typically Phe), and if some other less well understood criteria are met by the sequence downstream of the octa-peptide (Isaya *et al.*, 1992).

3.2. Outer Membrane-Targeting Signals

The targeting of two proteins of the outer mitochondrial membrane, porin and Mas70p, has been studied rather extensively. Porin does not have a cleavable targeting signal, and does not require a potential across the inner membrane. It does, however, require ATP for import, and utilizes at least a part of the same machinery as matrix proteins (Kleene *et al.*, 1987; Pfanner *et al.*, 1987; Hwang and Schatz, 1989). It is not clear what portion or portions of the chain contain the targeting information.

Mas70p, on the other hand, has a clearly demarcated N-terminal targeting signal, although it is not cleaved upon import. The first 12 residues have a net positive charge, and can, by themselves, mediate import into the matrix (Hurt *et al.*, 1985; Nakai *et al.*, 1989). The following $\sim$ 25 residues are quite apolar, and apparently provide a stop-transfer function, anchoring the protein to the outer membrane with the large C-terminal domain projecting outside the organelle (Reizman *et al.*, 1983; Hase *et al.*, 1984).

3.3. Intermembrane Space-Targeting Signals

There are at least three different pathways for import into the intermembrane space, exemplified by cytochrome b_2, cytochrome c heme lyase, and cytochrome c. Cytochrome b_2 has a bipartite targeting signal, with a typical N-terminal matrix-targeting part immediately followed by a second signal with many of the characteristics of a signal peptide for protein secretion (von Heijne *et al.*, 1989). The import pathway dictated by this combination of signals is first into the matrix, where the matrix-targeting part is removed, and then re-export across the inner membrane (Koll *et al.*, 1992). Because the latter step is similar to bacterial protein export (and thus presumably an evolutionary remnant from the pre-endosymbiont days of the mitochondrion), this pathway has been termed "conservative sorting" (Hartl and Neupert, 1990).

Cytochrome c heme lyase lacks a cleavable presequence, but is nevertheless targeted to the normal import machinery in the outer membrane. However, it never reaches the matrix compartment, but rather appears to deviate from the standard path after passage through the outer membrane, and it ends up in the intermembrane space (Lill *et al.*, 1992). It is possible that the lack of a typical matrix-targeting signal prevents its import into the matrix; it is not known what other feature or features of the nascent chain guide its targeting.

Cytochrome c, finally, is imported directly through the outer membrane into the intermembrane space. Apo-cytochrome c binds specifically to outer membrane lipids and appears to partly penetrate the membrane spontaneously (Jordi *et al.*, 1989). Full import is catalyzed by cytochrome c, and thus converts it to the soluble holo-enzyme (Nargang *et al.*, 1988; Nicholson and Neupert, 1989). In this case, no specific signal seems to be involved, and both targeting and import depend on global properties of the apo-cytochrome c molecule.

4. SIGNALS FOR CHLOROPLAST PROTEIN IMPORT

In terms of intra-organellar targeting, protein import into chloroplasts is even more complex than for mitochondria. With its two envelope membranes and the thylakoid stacks, there are at least six different compartments within this organelle (see Chapter 6).

4.1. Stromal Transit Peptides

Stromal transit peptides vary greatly in both length and sequence, yet some common characteristics can be defined (von Heijne *et al.*, 1989). Transit peptides are notably rich in hydroxylated amino acids (Ser in particular), and contain few or no acid residues. They do not appear to be designed to form well-defined secondary structures such as amphiphilic α-helices, and are in fact mostly predicted as random coils (von Heijne and Nishikawa, 1991). Model studies have shown that transit peptides can bind to lipid monolayers containing acidic phospholipids and galactolipids typically found in the chloroplast envelope (van't Hof *et al.*, 1991), although the binding is weaker than for mitochondrial targeting peptides. Competition experiments using an *in vitro* import system and synthetic peptides corresponding to different parts of a transit peptide have suggested that the central region may be involved in the initial binding to the chloroplast surface, and that the N-terminal and C-terminal ends rather affect a later step in the import pathway (Perry *et al.*, 1991).

No highly conserved cleavage-site motif for the stromal processing peptidase has been found, although ∼ 1/3 of all known transit peptides end with a loosely conserved VX(A/C)↓A pattern (Gavel and von Heijne, 1990b).

4.2. Thylakoid Transfer Domains

Similar to the mitochondrial intermembrane space proteins, proteins destined for the lumen of the thylakoids have a bipartite targeting signal composed of an N-terminal stromal transit peptide followed by a thylakoid transfer domain with all the characteristics of a secretory bacterial signal peptide (von Heijne *et al.*, 1989; Bassham *et al.*, 1991). Indeed, thylakoid transfer domains have been shown to function as signal peptides in *E. coli* (Meadows and Robinson, 1991). Upon import into the thylakoid, the transfer domain is cleaved by a peptidase with a substrate specificity very similar to the *E. coli* signal peptidase I (Halpin *et al.*, 1989; Anderson and Gray, 1991; Shackleton and Robinson, 1991). Whether the stromal transit peptide has to be removed prior to thylakoid import is somewhat controversial (Bauerle *et al.*, 1991; Bauerle and Keegstra, 1991; Mould *et al.*, 1991).

4.3. Envelope-Targeting Signals

Two proteins of the outer envelope membrane have been cloned, and their import has been partly characterized (Salomon *et al.*, 1990; Li *et al.*, 1991). Neither has a cleavable presequence, nor do they require ATP for proper membrane integration. The chloroplasts can be treated with thermolysine with no effect on the import, suggesting that insertion is "spontaneous" and independent of proteinaceous factors in the outer envelope membrane.

Two inner membrane proteins, the phosphate translocator and a 37-kDa protein of unknown function, have also been cloned (Flügge *et al.*, 1989; Dreseswerringloer *et al.*, 1991; Flügge *et al.*, 1991). These proteins appear to use the normal import pathway, although their presequences do not look like typical stroma-targeting transit peptides.

5. SIGNALS FOR NUCLEAR PROTEIN IMPORT

Nuclear import is mechanistically very different from the other import processes discussed so far (Silver, 1991; see Chapter 9). The nuclear pore complex allows passage of small macromolecules by passive diffusion, and only larger proteins require active, signal-sequence-dependent transport across the nuclear envelope. Nuclear-targeting sequences are integral parts of the fully folded, mature protein, can act cooperatively, and presumably need to be exposed on the surface of the molecule. They may bind to cytosolic receptors that in turn mediate the interactions with the nuclear pore (Adam and Gerace, 1991), and then piggy-back into the nucleus.

Nuclear-targeting sequences have long been known to be composed of clusters of basic amino acids (Lys and Arg), and it has recently been proposed that a common consensus motif is defined by a dibasic pair, a spacer segment of any ten amino acids, and a second basic cluster with at least three out of five basic residues (Dingwall and Laskey, 1991; Robbins *et al.*, 1991). About half of all known nuclear proteins contain this motif; in contrast, it is found in only 4% of all nonnuclear proteins.

6. SIGNALS FOR PEROXISOMAL PROTEIN IMPORT

The mechanism of protein import into peroxisomes is largely unknown, although at least three import-deficient mutants have been found that map to peroxisomal membrane proteins (Erdmann *et al.*, 1991; Hohfeld *et al.*, 1991; Tsukamoto *et al.*, 1991). Some genetic evidence also shows that at least two independent import pathways exist (Walton *et al.*, 1992); this conclusion is strengthened by the recent identification of two very different kinds of targeting signals: one C-terminal and one N-terminal (see Chapter 8).

6.1. The SKL Signal

A C-terminal Ser-Lys-Leu (SKL) motif is found on many peroxisomal proteins and has been shown to act as a targeting signal when transplanted onto reporter proteins (Gould *et al.*, 1989; Miyazawa *et al.*, 1989; Gould *et al.*, 1990a; Gould *et al.*, 1990b). Some conservative substitutions are allowed (Gould *et al.*, 1989), although the C-terminal Leu seems to be absolutely required. Furthermore, the signal only works from a C-terminal and not from internal positions.

6.2. Amino-Terminal Import Signals

Many peroxisomal proteins do not have SKL signals, and hence must be targeted in some other fashion. An N-terminal, cleavable-targeting signal was recently identified in rat 3-ketoacyl-CoA thiolase (Osumi *et al.*, 1991; Swinkels *et al.*, 1991). Deletion analysis suggested that the necessary information is carried by the 11-residue stretch MHRLQVVLGHL. Related N-terminal extensions are present in peroxisomal thiolases from other organisms, as well as in a plant glyoxysomal malate dehydrogenase. It is perhaps noteworthy that two tripeptides related to the SKL-motif (HRL and GHL) are found in the thiolase presequence, although the SKL pathway is probably not used by these proteins (Swinkels *et al.*, 1991).

7. SIGNALS FOR MEMBRANE PROTEIN ASSEMBLY

Just like soluble proteins, integral membrane proteins must be targeted to the appropriate organelle and suborganellar compartment, but in addition they must contain signals that guide their integration into the target membrane. These signals define the topology of the molecule, i.e., the number of transmembrane segments and the overall orientation relative to the membrane.

7.1. Topological Signals

Most membrane proteins depend on the cell's secretory machinery for their insertion, and hence use signal peptides for targeting and for triggering the initial translocation event. This is true in bacteria, where both inner and outer membrane proteins are made with (cleavable or uncleavable) N-terminal signal peptides, as well as in eukaryotic cells. Proteins of the plasma membrane and the membranes of the exocytic pathway are inserted into the ER membrane and further transported to their intended compartment. Mitochondrially encoded inner membrane proteins also seem to use signal peptide-like targeting peptides, as do thylakoid membrane proteins (see the relevant sections above).

The transmembrane topology is primarily determined by two features of the amino acid sequence: the number and positions of the stretches of apolar amino acids that form the transmembrane segments proper, and the distribution of positively charged amino acids (Lys and Arg) in the loops connecting the transmembrane segments (von Heijne and Manoil, 1990).

Typical transmembrane segments are 15–20 residues long and highly enriched in apolar amino acids. A stretch of as little as 8–10 consecutive leucines is sufficient to impart a transmembrane topology, whereas a much longer segment is needed if alanines are used instead (Kuroiwa *et al.*, 1991). Thus, overall hydrophobicity seems to be the main factor that distinguishes transmembrane from nonmembrane segments.

The orientation of the transmembrane segments is largely controlled by flanking positively charged amino acids. This was first suggested by a statistical analysis of bacterial inner membrane proteins, where it was shown that Arg and Lys residues were many times more abundant in cytoplasmically exposed parts compared to periplasmic parts (von Heijne, 1986a)—the "positive inside-rule." Similar observations have since been made for eukaryotic plasma membrane proteins (von Heijne and Gavel, 1988; Hartmann et al., 1989), for thylakoid membrane proteins (Gavel et al., 1991), and for mitochondrially encoded proteins of the inner mitochondrial membrane (Gavel and von Heijne, 1992).

The positive inside-rule has been tested experimentally in a number of systems. In E. coli, positively charged amino acids have a decisive influence on the orientation of inner membrane proteins (Boyd and Beckwith, 1989; von Heijne, 1989; Boyd and Beckwith, 1990; Nilsson and von Heijne, 1990; McGovern et al., 1991; Andersson et al., 1992), and they clearly play a role also for eukaryotic membrane proteins, although the results obtained so far are less clear-cut than for E. coli (Haeuptle et al., 1989; Szczesna-Skorupa and Kemper, 1989; Sato et al., 1990; Parks and Lamb, 1991; Andrews et al., 1992).

7.2. Degradation Signals

The ER is a site of "quality control" that only allows the further transport along the exocytic pathway of proteins that have folded and oligomerized properly (Desilva et al., 1990). Recently, it has been found that isolated subunits of oligomeric membrane proteins may have "degradation signals," possibly in the form of a charged residue, in their transmembrane domains; these are signals that become masked upon oligomerization (Bonifacino et al., 1990a,b; Klausner and Sitia, 1990; Bonifacino et al., 1991). The way in which these signals are recognized and the proteolytic system responsible for degradation have not yet been characterized.

8. CONCLUSIONS

Protein trafficking in the prokaryotic and eukaryotic cell depends on an intricate system of signals, receptors, and membrane-translocation machineries. Some of the signals are well characterized and can often be rather easily recognized in the primary amino acid sequence.

A measure of our understanding of these signals can be obtained by trying to formulate rules or patterns that discriminate among different signals, and then let a computer "direct" proteins to different compartments. This approach was recently tried on a sample of bacterial proteins, with a "sorting" algorithm based on a number of published (and unpublished) methods for detecting various targeting signals (Nakai and Kanehisa, 1991). The result was rather encouraging: About 80% of the proteins in each class (cytoplasm, inner membrane, periplasm, and outer membrane) were assigned to the correct compartment. A similar study focusing on eukaryotic protein sorting with a much larger number of possible compartments (nucleus, cytoplasm, ER membrane, ER lumen, Golgi, lysosome membrane, lysosome lumen, plasma membrane, peroxisome, mitochondrial matrix, inner membrane, intermembrane space, and outer membrane, and the extracellular medium) reached almost 60% correct predictions (Nakai, 1991).

Statistical and experimental studies of integral membrane proteins have also led to improvements in our ability to predict the transmembrane structure from the amino acid sequence; a recent method employing hydrophobicity analysis coupled with a screening procedure based on the positive inside-rule, in fact, managed to identify correctly the transmembrane segments and predict the correct orientation for 23 out of 24 bacterial inner membrane proteins of known sequence and topology (von Heijne, 1992).

In summary, then, we know quite a lot about the signals of protein sorting, and we are beginning to unravel the mechanisms of signal recognition and membrane translocation. Beyond their biological significance, these studies have also made it clear that we have a long way to go before our own mail services and overnight courier deliveries work as efficiently as Mother Nature's own inventions.

9. REFERENCES

Adam, S. A., and Gerace, L., 1991, Cytosolic proteins that specifically bind nuclear location signals are receptors for nuclear import, *Cell* **66**:837–847.

Akita, M., Sasaki, S., Matsuyama, S., Mizushima, S., 1990, SecA interacts with secretory proteins by recognizing the positive charge at the amino terminus of the signal peptide in *Escherichia coli*, *J. Biol. Chem.* **265**:8164–8169.

Anderson, C. M., and Gray, J., 1991, Cleavage of the precursor of pea chloroplast cytochrome-f by leader peptidase from *Escherichia coli*, *FEBS Lett.* **280**:383–386.

Andersson, H., Bakker, E., and von Heijne, G., 1992, Different positively charged amino acids have similar effects on the topology of a polytopic transmembrane protein in *Escherichia coli*, *J. Biol. Chem.* **267**:1491–1495.

Andersson, H., and von Heijne, G., 1991, A 30-residue-long "export initiation domain" adjacent to the signal sequence is critical for protein translocation across the inner membrane of *Escherichia coli*, *Proc. Natl. Acad. Sci. USA* **88**:9751–9754.

Andrews, D. W., Young, J. C., Mirels, L. F., and Czarnota, G. J., 1992, The role of the N-region in signal sequence and signal-anchor function, *J. Biol. Chem.* **267**:7761–7769.

Bansal, A., and Gierasch, L. M., 1991, The NPXY internalization signal of the LDL receptor adopts a reverse-turn conformation, *Cell* **67**:1195–1201.

Baranski, T. J., Faust, P. L., and Kornfeld, S., 1990, Generation of a lysosomal enzyme targeting signal in the secretory protein pepsinogen, *Cell* **63**:281–291.

Baranski, T. J., Koelsch, G., Hartsuck, J. A., and Kornfeld, S., 1991, Mapping and molecular modeling of a recognition domain for lysosomal enzyme targeting, *J. Biol. Chem.* **266**:23365–23372.

Barkocy-Gallagher, G. A., and Bassford, P. J., 1992, Synthesis of precursor maltose-binding protein with proline in the +1 position of the cleavage site interferes with the activity of *Escherichia coli* signal peptidase-I *in vivo*, *J. Biol. Chem.* **267**:1231–1238.

Bassham, D. C., Bartling, D., Mould, R. M., Dunbar, B., Weisbeek, P., Herrmann, R. G., and Robinson, C., 1991, Transport of proteins into chloroplasts—Delineation of envelope transit and thylakoid transfer signals within the pre-sequences of three imported thylakoid lumen proteins, *J. Biol. Chem.* **266**:23606–23610.

Bauerle, C., Dorl, J., and Keegstra, K., 1991, Kinetic analysis of the transport of thylakoid lumenal proteins in experiments using intact chloroplasts, *J. Biol. Chem.* **266**:5884–5890.

Bauerle, C., and Keegstra, K., 1991, Full-length plastocyanin precursor is translocated across isolated thylakoid membranes, *J. Biol. Chem.* **266**:5876–5883.

Bedwell, D. M., Strobel, S. A., Yun, K., Jongeward, G. D., and Emr, S. D., 1989, Sequence and structural requirements of a mitochondrial protein import signal defined by saturation cassette mutagenesis, *Mol. Cell. Biol.* **9**:1014–1025.

Black, M. T., Munn, J. G. R., and Allsop, A. E., 1992, On the catalytic mechanism of prokaryotic leader peptidase-I, *Biochem. J.* **282**:539–543.

Bonifacino, J. S., Cosson, P., and Klausner, R. D., 1990a, Colocalized transmembrane determinants for ER degradation and subunit assembly explain the intracellular fate of TCR chains, *Cell* **63**:503–513.

Bonifacino, J. S., Cosson, P., Shah, N., and Klausner, R. D., 1991, Role of potentially charged transmembrane residues in targeting proteins for retention and degradation within the endoplasmic reticulum, *EMBO J.* **10**:2783–2793.

Bonifacino, J. S., Suzuki, C. K., and Klausner, R. D., 1990b, A peptide sequence confers retention and rapid degradation in the endoplasmic reticulum, *Science* **247**:79–82.

Boyd, D., and Beckwith, J., 1989, Positively charged amino acid residues can act as topogenic determinants in membrane proteins, *Proc. Natl. Acad. Sci. USA* **86**:9446–9450.

Boyd, D., and Beckwith, J., 1990, The role of charged amino acids in the localization of secreted and membrane proteins, *Cell* **62**:1031–1033.

Braakman, I., Helenius, J., and Helenius, A., 1992, Role of ATP and disulphide bonds during protein folding in the endoplasmic reticulum, *Nature* **356**:260–262.

Chou, M. M., and Kendall, D. A., 1990, Polymeric sequences reveal a functional interrelationship between hydrophobicity and length of signal peptides, *J. Biol. Chem.* **265**:2873–2880.

Colley, K. J., Lee, E. U., and Paulson, J. C., 1992, The signal anchor and stem regions of the β-galactoside α-2,6-sialyltransferase may each act to localize the enzyme to the Golgi apparatus, *J. Biol. Chem.* **267**:7784–7793.

Demel, R. A., Goormaghtigh, E., and deKruijff, B., 1990, Lipid and peptide specificities in signal peptide lipid interactions in model membranes, *Biochem. Biophys. Acta* **1027**:155–162.

Desilva, A. M., Balch, W. E., and Helenius, A., 1990, Quality control in the endoplasmic reticulum: Folding and misfolding of vesicular stomatitis virus G-protein in cells and *in vitro*, *J. Cell Biol.* **111**:857–866.

Dingwall, C., and Laskey, R. A., 1991, Nuclear targeting sequences: A consensus, *Trends Biochem. Sci.* **16**:478–481.

Dreseswerringloer, U., Fischer, K., Wachter, E., Link, T. A., and Flügge, U. I., 1991, cDNA sequence and deduced amino acid sequence of the precursor of the 37-kDa inner envelope membrane polypeptide from spinach chloroplasts: Its transit peptide contains an amphiphilic α-helix as the only detectable structural element, *Eur. J. Biochem.* **195**:361–368.

Eberle, W., Sander, C., Klaus, W., Schmidt, B., von Figura, K., and Peters, C., 1991, The essential tyrosine of the internalization signal in lysosomal acid phosphatase is part of a β-turn, *Cell* **67**:1203–1209.

Endo, T., Shimada, I., Roise, D., and Inagaki, F., 1989, N-terminal half of a mitochondrial presequence peptide takes a helical conformation when bound to dodecylphosphocholine micelles: A proton nuclear magnetic resonance study, *J. Biochem.* **106**:396–400.

Erdmann, R., Wiebel, F. F., Flessau, A., Rytka, J., Beyer, A., Frohlich, K. U., and Kunau, W. H., 1991, PAS1, a yeast gene required for peroxisome biogenesis, encodes a member of a novel family of putative ATPases, *Cell* **64**:499–510.

Fikes, J. D., Barkocy-Gallagher, G. A., Klapper, D. G., and Bassford, P. J., 1990, Maturation of *Escherichia coli* maltose-binding protein by signal peptidase-I *in vivo:* Sequence requirements for efficient processing and demonstration of an alternate cleavage site, *J. Biol. Chem.* **265**:3417–3423.

Flügge, U. I., Fischer, K., Gross, A., Sebald, W., Lottspeich, F., and Eckerskorn, C., 1989, The triose phosphate-3-phosphoglycerate-phosphate translocator from spinach chloroplasts: Nucleotide sequence of a full-length cDNA clone and import of the *in vitro* synthesized precursor protein into chloroplasts, *EMBO J.* **8**:39–46.

Flügge, U. I., Weber, A., Fischer, K., Lottspeich, F., Eckerskorn, C., Waegemann, K., and Soll, J., 1991, The major chloroplast envelope polypeptide is the phosphate translocator and not the protein import receptor, *Nature* **353**:364–367.

Folz, R. J., Nothwehr, S. F., and Gordon, J. I., 1988, Substrate specificity of eukaryotic signal peptidase, *J. Biol. Chem.* **263**:2070–2078.

Gabathuler, R., and Kvist, S., 1990, The endoplasmic reticulum retention signal of the E3/19K protein of adenovirus type-2 consists of 3 separate amino acid segments at the carboxy terminus, *J. Cell Biol.* **111**:1803–1810.

Gavel, Y., Steppuhn, J., Herrmann, R., and von Heijne, G., 1991, The "positive-inside" rule applies to thylakoid membrane proteins, *FEBS Lett.* **282**:41–46.

Gavel, Y., and von Heijne, G., 1990a, Cleavage-site motifs in mitochondrial targeting peptides, *Protein Eng.* **4**:33–37.

Gavel, Y., and von Heijne, G., 1990b, A conserved cleavage-site motif in chloroplast transit peptides, *FEBS Lett.* **261**:455–458.

Gavel, Y., and von Heijne, G., 1992, The distribution of charged amino acids in mitochondrial inner membrane proteins suggests different modes of membrane integration for nuclearly and mitochondrially encoded proteins, *Eur. J. Biochem.* **205**:1207–1215.

Gething, M. J., and Sambrook, J., 1992, Protein folding in the cell, *Nature* **355**:33–45.

Gierasch, L. M., 1989, Signal sequences, *Biochemistry* **28**:923–930.

Glick, B., and Schatz, G., 1991, Import of proteins into mitochondria, *Annu. Rev. Genet.* **25**:21–44.

Gould, S. J., Keller, G.-A., Hosken, N., Wilkinson, J., and Subramani, S., 1989, A conserved tripeptide sorts proteins to peroxisomes, *J. Cell Biol.* **108**:1657–1664.

Gould, S. J., Keller, G. A., Schneider, M., Howell, S. H., Garrard, L. J., Goodman, J. M., Distel, B., Tabak, H., and Subramani, S., 1990a, Peroxisomal protein import is conserved between yeast, plants, insects and mammals, *EMBO J.* **9**:85–90.

Gould, S. J., Krisans, S., Keller, G. A., and Subramani, S., 1990b, Antibodies directed against the peroxisomal targeting signal of firefly luciferase recognize multiple mammalian peroxisomal proteins, *J. Cell Biol.* **110**:27–34.

Green, R., Kramer, R. A., and Shields, D., 1989, Misplacement of the amino-terminal positive charge in the prepro-α-factor signal peptide disrupts membrane translocation *in vivo, J. Biol. Chem.* **264**:2963–2968.

Haeuptle, M. T., Flint, N., Gough, N. M., and Dobberstein, B., 1989, A tripartite structure of the signals that determine protein insertion into the endoplasmic reticulum membrane, *J. Cell Biol.* **108**:1227–1236.

Halpin, C. Elderfield, P. D., James, H. E., Zimmermann, R., Dunbar, B., and Robinson, C., 1989, The reaction specificities of the thylakoidal processing peptidase and *Escherichia coli* leader peptidase are identical, *EMBO J.* **8**:3917–3921.

Hartl, F. U., and Neupert, W., 1990, Protein sorting to mitochondria: Evolutionary conservations of folding and assembly, *Science* **247**:930–938.

Hartmann, E., Rapoport, T. A., and Lodish, H. F., 1989, Predicting the orientation of eukaryotic membrane proteins, *Proc. Natl. Acad. Sci. USA* **86**:5786–5790.

Hase, T., Müller, U., Riezman, H., and Schatz, G., 1984, A 70-kd protein of the yeast mitochondrial outer membrane is targeted and anchored via its extreme amino terminus, *EMBO J.* **3**:3157–3164.

Hendrick, J. P., Hodges, P. E., and Rosenberg, L. E., 1989, Survey of amino-terminal proteolytic cleavage sites in mitochondrial precursor proteins: Leader peptides cleaved by two matrix proteases share a three-amino acid motif, *Proc. Natl. Acad. Sci. USA* **86**:4056–4060.

High, S., and Dobberstein, B., 1991, The signal sequence interacts with the methionine-rich domain of the 54-kD protein of signal recognition particle, *J. Cell Biol.* **113**:229–233.

Hikita, C., and Mizushima, S., 1992, Effects of total hydrophobicity and length of the hydrophobic domain of a signal peptide on *in vitro* translocation efficiency, *J. Biol. Chem.* **267**:4882–4888.

Hohfeld, J., Veenhuis, M., and Kunau, W. H., 1991, PAS3, a *Saccharomyces* cerevisiae gene encoding a peroxisomal integral membrane protein essential for peroxisome biogenesis, *J. Cell Biol.* **114**:1167–1178.

Hoyt, D. W., and Gierasch, L. M., 1991, Hydrophobic content and lipid interactions of wild-type and mutant OmpA signal peptides correlate with their *in vivo* function, *Biochemistry* **30**:10155–10163.

Hurt, E. C., Pesold, H. B., Suda, K., Oppliger, W., and Schatz, G., 1985, The first twelve amino acids (less than half of the pre-sequence) of an imported mitochondrial protein can direct mouse cytosolic dihydrofolate reductase into the yeast mitochondrial matrix, *EMBO J.* **4**:2061–2068.

Hurtley, S. M., 1992, Golgi localization signals, *Trends Biochem. Sci.* **17**:2–3.

Hwang, S. T., and Schatz, G., 1989, Translocation of proteins across the mitochondrial inner membrane, but not into the outer membrane, requires nucleoside triphosphates in the matrix, *Proc. Natl. Acad. Sci. USA* **86**:8432–8436.

Isaya, G., Kalousek, F., and Rosenberg, L. E., 1992, Amino-terminal octapeptides function as recognition signals for the mitochondrial intermediate peptidase, *J. Biol. Chem.* **267**:7904–7910.

Jackson, M. R., Nilsson, T., and Peterson, P. A., 1990, Identification of a consensus motif for retention of transmembrane proteins in the endoplasmic reticulum, *EMBO J.* **9**:3153–3162.

Johansson, M., Nilsson, I., and von Heijne, G., 1993, Positively charged amino acids placed next to a signal sequence block protein translocation more efficiently in *Escherichia coli* than in mammalian microsomes, *Mol. Gen. Genet.*, **239**:251–256.

Jordi, W., de Kruijff, B., and Marsh, D., 1989, Specificity of the interaction of amino-terminal and carboxy-terminal fragments of the mitochondrial precursor protein apocytochrome-*c* with negatively charged phospholipids: A spin-label electron spin resonance study, *Biochemistry* **28**:8998–9005.

Kalousek, F., Hendrick, J. P., and Rosenberg, L. E., 1988, Two mitochondrial matrix proteases act sequentially in the processing of mammalian matrix enzymes, *Proc. Natl. Acad. Sci. USA* **85**:7536–7540.

Klausner, R. D., and Sitia, R., 1990, Protein degradation in the endoplasmic reticulum, *Cell* **62**:611–614.

Kleene, R., Pfanner, N., Pfaller, R., Link, T. A., Sebald, W., Neupert, W., and Tropschug, M.,

1987, Mitochondrial porin of *Neurospora crassa:* cDNA cloning, *in vitro* expression and import into mitochondria, *EMBO J.* **6**:2627–2683.

Kohara, A., Yamamoto, Y., and Kikuchi, M., 1991, Alteration of N-terminal residues of mature human lysozyme affects its secretion in yeast and translocation into canine microsomal vesicles, *J. Biol. Chem.* **266**:20363–20368.

Koll, H., Guiard, B., Rassow, J., Ostermann, J., Horwich, A. L., Neupert, W., and Hartl, F. U., 1992, Antifolding activity of hsp60 couples protein import into the mitochondrial matrix with export to the intermembrane space, *Cell* **68**:1163–1175.

Kornfeld, S., and Mellman, I., 1989, The biogenesis of lysosomes, *Annu. Rev. Cell. Biol.* **5**:483–525.

Kuroiwa, T., Sakaguchi, M., Mihara, K., and Omura, T., 1991, Systematic analysis of stop-transfer sequence for microsomal membrane, *J. Biol. Chem.* **266**:9251–9255.

Kyte, J., and Doolittle, R. F., 1982, A simple method for displaying the hydropathic character of a protein, *J. Mol. Biol.* **157**:105–132.

Laws, J. K., and Dalbey, R. E., 1989, Positive charges in the cytoplasmic domain of *Escherichia coli* leader peptidase prevent an apolar domain from functioning as a signal, *EMBO J.* **8**:2095–2099.

Lehnhardt, S., Pollitt, N. S., Goldstein, J., and Inouye, M., 1988, Modulation of the effects of mutations in the basic region of the OmpA signal peptide by the mature portion of the protein, *J. Biol. Chem.* **263**:10300–10303.

Lemire, B. D., Fankhauser, C., Baker, A., and Schatz, G., 1989, The mitochondrial targeting function of randomly generated peptide sequences correlates with predicted helical amphiphilicity, *J. Biol. Chem.* **264**:20206–20215.

Li, H. M., Moore, T., and Keegstra, K., 1991, Targeting of proteins to the outer envelope membrane uses a different pathway than transport into chloroplasts, *Plant Cell* **3**:709–717.

Li, P., Beckwith, J., and Inouye, H., 1988, Alteration of the amino terminus of the mature sequence of a periplasmic protein can severely affect protein export in *Escherichia coli, Proc. Natl. Acad. Sci. USA* **85**:7685–7689.

Lill, R., Stuart, R. A., Drygas, M. E., Nargang, F. E., and Neupert, W., 1992, Import of cytochrome-c heme lyase into Mitochondria: A novel pathway into the intermembrane space, *EMBO J.* **11**:449–456.

Lutcke, H., High, S., Romisch, K., Ashford, A. J., and Dobberstein, B., 1992, The methionine-rich domain of the 54 kDa subunit of signal recognition particle is sufficient for the interaction with signal sequences, *EMBO J.* **11**:1543–1551.

MacIntyre, S., Eschbach, M. L., and Mutschler, B., 1990, Export incompatibility of N-terminal basic residues in a mature polypeptide of *Escherichia coli* can be alleviated by optimizing the signal peptide, *Mol. Gen. Genet.* **221**:466–474.

Martin, J., Mahlke, K., and Pfanner, N., 1991, Role of an energized inner membrane in mitochondrial protein import: $\Delta\Psi$ drives the movement of presequences, *J. Biol. Chem.* **266**:18051–18057.

McGovern, K., Ehrmann, M., and Beckwith, J., 1991, Decoding signals for membrane protein assembly using alkaline phosphatase fusions, *EMBO J.* **10**:2773–2782.

Meadows, J. W., and Robinson, C., 1991, The full precursor of the 33 kDa oxygen-evolving complex protein of wheat is exported by *Escherichia coli* and processed to the mature size, *Plant Mol. Biol.* **17**:1241–1243.

Miyazawa, S., Osumi, T., Hashimoto, T., Ohno, K., Miura, S., and Fujiki, Y., 1989, Peroxisome targeting signal of rat liver acyl-coenzyme A oxidase resides at the carboxy terminus, *Mol. Cell. Biol.* **9**:83–91.

Mould, R. M., Shackleton, J. B., and Robinson, C., 1991, Transport of proteins into chloroplasts: Requirements for the efficient import of two lumenal oxygen-evolving complex proteins into isolated thylakoids, *J. Biol. Chem.* **266**:17286–17289.

Muesch, A., Hartmann, E., Rohde, K., Rubartelli, A., Sitia, R., and Rapoport, T. A., 1990, A novel pathway for secretory proteins, *Trends Biochem. Sci.* **15**:86–88.

Munro, S., 1991, Sequences within and adjacent to the transmembrane segment of α-2,6-sialyltransferase specify Golgi retention, *EMBO J.* **10**:3577–3588.

Nakai, K., 1991, Predicting various targeting signals in amino acid sequences, *Bull. Inst. Chem. Res. Kyoto Univ.* **69**:269–291.

Nakai, K., and Kanehisa, M., 1991, Expert system for predicting protein localization sites in gram-negative bacteria, *Proteins: Struct. Funct. Genet.* **11**:95–110.

Nakai, M., Hase, T., and Matsubara, H., 1989, Precise determination of the mitochondrial import signal contained in a 70 kDa protein of yeast mitochondrial outer membrane, *J. Biochem.* **105**:513–519.

Nargang, F. E., Drygas, M. E., Kwong, P. L., Nicholson, D. W., and Neupert, W., 1988, A mutant of *Neurospora crassa* deficient in cytochrome *c* heme lyase activity cannot import cytochrome *c* into mitochondria, *J. Biol. Chem.* **263**:9388–9394.

Nicholson, D. W., and Neupert, W., 1989, Import of cytochrome *c* into mitochondria: Reduction of heme, mediated by NADH and flavin nucleotides, is obligatory for its covalent linkage to apocytochrome *c*, *Proc. Natl. Acad. Sci. USA* **86**:4340–4344.

Nilsson, I., and von Heijne, G., 1992, A signal peptide with a proline next to the cleavage site inhibits leader peptidase when present in a *sec*-independent protein, *FEBS Lett.* **299**:243–246.

Nilsson, I. M., and von Heijne, G., 1990, Fine-tuning the topology of a polytopic membrane protein: Role of positively and negatively charged residues, *Cell* **62**:1135–1141.

Nilsson, T., Lucocq, J. M., Mackay, D., and Warren, G., 1991, The membrane spanning domain of β-1,4-galactosyltransferase specifies *trans* Golgi localization, *EMBO J.* **10**:3567–3575.

Osumi, T., Tsukamoto, T., Hata, S., Yokota, S., Miura, S., Fujiki, Y., Hijikata, M., Miyazawa, S., and Hashimoto, T., 1991, Amino-terminal presequence of the precursor of peroxisomal 3-ketoacyl-CoA thiolase is a cleavable signal peptide for peroxisomal targeting, *Biochem. Biophys. Res. Commun.* **181**:947–954.

Parks, G. D., and Lamb, R. A., 1991, Topology of eukaryotic type-II membrane proteins: Importance of N-terminal positively charged residues flanking the hydrophobic domain, *Cell* **64**:777–787.

Pelham, H. R. B., 1990, The retention signal for soluble proteins of the endoplasmic reticulum, *Trends Biochem. Sci.* **15**:483–486.

Perlman, D., and Halvorson, H. O., 1983, A putative signal peptidase recognition site and sequence in eukaryotic and prokaryotic signal peptides, *J. Mol. Biol.* **167**:391–409.

Perry, S. E., Buvinger, W. E., Bennett, J., and Keegstra, K., 1991, Synthetic analogues of a transit peptide inhibit binding or translocation of chloroplastic precursor proteins, *J. Biol. Chem.* **266**:11882–11889.

Peters, C., Braun, M., Weber, B., Wendland, M., Schmidt, B., Pohlmann, R., Waheed, A., and von Figura, K., 1990, Targeting of a lysosomal membrane protein: A tyrosine-containing endocytosis signal in the cytoplasmic tail of lysosomal acid phosphatase is necessary and sufficient for targeting to lysosomes, *EMBO J.* **9**:3497–3506.

Pfanner, N., Rassow, J., Vanderklei, I. J., and Neupert, W., 1992, A dynamic model of the mitochondrial protein import machinery, *Cell* **68**:999–1002.

Pfanner, N., Tropschug, M., and Neupert, W., 1987, Mitochondrial protein import: Nucleoside triphosphates are involved in conferring import-competence to precursors, *Cell* **49**:815–823.

Pollock, R. A., Hartl, F. U., Cheng, M. Y., Ostermann, J., Horwich, A., and Neupert, W., 1988, The processing peptidase of yeast mitochondria: The two co-operating components MPP and PEP are structurally related, *EMBO J.* **7**:3493–3500.

Puziss, J. W., Fikes, J. D., and Bassford, P. J., 1989, Analysis of mutational alterations in the hydrophilic segment of the maltose-binding protein signal peptide, *J. Bacteriol.* **171**:2303–2311.

Puziss, J. W., Strobel, S. M., and Bassford, P. J., 1992, Export of maltose-binding protein species with altered charge distribution surrounding the signal peptide hydrophobic core in *Escherichia coli* cells harboring prl suppressor mutations, *J. Bacteriol.* **174**:92–101.

Riezman, H., Hase, T., van Loon, A., Grivell, L. A., Suda, K., and Schatz, G., 1983, Import of proteins into mitochondria: A 70 kilodalton outer membrane protein with a large carboxy-terminal deletion is still transported to the outer membrane, *EMBO J.* **2**:2161–2168.

Robbins, J., Dilworth, S. M., Laskey, R. A., and Dingwall, C., 1991, Two interdependent basic domains in nucleoplasmin nuclear targeting sequence: Identification of a class of bipartite nuclear targeting sequence, *Cell* **64**:615–623.

Roise, D., Horvath, S. J., Tomich, J. M., Richards, J. H., and Schatz, G., 1986, A chemically synthesized pre-sequence of an imported mitochondrial protein can form an amphiphilic helix and perturb natural and artificial phospholipid bilayers, *EMBO J.* **5**:1327–1334.

Rusch, S. L., and Kendall, D. A., 1992, Signal sequences containing multiple aromatic residues, *J. Mol. Biol.* **224**:77–85.

Salomon, M., Fischer, K., Flügge, U. I., and Soll, J., 1990, Sequence analysis and protein import studies of an outer chloroplast envelope polypeptide, *Proc. Natl. Acad. Sci. USA* **87**:5778–5782.

Sato, T., Sakaguchi, M., Mihara, K., and Omura, T., 1990, The amino-terminal structures that determine topological orientation of cytochrome-P-450 in microsomal membrane, *EMBO J.* **9**:2391–2397.

Schatz, P. J., and Beckwith, J., 1990, Genetic analysis of protein export in *Escherichia coli*, *Annu. Rev. Genet.* **24**:215–248.

Shackleton, J. B., and Robinson, C., 1991, Transport of proteins into chloroplasts: The thylakoidal processing peptidase is a signal-type peptidase with stringent substrate requirements at the −3-position and −1-position, *J. Biol. Chem.* **266**:12152–12156.

Shen, L. M., Lee, J.- I., Cheng, S., Jutte, H., Kuhn, A., and Dalbey, R. E., 1991, Use of site-directed mutagenesis to define the limits of sequence variation tolerated for processing of the M13 procoat protein by the *Escherichia coli* leader peptidase, *Biochemistry* **30**:11775–11781.

Silver, P. A., 1991, How proteins enter the nucleus, *Cell* **64**:489–497.

Summers, R. G., Harris, C. R., and Knowles, J. R., 1989, A conservative amino acid substitution, arginine for lysine, abolishes export of a hybrid protein in *Escherichia coli:* Implications for the mechanism of protein secretion, *J. Biol. Chem.* **264**:20082–20088.

Sung, M., and Dalbey, R., 1992, Identification of potential active-site residues in the *Escherichia coli* leader peptidase, *J. Biol. Chem.* **267**:13154–13159.

Sweet, D. J., and Pelham, H. R. B., 1992, The *Saccharomyces cerevisiae* SEC20 gene encodes a membrane glycoprotein which is sorted by the HDEL retrieval system, *EMBO J.* **11**:423–432.

Swift, A. M., and Machamer, C. E., 1991, A Golgi retention signal in a membrane-spanning domain of coronavirus-E1 protein, *J. Cell Biol.* **115**:19–30.

Swinkels, B. W., Gould, S. J., Bodnar, A. G., Rachubinski, R. A., and Subramani, S., 1991, A novel, cleavable peroxisomal targeting signal at the amino-terminus of the rat 3-ketoacyl-CoA thiolase, *EMBO J.* **10**:3255–3262.

Szczesna-Skorupa, E., and Kemper, B., 1989, NH2-terminal substitutions of basic amino acids induce translocation across the microsomal membrane and glycosylation of rabbit cytochrome P450C2, *J. Cell Biol.* **108**:1237–1243.

Teasdale, R. D., Dagostaro, G., and Gleeson, P. A., 1992, The signal for Golgi retention of bovine-β1,4-galactosyltransferase is in the transmembrane domain, *J. Biol. Chem.* **267**:4084–4096.

Tsukamoto, T., Miura, S., and Fujiki, Y., 1991, Restoration by a 35K membrane protein of peroxisome assembly in a peroxisome-deficient mammalian cell mutant, *Nature* **350**:77–81.

van't Hof, R., Demel, R. A., Keegstra, K., and de Kruijff, B., 1991, Lipid peptide interactions between fragments of the transit peptide of ribulose-1,5-bisphosphate carboxylase oxygenase and chloroplast membrane lipids, *FEBS Lett.* **291**:350–354.

Vega, M. A., Rodriguez, F., Segui, B., Cales, C., Alcalde, J., and Sandoval, I. V., 1991, Targeting of lysosomal integral membrane protein LIMP-II: The tyrosine-lacking carboxyl cytoplasmic tail of LIMP-II is sufficient for direct targeting to lysosomes, *J. Biol. Chem.* **266**:16269–16272.

von Heijne, G., 1983, Patterns of amino acids near signal-sequence cleavage sites, *Eur. J. Biochem.* **133**:17–21.

von Heijne, G., 1984, Analysis of the distribution of charged residues in the N-terminal region of signal sequences: Implications for protein export in prokaryotic and eukaryotic cells, *EMBO J.* **3**:2315–2318.

von Heijne, G., 1985, Signal sequences: The limits of variation, *J. Mol. Biol.* **184**:99–105.

von Heijne, G., 1986a, The distribution of positively charged residues in bacterial inner membrane proteins correlates with the transmembrane topology, *EMBO J.* **5**:3021–3027.

von Heijne, G., 1986b, Mitochondrial targeting sequences may form amphiphilic helices, *EMBO J.* **5**:1335–1342.

von Heijne, G., 1986c, Net N-C charge imbalance may be important for signal sequence function in bacteria, *J. Mol. Biol.* **192**:287–290.

von Heijne, G., 1986d, A new method for predicting signal sequence cleavage sites, *Nucleic Acids Res.* **14**:4683–4690.

von Heijne, G., 1989, Control of topology and mode of assembly of a polytopic membrane protein by positively charged residues, *Nature* **341**:456–458.

von Heijne, G., 1990, The signal peptide, *J. Membr. Biol.* **115**:195–201.

von Heijne, G., 1992, Membrane protein structure prediction: Hydrophobicity analysis and the 'positive inside' rule, *J. Mol. Biol.* **225**:487–494.

von Heijne, G., and Gavel, Y., 1988, Topogenic signals in integral membrane proteins, *Eur. J. Biochem.* **174**:671–678.

von Heijne, G., and Manoil, C., 1990, Membrane proteins: From sequence to structure, *Protein Eng.* **4**:109–112.

von Heijne, G., and Nishikawa, K., 1991, Chloroplast transit peptides: The perfect random coil?, *FEBS Lett.* **278**:1–3.

von Heijne, G., Steppuhn, J., and Herrmann, R. G., 1989, Domain structure of mitochondrial and chloroplast targeting peptides, *Eur. J. Biochem.* **180**:535–545.

Walton, P. A., Gould, S. J., Feramisco, J. R., and Subramani, S., 1992, Transport of microinjected proteins into peroxisomes of mammalian cells: Inability to Zellweger cell lines to import proteins with the SKL tripeptide peroxisomal targeting signal, *Mol. Cell. Biol.* **12**:531–541.

Wickner, W., Driessen, A. J. M., and Hartl, F. U., 1991, The enzymology of protein translocation across the *Escherichia coli* plasma membrane, *Annu. Rev. Biochem.* **60**:101–124.

Yamamoto, Y., and Kikuchi, M., 1989, Synthesis, processing and degradation in yeast of precursor human lysozyme with newly designed signal sequences, *Eur. J. Biochem.* **184**:233–236.

Yamamoto, Y., Taniyama, Y., and Kikuchi, M., 1989, Important role of the proline residue in the signal sequence that directs the secretion of human lysozyme in *Saccharomyces cerevisiae*, *Biochemistry* **28**:2728–2732.

Yamane, K., and Mizushima, S., 1988, Introduction of basic amino acids residues after the signal peptide inhibits protein translocation across the cytoplasmic membrane of *Escherichia coli*, *J. Biol. Chem.* **263**:19690–19696.

Yang, M., Jensen, R. E., Yaffe, M. P., Oppliger, W., and Schatz, G., 1988, Import of proteins into yeast mitochondria: The purified matrix processing protease contains two subunits which are encoded by the nuclear MAS1 and MAS2 genes, *EMBO J.* **7**:3857–3862.

Zhu, H. Y., and Dalbey, R. E., 1989, Both a short hydrophobic domain and a carboxyl-terminal hydrophilic region are important for signal function in the *Escherichia coli* leader peptidase, *J. Biol. Chem.* **264**:11833–11838.

Chapter 2

Insertion of Proteins into Membranes
A Survey

Vincent Géli and Hélène Bénédetti

1. INTRODUCTION

Integral membrane proteins are defined as proteins that span the membrane at least once. Until now, hundreds of coding sequences have been obtained for integral membrane proteins, but by contrast only a limited amount of information about the atomic structure of detergent solubilized proteins has been reported. So far, four kinds of structures have been observed for integral membrane proteins whose structures have been determined either by X-ray crystallography or electron crystallography. The structures known with high resolution are the photosynthetic reaction centers, the porins, bacteriorhodopsin, and the light harvesting complex II. Determination of these three-dimensional (3-D) structures has provided the information upon which the extensively used prediction methods for the arrangement of membrane proteins have been based. In the absence of three-dimensional structure information, computational methods based on the analysis and comparison of amino-acid sequences have been used to predict the topology of membrane proteins. These methods give a two-dimensional picture of the arrangement of the protein in the membrane. In the meantime, new experimental

Vincent Géli and Hélène Bénédetti Laboratoire d'Ingéniérie et de Dynamique des Systèmes Membranaires, F-13402 Marseille Cedex 20, France.

Subcellular Biochemistry, Volume 22: Membrane Biogenesis, edited by A. H. Maddy and J. R. Harris. Plenum Press, New York, 1994.

procedures have been developed, increasing the possibilities to probe membrane topology, and thus the validity of the computational methods.

On the other hand, a large body of information has been available on the mechanism by which proteins are inserted/transported into/across biological membranes. How is the structure of integral proteins in the membrane related to the mechanism of membrane insertion? Do integral membrane proteins contain in their sequence the information necessary for their folding in the membrane and what governs their assembly? What are the topogenic elements that guide the membrane insertion of the protein and determine its orientation? Is insertion into the membranes a process helped by specific proteins acting as catalysts or a spontaneous reaction? To what extent do inserted proteins use the same cellular machineries as the transported ones? Does membrane insertion require energy input? What can we learn regarding the mechanisms of membrane protein insertion from proteins of known structure that are converted from a water-soluble form to a membrane-bound form?

We have tried to summarize information about the structure of integral proteins and to review some of the recent advances dealing with the points just mentioned.

2. STRUCTURE OF MEMBRANE PROTEINS

The three-dimensional structure of a great number of water-soluble proteins has been determined. In contrast, little structural information is available concerning membrane proteins. The explanation for this resides in the fact that well-ordered 3-D crystals of membrane proteins are very difficult to obtain, and such crystals are prerequisite to perform high-resolution crystallographic analyses that would give an atomic resolution (< 3Å) of the tertiary structure of the protein.

To be purified and then crystallized, membrane proteins need to be solubilized in particular detergents that will keep them in their native conformation and shield their hydrophobic surface parts (which are in contact with the apolar chains of the lipids inside the membrane) from contact with water by forming micelles. Therefore, crystals will have to form in the presence of a detergent, and there is no standard method to find the right detergent and the right crystallization conditions (involving several parameters) for a given membrane protein. Furthermore, membrane proteins tend to form less-ordered 3-D crystals than do water-soluble proteins, probably because polar interactions are then less important between the molecules (since a portion of their surface is hydrophobic) (Kühlbrandt, 1988).

Although X-ray crystallography on well-ordered 3-D crystals is the method that provides highest resolution, another method based on image analysis of electron micrographs of 2-D crystals (called electron crystallography) has provided valuable structural information and allowed the determination of the 3-D structure of two integral proteins at a resolution level as high as 6Å and 3.5Å.

Up to now, the 3-D structure of four kinds of membrane proteins has been determined. For two of them, the 3-D structure has been achieved by X-ray crystallography (the photosynthetic reaction centers and the porins); for the other two it has been resolved by electron crystallography (light harvesting complex II and bacteriorhodopsin).

2.1. Structures Obtained by X-ray Crystallography

2.1.1. The Photosynthetic Reaction Centers

The 3-D structure of the photosynthetic reaction center (RC) of *Rhodopseudomonas viridis* was first determined at 3Å (Deisenhofer *et al.*, 1984) and then at 2.3Å resolution (Deisenhofer and Michel, 1989). The structure of this center has also been determined in another genus of purple bacteria, *Rhodopseudomonas sphaeroides*, at a resolution of 2.8Å (Allen *et al.*, 1986, 1987; Chang *et al.*, 1986). This center is a complex between proteins and pigments that carries out the initial photosynthetic steps of light-induced electron transfer from a donor to a series of acceptor species. This complex is formed by three subunits. Two of them, L and M, are polytopic membrane proteins that possess five membrane-spanning helices. Their folding is similar, consistent with their sequence homologies, and they are arranged almost symmetrically in the membrane (related by a twofold rotation axis). They ligate different cofactors (4 bacteriochlorophylls, 2 bacteriopheophytins, 2 quinones, and a nonheme atom of iron).

The third subunit, H, is a bitopic protein. It has a single membrane-spanning helix, and most of its bulk is located in the cytoplasm.

In *R. viridis,* a fourth subunit (a 4-heme cytochrome), which is lipid-anchored on the periplasmic side of the membrane, interacts with the periplasmic parts of the other three subunits.

The membrane-spanning region of the photosynthetic reaction centers is then constituted of 11 transmembrane α-helices that create a sort of framework around the cofactors. These helices comprise from 24 to 28 residues, and as 20 residues are enough to cross the membrane, a portion of them extends out of the membrane. They are very hydrophobic and are devoid of charged residues, therefore constituting apolar ends of the membrane-spanning region.

2.1.2. The Porins

Porins are found in the outer membrane of gram-negative bacteria, mitochondria, and chloroplasts. They are organized in homotrimers that are quite resistant to proteases and detergent denaturation.

In gram-bacteria, they form weakly selective ion channels through which small hydrophilic molecules (not bigger than 600D) can penetrate into the periplasmic space (Nikaido and Vaara, 1985; Benz and Bauer, 1988). They belong to a particular class of membrane proteins since their secondary structure is mainly organized in β-sheets (Nabedzyk *et al.*, 1988).

Both crystallographic methods (X-ray crystallography and electron crystallography) have been used to obtain information on the 3-D structure of these proteins.

The porin of *Rhodobacter capsulatus* has been successfully crystallized (Nestel *et al.*, 1989) and a high resolution of its 3-D structure (at 1.8Å) has now been determined (Weiss *et al.*, 1991). Electron crystallography studies on OmpF (devoid of specificity for the penetrating compounds) and PhoE (specific for phosphorylated compounds) porins of *E. coli* have given high-resolution projection maps of 3.5Å (Sass *et al.*, 1989) and 6.5Å (Jap, 1989), respectively. However, very recently, crystals of these two porins have given a higher resolution, at the atomic level, of the 3-D structure of these proteins (2.4Å for OmpF and 3Å for PhoE) (Cowan *et al.*, 1993). These data are in agreement and demonstrate that each monomer of a porin is formed by 16 antiparallel β-strands that form a right-handed twisted barrel. The lower rim of the barrel is constructed in a rather regular way since the loops connecting the β-strands contain only a few residues. In contrast, the upper rim is less regularly organized and contains some big loops (10 to 40 residues). The biggest one forms an α-helix of about 40 residues in length and runs into the inside of the barrel, thereby delimiting the size and the shape of the pore. The second largest loop (≈ 20 residues), also organized in α-helix, protrudes significantly out of the membrane, and as this region presents the bigger differences between PhoE and the other porins, it is thought to carry the phosphate-binding site of PhoE. The outside of the barrel is hydrophobic and the inside hydrophilic. PhoE porin contains two additional lysine residues inside its pore, and one of them has been shown to be responsible for the anion selectivity of PhoE.

For OmpF and PhoE, the occurrence of aromatic residues is remarkable at the interface between the lipid bilayer and the extracellular medium or the periplasmic space. This is in accordance with the hypothesis offered by Schiffer *et al.* (1992) about the role that tryptophane residues could play in anchoring the membrane protein on the side of the membrane by forming hydrogen bonds with the head groups of the lipids.

The monomers associate by juxtaposing the region where the rim of the barrel is shorter and comparatively more polar. For PhoE and OmpF, 35% of the residues of the monomers are thought to participate in the formation of the trimer by both hydrophobic and polar interactions. Thus, various types of interactions probably account for the stability of the trimers.

2.2. Structures Obtained by Electron Crystallography

2.2.1. Bacteriorhodopsin

Bacteriorhodopsin (BR) functions as a light-driven proton pump in *Halobacterium halobium* (Oesterhelt and Stoeckenius, 1973). It is associated with a retinal chromophore that gives it its purple color. Furthermore, it forms 2-D crystals in the so-called membrane of these bacteria.

Attempts to obtain 3-D crystals of bacteriorhodopsin have been made for several years by Michel and others (see Deisenhofer and Michel, 1989, and references therein). Crystals have been obtained but they do not diffract well.

Since 1975, it is known that BR is made up of seven α-helices (Henderson and Unwin, 1975), but the determination of the way in which they are connected, where and how the retinal is associated, and which parts of the structure are directly involved in the proton-pumping function have needed a great amount of work during these last 15 years involving spectroscopy, neutron diffraction, biochemical and genetic studies. All this information has facilitated the interpretation of a high-resolution electron-density map (3.5Å in a direction parallel to the membrane and about 10Å in the perpendicular direction) obtained by electron microscopy of 2-D crystals (Henderson *et al.*, 1990).

The α-helices constituting BR are 20 to 27 residues long (≈ 30Å) and completely span the membrane, which is thought to be 30Å thick. Three of them are perpendicular to the plan of the membrane, whereas the others form an angle of 20° with it. Three of the helices also show a slight kinking at the proline residues near their center.

These seven helices constitute a sort of hollow but flattened cylinder in the center of which the retinal is bound (via a Schiff base). The inside part of the cylinder is made up by the most polar faces of the helices, which contain some charged amino-acids (4 Asp, 1 Lys) and hydrogen-bonding residues. Its outside part, in contact with the lipids, is much more hydrophobic. Like porins, BR is organized in trimers in the membrane.

2.2.2. The Pea Light Harvesting Complex II

The pea light harvesting complex II (LHCII) collects the solar energy and transfers it to photosystem II where it is converted into a potential gradient across the thylakoid membrane of chloroplasts. It is complexed with 15 chromophores and several carotenoids (Butler and Kuhlbrandt, 1988; Siefermann-Harms *et al.*, 1985) and is organized in trimers in the membrane (Butler and Kuhlbrandt, 1988), which form well-ordered 2-D crystals. Electron crystallography studies of these crystals have allowed Kuhlbrandt and Wang (1991) to determine the structure of LHCII at 6Å resolution. This complex, as RC and BR, belongs to the α-helix bundle scheme. Each monomer has three α-helices. Two of them are longer (46Å to 49Å) than the length required to cross the membrane (30Å) and protrude somewhat beyond the membrane surface. They are inclined by 25° and 31° with respect to the normal membrane and are slightly twisted around one another. The third helix is shorter ($\approx$ 30Å) and its tilt angle with respect to the normal membrane is about 11°. These three helices seem to form a scaffold on which chlorophylls attach in such a way that optimal spacing and orientation can be reached for energy transfer.

2.3. Nuclear Magnetic Resonance Imaging of Low Molecular Weight Membrane Proteins

High-resolution nuclear magnetic resonance (NMR) spectroscopy has not only been used successfully to study the structure of small molecules in solution or small water-soluble proteins but also to investigate the structure of small membrane proteins in organic solvent or detergent micelles. This is the case for gramicidin A (which has 15 aa and dimerizes to form an ionic channel into the membrane) (Smith *et al.*, 1989), melittin (26 aa), a tetramer of which is thought to form a channel in membranes (Bazzo *et al.*, 1988), and M13 phage coat protein (50-aa long, which integrates into the inner membrane of its host, *E. coli*, during the reproductive cycle of the phage) (Henry and Sykes, 1990).

This method could be theoretically applied to bigger membrane proteins provided that they could be isotopically labeled with ^{13}C and ^{15}N.

2.4. Structure of the Soluble Form of Membrane Proteins

Some proteins can exist under two forms: a soluble one and a membrane one. The 3-D structure of the soluble form of three of these proteins has been determined at a high resolution. All of them are thought spontaneously to insert into the membrane in order to form channels.

Colicin A is a bacteriocin produced by and active against *E. coli*. The molecule can be divided into three domains, each one being implicated in one step of the mode of action. A central domain targets the toxin to its receptor on the outer membrane of the bacteria; an N-terminal domain allows it to translocate across the outer membrane and the periplasmic space, and a C-terminal domain is involved in the formation of voltage-dependent channels in the inner membrane that will destroy the energy potential of the cells (for reviews see Lazdunski *et al.*, 1988, and Pattus *et al.*, 1990). It is the 3-D structure of the soluble form of this C-terminal domain that has been determined (Parker *et al.*, 1989). This domain can be described as a bundle of ten α-helices organized in a three-layer structure. Two of these helices, totally hydrophobic, are completely buried inside the bundle.

The δ-endotoxin is an insecticidal protein produced by *Bacillus thuringiensis*. After binding to a receptor at the surface of targeted membranes, it forms channels in the cell membrane. The domain of the protein implicated in the formation of the channel is built on the same model as the C-terminal domain of colicin A: One hydrophobic helix is buried inside the bundle formed by six other amphipathic α-helices (Li *et al.*, 1991).

Diphtheria toxin (DT) is produced by *Corynebacterium diphtheriae*. Just before or after binding to a receptor at the cell surface, it is cleaved into two fragments that are endocytosed. Under the low pH conditions of the endosomes, one of the fragments undergoes a conformational change that promotes the insertion and formation of a channel in the membrane. The other fragment, which carries the lethal activity, is thought to gain access to its target in the cytoplasm upon translocating across this pore (Moskaug *et al.*, 1989). Again, the domain that will become membrane anchored displays a structure similar to that of the C-terminal domain of colicin A in its soluble form. Indeed, it comprises nine α-helices more or less arranged in three layers, with the two pairs of hydrophobic helices, the most hydrophobic pair constituting the central core layer (Choe *et al.*, 1992). A particular feature of the structure of the pore-forming fragment is the acidic composition of the connecting loops that render them highly charged and water-soluble at neutral pH but nearly neutral and lipid soluble at acidic pH. This could account for the conformational change of this fragment in an acidic environment.

The elucidation of these structures is of great interest because they allow us to distinguish precisely between the different individual elements of secondary structure that will rearrange in order to build the new 3-D membrane structure. Models for the insertion into the membrane of these polypeptides can be drawn up and further confirmed or invalidated by experiments (Parker *et al.*, 1990, Li *et al.*, 1991; Choe *et al.*, 1992) (see Section 6 this volume).

What can we learn from these data about the 3-D structure of membrane proteins in general? First, the surface area and the internal atomic packing of membrane proteins seem to be similar to those of soluble proteins of the same size. Along the same line, residues buried in the interior of both types of proteins are hydrophobic and exhibit comparable hydrophobicities.

However, these similarities do not extend to the surface residues. Indeed, to minimize surface energies most polar residues of water-soluble proteins segregate to their surface while the bilayer exposed residues at the surface of membrane proteins are even more hydrophobic than the ones in the interior.

Except these differences in the polarity of the surface residues, related to the different environment in which the proteins are soluble, the membrane proteins whose structure has been resolved seem to have the same packing rules and therefore the same structural organization as water-soluble proteins (α-helical bundles or β-barrels) although β-strand structures seem to be less widespread.

3.   FOLDING OF INTEGRAL MEMBRANE PROTEINS

3.1.   Secondary Structure of Transmembrane Elements and Energetics of Protein Insertion

The fatty-acyl region of a lipid bilayer is unable to establish any hydrogen bond. Then, upon insertion of a polypeptide chain into the lipid bilayer, the hydrogen bonding groups, formerly interacting with water molecules, would have to form hydrogen bonds with one another. This will result in the formation of periodic structures like α-helices or β-sheets. For membrane-spanning polypeptides, such structures are far more stable than those they would adopt in an aqueous solvent (Engelman and Steitz 1981). The maximum number of hydrogen bonds is formed in the α-helix structure, and this observation led Singer (1971) to assume that the membrane interior of integral proteins might be largely organized in α-helices. This hypothesis is now confirmed by the experimental information given by the determination of the 3-D structure of some membrane proteins, described in Chapter 2 of Singer (1971).

The two-stranded antiparallel β-structure is theoretically able to establish only half of the possible hydrogen bonds. However, in the particular case of the porins, the organization of the β-strands in a twisted barrel allows the formation of all hydrogen bonds. Furthermore, some hydrogen bonds could be formed with water molecules inside the lumen of the pore.

Except for the particular case of porins, no other known membrane protein has been found to have a secondary structure in β-sheets. In contrast to α-helical transmembrane domains, which are characterized by an uninterrupted stretch of about 20 hydrophobic residues, 10 to 12 amino acids are enough to span the membrane in a β-strand conformation. It seems that this particular organization is related to the localization of the porins in the outer membrane (of bacteria and mitochondria) (MacIntyre et al., 1988; Popot et al., 1992). Perhaps such a structure has evolved in order to allow the proteins to be efficiently translocated across the inner membrane and inserted only in the outer membrane, whereas an α-helical conformation would have resulted in anchoring them in the inner membrane.

Another characteristic of the fatty-acyl region of lipid bilayers, which has consequences on protein insertion, is its poor solvent capacity for polar groups due to its low dielectric constant. The cost in energy to bury charged residues, ion-pair residues, and discharged but polar residues has been estimated (Singer 1971, 1976) and it proves quite costly, the latter case being the less costly. From mutagenesis experiments and the example of bacteriorhodopsin, it seems, however, that a hydrophobic polypeptide chain can stably insert into the membrane even if it bears one ionic residue (for a review see Singer, 1990).

This thermodynamic information suggests that membrane proteins that do not form aqueous channels are amphipathic with almost all the charged and polar residues lining outside the membrane. These energetics assumptions are again confirmed by the experimental data since polar and ionic residues are very rare in the transmembrane segments of non-channel-forming membrane proteins.

3.2. The α-Helix as an Autonomous Folding Domain

Different lines of evidence tend to suggest that hydrophobic transmembrane α-helix bundles of membrane proteins does not require that the helices be part of the same polypeptide chain. Short peptides of less than 50 residues can form single hydrophobic transmembrane α-helices and constitute small subunits of mitochondria and chloroplast complexes (for a review, see Popot and de Vitry, 1990).

Single synthetic transmembrane helices have been shown to be able to associate in order to form ion-channels in model membranes (Lear et al., 1988; Oiki et al., 1988).

Upon insertion into lipid vesicles, both of the peptides corresponding to the first two α-helices of BR refold into a structure close to the native one in the BR molecule, as assessed by spectroscopic methods (Popot et al., 1987). Furthermore, the 7-helix BR molecule can be reconstituted in a functional way by fusing separate populations of lipid vesicles carrying fragments comprising its first two or its last five α-helices (Popot et al., 1987).

A functional β2-adrenergic receptor can be reconstituted in vivo upon the assembly of independently expressed portions of the molecule (Kobilka et al., 1988).

Some families of membrane proteins have a sequence constituted of repeated elements (see Popot et al., 1992, and references therein) that could be accounted for by gene duplication and fusion.

Furthermore, some multispanning membrane proteins can exist as single polypeptides or as heterodimers depending on the species (Heinemeyer et al., 1984; Kallas et al., 1988; Yamagushi et al., 1988) and some others can have evolved from an ancestor multispanning membrane protein by addition, deletion, or duplication of transmembrane helices (examples reviewed in Popot et al., 1992).

4. TRANSMEMBRANE TOPOLOGY

4.1. Classification of Integral Membrane Proteins

Because β-barrel-type membrane-proteins seem to be less widespread and because much less is known about their insertion and assembly into membranes than in the case of α-helical membrane proteins, this chapter will only deal with α-helical membrane proteins.

4.1.1. Single-Helix Membrane Proteins

The simplest transmembrane proteins span the membrane once and are usually called *bitopic proteins* (Blobel, 1980). Many of these proteins contain large extramembrane domains that can be removed from their membrane anchor by proteolytic cleavage. Short single-helix proteins for which the transmembrane region is the essential part of the protein are abundant in mitochondria and chloroplasts (de Vitry *et al.*, 1991).

According to the location of their N- and C-termini, facing the cytoplasm or the extracytoplasmic region, they can be classified into different groups. For the group I, the N-terminus faces the extracytoplasmic surface of the lipid bilayer. This group includes a wide range of proteins involved in recognition, adhesion, and signal transduction system (reviewed by Singer, 1990). For the group II, the N-terminus faces the cytoplasm. Usually these proteins have their active domains on the cytoplasmic side of the membrane. However, some proteins of group II have their active sites on the exterior surface of the membrane (reviewed by Singer, 1990).

4.1.2. Polytopic Membrane Proteins

Polytopic membrane proteins span the bilayer more than once. To this group belong ATPases, sugar transporters, various types of receptors, including G-proteins-coupled receptors, membrane-bound proteases, pumps and voltage-sensitive Ca^{2+}, Na^+, and K^+ channels (Singer, 1990; Popot and de Vitry, 1990). According to the degree of integration into the membrane of the various types of proteins, a correlation with their function can be established; for instance, proteins involved in electron transfer are strongly embedded into the membrane in contrast to ligand receptors, which are poorly embedded (Popot and de Vitry, 1990; Popot *et al.*, 1992).

4.2. Topology of Integral Membrane Proteins

Prediction methods based on hydrophobicity and energetic constraints provide valuable models on the arrangement of proteins into membranes (White and Jacobs, 1990). To determine the transmembrane orientation of the membrane proteins, these computational methods still need to be combined with experimental approaches such as surface labeling, protease susceptibility, antibody recognition, protein-fusion to enzymatic reporters, and protein engineering (reviewed by Jennings, 1989). Such studies have led to the recognition of topogenic signals that guide the insertion and determine the orientation of the protein in the membrane.

4.2.1. Topogenic Signals

Segments of helical membrane proteins have been identified that specify transmembrane topology, and they can be classified as follows: (1) amino-terminal cleavable leader peptides that target the proteins and initiate insertion into the membrane; (2) amino-terminal or internal uncleaved signals able to target and promote translocation across the membrane of the N- or C-terminus of the polypeptide; (3) stop-transfer sequences that halt the translocation of the polypeptide after it is initiated by a signal sequence. These so-called signal-anchor sequences contain hydrophobic cores of about 20 residues (Adams and Rose, 1985; Davis and Model, 1985; Garoff, 1985) and are flanked by hydrophilic regions. Accumulating evidence indicates that the transmembrane topology is not determined by the apolar regions but rather by the charged amino acids of the neighboring hydrophilic regions. Cleavable signal peptides can be viewed as specialized variations of uncleaved signal peptides.

4.2.2. The Role of the Polar Flanking Regions

Based on a statistical analysis of the sequences of integral membrane proteins, von Heijne (1986a) proposed, for prokaryotic proteins, the "positive inside" rule, suggesting that the most positively charged of the regions flanking the membrane-spanning segments remain in the cytoplasm. This hypothesis has been experimentally checked for several prokaryotic and eukaryotic proteins.

4.2.2a. Prokaryotic Proteins. The *E. coli* signal peptidase, a membrane-bound protease spanning the membrane twice and that has small N- and large C-terminal domains facing the periplasm, has served as a model for studying sequence determinants of membrane protein topology (Dalbey and Wickner, 1986, 1987; Dalbey *et al.*, 1987; von Heijne *et al.*, 1988; Laws and Dalbey, 1989). Indeed, the topology of the signal peptidase has been inverted by reducing the cytoplasmic region and by adding positive charges in the N-terminus (von Heijne, 1989). It was further shown in another study, that the balance of positive charges upstream and downstream a transmembrane segment was a determinant for its orientation (Nilsson and von Heijne, 1990). The influence of the positive charges is stronger when they are located close to the membrane-spanning segments (Anderson and von Heijne, 1991).

Similar conclusions on the role of the cytoplasmic regions of MalF, one of the components of the maltose transport system, were obtained by using protein fusions with enzymatic reporters (Manoil and Beckwith, 1986; Boyd *et al.*, 1987; Froshauer *et al.*, 1988; Boyd and Beckwith, 1989). Deletions in the MalF-alkaline phosphatase were introduced in order to identify topogenic determinants. Cytoplasmic domains were identified as such and their strength was correlated with the density of positive charges (McGovern *et al.*, 1991). In addition, a MalF protein carrying a deletion in its first membrane-spanning segment still inserts functionally in the membrane, indicating that membrane-spanning segments other than the first one are able to promote proper insertion of the protein (Ehrmann and Beckwith, 1991).

Another amazing example lies in the pathway of assembly of components of the cell-surface in prokaryotic organisms. Gram-negative bacteria assemble their pili from prepilin (Storm and Lory, 1986). Sequence analysis has revealed that the amino-terminal extension, which is cleaved by a specific protease, contains positive charges (Nunn and Lory, 1991; Dupuy *et al.*, 1991). These charges may be important in promoting a proper transmembrane orientation across the inner membrane before their subsequent cleavage by a cytoplasmic-specific protease and assembly (for a discussion see Tomassen *et al.*, 1992).

Many explanations can account for the topological control made by basic residues. Cytoplasmic-positive charges provide a barrier for translocation. It has been previously observed for the *E. coli* alkaline phosphatase, a translocated preprotein, that introduction of positive charges in the amino-terminus of the mature sequence severely affects its export (Li *et al.*, 1988). Neutralization of the basic residues prior to translocation may be energetically costly if translocation had to occur through an apolar environment. Therefore, cytoplasmic-positive charges could prevent membrane-spanning segments from flipping through the membrane (Boyd and Beckwith, 1990; Dalbey, 1990). On the other hand, cytoplasmic-positive charges may orientate the membrane-spanning segment according to the polarity of the membrane potential. Another possibility lies in the electrostatic interaction between the cytoplasmic-positive charges and the head groups of the acidic phospholipids. This binding was proposed as a key step in the insertion of the M13 procoat protein in the *E. coli* inner membrane (Kuhn *et al.*, 1986; Gallusser and Kuhn, 1990). Interestingly, the well-characterized bacteriorhodopsin does not obey the positive-inside rule (Popot *et al.*, 1992).

4.2.2b. Eukaryotic Proteins. Proteins sorted to various compartments acquire their orientation in the endoplasmic reticulum. Hydrophobicity of signal-anchor sequences is essential for membrane insertion, but the final membrane orientation is also strongly influenced by the hydrophilic flanking regions. For example, the secreted multilineage colony-stimulating factor was converted into an uncleaved anchored protein with an inverted orientation by exchanging N- and C-terminal sequences flanking the hydrophobic core of the signal sequence (Haeuptle *et al.*, 1989). These results indicated that hydrophilic sequences determine both cleavage by signal peptidase and membrane orientation. Conversely, the anchored protein cytochrome P450 was converted into a secretory protein by exchanging N-terminal acidic residues to basic residues (Szczesna-Skorupa *et al.*, 1988, 1989).

In many cases, sequences initiating and halting translocation have been shown to be interchangeable (Zerial *et al.*, 1987; Audigier *et al.*, 1987; Rothman *et al.*, 1988; Wessels and Spiess, 1988). Thus, the so-called positive-inside rule (von Heijne and Gavel, 1988) seems to apply also for proteins inserted into the endoplasmic reticulum, but a number of exceptions exist (Jennings, 1989). A statistical analysis of eukaryotic transmembrane proteins synthesized on the rough endoplasmic reticulum indicates that the charge differences between the 15 residues flanking the first internal signal-anchor sequence upstream and downstream determine their transmembrane orientation (Hartmann *et al.*, 1989a). A positive difference (Δ(C-N) positive) correlates with an Nexo/Ccyt orientation. Thus, the orientation of identical hydrophobic sequences is determined by their position within the polypeptide chain.

The role of charges in the flanking regions was tested for the asialoglycoprotein receptor, a single-spanning protein with a cytoplasmic amino terminus (Beltzer *et al.*, 1991). When charged residues flanking upstream and downstream the hydrophobic signal were mutated to opposite charges, inverted orientation was obtained for 50% of the polypeptides. If only one flanking region was mutated, the effect on the orientation was not significant. In contrast, orientation analysis on the same mutants, but carrying in addition a deletion of the amino-terminal region preceding the mutated anchor sequence, showed that such deleted mutants were more severely affected. Inverted topology is easier to obtain if the domain to be translocated is shortened (Beltzer *et al.*, 1991). In another study, mutations flanking the hydrophobic core of the preprolactin signal sequence were analyzed for their ability to convert the signal into a type I or type II signal-anchor sequence (Andrews *et al.*, 1992). Surprisingly, the adopted topology was opposite to that expected according to the charge distribution, suggesting that factors other than charge distribution determine the transmembrane orientation (Andrews *et al.*, 1992).

4.3. Vectoriality of Membrane Protein Insertion

Prokaryotic proteins such as lactose permease, SecY, and MalF have been shown to insert into the cytoplasmic membrane with a proper topology in absence of some of their membrane-spanning segments (Bibi *et al.*, 1991; Schatz *et al.*, 1991; Ehrmann and Beckwith, 1991). Thus, for prokaryotic cytoplasmic membrane proteins, insertion of topological elements, which are distributed all along the sequence, can occur nonsequentially. Current models for membrane insertion support independent insertion of the topogenic elements (Popot and Engelman, 1990; von Heijne and Manoil, 1990). However, the most hydrophilic transmembrane segment of the lactose permease exports with low efficiency an

alkaline phosphatase reporter fusion (Calamia and Manoil, 1992). Thus, as proposed by Calamia and Manoil, translocation of membrane-spanning segments
containing charged residues may require ion pairing between charged residues
belonging to different membrane-spanning segments (Calamia and Manoil, 1992).

For eukaryotic proteins, membrane topology has been proposed to depend
on the orientation of their most N-terminal signal-anchor sequence (Hartmann *et
al.*, 1989a), which interacts with the signal recognition particle (SRP) and the
SRP receptor (Wessels and Spiess, 1988). The general view is that each transmembrane segment would orientate in the opposite direction of the one preceding
it (Hartmann *et al.*, 1989a). However, this sequential model for insertion could
be related to the obligatory co-translational nature of the insertion process (von
Heijne and Manoil, 1990). It is noticeable that all known membrane proteins
with N-terminal cleavable signal sequences have their N-termini directed to the
exoplasmic side of the membrane.

4.4.   Intramembrane Interactions

4.4.1.   Intramembrane Helix–Helix Association

Topogenic signals have been defined above; however, predictions of transmembrane orientation of membrane-spanning helices do not indicate how the
membrane helices of multiple-helix proteins are packed within the bilayer. Many
proteins have been reconstituted in active forms after reconstitution from fragments, suggesting that interactions between different regions of a protein are
responsible for the three-dimensional arrangement of the protein within the bilayer. These reconstitutions have been made *in vitro* (e.g., bacteriorhodopsin) (Liao
et al., 1983; Popot *et al.*, 1987) or *in vivo* (e.g., lactose permease) (Bibi and
Kaback, 1990) and the chimeric α2-, β2-adrenergic receptors (Kobilka *et al.*,
1988). Other data indicate that the transmembrane segment of several single-
helix proteins like the glycophorin A (Bormann *et al.*, 1989; Lemmon *et al.*,
1992) or the T-cell receptor (Manolios *et al.*, 1990; Cosson *et al.*, 1991) might
have a role in their oligomerization. Probably, α-helical transmembrane regions
insert independently and then assemble within the bilayer to form oligomers;
however, this point requires further investigation. To explain such an ability for
refolding of the proteins mentioned above or to account for the role of the
transmembrane domain in mediating oligomerization, intramembrane helix-helix
association has been proposed (Popot and Engelman, 1990; Bormann and Engelman, 1992).

4.4.2.   Two-Stage Folding

These studies provide strong evidence in favor of the "two-stage model" for
membrane protein folding and oligomerization (Popot and Engelman, 1990). In

this model, the first step would consist in the insertion of the α-helical transmembrane regions independently of each other. The second step would consist in the assembly of these α-helices with minimal rearrangement in order to form the final and functional structure.

According to this model, the insertion and folding of multispanning membrane proteins could be governed by the insertion of single transmembrane α-helical segments whose topology is determined by signals depicted in Section 3.2.

The insertion and assembly of β-barrel proteins cannot proceed along the same line. Indeed, β-strands cannot insert autonomously into the lipid bilayer since most of their peptide hydrogen bonds cannot form in the nonpolar membrane interior. Furthermore, it has been shown that the deletion of a single membrane-spanning β-strand prevented the correct insertion of OmpA (Klose *et al.*, 1988) and PhoE proteins (Bosch *et al.*, 1989) into the outer membrane. Therefore, it seems that the β-strands insert in the outer membrane in a cooperative way after or during the β-barrel formation.

5. MECHANISMS OF INSERTION OF MEMBRANE PROTEINS

5.1. Insertion as an Assisted Process

In vivo, most membrane proteins with large extramembraneous domains do not insert spontaneously into the membranes, but use multicomponent systems for this purpose. Such systems are also responsible for the translocation of proteins across membranes. The fact that some proteins stay embedded in the membrane is only due to their particular characteristics (hydrophobic segments, particular distribution of charged amino-acid residues). It seems that proteins with multiple membrane segments behave differently for their membrane insertion (see Section 5.2).

These multicomponent systems, although only partly known, seem to differ between prokaryotes and eukaryotes and between the different target organelles in eukaryotes (mitochondria, chloroplasts, peroxysomes, ER). However, the processes of insertion/translocation seem to occur with homologous mechanisms:

Proteins are targeted to these systems by the presequences or internal sequences.

Particular proteins, called chaperones, maintain proteins translocation-competent, in an unfolded conformation unless the translocation takes place cotranslationally,

At their cytosolic surface, the membranes exhibit receptors for targeted polypeptides.

Energy is required in the form of ATP or GTP and sometimes of $\Delta\Psi$.

Most of the time, the signal sequence is cleaved off on the other side of the membrane.

However, this process is dispensable for translocation or insertion of proteins into membranes. Finally, newly translocated or inserted proteins have to fold properly in their new compartment, and this is achieved with the help of molecular chaperones.

5.1.1. Targeting Sequences

Signal sequences are generally found at the amino terminus of proteins as cleavable presequences, but their cleavage by specific proteases (reviewed by M. Müller, 1992) is not required for translocation (Müller et al., 1987; Kaiser et al., 1987; Fikes and Bassford, 1987; Géli et al., 1990). There is no sequence homology among the signal sequences that target peptides to a given compartment; hence, the function of signal sequences does not result from a particular primary structure but rather from structural characteristics (for reviews see von Heijne, 1985, 1990; Watson, 1984).

 Bacterial and ER (endoplasmic reticulum) signal sequences share the same structural characteristics and are functionally interchangeable (von Heijne, 1988). They can be divided into three different domains: an N-terminal domain that is positively charged; a central hydrophobic region; and a helical carboxy-terminal domain. Genetic studies have indicated that the N-terminal and the central regions are the only ones to be required for efficient targeting, the C-terminal domain being involved in the recognition by the signal peptidase enzyme.

Mitochondrial targeting peptides are structurally different from ER or bacteria signal sequences and have the ability to form an amphiphilic α-helix with charged residues on one side and hydrophobic residues on the other (von Heijne, 1986b). These features are important for efficient targeting (Bedwell et al., 1989; Endo et al., 1989a; Lemire et al., 1989).

Chloroplast targeting peptides also have particular structural features with three different domains and a specific cleavage site (von Heijne et al., 1989).

5.1.2. Antifolding Proteins

The so-called chaperone proteins (Ellis and van der Vies, 1991) seem to play a role at two different stages in protein insertion or translocation. First, they keep the preproteins in a conformation competent for translocation, preventing them from acquiring their final tertiary structure. Second, they act after the process of translocation, allowing the entire protein or part of the protein that has crossed the membrane to correctly fold on the other side. Interactions of chaperones with the nascent or completed proteins and energy derived from ATP hydrolysis seem to be prerequisite for these events, although this has not yet been demonstrated in every case.

Identified chaperones belong to different families of heat-shock proteins (hsp) classified according to their molecular weight, hsp10, 60, 70, and 90. Chaperones interact with the mature part of the precursors (Randall et al., 1990) but the nature of the interaction is still unknown.

5.1.2a. In Eukaryotic Cells. In eukaryotic cells, chaperones are located in different cellular compartments. Translocation of many proteins destined for the mitochondria or for the endoplasmic reticulum (ER) are stimulated by the same cytosolic hsp70 protein (Chirico et al., 1988; Deshaies et al., 1988). These proteins would maintain the precursor in a loosely folded state.

Other chaperones are located inside the mitochondrial matrix or in the ER lumen. Three mitochondrial matrix chaperones, hsp70, hsp60, and hsp10, have been identified so far (McMullin and Hallberg, 1988; Craig et al., 1989; Hartmann et al., 1992); hsp60 and hsp10 are related to the E. coli GroEL/ES proteins, In vivo, hsp60 is essential for the assembly of oligomeric proteins (Cheng et al., 1989; Osterman et al., 1989). Its antifolding activity couples protein import into the matrix with the subsequent export to the intermembrane space (Koll et al., 1992); however, this so-called conservative sorting as a pathway for intermembrane space localization (Hartl and Neupert, 1990) is controversial. Glick et al. (1992) reported that intermembrane space localized precursors are sorted by a stop-transfer mechanism. Another matrix chaperone is a member of the hsp70 family (Craig et al., 1989). This chaperone is necessary for the completion of translocation (Scherer et al., 1990; Kang et al., 1990) and precedes any interaction with hsp60 (Manning-Krieg et al., 1991). These sequential interactions with the matrix chaperones require ATP (Manning-Krieg et al., 1991; Baker and Schatz, 1991). It is reasonable to believe that chaperones which prevent tight folding of newly synthesized proteins and aggregation and which promote refolding and even disaggregation are involved in the efficient insertion of membrane proteins and in the efficient translocation of the transported proteins across membranes.

In the case of import or insertion into the ER membrane, the interaction of the signal recognition particle (SRP) with the signal sequence of the partially translated secretory proteins induces a translation arrest. Then, the protein synthesis becomes linked to the initiation of the translocation process, thereby avoiding any incompetent folding of the precursor. Cytoplasmic chaperone proteins of the hsp70 family have been discovered in yeast and mammalian cells (Gething and Sambrook, 1992, and references therein). In yeast, two of these chaperone proteins (Ssa1p and Ssa2p) have been shown to be involved both in the in vivo and in vitro posttranslational import of precursor proteins in the lumen of the ER (Deshaies *et al.*, 1988; Chirico *et al.*, 1988). They could constitute elements of an alternative, SRP-independent translocation pathway (some data imply the existence of such a pathway in yeast) or could salvage precursors that prematurely fold into a translocation-incompetent conformation during translation (the rate of elongation of protein could be higher than the rate of translocation).

First in mammalian cells and subsequently in yeast, a chaperone protein belonging to the hsp70 type and called Bip has been found in the lumen of the ER (Gething and Sambrook, 1992, and references therein). It is constitutively expressed but is induced by the accumulation of unfolded precursor in the cytoplasm (Kozutsumi *et al.*, 1988). It has been demonstrated to bind transiently to different nascent wild-type exported proteins and permanently to unfolded proteins, thereby blocking their transport out of the ER (Bole *et al.*, 1986; Gething *et al.*, 1986; Dorner *et al.*, 1987).

The addition of ATP can induce *in vitro* the dissociation of complexes between Bip and mammalian nascent polypeptide chains (Munro and Pelham, 1986). Furthermore, Flynn *et al.* (1991) have shown that a peptide-dependent ATPase activity is associated with Bip. These lines of evidence led to the conclusion that Bip plays a role in the folding and assembly of newly synthesized proteins in the ER lumen (Gething *et al.*, 1986; Bole *et al.*, 1986; Pelham, 1986). Vogel *et al.* (1990) also demonstrated that Bip could be involved in translocation since when it was altered by mutation, precursors accumulated in the yeast cytoplasm. Recently, the isolation of different mutants in the yeast gene KAR2, which encodes Bip, and cross-linking experiments have allowed Sanders *et al.* (1992) to show that Bip functions at two different stages in protein translocation: (1) It seems to facilitate the interaction of precursor proteins with Sec61p (a component of the translocation machinery in yeast) prior to the initiation of translocation and (2) it seems to participate directly in translocation since it has been cross-linked to partially translocated proteins and since one particular Kar2 mutant exhibits a defect in import, but not in precursor interaction with Sec61p.

By the means of a new reconstitution method, Brodsky *et al.* (1993) have recently shown that the cytosolic (Ssa1p) and the lumenal hsp70 isoenzymes in yeast were not interchangeable and played specific roles during translocation.

5.1.2b.   In Prokaryotes.   Identified *E. coli* chaperones are SecB, DnaK, and the GroEL/ES complex. SecB (Kumamoto and Beckwith, 1983) interacts with most of the translocated proteins (Collier *et al.*, 1988; Kumamoto, 1989; Lecker *et al.*, 1989) and therefore probably with most membrane-inserted proteins. Interestingly, heat-shock proteins other than GroEL and DnaK were shown to suppress a SecB defect, indicating that many proteins can perform the same function (Altman *et al.*, 1991).

Cross-linking experiments have provided evidence that GroEL and unfolded newly synthesized pre-β-lactamase interact (Bokchareva *et al.*, 1988). The complex GroEL/ES was shown to inhibit the folding of pre-β-lactamase (Laminet *et al.*, 1990), and temperature-sensitive mutations in the genes encoding GroES and GroEL affect its export (Kusukawa *et al.*, 1989).

DnaK, the *E. coli* hsp70 homologue, in conjunction with GroEL/ES, facilitates the export of hybrid proteins comprising the amino terminus of secreted proteins fused to the cytoplasmic protein β-galactosidase (Phillips and Silhavy, 1990). In addition, DnaK is able to disaggregate and reactivate, in an ATP-dependent manner, the heat-inactivated RNA polymerase (Skowyra *et al.*, 1990).

Once the translocation/insertion through or into the plasma membrane is achieved, SecD and SecF, proteins of unknown function that extend into the periplasmic space, might catalyze the folding of proteins in the periplasm as Bip and hsp70, respectively, do in the lumen of the ER and the matrix of mitochondria. Indeed, SecD was found to have a role in the release of the translocated mature maltose binding protein (Matsuyama *et al.*, 1993). However, since there is no ATP in the periplasmic space, the resulting mechanism must be different from the others.

5.1.3. Components of the Transport/Insertion Machineries

The machineries facilitating the process of insertion and translocation of proteins are composed of different cytosolic, peripheral, and integral membrane proteins. Cytosolic proteins often correspond to the previously depicted chaperone proteins; some peripheral and integral membrane proteins constitute receptors that help to target the precursors to the right membrane, and membrane-embedded complexes would directly participate in the translocation or insertion process since translocation is thought to occur through a proteinaceous pore (Gilmore and Blobel, 1985; Pfanner *et al.*, 1987a; Simon and Blobel, 1991, 1992; Schiebel and Wickner, 1992). Genetic studies, cross-linking experiments and the development of *in vitro* systems of import have been used to identify some of these proteins and deduce or demonstrate the role of some of them. We will review the progress in this work up to now and will attempt to offer a model for the events occurring during the translocation of proteins across the ER membrane of eukaryotic cells, the membranes of mitochondria, and the plasma membrane of bacteria.

5.1.3a.   Translocation/Insertion across/into the *E. coli* Inner Membrane. *E. coli* export machinery has been characterized biochemically and genetically (Wickner *et al.*, 1991; Stader *et al.*, 1989; Schatz and Beckwith, 1990). *In vitro*, translocation (Müller and Blobel, 1984; Cabelli *et al.*, 1988; Driessen and Wickner, 1990; Lill *et al.*, 1989; Kumamoto, 1989; Brundage *et al.*, 1990) requires the peripherical cytoplasmic membrane SecA protein (Oliver and Beckwith, 1982), which binds the complex precursor/SecB (Hartl *et al.*, 1990). SecA interacts by its N-terminal domain (Kimura *et al.*, 1991) with positive charges located at the amino terminus of signal peptides (Akita *et al.*, 1990). SecA is bound to the membrane by interacting with the presumed trans-locator complex formed by the two polytopic membrane proteins SecE and SecY (Shiba *et al.*, 1984; Akiyama and Ito, 1987; Schatz *et al.*, 1989; Bieker and Silhavy, 1990; Brundage *et al.*, 1990; Hartl *et al.*, 1990; Nishiyama *et al.*, 1992). Cross-linking experiments suggest that the complex formed by SecA, SecE, and SecY shield the translocating preprotein from the phospholipids (Joly and Wickner, 1993). Another gene (SecD) was genetically shown to be involved in protein export (Gardel *et al.*, 1987); perhaps it is necessary for completion of transloca-tion (Wickner *et al.*, 1991; Matsuyama *et al.*, 1993). It has also been proposed that signal sequences strongly increase the efficiency of the export process but are not always essential for export (Derman *et al.*, 1993).

The following polytopic membrane proteins SecY (Akiyama and Ito, 1989), leader peptidase (Wolfe *et al.*, 1985), and lactose permease (Ito and Akiyama, 1991) depend on SecA and SecY for their insertion. In addition, when the total incorporation of radiolabeled membrane proteins into the inner and outer mem-branes of SecA and SecY temperature-sensitive *E. coli* mutants was measured, only one third of the total membrane proteins were recovered in the envelope (Baker *et al.*, 1987). These results indicate that most of the membrane proteins use the general export pathway for their insertion (Saier *et al.*, 1989). However, several proteins are Sec-independent for their integration and may depend on unidentified components, suggesting that some membrane proteins behave differ-ently.

5.1.3b. Translocation/Insertion across/into the Mitochondrial Mem-branes. The mitochondrial import apparatus is constituted of import receptors (Sollner *et al.*, 1989, 1990; Hines *et al.*, 1990) that bring the precursors to an hetero-oligomeric channel (Vestweber *et al.*, 1989; Baker *et al.*, 1990; Kiebler *et al.*, 1990; Baker and Schatz, 1991; Glick and Schatz, 1991; Pfanner *et al.*, 1992; Scherer *et al.*, 1992). This channel spans both membranes at regions defined as contact sites (Schleyer and Neupert, 1985). However, the mitochondrial inner membrane contains a protein translocation activity identical to that observed for intact mitochondria (Hwang *et al.*, 1989; Géli and Glick, 1990). Pro-teins destined to the outer membrane probably share the early import steps with inner-membrane proteins, but are rerouted before they reach the inner mem-brane.

5.1.3c. Translocation/Insertion across/into the ER Membrane. As soon as they emerge from ribosomes (after polymerization of about 70 residues), the signal sequences of secreted proteins are recognized by a cytoplasmic ribonucleoprotein particle called signal recognition particle (SRP). In mam-malian cells, SRP has been shown to consist of one 7S RNA molecule and six proteins noncovalently bound to it. The signal sequence recognition activity has been localized in the 54-kDa subunit, since this protein has been successfully cross-linked to the signal sequence of different polypeptide chains (Kurzchalia *et al.*, 1986; Krieg *et al.*, 1986).

Concomitantly with signal sequence binding, SRP arrests or delays the elongation of the protein by accentuating the naturally occurring translational pauses (Wolin and Walter, 1989) probably by a subsequent interaction with the ribosome (Walter *et al.*, 1981). Evidence suggests that the SRP would be pre-bound to the ribosomes (Meyer, 1991). This effect on elongation is released when SRP (engaged in the ribosome-nascent polypeptide complex) interacts with its receptor at the surface of the ER (Meyer, 1991; Gilmore *et al.*, 1982).

This receptor is a complex of two proteins *a* and *b* (Tajima *et al.*, 1986), and it is the *a* subunit (also called docking protein), the primary structure of which has been determined (Hortsch *et al.*, 1988; Lauffer *et al.*, 1985), that carries the SRP binding activity. The release of SRP from the signal sequence upon binding to its receptor is GTP dependent (Connolly and Gilmore, 1989; Gilmore and Blobel, 1983).

At this stage, ribosomes are thought to become associated with a putative receptor at the surface of the ER membrane (Borgese *et al.*, 1974). Originally, there was some evidence that the two major integral membrane glycoproteins, ribophorins I and II, would be these ribosome receptors (Kreibich *et al.*, 1978a,b; Marcantonio *et al.*, 1984), the strongest of which being that antibodies directed against the cytoplasmic domain of ribophorin I can block the transloca-tion and the elongation arrest activity *in vitro* (Yu *et al.*, 1990). However, other evidence argues against this composition for the ribosome receptor. For instance, Hortsch *et al.* (1986) showed that the ribosome receptor involved components more sensitive to proteolysis than ribophorins were. In addition, ribophorins were recently shown to be part of a trimeric complex (with a protein of 45 kDa) that corresponds to the oligosaccharyltransferase activity present in the lumen of the ER (Kelleher *et al.*, 1992). Two other candidates have been described to have a ribosome binding activity, a protein of 180 kDa (Savitz and Meyer, 1990) and one of 34 kDa (Tazawa *et al.*, 1991). But two recent reports provide compelling evidence that even if the 180 kDa protein is in close proximity to ER membrane-bound ribosomes, it does not fractionate with the ribosome-binding activity and subsequently cannot be, alone, the bona fide ribosome receptor (Nunnari *et al.*, 1991; Collins and Gilmore, 1991). The existence of the ribosome receptor is still under debate.

Once the precursor protein has been targeted to the membrane of the ER and its translocation initiated, cross-linking experiments have shown that transloca-tion intermediates are in close proximity to a 34 kDa integral glycoprotein of the ER membrane, which has been called signal sequence receptor (SSR) (Wied-mann *et al.*, 1987; Krieg *et al.*, 1989; Wiedmann *et al.*, 1989; High *et al.*, 1991; Thrift *et al.*, 1991). Until recently, this major protein of the ER membrane (SSRa) was thought to form a dimeric complex with another integral glycoprotein of 22 kDa (SSRb) (Gölich *et al.*, 1990), but in fact these two proteins belong to a tetrameric complex that includes two other integral membrane glycoproteins: pp90 (a 90 kDa phosphoglycoprotein) and gp25L (25 kDa) (Wada *et al.*, 1991). Hartmann *et al.* (1989b) have demonstrated that antibodies or Fab directed against the cytoplasmic part of SSRa blocked the *in vitro* translocation and the elongation arrest release activity, which could account for an essential role of SSRa in the translocation process. Furthermore, as SSRa could be cross-linked not only to the signal sequence but also to the mature part of precursors, it was thought to participate directly in the putative protein-transporting pore in the ER membrane (Wiedmann *et al.*, 1989; Thrift *et al.*, 1991). However, a different experimental approach consisting of depleting detergent solubilized microsomal membranes in SSRa and then using the reconstituted vesicles in an *in vitro* translocation assay led to the opposite conclusion (Migliaccio *et al.*, 1992). The role of SSRa in translocation is also investigated now on the basis of the phosphorylation of pp90 and SSRa and their ability to bind calcium. Wada *et al.* (1991) have proposed an alternative role for SSRa in association with pp90; that is, the regulation of calcium-dependent events in the luminal side of the ER membrane.

Kellaris *et al.* (1991) have recently cross-linked another integral membrane protein of 34 kDa to the signal sequence of a secretory protein. It has been called IMP34 and is not glycosylated. Although this cross-link is specific for nascent polypeptides, it remains to be determined whether this factor is really a part of the translocation/insertion machinery.

Two other integral membrane proteins that have been cross-linked to translocating nascent chains are good candidates to participate in the transfer of the polypeptide chains through the ER membrane. One of them is a homologue of the bacterial SecYp and the yeast Sec61p (see below), respectively, implicated in the translocation of polypeptides through the cytoplasmic membrane of bacteria and the ER membrane of yeast cells (Görlich *et al.*, 1992b). The other one, called TRAM (for translocating chain associating membrane protein) has eight putative transmembrane segments and would interact with the nascent chains early during their passage through the membrane (Görlich *et al.*, 1992a).

In the yeast *Saccharomyces cerevisiae*, a protein homologous to the 54 kDa subunit of the mammalian SRP has been identified (Hann *et al.*, 1989). It is associated with a small cytoplasmic RNA (scR1) and presumably, together with other yet to be identified proteins, forms a 16S particle that could be called a yeast SRP (Amaya and Nakano, 1991; Hann and Walter, 1991). Hann and Walter (1991) also mention that a putative subunit of the SRP receptor has been isolated in yeast, which provides further evidence that an SRP targeting pathway exists in *S. cerevisiae*. However, the genes encoding these components are not essential for cell growth, and the translocation of some proteins is only slightly affected in the corresponding mutants. These observations suggest that an alternative targeting pathway might exist in *S. cerevisiae*.

Genetic studies in yeast have also led to the identification of three other ER membrane proteins, Sec61p, Sec62p, and Sec63p, whose genes are essential for growth and translocation *in vivo* and *in vitro* (Deshaies and Schekman, 1989; Rothblatt *et al.*, 1989; Stirling *et al.*, 1992; Toyn *et al.*, 1988; Feldheim *et al.*, 1992). They exist together in a complex with a glycoprotein of 31.5 kDa and another protein of 23 kDa (Deshaies *et al.*, 1991). Recently, Sanders *et al.* (1992) and Müsch *et al.* (1992) have demonstrated that Sec61p can be cross-linked to polypeptides that have initiated but not completed their membrane transport, providing evidence for a direct involvement of Sec61p in the translocation process. Furthermore, Müsch *et al.* (1992) could transiently cross-link Sec62p to proteins initiating their translocation—that is, before they interact with Sec61p—whereas Sanders *et al.* (1992) showed that the functions of Sec62p and Sec63p were required for the interaction of Sec61p with the translocating protein. Thus, these Sec proteins also seem to play a direct role in translocation/insertion of proteins into the ER.

These findings demonstrate that secreted polypeptides are in close proximity to integral membrane proteins of the ER all along their process of translocation. They are in good agreement with earlier studies that suggested that early stages of ER translocation occur in a proteinaceous environment (Gilmore and Blobel, 1985). The existence of a large ion-conducting channel was demonstrated in the ER membrane of mammalian cells when polypeptide chain translocation was blocked (Simon and Blobel, 1991).

It is widely accepted that transmembrane proteins use the same machineries as transported proteins. However, it is not yet understood how membrane-embedded sequences of membrane integral proteins leave the aqueous environment of the putative proteinaceous channel to insert into the bilayer. The currently accepted view favors the idea that translocation takes place across a proteinaceous channel (called translocon) through which proteins would be threaded.

Recently, Ooi and Weiss (1992) have shown that a 221-residue nascent polypeptide could not only move inward but also backward through this putative translocation channel. The retrograde movement of the peptide appears upon cessation of chain elongation. These data lead the authors to propose that this putative translocation channel would be a passive pore through which vectorial transport of the nascent polypeptide would be driven by the chain elongation and the further modifications of the peptide in the lumen of the ER.

According to alternative models, the transport of polypeptides would occur through a hexagonal lipid phase (favored by certain phospholipids). Indeed, the ability of signal sequences to insert into protein-free phospholipid bilayers strongly correlates with their functionality in the process of translocation.

5.1.4. Energy Requirements

Transport across bacterial and mitochondrial membranes requires ATP and a potential across the membrane (Chen and Tai, 1985; Geller *et al.*, 1986; Eilers *et al.*, 1987; Pfanner *et al.*, 1987b; Eilers *et al.*, 1988). For the import of matrix proteins, ATP is required in the matrix (Hwang and Schatz, 1989). It was recently reported that cytochrome c1 and the adenine nucleotide translocator were not affected by matrix ATP depletion for their transport to the inter-membrane space and the inner membrane, respectively, suggesting that these proteins do not cross the inner membrane during their import pathway (Wachter *et al.*, 1992).

With bacteria, translocation occurs through consecutive steps with different energy requirements (Geller ahd Green, 1989; Tani *et al.*, 1989; Schiebel *et al.*, 1991). ATP binds to SecA and allows translocation of a small domain of the precursor. Hydrolysis of ATP is required for the release of the preprotein from SecA (Geller and Green, 1989; Lill *et al.*, 1989; Schiebel *et al.*, 1991). The SecA-dependent ATP hydrolysis appears to be the only requirement for ATP (Lill *et al.*, 1989). The transmembrane electrical potential and the proton-motive force were proposed to be both necessary for translocation (Geller *et al.*, 1986; Yamane *et al.*, 1987; Yamada *et al.*, 1989a; Tani *et al.*, 1989; Geller and Green, 1989; Driessen and Wickner, 1991; Schiebel *et al.*, 1991; Driessen, 1992). Results of these studies indicate that the proton motive force is required for the completion of translocation (since only a small domain of the precursor is trans-located after ATP had bound to SecA). The overproduction of SecA allows a proton motive force independent translocation (Yamada *et al.*, 1989b), suggest-ing that in the absence of a proton motive force, cycles of interaction of the substrate with SecA allow the complete transit of the protein (Schiebel *et al.*, 1991; Driessen, 1992). Recently it was proposed that the translocation itself can drive the unfolding of a preprotein domain in absence of any exogenous energy input (Arkowitz *et al.*, 1993).

The requirement for the membrane potential is shared by all exported proteins (Daniels *et al.*, 1981; Enequist *et al.*, 1981). Interestingly, the polytopic integral lactose permease, when synthesized in a cell-free transcription-translation sys-tem, integrates into inverted inner membrane vesicles independently from the proton-motive force (Ahrem *et al.*, 1989). A relationship between the lengths of the polar regions to be translocated and the requirement for a transmembrane ion gradient was proposed (Ahrem *et al.*, 1989; Saier *et al.*, 1989).

With mitochondria, the import is mainly dependent on the electrochemical potential rather than the proton-motive force. The potential is necessary to move the N-terminal part of the precursor across both mitochondrial membranes but is not necessary for subsequent translocation (Eilers *et al.*, 1988). It is not excluded, however, that membrane potential affects a voltage-sensitive component of the import machinery and thus does not act as a driving force for import.

5.2. "Spontaneous" Insertion of Membrane Proteins

5.2.1. *In Vitro* Reconstitution of Integral Membrane Proteins into Membranes

To date, only two membrane proteins have been fully renatured *in vitro*, starting from a completely unfolded polypeptide: bacteriorhodopsin (and two of its proteolytic fragments) and the OmpF porin. Other proteins, like purified lactose permease, have been reconstituted in proteoliposomes in a fully active form.

In the case of the bacteriorhodopsin (BR), Huang *et al.* (1981) first demonstrated that upon denaturation in organic solvents, bacterio-opsin (BR without the retinal) could be efficiently renatured after transfer to an SDS solution and a subsequent dilution by the addition of lipid/detergent mixed micelles. The protein bound the retinal, it recovered the spectrum of BR, and, upon detergent removal, pumped protons into the vesicles.

The two proteolytic fragments obtained after treatment of bacterio-opsin with chymotrypsin (and which contain, respectively, five and two transmembrane helices of the entire protein) could also be denatured and, by following the same protocol, be made to renature (separately or not) and reassociate in order to form an active BR (Huang *et al.*, 1981; Liao *et al.*, 1983).

Subsequently, using a modified version of this protocol, renaturation of BR from its entire denatured polypeptide chain or from its two denatured proteolytic fragments was obtained at such low lipid-to-protein ratios that reformation of a two-dimensional lattice was induced. This is similar to that present in purple membranes (Popot *et al.*, 1986, 1987). Then, more stringent refolding studies based on a structural criterion became possible using crystallographic methods. They revealed that the denatured and reassembled BR molecules were indistinguishable from the native ones (Popot *et al.*, 1986, 1987).

After denaturation of the OmpF porin with a chaotropic agent, it could renature and recover its native trimeric stage in the presence of amphiphilic molecules (as checked by analytical centrifugation, gel electrophoresis, protease resistance, CD data, and channel formation in planar lipid bilayers) (Eisele and Rosenbusch, 1990). Furthermore, this renaturation can take place without pre-existing bilayers.

Purified lactose permease, which spans the bacterial inner membrane 12 times (Calamia and Manoil, 1990), transports β-galactosides and H^+ with a 1:1 stochiometry (Kaback *et al.*, 1990). The protein has been inserted into proteoliposomes (Newman *et al.*, 1981; Vitanen *et al.*, 1984) and the transport activity was reconstituted, suggesting that a significant proportion of the protein had the orientation found in the bacterial membrane (Vitanen *et al.*, 1984). It is noticeable that *in vivo* the protein assembles and is functional when expressed in fragments (Bibi and Kaback, 1990; Wrubel *et al.*, 1990).

The ability of these proteins to renaturate efficiently *in vitro* clearly shows that the information required for their folding is entirely contained within their primary sequence. However, these results do not mean that *in vivo* these proteins fold without the help of any other proteins—for example, chaperone proteins—since intracellularly the physicochemical conditions are different. But these *in vitro* experiments contribute to a better understanding of the native state of a protein and raise questions about the folding pathway and the intramolecular interactions that drive polypeptide folding. For example, the efficient refolding of BR confirms that the native structure of BR, which could have been forced to lie at a higher energy stage (because of the presence of lipids and its way of insertion) has in fact a free energy minimum. Furthermore, the reassociation of fragments in a fully active BR and the functional complementation of fragments from the lactose permease argue in favor of the "two-stage model" of membrane protein folding and oligomerization (Popot and de Vitry, 1990), discussed in Section 4.

5.2.2. The Sec-Independent Insertion into the *E. coli* Inner Membrane

Insertion of several proteins has been found to be SecA/SecY independent. These proteins are small and include proteins as the M13 procoat protein (Wolfe *et al.*, 1985) and the secretory prepromelittin (Cobet *et al.*, 1989). M13 procoat with an OmpA leader peptide retained its independence from the Sec function, indicating that the mature sequence rather than the signal sequence was responsible for that property (Kuhn *et al.*, 1987). When proper fusion proteins were constructed in order to extend their extracellular domains, both proteins became Sec dependent for their insertion (Cobet *et al.*, 1989; Kuhn, 1988). Thus, it seems that if the domain to be translocated is large, the Sec genes are required for the membrane insertion. Some larger proteins were also found to be SecA/SecY independent. The colicin A immunity protein (178 residues with periplasmic domains of about 20 residues), which spans the membrane four times (Géli *et al.*, 1988, 1989), was found to insert without the requirement of the Sec genes (Géli, unpublished results). Here again, when the reporter molecule alkaline phosphatase was fused within the second periplasmic region (i.e., after the third membrane-spanning segment), the fusion protein regained a Sec dependence for its insertion. These results seem to agree with those obtained for the M13 procoat protein and the prepromelittin and with the proposal from von Heijne who suggested that translocation of periplasmic regions longer than 60–70 residues are dependent on the Sec gene products (Andersson and von Heijne, 1993). However, the MalF protein, which contains a periplasmic region of approximately 180 residues, was found to insert without the requirement of Sec genes (McGovern and Beckwith, 1991). It is possible that the membrane-spanning segments flanking this 180-residue periplasmic domain cooperate to promote its translocation in a way resembling the proposed model for the M13 procoat insertion (Kuhn, 1987). Further investigation is required to determine the factors responsible for the Sec dependence.

5.2.3. Translocation of Proteins without the Requirement of the SRP-Induced Translation Arrest

It has been demonstrated for several proteins that the translocation across the ER membrane is independent on the elongation process and therefore of the SRP-induced translation arrest. These proteins fall into two different classes: (1) Some of them undergo translocation posttranslationally and are independent on the SRP targeting pathway; (2) the others exhibit translocation that is independent of elongation, but they need to be still attached (as peptidyl-tRNA) to the ribosomes in order to be imported in an SRP-dependent way.

Small polypeptides having fewer than 75 amino acid residues, such as prepromelitin (Zimmermann and Mollay, 1986), GLa peptide (Schlenstedt and Zimmermann, 1987), and preprocecropin A (Schlenstedt et al., 1990), belong to the first class. These properties come from their size since the translocation of artificially long derivatives of these polypeptides becomes dependent of the SRP targeting pathway (Müller and Zimmermann, 1987, 1988; Schlenstedt and Zimmermann, 1987). Preprocecropin A can translocate via both pathways (Schlenstedt et al., 1990).

This phenomenon can be explained if we consider that 30 to 40 amino acids of a nascent polypeptide are buried within the ribosome (Blobel and Sabatini, 1970) and that a signal sequence is 15 to 25 residues long. Then, in the case of short polypeptides, their translation is already achieved when the signal sequence emerges from the ribosome. In other words, the polypeptide is released from the ribosome before SRP has been able to interact with the signal sequence and before the ribosome could have bound to its receptor at the surface of the membrane. ATP and a cytosolic protein are involved in the translocation of these polypeptides (Müller and Zimmermann, 1987). Furthermore, the primary structure of these short polypeptides is important for their ability to be translocated, and they have probably evolved in a way that allows them to stay competent for membrane interaction without the help of the SRP targeting pathway components (Müller and Zimmermann, 1987, 1988). Further support for this hypothesis comes from the fact that the shortening of long precursors to less than 75 amino acids prevents them from translocating, probably because their primary structure does not fulfill the criteria necessary to allow them to interact directly with the membrane (Ibrahimi *et al.*, 1986; Siegel and Walter, 1988).

As already mentioned, yeast cells are able to grow in the absence of SRP (Hann and Walter, 1991). Presumably, all the essential proteins that have to cross or insert into the ER membrane are able to do so independently of SRP with an efficiency great enough to allow cell survival. Indeed, using *in vitro* translocation assays, it has been demonstrated that many precursor proteins are able to cross membranes posttranslationally, in an SRP-independent pathway (Hansen *et al.*, 1986; Rothblatt and Meyer, 1986; Walters and Blobel, 1986). Furthermore, Hann and Walter (1991) have shown that in the absence of SRP, the *in vivo* integration or translocation of some proteins is hardly affected (preinvertase and preproalphafactor) or not affected at all (preproCPY). These observations imply that an alternative, SRP-independent targeting pathway exists in yeast. According to the *in vitro* results, this pathway would be posttranslational and would require that ATP and precursors interact with the cytoplasmic chaperone protein hsp70 (Chirico *et al.*, 1988; Deshaies *et al.*, 1988) in order to acquire a conformation competent for translocation. Hann and Walter (1991) suggest that the degree to which a given protein is independent of SRP is determined by the degree to which it is able to stay in a competent conformation for import in the cytoplasm.

It is not known how such proteins are targeted to the ER membrane (whether it is directly via the interaction between the signal sequence and its receptor or if other components are involved), and once associated with the membrane, it is not clear whether they use the same multicomponent membrane complex (translocon) to cross or integrate into the ER membrane.

Finally, it has been reported that some precursors can be imported, *in vitro*, into mammalian cell microsomes without any coupling between elongation and translocation (Hansen *et al.*, 1986; Caulfield *et al.*, 1986; Mueckler and Lodish, 1986a,b; Perara *et al.*, 1986; Chao *et al.*, 1987). In these experiments, translation of precursors has been initiated, and then elongation has been stopped prematurely with cycloheximide. Then, microsomal membranes (with endogenous SRP) have been added and the translocation assayed. Because in these experiments translocation was observed, it has been concluded that translocation of proteins across mammalian ER membranes is not necessarily coupled to translation. However, SRP is still required for this process (Hansen *et al.*, 1986; Mueckler and Lodish, 1986a) and the SRP-dependent targeting occurs only if the preprotein is still attached to the ribosome (Garcia and Walter, 1988). These authors propose that ribosomes could be necessary for the interaction between SRP and the signal sequence to occur since no affinity of SRP for isolated signal sequences or ribosome-free preproteins has yet been demonstrated.

ATP is required during this elongation-independent translocation (Garcia and Walter, 1988; Mueckler and Lodish, 1986b), probably in relation with the cytoplasmic hsp70 proteins, in order to allow the precursors to keep a translocation-competent conformation and to the signal sequences to remain accessible to SRP.

5.3. Role of Lipids

The possible role of phospholipids in protein translocation was investigated using *E. coli* strains defective in the synthesis of anionic phospholipids. Translocation of exported proteins was shown to be severely impaired (de Vrije *et al.*, 1988). An interaction between signal peptides and negatively charged phospholipids was proposed to be essential for the initiation of *E. coli* protein translocation (de Vrije, 1989). It has also been demonstrated that both SecA ATPase activity and translocation across the inner membrane require acidic phospholipids (Lill *et al.*, 1990). Moreover, translocation was shown to be dependent on the phosphatidylglycerol content (Kusters *et al.*, 1991). The functional binding of SecA with the inner membrane was then shown to need acidic phospholipids (Hendrick and Wickner, 1991). On the other hand, it has been suggested that signal sequences of precursors could directly interact with acidic phospholipids in the translocation process (Kusters *et al.*, 1991). Indeed, *E. coli* signal peptides affect local lipid structure by inducing the formation of nonbilayer lipid structure in model membranes (Killian *et al.*, 1990). These modifications were proposed to allow the protein to cross the membrane (Killian *et al.*, 1990). Further advances will require the *in vivo* localization of signal sequences during the process of translocation.

The possible involvement of lipids in mitochondrial import was investigated by using a drug interacting with acidic phospholipids. The import was blocked in isolated yeast mitochondria (Eilers *et al.*, 1989). Acidic phospholipids on the mitochondrial surface partly unfolded the precursor used in these experiments (Eilers *et al.*, 1989; Endo *et al.*, 1989b). In that case, lipids were proposed to be involved in the attainment of an import-competent conformation. Negatively charged phospholipids are also involved in the import pathway of the apocytochrome *c* (Jordi, 1990). For this particular import pathway (Hartl *et al.*, 1989), spontaneous membrane insertion into and partial translocation across lipid bilayers have allowed the apocytochrome *c* to bypass early steps in the general import pathway (Dumont and Richards, 1984; Jordi, 1990).

6. MEMBRANE INSERTION OF TOXINS

To reach their target and to produce their effect, secreted bacterial toxins often have to translocate across or to insert into membranes. Diphtheria toxin and pore-forming colicins are the most well-characterized membrane-translocating

toxins. To illustrate the contribution of work on bacterial toxins in the understanding of protein-insertion mechanisms, we will focus on the interaction of *E.
coli* membranes with colicin A, one of the pore-forming colicins.

6.1. Colicin A Unfolds during Its Translocation

Colicin A (Mr 63000) is a bactericidal protein able to form voltage-
dependent channels in *E. coli* cytoplasmic membranes. Like many toxins, colicin
A is organized in structural domains involved in entry into the target cell and
membrane insertion. The central domain binds to a receptor at the surface of the
target cell; the N-terminal domain interacts with a proteinaceous complex (Tol
proteins, Webster, 1991) responsible for the so-called translocation step; and the
C-terminal domain carries the ionophoric activity. Thus pore formation implies
that a portion of the colicin molecule crosses the outer membrane and the periplasmic space, allowing the pore-forming domain to insert into the membrane.
Since (1) the N-terminal domain of colicin A interacts with at least one
of the components of its translocation system (TolA) (Bénédetti *et al.*, 1991),
(2) urea denaturation of colicin A decreases the time required for the protein to
form its pore, and (3) colicin A spans the whole envelope when its pore has
formed (Bénédetti *et al.*, 1992), it is proposed that the components of the translocation system are responsible for colicin A unfolding, perhaps by breaking
interdomain interactions. The molecule being in an extended form, the pore-
forming domain can then penetrate into the inner membrane to form a channel.
Insights about the insertion of the pore-forming domain come from *in vitro*
experiments with the purified domain and membrane model systems.

6.2. Membrane Insertion of the Colicin A Pore-Forming Domain

The pore-forming domain of colicin A can be purified after cleavage of the
entire molecule by proteases (Martinez *et al.*, 1983). The pore-forming domain is
water soluble and has a ionophoric activity similar to the entire molecule when
lipid model systems are used (Pattus *et al.*, 1990). The crystal structure of the
pore-forming domain of colicin A has been refined at 2.4Å resolution. The
polypeptide chain consists of ten α-helices with a hydrophobic core made of a
helical hairpin, which is buried in the soluble form of the protein (Parker *et al.*,
1989). *In vitro*, the channel formation can be divided into two distinct steps: The
protein spontaneously inserts into the lipid bilayer and then the application of a
transmembrane potential causes the insertion of some of the helices of the molecule and channel opening. The potential independent insertion of the colicin
pore-forming domain requires acidic pH and negatively charged phospholipids
(for reviews, cf Slatin, 1988; Lazdunski *et al.*, 1988; Cramer *et al.*, 1990; Pattus
et al., 1990). At acidic pH, the protein has a molten globular conformation that

has been shown to be the intermediate form for membrane insertion (van der Goot *et al.*, 1991). It was the first time that a relation between a molten globule conformation and a competent state for membrane insertion was established. The role of the positively charged phospholipids would be either to orientate the molecule at the surface of the bilayer (Parker *et al.*, 1989) or to decrease locally the surface pH in order to facilitate the attainment of the molten globular conformation (Gonzalez-Manas *et al.*, 1992). Potential-independent insertion of the protein is accompanied by a rearrangement of the molecule, the elements of secondary structure being conserved (Lakey *et al.*, 1990). The protein, although in contact with the acyl chain of the phospholipids (Gonzalez-Manas *et al.*, 1992), remains surface localized (Géli *et al.*, 1992). When a membrane potential is applied, a portion of the molecule deeply inserts in the bilayer, but the conformational changes occurring at this step are not documented. Voltage-sensitive fragments required for insertion were defined for the pore-forming domain of colicin El (Merrill and Cramer, 1990).

As mentioned above, *in vivo* channel formation requires the Tol proteins. Questions arise about the relations between the pore seen *in vivo* and *in vitro* (Letellier, 1992). Do the Tol proteins unfold the colicin in order to allow the appearance of a molten globular conformation? Are the pores seen *in vivo* and *in vitro* identical? These questions remain unsolved.

As for colicin pore-forming domains, membrane penetration of the diphtheria toxin is triggered by a low pH that induces a conformational change (for a review, cf London, 1992). The transmembrane fragment of diphtheria toxin has a rather similar structure as the pore-forming domain of colicin A (Choe *et al.*, 1992). The proposed model for its insertion involves two pairs of hydrophobic helices (Choe *et al.*, 1992). The hydrophobic interaction between the two pairs of hydrophobic helices with the lipids might be stronger than the colicin A surface interaction, which occurs in absence of potential. This could explain their differences with respect to the requirement for a membrane potential in order to insert into the bilayer. With the availability of their three-dimensional-structures, bacterial toxins are proving useful in understanding mechanisms of protein translocation and insertion.

7. CONCLUSIONS

The determination of the 3-D structure of different membrane proteins has confirmed some previous structural proposals based on thermodynamic considerations. On this basis, integral membrane proteins can be classified in two different groups, the α-bundle group and the β-barrel group.

Until now, the membrane proteins known to have a predominantly β-sheet secondary structure have all been shown to belong to the bacterial or mitochondrial outer membrane. However, our knowledge is not wide enough to rule out the possibility of other localizations. The polypeptide chain of this type of protein is overall hydrophilic, but the face of the barrel in contact with lipids is hydrophobic while the face constituting the lumen of the pore is hydrophilic.

The proteins built on the α-bundle scheme are inserted into the membrane via their α-helical segments, which turn their most hydrophobic face toward the lipids and their most hydrophilic one toward the core of the protein. These transr embrane α-helices turn out to be autonomous folding domains, and the formation of the bundles does not require that the helices be part of the same polypeptide chain.

While the overall characteristics of the 3-D structure of membrane proteins now seem to stand out, the specific role that some residues could play in stabilizing their structure is still unknown. However, the study of the known 3-D structures of membrane proteins should help in addressing this question.

Even if in some cases integral membrane proteins can be efficiently refolded and inserted into lipids *in vitro* without the help of any other protein; the insertion mechanism of most membrane proteins *in vivo* involves a complex multicomponent protein machinery and energy. Exceptions to this multicomponent-dependent insertion mechanism are constituted by small proteins such as the M13 procoat protein for prokaryotic cells and prepromelittin, preprocecropin A, and GLa peptide for eukaryotic cells. However, larger prokaryotic integral membrane proteins such as MalF or the colicin A immunity protein have also been found to insert independently from the identified Sec machinery.

The mechanisms followed for the translocation or insertion of membrane proteins with large extramembraneous domains through or into different membranes appear to be similar in all living organisms. Indeed, (1) proteins to be inserted or translocated need to be kept in a loose conformation by chaperone proteins before being targeted to the appropriate membrane unless the insertion/translocation process is coupled to translation; (2) these proteins are targeted to the membrane by presequences or internal sequences; (3) targeting to the membrane is made via soluble cytoplasmic factors and membrane-bound receptors; (4) the polypeptide chains are thought to cross the membrane in a proteinaceous environment, through an aqueous pore protein complex; and (5) some proteins implicated in the translocation/insertion mechanism have even been shown to be homologous between species.

However, a great amount of work combining both genetic and biochemical approaches has still to be done to identify all the components involved in such mechanisms and to discover if they share other homologies between species, thus providing further arguments in favor of an evolutionary conservation of the translocation mechanism.

Different experiments have shown that the difference between the insertion and the translocation of a protein into or across a membrane resides in the presence or not, in the polypeptide chain, of a hydrophobic stretch of residues long enough to cross the membrane (20 aa). During the transport of polypeptides through the putative aqueous pore, the mechanisms responsible for the lateral diffusion of hydrophobic stretches into the lipid phase and the transport across the membrane of the other types of sequences are being investigated.

Another interesting question is the type of mechanism that drives polypeptide translocation into or across membranes. Different suggestions have been made about the possible mechanisms implicated such as: (1) the force and the energy given by the polypeptide chain elongation, (2) the possible role of the electrochemical gradient, (3) the energy associated with the folding of the already translocated portions of the protein, which may pull the polypeptide across the membrane, and (4) the existence of a pump inside the membrane (see Simon *et al.*, 1992; Ooi and Weiss, 1992, and references therein). The most recent suggestion was made by Simon *et al.* (1992), who proposed that the Brownian ratchet mechanism could be the translocation driving force. Such a mechanism has the advantages of being nonspecific, fast, and powerful enough to be able to achieve this task, but its validity still needs to be demonstrated experimentally.

The mechanism of insertion of membrane proteins into membranes *in vivo* is still far from being completely understood. The machinery used for the insertion of some of these membrane proteins is the same as that used for the translocation process, but not all of its constituents are known. Membrane proteins that span the membrane many times seen to be inserted by different mechanisms. Furthermore, the driving force of both processes is still under investigation.

ACKNOWLEDGMENTS. We thank Dr. Alan Munn and David Espesset for careful reading of the manuscript.

8. REFERENCES

Adams, G. A., and Rose, J. K., 1985, Structural requirements of a membrane-spanning domain for protein anchoring and cell surface transport, *Cell* **41**:1007–1015.

Ahrem, B., Hoffschulted, H. K., and Müller, M., 1989, *In vitro* membrane assembly of a polytopic, transmembrane protein results in an enzymatically active conformation, *J. Cell Biol.* **108**:1637–1646.

Akita, M., Sasaki, S., Matsuyama, S., and Mizushima, S., 1990, SecA interacts with secretory proteins by recognizing the positive charge at the amino terminus of the signal peptide in *Escherichia coli*, *J. Biol. Chem.* **265**:8164–8169.

Akiyama, Y., and Ito, K., 1987, Topology analysis of the SecY protein, an integral membrane protein involved in protein export in *Escherichia coli*, *EMBO J.* **6**:3456–3470.

Akiyama, Y., and Ito, K., 1989, Export of *Escherichia coli* alkaline phosphatase attached to an integral membrane protein, SecY, *J. Biol. Chem.* **264**:437–442.

Allen, J. P., Feher, G., Yeates, T. O., Rees, D. C., and Deisenhofer, J., 1986, Structural homology of reaction centers from *Rhodopseudomonas sphaeroides* and *Rhodopseudomonas viridis* as determined by X-ray diffraction, *Proc. Natl. Acad. Sci. USA* **83:**8589–8593.

Allen, J. P., Feher, G., Yeates, T. O., Komiya, H., and Rees, D. C., 1987, Structure of the reaction center from *Rhodobacter sphaeroides* R-26: The cofactors, *Proc. Natl. Acad. Sci. USA* **84:**5730–5734.

Altman, E., Kumamoto, C. A., and Emr, S. D., 1991, Heat-shock proteins can substitute for SecB function during protein export in *Escherichia coli, EMBO J.* **10:**239–245.

Amaya, Y., and Nakano, A., 1991, SRH1 protein, the yeast homologue of the 54kDa subunit of signal recognition particle, is involved in ER translocation of secretory proteins, *FEBS Lett.* **283:**325–328.

Andersson, H., and von Heijne, G., 1991, a 30-residue-long "export initiation domain" adjacent to the signal sequence is critical for protein translocation across the inner membrane of *Escherichia coli, Proc. Natl. Acad. Sci. USA* **88:**9751–9755.

Andersson, H., and von Heijne, G., 1993, Sec-dependent and Sec-independent assembly of *E. coli* inner membrane proteins: The topological rules depend on chain length, *EMBO J.* **12:**683–691.

Andrews, D. W., Young, J. C., Mirels, L. F., and Czarnota, G. J. 1992, The role of the N-region in signal sequence and signal-anchor function, *J. Bio. Chem.* **267:**7761–7769.

Arkowitz, R. A., Joly, J. C., and Wickner, W., 1993, Translocation can drive the unfolding of a preprotein domain, *EMBO J.* **12:**243–253.

Audigier, Y., Friedlander, M., and Blobel, G., 1987, Multiple topogenic sequences in bovine opsin, *Proc. Natl. Acad. Sci. USA* **84:**5783–5787.

Baker, K. N., Machman, N., Jackson, M., and Holland, I. B., 1987, Role of SecA and SecY in protein export as revealed by studies of TonA assembly into the other membrane of *Escherichia coli, J. Mol. Biol.* **198:**6932–703.

Baker, K. P., Schaniel, A., Vestweber, D., and Schatz, G., 1990, A yeast mitochondrial outer membrane protein essential for protein import and cell viability, *Nature* **348:**605–609.

Baker, K. P., and Schatz, G., 1991, Mitochondrial proteins essential for viability mediate protein import into yeast mitochondria, *Nature* **349:**205–208.

Bazzo, R., Tappin, M. J., Pastore, A., Harvey, T. S., Carver, J. A., and Campbell, I. D., 1988, The structure of melittin: A 1H study in methanol, *Eur. J. Biochem.* **173:**139–146.

Bedwell, D. M., Stobel, S. A., Yun, K., Jongeward, G. D., and Emr, S. D., 1989, Sequence and structural requirements of a mitochondrial protein import signal defined by saturation cassette mutagenesis, *Mol. Cell Biol.* **9:**1014–1025.

Beltzer, J. P., Fiedler, K., Fuhrer, C., Geffen, I., Handschin, C., Wessels, H. P., and Spiess, M., 1991, Charged residues are major determinants of the transmembrane orientation of a signal-anchor sequence, *J. Biol. Chem.* **266:**973–978.

Bénédetti, H., Lazdunski, C., and Lloubés, R., 1991, Protein import into *Escherichia coli*: Colicins A and E1 interact with a component of their translocation system, *EMBO J.* **10:**1989–1995.

Bénédetti, H., Lloubés, R. Lazdunski, C., and Letellier, L., 1992, Colicin A unfolds during its translocation in *Escherichia coli* cells and spans the whole cell envelope when its pore has formed, *EMBO J.* **11:**441–447.

Benz, R., and Bauer, K., 1988, Permeation of hydrophilic molecules through the outer membrane of gram-negative bacteria: Review on bacterial porins, *Eur. J. Biochem.* **176:**1–19.

Bibi, E., and Kaback, H. R., 1990, *In vivo* expression of the lacY gene in two segments leads to functional lac permease, *Proc. Natl. Acad. Sci. USA* **87:**4325–4329.

Bibi, E., Verner, G., Chang, C-Y., and Kaback, H. R., 1991, Organization and stability of a polytopic membrane protein: Deletion analysis of the lactose permease of *Escherichia coli, Proc. Natl. Acad. Sci. USA* **88:**7271–7275.

Bieker, K. L., and Silhavy, T. J., 1990, The genetics of protein secretion in *E. coli, Trends Genet. Sci.* **6**:329–334.

Blobel, G., 1980, Intracellular protein topogenesis, *Proc Natl. Acad. Sci. USA* **77**:1496–1501.

Blobel, G., and Sabatini, D. D., 1970, Controlled proteolysis of nascent polypeptides in rat liver cell fractions: Location of the polypeptides within ribosomes, *J. Cell Biol.* **45**:130–145.

Bochkareva, E. S., Lissin, N. M., and Girshovich, A. S., 1988, Transient association of newly synthesized unfolded proteins with the heat-shock GroEL protein, *Nature* **336**:254–257.

Bole, D. G., Hendershot, L. M., and Keaney, J. F., 1986, Post-translational association of immunoglobulin heavy chain binding protein with nascent heavy chains in nonsecreting and secretory hybridomas, *J. Cell Biol.* **102**:1558–1566.

Borgese, N., Mok, W., Kreibich, G., and Sabatini, D. D., 1974, Ribosomal-membrane interaction: *In vitro* binding of ribosomes to microsomal membranes, *J. Mol. Biol.* **88**:559–580.

Bormann, B. J., Knowles, W. J., and Marchesi, V. T., 1989, Synthetic polypeptides mimic the assembly of transmembrane glycoproteins, *J. Biol. Chem.* **264**:4033–4037.

Bormann, B. J., and Engelman, D. M., 1992, Intramembrane helix-helix association in oligomerization and transmembrane signaling, *Ann. Rev. Biophys. Biomol. Struct.* **21**:223–242.

Bosch, D., Scholten, M., Verhagen, C., and Tommassen, J. 1989, The role of the carboxy-terminal membrane-spanning fragment in the biogenesis of *Escherichia coli* K12 outer membrane protein PhoE, *Mol. Gen. Genet.* **216**:144–148.

Boyd, D., and Beckwith, J., 1989, Positively charged amino-acid residues can act as topogenic determinants in membrane proteins, *Proc. Natl. Acad. Sci. USA* **86**:9446–9450.

Boyd, D., and Beckwith, J., 1990, The role of charged amino acids in the localization of secreted and membrane proteins, *Cell* **62**:1031–1033.

Boyd, D., Manoil, C., and Beckwith, J., 1987, Determinants of membrane protein topology, *Proc. Natl. Acad. Sci. USA* **84**:8525–8529.

Brodsky, J. L., Hamamoto, S., Feeldheim, D., and Schekman, R., 1993, Reconstitution of protein translocation from solubilized yeast membranes reveals topologically distinct roles for Bip and cytosolic hsc70, *J. Cell Biol.* **120**:95–102.

Brundage, L., Hendrick, J. P., Schiebel, E., Driessen, A. J. M., and Wickner, W., 1990, The purified *E. coli* integral membrane protein SecY/E is sufficient for reconstitution of SecA-dependent precursor protein translocation, *Cell* **62**:649–657.

Butler, P. J. G., and Kühlbrandt, W., 1988, Determination of the aggregate size in detergent solution of the light-harvesting chlorophyll a/b-protein complex from chloroplast membranes, *Proc. Natl. Acad. Sci. USA* **85**:3797–3801.

Cabelli, R. J., Chen, L., Tai, P. C., and Oliver, D. B., 1988, SecA protein is required for secretory protein translocation into *E. coli* membrane vesicles, *Cell* **55**:683–692.

Calamia, J., and Manoil, C., 1990, Lac permease of *Escherichia coli:* Topology and sequence elements promoting membrane insertion, *Proc. Natl. Acad. Sci. USA* **87**:4937–4941.

Calamia, J., and Manoil, C., 1992, Membrane protein spanning segments as export signals, *J. Mol. Biol.* **224**:539–543.

Caulfield, M. P., Duong, L. T., and Rosenblatt, M., 1986, Demonstration of post-translational secretion of human placental lactogen by a mammalian *in vitro* translation system, *J. Biol. Chem.* **261**:10953–10956.

Chang, C. H., Tiede, D., Tang, J., Smith, U., Norris, J. R., and Schiffer, M., 1986, Structure of *Rhodopseudomonas sphaeroids* R-26 reaction center, *FEBS Lett.* **205**:82–86.

Chao, C. C. K., Bird, P., Gething, M. J., and Sambrook, J., 1987, Post-translational translocation of influenza virus hemagglutinin across microsomal membranes, *Mol. Cell Biol.* **7**:3842–3845.

Chen, L., and Tai, P., 1985, ATP is essential for protein translocation into *Escherichia coli* membrane vesicles, *Proc. Natl. Acad. Sci. USA* **82**:4384–4388.

Cheng, M. Y., Hartl, F. U., Martin, J., Pollock, R. A., Kalousek, F., Neupert, W., Hallberg, E. M.,

Hallberg, R. L., and Horwich, A. L., 1989, Mitochondrial heat-shock protein hsp60 is essential for assembly of proteins imported into yeast mitochondria, *Nature* **337**:620–625.

Chirico, W. J., Waters, M. G., and Blobel, G., 1988, 70 K heat-shock-related proteins stimulate protein translocation into microsomes, *Nature* **332**:805–810.

Choe, S., Bennett, M. J., Fujii, G., Curmi, P. M. G., Kantardjieff, K. A., Collier, R. J., and Eisenberg, D., 1992, The crystal structure of diphtheria toxin, *Nature* **357**:216–222.

Cobet, W., Mollay, C., Müller, G., and Zimmermann, R., 1989, Export of honeybee prepromelittin in *E. coli* depends on the membrane potential but does not depend on proteins SecA and SecY, *J. Biol. Chem.* **264**:10169–10176.

Collier, D. N., Bankaitis, V. A., Weiss, J. B., and Bassford, P. J., Jr., 1988, The antifolding activity of SecB promotes the export of the *E. coli* maltose-binding protein, *Cell* **53**:273–283.

Collins, O. G., and Gilmore, R., 1991, Ribosome binding to the endoplasmic reticulum: A 180 kD protein identified by crosslinking to membrane bound ribosomes is not required for ribosome binding activity, *J. Cell Biol.* **114**:639–649.

Connolly, T., and Gilmore, R., 1989, The signal recognition particle receptor mediates the GTP-dependent displacement of SRP from the signal sequence of the nascent polypeptide, *Cell* **57**:599–610.

Cosson, P., Lankford, S. P., Bonifacio, J. S., and Klausner, R. D., 1991, Membrane protein association by potential intermembrane charge pairs, *Nature* **351**:414–416.

Cowan, S. W., Schirmer, T., Rummel, G., Steiert, M., Ghosh, R., Pauptit, R. A., Jansonius, J. N., and Rosenbush, J. P., 1993, Crystal structures explain functional properties of two *E. coli* porins, *Nature* **358**:727–733.

Craig, A. E., Kramer, J., Shilling, J., Werner-Washburne, M., Holmes, S., *et al.*, 1989, SSC1, an essential member of the yeast HSP70 multigene family, encodes a mitochondrial protein, *Mol. Cell Biol.* **9**:3000–3008.

Cramer, W. A., Cohen, F. S., Merrill, A. R., and Song, H. Y., 1990, Structure and dynamics of the colicin E1 channel, *Mol. Microbiol.* **4**:519–526.

Dalbey, R. E., 1990, Positively charged residues are important determinants of membrane-protein topology, *Trends Biochem. Sci* **15**:253–257.

Dalbey, R. E., and Wickner, W., 1986, The role of the polar carboxy-terminal domain of *Escherichia coli* leader peptidase in its translocation across the plasma membrane, *J. Biol. Chem.* **261**:13844–13849.

Dalbey, R. E., and Wickner, W., 1987, Leader peptidase of *Escherichia coli:* Critical role of a small domain in membrane assembly, *Science* **235**:783–787.

Dalbey, R. E., Kuhn, A., and Wickner, W., 1987, The internal signal sequence of *Escherichia coli* leader peptidase is necessary, but not sufficient, for its rapid membrane assembly, *J. Biol. Chem.* **262**:13241–13245.

Daniels, C. J., Bole, D. G., Quay, S. C., and Oxender, D. L., 1981, Role for membrane potential in the secretion of protein into the periplasm of *Escherichia coli*, *Proc. Natl. Acad. Sci. USA* **78**:5396–5400.

Davis, N. G., and Model, P., 1985, An artificial anchor domain: Hydrophobicity suffices to stop transfer, *Cell* **41**:607–614.

Deisenhofer, J., Epp, O., Miki, K., Huber, R., and Michel, H., 1984, X-ray structure analysis of a membrane protein complex: Electron density at 3 Å resolution and a model of the chromophores of the photosynthetic reaction center from *Rhodopseudomonas viridis*, *J. Mol. Biol.* **180**:385–398.

Deisenhofer, J., and Michel, H., 1989, The photosynthetic reaction center from the purple bacterium *Rhdopseudomonas viridis*, *Science* **245**:1463–1473.

Derman, A. I., Puziss, J. W., Bassford, P. J., Jr., and Beckwith, J., 1993, A signal sequence is not required for protein export in *prlA* mutants of *Escherichia coli*, *EMBO J.* **12**:879–888.

Deshaies, R. J., Koch, B. D., Werner-Washburne, M., Craig, E. A., and Schekman, R., 1988, A subfamily of stress proteins facilitates translocation of secretory and mitochondrial precursor polypeptides, *Nature* **332**:800–805.

Deshaies, R. J., Sanders, S. L., Feldheim, D. A., and Schekman, R., 1991, Assembly of yeast Sec proteins involved in translocation into the endoplasmic reticulum into a membrane-bound multi-subunit complex, *Nature* **349**:806–808.

Deshaies, R., and Schekman, R., 1989, Sec62 encodes a putative membrane protein required for protein translocation into the yeast endoplasmic reticulum, *J. Cell Biol.* **109**:2653–2664.

de Vitry, C., Diner, B. A., and Popot, J. L., 1991, Photosystem II particles from *Chlamydomonas reinhardtii* purification, molecular weight, small subunit composition and protein phosphorylation, *J. Biol. Chem.* **266**:16614–16620.

de Vrije, T., 1989, Studies on the role of phospholipids in the translocation process of outer membrane protein PhoE across *Eschedrichia coli* inner membranes, Ph.D. thesis, University of Utrecht.

de Vrije, G. J., de Swart, R. L., Dowhan, W., Tommassen, J., and de Kruijff, B., 1988, Phosphatidylglycerol is involved in protein translocation across *Escherichia coli* inner membranes, *Nature* **334**:173–175.

Dorner, A. J., Bole, D. G., and Kaufman, R. J., 1987, The relationship of N-linked glycosylation and heavy chain-binding protein association with the secretion of glycoproteins, *J. Cell Biol.* **105**:2665–2674.

Driessen, A. J. M., and Wickner, W., 1990, Solubilization and functional reconstitution of the protein-translocation enzymes of *Escherichia coli*, *Proc. Natl. Acad. Sci. USA* **87**:3107–3111.

Driessen, A. J. M., and Wickner, W., 1991, Proton transfer is rate-limiting for translocation of precursor proteins by the *Escherichia coli* translocase, *Proc. Natl. Acad. Sci. USA* **88**:2471–2475.

Driessen, A. J. M., 1992, Precursor protein translocation by the *Escherichia coli* translocase is directed by the protonmotive force, *EMBO J.* **11**:847–853.

Dumont, M. E., and Richards, F. M., 1984, Insertion of apocytochrome *c* into lipid vesicles, *J. Biol. Chem.* **259**:4147–4156.

Dupuy, B., Taha, M. K., Pugsley, A. P., and Marshall, C. A., 1991, *Neisseria gonorrhoeae* prepilin export studied in *Escherichia coli*, *J. Bacteriol.* **173**:7589–7598.

Ehrmann, M., and Beckwith, J., 1991, Proper insertion of a complex membrane protein in the absence of its amino-terminal export signal, *J. Biol. Chem.* **266**:16530–16534.

Eilers, M., Endo, T., and Schatz, G., 1989, Adriamycin, a drug interacting with acidic phospholipids, blocks import of precursor proteins by isolated yeast mitochondria, *J. Biol. Chem.* **264**:2945–2950.

Eilers, M., Hwang, S., and Schatz, G., 1988, Unfolding and refolding of a purified precursor protein during import into isolated mitochondria, *EMBO J.* **7**:1139–1145.

Eilers, M., Oppliger, W., and Schatz, G., 1987, Both ATP and an energized inner membrane are required to import a purified precursor protein into mitochondria, *EMBO J.* **6**:1073–1077.

Eisele, J. L., and Rosenbusch, J. P., 1990, *In vitro* folding and oligomerization of a membrane protein: Transition of bacterial porin from random coil to native conformation, *J. Biol. Chem.* **265**:10217–10220.

Ellis, R. J., and van der Vies, S. M., 1991, Molecular chaperones, *Annu. Rev. Biochem.* **60**:321–347.

Endo, T., Shimada, I., Roise, D., and Inagaki, F., 1989a, N-terminal half of a mitochondrial presequence peptide takes a helical conformation when bound to dodecylphosphocholine micelles—a proton magnetic resonance study, *J. Biochem.* **106**:396–400.

Endo, T., Eilers, M., and Schatz, R. G., 1989b, Binding of tightly folder artificial mitochondrial precursor protein to the mitochondrial membrane involves a lipid conformational change, *J. Biol. Chem.* **264**:2951–2956.

Enequist, H. G., Hirst, T. R., Hardy, S. J. S., Harayama, S., and Randall, L. L., 1981, Energy is required for maturation of exported proteins in *Escherichia coli, Eur. J. Biochem.* **116**:227–233.

Engelman, D. M., and Steitz, T. A., 1981, The spontaneous insertion of proteins into and across membranes, the helical hairpin hypothesis, *Cell* **23**:411–422.

Feldheim, D., Rothblatt, J., and Schekman, R., 1992, Topology and functional domains of Sec63p, an ER membrane protein required for secretory protein translocation, *Mol. Cell Biol.* **12**:3288–3296.

Fikes, J. D., and Bassford, P. J., 1987, Export of unprocessed precursor maltose-binding protein to the periplasm of *Escherichia coli* cells, *J. Bacteriol.* **169**:2352–2359.

Flynn, G. C., Pohl, J., Flocco, M. T., and Rothman, J. E., 1991, Peptide-binding specificity of the molecular chaperone bip, *Nature* **353**:726–728.

Froshauer, S., Green, G. N., Boyd, D., McGovern, K., and Beckwith, J., 1988, Genetic analysis of the membrane insertion and topology of MalF, a cytoplasmic membrane protein of *Escherichia coli, J. Mol. Biol.* **200**:501–511.

Gallusser, A., and Kuhn, A., 1990, Initial steps in protein membrane insertion: Bacteriophage M13 procoat protein binds to the membrane surface by electrostatic interaction, *EMBO J.* **9**:2723–2729.

Garcia, P. D., and Walter, P., 1988, Full-length prepro-α-factor can be translocated across the mammalian microsomal membrane only if translation has not terminated, *J. Cell Biol.* **106**:1043–1048.

Gardell, C., Benson, S., Hunt, J., Michaelis, S., and Beckwith, J., 1987, *secD*, a new gene involved in protein export in *Escherichia coli, J. Bacteriol.* **169**:1286–1290.

Garoff, H., 1985, Using recombinant DNA techniques to study protein targetin in the eukaryotic cell, *Annu. Rev. Cell Biol.* **1**:403–445.

Géli, V., Baty, D., and Lazdunski, C., 1988, Use of a foreign epitope as a "tag" for the localization of minor proteins within a cell: The case of the immunity protein to colicin A, *Proc. Natl. Acad. Sci. USA* **85**:689–693.

Géli, V., Baty, D., Pattus, F. P., and Lazdunski, C., 1989, Topology and function of the integral membrane protein conferring immunity to colicin A, *Mol. Microbiol.* **3**:679–687.

Géli, V., and Glick, B., 1990, Mitochondrial protein import, *J. Bioenerg. Biomembr.* **22**:725–751.

Géli, V., Yang, M., Suda, K., Lustig, A., and Schatz, G., 1990, The MAS-encoded processing protease of yeast mitochondria: Overproduction and characterization of its two nonidentical subunits, *J. Biol. Chem.* **265**:19216–19222.

Géli, V., Koorengevel, M. C., Demel, R. A., Lazdunski, C., and Killian, A., 1992, Acidic interaction of the colicin A pore-forming domain with model membranes of *E. coli* lipids results in a large perturbation of acyl chain order and stabilization of the bilayer, *Biochemistry* **31**:11089–11094.

Geller, B. L., Movva, N. R., and Wickner, W., 1986, Both ATP and the electrochemical potential are required for optimal assembly of pro-OmpA into *Escherichia coli* inner membrane vesicles, *Proc. Natl. Acad. Sci. USA* **83**:4219–4222.

Geller, B. L., and Green, H. M., 1989, Translocation of proOmpA across inner membrane vesicles of *Escherichia coli* occurs in two consecutive energetically distinct steps, *J. Biol. Chem.* **264**:16465–16469.

Gething, M. J., McCammon, K., and Sambrook, J., 1986, Expression of wild-type and mutant forms of influenza hemagglutinin: The role of folding in intracellular transport, *Cell* **46**:939–950.

Gething, H. J., and Sambrook, J., 1992, Protein folding in the cell, *Nature* **355**:33–45.

Gilmore, R., and Blobel, G., 1983, Transient involvement of signal recognition particle and its reception in the microsomal membrane prior to protein translocation, *Cell* **35:**677–685.

Gilmore, R., and Blobel, G., 1985, Translocation of secretory proteins across the microsomal membrane occurs through an environment accessible to aqueous perturbants, *Cell* **42:**497–505.

Gilmore, R., Blobel, G., and Walter, P., 1982, Protein translocation across the endoplasmic reticulum. I: Detection in the microsomal membrane of a receptor for the signal recognition particle, *J. Cell Biol.* **95:**463–469.

Glick, B., and Schatz, G., 1991, Import of proteins into mitochondria, *Annu. Rev. Genet.* **25:**21–44.

Glick, B., Brandt, A., Cunningham, K., Muller, S., Hallberg, R., and Schatz, G., 1992, Cytochrome c1 and b2 are sorted to the intermembrane space of yeast mitochondria by a stop-transfer mechanism, *Cell* **69:**809–822.

Gonzàlez-Manas, J. M., Lakey, J. H., and Pattus, F., 1992, Brominated phospholipids as a tool for monitoring the membrane insertion of colicin A, *Biochemistry* **31:**7294–7300.

Görlich, D., Prehn, S., Hartmann, E., Herz, J., Otto, A., Kraft, R., Wiedmann, M., Knespel, S., Dobberstein, B., and Rapoport, T. A., 1990, The signal sequence receptor has a second subunit and is part of a translocation complex in the endoplasmic reticulum as probed by bifunctional reagents, *J. Cell Biol.* **111:**2283–2294.

Görlich, D., Hartmann, E., Prehn, S., and Rapoport, T. A., 1992a, A protein of the endoplasmic reticulum involved early in polypeptide translocation, *Nature* **357:**47–52.

Görlich, D., Prehn, S., Hartmann, E., Kalies, K. U., and Rapoport, T. A., 1992b, A mammalian homolog of Sec61p and SecYp is associated with ribosomes and nascent polypeptides during translocation, *Cell* **71:**489–503.

Haeuptle, M. T., Flint, N., Gough, N. M., and Dobberstein, B., 1989, A tripartite structure of the signals that determine protein insertion into the endoplasmic reticulum membrane, *J. Cell Biol.* **108:**1227–1236.

Hann, B. C., Poritz, M. A., and Walter, P., 1989, *Saccharomyces cerevisiae* and *Schizosaccharomyces pombe* contain a homologue to the 54-kD subunit of the signal recognition particle that in *S. cerevisiae* is essential for growth, *J. Cell Biol.* **109:**3223–3230.

Hann, B. C., and Walter, P., 1991, The signal recognition particle in *S. cerevisiae*, *Cell* **67:**131–144.

Hansen, W., Garcia, P. D., and Walter, P., 1986, *In vitro* protein translocation across the yeast endoplasmic reticulum: ATP-dependent post-translational translocation of the prepro-α-factor, *Cell* **45:**397–406.

Hartl, F. U., Pfanner, N., Nicholson, D. W., and Neupert, W., 1989, Mitochondrial protein import, *Biochem. Biophys. Acta* **998:**1–45.

Hartl, F. U., and Neupert, W., 1990, Protein sorting to mitochondria: Evolutionary conservations of folding and assembly, *Science* **247:**930–938.

Hartl, F. U., Lecker, S., Schiebel, E., Hendrick, J. P., and Wickner, W., 1990, The binding cascade of SecB to SecA to SecY/E mediates preprotein targeting to the *E. coli* plasma membrane, *Cell* **63:**269–279.

Hartmann, D. J., Hoogenraad, N. J., Condron, R., and Hoj, P. B., 1992, Identification of a mammalian 10-kDa heat shock protein, a mitochondrial chaperonin 10 homologue essential for assisted folding of trimeric ornithine transcarbamoylase *in vitro*, *Proc. Natl. Acad. Sci. USA* **89:**3394–3398.

Hartmann, E., Rapoport, T. A., and Lodish, H. F., 1989a, Predicting the orientation of membrane-spanning proteins, *Proc. Natl. Acad. Sci. USA* **86:**5786–5790.

Hartmann, E., Wiedmann, M., and Rapoport, T. A., 1989b, A membrane component of the endoplasmic reticulum that may be essential for protein translocation, *EMBO J.* **8:**2225–2229.

Heinemeyer, W., Alt, J., and Herrmann, R. G., 1984, Nucleotide sequence of the clustered genes for apocytochrome b6 and subunit 4 of the cytochrome b/f complex in the spinach plastid chromosome, *Curr. Genet.* **8:**543–549.

Henderson, R., and Unwin, P. N. T., 1975, Three-dimensional model of purple membrane obtained by electron microscopy, *Nature* **257**:28–32.

Henderson, R., Baldwin, J. M., Ceska, T. A., Zemlin, F., Beckmann, E., and Downing, K. H., 1990, Model for the structure of bacteriorhodopsin based on light-resolution electron cryomicroscopy, *J. Mol. Biol.* **213**:899–929.

Hendrick, J. P., and Wickner, W., 1991, SecA protein needs both acidic phospholipids and SecY/E protein for functional high-affinity binding to the *Escherichia coli* plasma membrane, *J. Biol. Chem.* **266**:24596–24600.

Henry, G. D., and Sykes, B. D., 1990, Detergent solubilized M13 coat protein exists as an asymmetric dimer: Observation of individual monomers by ^{15}N, ^{13}C and ^{1}H nuclear magnetic resonance spectroscopy, *J. Mol. Biol.* **212**:11–14.

High, S., Görlich, D., Wiedmann, M., Rapoport, T. A., and Dobberstein, B., 1991, The identification of proteins in the proximity of signal-anchor sequences during their targeting to and insertion into the membrane of the ER, *J. Cell Biol.* **113**:35–44.

Hines, V., Grandt, A., Griffiths, G., Horstmann, H., Brütsch, H., and Schatz, G., 1990, Protein import into yeast mitochondria is accelerated by the outer membrane protein MAS70, *EMBO J.* **9**:3191–3200.

Hortsch, M., Avossa, D., and Meyer, D. I., 1986, Characterization of secretory protein translocation: Ribosome-membrane interaction in endoplasmic reticulum, *J. Cell Biol.* **103**:241–253.

Hortsch, M., Labeit, S., and Meyer, D. I., 1988, Complete cDNA sequence coding for human docking protein, *Nucleic Acids Res.* **16**:361–362.

Huang, K. S., Bayley, H., Liao, M. J., London, E., and Khorana, H. G., 1981, Refolding of an integral membrane protein: Denaturation, renaturation, and reconstitution of intact bacteriorhodopsin and two proteolytic fragments, *J. Biol. Chem.* **256**:3802–3809.

Hwang, S., Jascur, T., Vestweber, D., Pon, L., and Schatz, G., 1989, Disrupted yeast mitochondria can import precursor proteins directly through their inner membrane, *J. Cell Biol.* **109**:487–493.

Hwang, S., and Schatz, G., 1989, Translocation of protein across mitochondrial inner membrane, but not into the outer membrane, requires necleoside triphosphates in matrix, *Proc. Natl. Acad. Sci. USA* **86**:8432–8436.

Ibrahimi, I. M., Cutler, D., Stueber, D., and Bujard, H., 1986, Determinant for protein translocation across mammalian endoplasmic reticulum: Membrane insertion of truncated and full-length prelysozyme molecules, *Eur. J. Biochem.* **155**:571–576.

Ito, K., and Akyama, Y., 1991, *In vivo* analysis of integration of membrane proteins in *Escherichia coli*, *Mol. Microbiol.* **5**:2243–2253.

Jap, B. K., 1989, Molecular design of PhoE porin and its functional consequences, *J. Mol. Biol.* **205**:407–419.

Jennings, M. L., 1989, Topography of membrane proteins, *Annu. Rev. Biochem.* **58**:999–1027.

Joly, J. C., and Wickner, W., 1993, The SecA and SecY subunits of translocase are the nearest neighbors of a translocating preprotein, shielding it from phospholipds, *EMBO J.* **12**:255–263.

Jordi, W., 1990, The molecular mechanism of import of the precursor protein apocytorhrome *c* into mitochondria, Ph.D. thesis, University of Utrecht.

Kaback, H. R., Bibi, E., and Roepe, P., 1990, β-galactoside transport in *E. coli*: A functional dissection of lac permease, *Trends Biochem. Sci.* **15**:309–314.

Kaiser, C. A., Preuss, D., Grisafi, P., and Botstein, D., 1987, Many random sequences functionally replace the secretion signal sequence of yeast invertase, *Science* **235**:312–317.

Kallas, T., Spiller, S., and Malkin, R., 1988f, Characterization of two operons encoding the cytochrome b6-f complex of the *cyanobacterium Nostoc* PCC 7906, *J. Biol. Chem.* **263**:14334–14342.

Kang, P. J., Ostermann, J., Shilling, J., Neupert, W., Craig, E. A., and Pfanner, N., 1990,

Requirement for hsp70 in the mitochondrial matrix for translocation and folding of precursor proteins., *Nature* **348**:137–142.

Kellaris, K. V., Bowen, S., and Gilmore, R., 1991, ER translocation intermediates are adjacent to a non-glycosylated 34 kD integral membrane protein, *J. Cell Biol.* **114**:21–33.

Kelleher, D. J., Kreibich, G., and Gilmore, R., 1992, Oligosaccharytransferase activity is associated with a protein complex composed of Ribophorins I and II and a 48kd protein, *Cell* **69**:55–65.

Kiebler, M., Pfaller, R., Sollner, T., Griffiths, G., Horstmann, H., Pfanner, N., and Neupert, W., 1990, Identification of a mitochondrial receptor complex required for recognition and membrane insertion of precursor protein, *Nature* **348**:610–616.

Killian, J. A., de Jong, A. M. P., Bijvelt, J., Verkleij, A. J., and de Kruiff, B., 1990, Induction of non-bilayer lipid structures by functional signal peptides, *EMBO J.* **9**:815–819.

Kimura, E., Akita, M., Matsuyama, S., and Mizushima, S., 1991, Determination of a region in SecA that interacts with presecretory proteins in *Escherichia coli*, *J. Biol. Chem.* **266**:6600–6606.

Klose, M., Schwart, H., MacIntyre, S., Frendl, R., Eschbach, M. L., and Henning, U., 1988, Internal deletions in the gene for an *Escherichia coli* outer membrane protein define an area possibly important for recognition of the outer membrane by this polypeptide, *J. Biol. Chem.* **263**:13291–13296.

Kobilka, B., Kobilka, T., Daniel, K., Regan, J., Caron, M., and Lefkowitz, R., 1988, Chimeric α2,β2-adrenergic receptors: Delineation of domains involved in effector coupling and ligand binding specificity, *Science* **240**:1310–1316.

Koll, H., Guiard, B., Rassow, J., Ostermann, J., Horwich, A. L., Neupert, W., and Hartl, F. U., 1992, Antifolding activity of hsp60 couples protein import into the mitochondrial matrix with export to the intermembrane space, *Cell* **68**:1163–1175.

Kozutsumi, Y., Segal, M., Normington, K., Gething, M-J., and Sambrook, J., 1988, The presence of malfolded proteins in the endoplasmic reticulum signals the induction of glucose-regulated proteins, *Nature* **332**:462–464.

Kreibich, G., Ulrich, B. L., and Sabatini, D. D., 1978a, Proteins of rough microsomal membranes related to ribosome binding: Identification of ribophorins I and II, membrane proteins characteristic of rough microsomes, *J. Cell Biol.* **77**:464–487.

Kreibich, G., Freienstein, C. M., Pereyra, B. N., Ulrich, B. L., and Sabatini, D. D., 1978b, Proteins of rough microsomal membranes related to ribosome binding: Cross-linking of bound ribosomes to specific membrane proteins exposed at the binding sites, *J. Cell Biol.* **77**:488–506.

Krieg, U. C., Johnson, A. E., and Walter, P., 1989, Protein translocation across the endoplasmic reticulum membrane: Identification by photocrosslinking of the 39-kD integral membrane glycoprotein as part of a putative translocation tunnel, *J. Cell Biol.* **109**:2033–2043.

Krieg, U. C., Walter, P., and Johnson, A. E., 1986, Photocrosslinking of the signal sequence of nascent preprolactin to the 54 kilodalton polypeptide of the signal recognition particle, *Proc. Natl. Acad. Sci. USA* **83**:8604–8608.

Kuhn, A., 1987, Bacteriophage M13 procoat proteins inserts into the plasma membrane as a loop structure, *Science* **238**:1413–1415.

Kuhn, A., 1988, Alterations in the extracellular domain of M13 procoat protein make its membrane insertion dependent on SecA and SecY, *Eur. J. Biochem.* **177**:267–271.

Kuhn, A., Kreil, G., and Wickner, W., 1986, Both hydrophobic domains of M13 procoat are required to initiate membrane insertion, *EMBO J.* **5**:3681–3685.

Kuhn, A., Kreil, G., and Wickner, W., 1987, Recombinant forms of M13 procoat with an OmpA leader sequence or a large carboxy-terminal extension retain their independence of SecY function, *EMBO J.* **6**:501–505.

Kühlbrandt, W., 1988, Three-dimensional crystallization of membrane proteins, *Rev. Biophys.* **21**:429–477.

Kühlbrandt, W., and Wang, D. N., 1991, Three-dimensional structure of plant light-harvesting complex determined by electron crystallography, *Nature* **350**:130–134.

Kumamoto, C. A., 1989, *Escherichia coli* SecB protein associates with exported protein precursors *in vivo*, *Proc. Natl. Acad. Sci. USA* **86**:5320–5324.

Kumamoto, C. A., and Beckwith, J., 1983, Mutations in a new genee, SecB, cause defective protein localization in *Escherichia coli*, *J. Bacteriol.* **154**:253–260.

Kurzchalia, T. V., Wiedmann, M., Girshowich, A. S., Bochkareva, E. S., Bielka, H., and Rapoport, T. A., 1986, The signal sequence of nascent preprolactin interacts with the 54 K polypeptide of the signal recognition particle, *Nature* **320**:634–636.

Kusakawa, N., Yura, T., Uequichi, C., Akiyama, Y., and Ito, K., 1989, Effects of mutations in heat-shock genes *groES* and *groEL* on protein export in *Escherichia coli*, *EMBO J.* **8**:3517–3521.

Kusters, R., Dowhan, W., and de Kruijff, B., 1991, Negatively charged phospholipids restore prePhoE translocation across phosphatidylglycerol-depleted *Escherichia coli* inner membranes, *J. Biol. Chem.* **266**:8659–8662.

Lakey, J., Massote, D., Heitz, F., Dasseux, J. L., Faucon, J. F., Parker, M. W., and Pattus, F., 1990, Membrane insertion of the pore-forming domain of colicin A, *Eur. J. Biochem.* **196**:599–607.

Laminet, A. A., Ziegelhoffer, T., Georgopoulos, C., and Plückthun, A., 1990, The *Escherichia coli* heat shock proteins GroEL and GroES modulate the folding of the β-lactamase precursor, *EMBO J.* **9**:2315–2319.

Lauffer, L., Garcia, P. B., Harkins, R. N., Coussens, L., Ullrich, A., and Walter, P., 1985, Topology of signal recognition particle receptor in endoplasmic reticulum membrane, *Nature* **318**:334–338.

Laws, J. K., and Dalbey, R. E., 1989, Positive charges in the cytoplasmic domain of *Escherichia coli* leader peptidase prevent an apolar domain from functioning as a signal, *EMBO J.* **8**:2095–2099.

Lazdunski, C., Baty, D., Géli, V., Cavard, D., Morlon, J., Lloubès, R., Howard, P., Knibiehler, M., Chartier, M., Varenne, S., Frenette, M., Dasseux, J. L., and Pattus, F., 1988, The membrane channel-forming colicin A: Synthesis, secretion, structure, action and immunity, *Biochem. Biophys. Acta* **947**:445–464.

Lear, J. D., Wassermann, Z. R., and Degrado, W. F., 1988, Synthetic amphiphilic peptide models for protein ion channels, *Science* 1177–1181.

Lecker, S., Lill, R., Ziegelhofer, T., Bassford, P. J. J., Kumamoto, C. A., and Wickner, W., 1989, Three pure chaperone proteins of *Escherichia coli*, SecB trigger factor, and GroEL, form soluble complexes with precursor proteins *in vivo*, *EMBO J.* **8**:2703–2709.

Lemire, B. D., Frankhauser, C., Baker, A., and Schatz, G., 1989, The mitochondrial targeting function of randomly generated peptide sequences correlates with predicted helical amphiphilicity, *J. Biol. Chem.* **264**:20206–20215.

Lemmon, M. A., Flanagan, J. M., Hunt, J. F., Adair, B. D., Bormann, B. J., Dempsey, C. E., and Engelman, D. M., 1992, Glycophorin A dimerization is driven by specific interactions between transmembrane α-helices, *J. Biol. Chem.* **267**:7683–7689.

Letellier, L., 1992, Bacteriocin and bacteriophage channels in prokaryotes in *Alkali Cation Transport in Prokaryotes* (E. Bakker, ed.) CRC Press, Boca Raton, Florida, pp. 359–376.

Li, P., Beckwith, J., and Inouye, H., 1988, Alteration of the amino terminus of the mature sequence of a periplasmic protein can severely affect protein export in *Escherichia coli*, *Proc. Natl. Acad. Sci. USA* **85**:7685–7689.

Li, J. Carroll, J., and Ellar, D. J., 1991, Crystal structure of insectiditial δ-endotoxin from *Bacillus thuringiensis* at 2.5 Å resolution, *Nature* **353**:815–821.

Liao, M. J., London, E., and Khorana, H. G., 1983, Regeneration of native bacteriorhodopsin from two chymotryptic fragments, *J. Biol. Chem.* **258**:9949–9955.

Lill, R., Cunningham, K., Brundage, L., Ito, K., Oliver, D., and Wickner, W., 1989, SecA protein hydrolyzes ATP and is an essential component of the protein translocation ATPase of *Escherichia coli*, *EMBO J.* **8**:961–966.

Lill, R., Dowhan, W., and Wickner, W., 1990, The ATPase activity of SecA is regulated by acidic phospholipids, SecY, and the leader and mature domains of precursor proteins, *Cell* **60**:271–280.

London, E., 1992, Diphtheria toxin: Membrane interaction and membrane translocation, *Biochem. Biophys. Acta* **113**:25–51.

MacIntyre, S., Freudl, R., Eschbach, M. L., and Henning, U., 1988, An artificial hydrophobic sequence functions as either an anchor or a signal sequence at only one of two proteins within the *Escherichia coli* outer membrane protein OmpA, *J. Biol. Chem.* **263**:19053–19059.

Manning-Krieg, U. C., Scherer, P. E., and Schatz, G., 1991, Sequential action of mitochondrial chaperones in protein import into the matrix, *EMBO J.* **10**:3273–3280.

Manoil, C., and Beckwith, J., 1986, A genetic approach to analyzing membrane protein topology, *Science* **233**:1403–1408.

Manolios, N., Bonifacino, J. S., and Klausner, R. S., 1990, Transmembrane helical interactions and the assembly of the T-cell receptor complex, *Science* **249**:274–277.

Marcantonio, E. E., Amar-Costesec, A., and Kreibich, G., 1984, Segregation of the polypeptide translocation apparatus to regions of the endoplasmic reticulum containing ribophorins and ribosomes: Rat liver microsomal subfractions contain equimolar amounts of ribophorins and ribosomes, *J. Cell Biol.* **99**:2254–2259.

Martinez, M. C., Lazdunski, C., and Pattus, F., 1983, Isolation, molecular and functional properties of the C-terminal domain of colicin A, *EMBO J.* **2**:1501–1507.

Matsuyama, S. E., Fujita, Y., and Mizushima, S., 1992, SecD is involved in the release of translocated secretory proteins from the cytoplasmic membrane of *Escherichia coli*, *EMBO J.* **12**:265–270.

McGovern, K., Ehrmann, M., and Beckwith, J., 1991, Decoding signals for membrane protein assembly using alkaline phosphatase fusions, *EMBO J.* **10**:2773–2782.

McGovern, K., and Beckwith, J., 1991, Membrane insertion of the *Escherichia coli* MalF protein in cells with impaired secretion machinery, *J. Biol. Chem.* **266**:20870–20876.

McMullin, T. W., and Hallberg, R. L., 1988, A highly evolutionarily conserved mitochondrial protein is structurally related to the protein encoded by the *Escherichia coli groEL* gene, *Mol. Cell Biol.* **8**:371–380.

Merrill, A. R., and Cramer, W. A., 1990, Identification of a voltage-responsive segment of the potential-gated colicin E1 ion channel, *Biochemistry* **29**:8529–8534.

Meyer, D. E., 1991, Protein translocation into the endoplasmic reticulum: A light at the end of the tunnel, *Trends Cell Biol.* **1**:154–159.

Migliaccio, G., Nicchitta, C. V., and Blobel, G., 1992, The signal sequence receptor, unlike the signal recognition particle receptor, is not essential for protein translocation, *J. Cell Biol.* **117**:15–25.

Moskaug, J. O., Sletten, K., Sandvig, K., and Olnes, S., 1989, Translocation of diphtheria toxin A-fragment to the cytosol: Role of the site of interfragment cleavage, *J. Biol. Chem.* **264**:15709–15713.

Mueckler, M., and Lodish, H. F., 1986a, Post-translational insertion of a fragment of the glucose transporter into microsomes requires phosphoanhydride bond cleavage, *Nature* **322**:549–552.

Mueckler, M., and Lodish, H. F., 1986b, The human glucose transporter can insert post-translationally into microsomes, *Cell* **44**:629–637.

Müller, G., and Zimmermann, R., 1987, Import of honeybee prepromelittin into the endoplasmic reticulum: Structural basis for independence of SRP and docking protein, *EMBO J.* **6**:2099–2107.

Müller, G., and Zimmermann, R., 1988, Import of honeybee prepromelittin into the endoplasmic reticulum: Energy requirements for membrane insertion, *EMBO J.* **7**:639–648.

Müller, M., 1992, Proteolysis in protein important export: Signal peptide processing in eu- and prokaryotes, *Experentia* **48**:118–129.

Müller, M., and Blobel, G., 1984, In vitro translocation of bacterial proteins across the plasma membrane of *Escherichia coli, Proc. Natl. Acad. Sci. USA* **81**:7421–7425.

Müller, M., Fisher, R. P., Rienhöfer-Schweer, A., and Hoffschulte, H. K., 1987, DCCD inhibits protein translocation into plasma membrane vesicles from *Escherichia coli* at two different steps, *EMBO J.* **6**:3855–3861.

Munro, S., and Pelham, H. R. B., 1986, An hsp70-like protein in the ER: Identity with the 78 kD glucose regulated protein and immunoglobulin heavy chain binding protein, *Cell* **46**:291–300.

Müsch, A., Wiedmann, M., and Rapoport, T. A., 1992, Yeast sec proteins interact with polypeptides traversing the endoplasmic reticulum membrane, *Cell* **69**:343–352.

Nabedryk, E., Garavito, R. M., and Breton, J., 1988, The orientation of β-sheets in porin: A polarized fourier transform infared spectroscopic investigation, *Biophys. J.* **53**:671–676.

Nestel, U., Wacker, T., Waitzik, D., Weckesser, J., Kreutz, W., and Welte, W., 1989, Crystallization and preliminary X-ray analysis of porin from *Rhodobacter capsulatus, FEBS Lett.* **242**:405–408.

Newman, M. J., Foster, D. C., Wilson, T. H., and Kaback, H. R., 1981, Purification and reconstitution of functional lactose carrier from *Escherichia coli, J. Biol. Chem.* **256**:11804–11808.

Nikaïdo, H., and Vaara, M., 1985, Molecular basis of bacterial outer membrane permeability, *Microbiol. Rev.* **49**:1–32.

Nilsson, I. M., and von Heijne, G., 1990, Fine-tuning the topology of a polytopic membrane protein: Role of positively and negatively charged amino acids, *Cells* **62**:1135–1141.

Nishiyama, K., Mizushima, S., and Tokuda, H., 1992, The carboxyl-terminal region of SecE interacts with SecY and is functional in the reconstitution of protein translocation activity in *Escherichia coli, J. Biol. Chem.* **267**:7170–7176.

Nunn, D., and Lory, S., 1991, Product of the *Pseudomonas aeruginosa* gene pilD is a prepilin leader peptidase, *Proc. Natl. Acad. Sci. USA* **88**:3281–3285.

Nunnari, J. M., Zimmerman, O. L., Ogg, S. C., and Walter, P., 1991, Characterization of the rough endoplasmic reticulum ribosome binding activity, *Nature* **352**:638–640.

Oesterhelt, D., and Stoeckenius, W., 1973, Functions of a new photoreceptor membrane, *Proc. Natl. Acad. Sci. USA* **70**:2853–2857.

Oiki, S., Danho, W., and Montal, M., 1988, Channel protein engineering: Synthetic 22-mer peptide from the primary structure of the voltage-sensitive sodium channel forms ionic channels in lipid bilayers, *Proc. Natl. Acad. Sci. USA* **85**:2392–2397.

Oliver, D. B., and Beckwith, J., 1982, Identification of a new gene (*secA*) and gene product involved in the secretion of envelope proteins in *Escherichia coli, J. Bacteriol.* **150**:686–691.

Ooi, C. E., and Weiss, J., 1992, Bidirectional movement of a nascent polypetide across microsomal membranes reveals requirements for vectorial translocation of proteins, *Cell* **71**:87–96.

Osterman, J., Horwich, A. L., Neupert, W., and Hartl, F. U., 1989, Protein folding in mitochondria requires complex formation with hsp60 and ATP hydrolysis, *Nature* **341**:125–130.

Parker, M. W., Pattus, F., Tucker, A. D., and Tsernoglou, D., 1989, Structure of the membrane-pore-forming fragment of colicin A, *Nature* **337**:93–96.

Parker, M. W., Tucker, A. D., Tsernoglou, D., and Pattus, F., 1990, Insights into membrane insertion based on studies of colicins, *Trends Biochem. Sci.* **15**:126–129.

Pattus, F., Massotte, D., Wiulmsen, H. U., Lakey, J., Tsernoglou, D., Tucker, A., and Parker, M., 1990, Colicins: Prokaryotic killer pores, *Experientia* **46**:180–192.

Pelham, H. R. B., 1986, Speculations on the functions of the major heat shock and glucose-regulated proteins, *Cell* **46**:959–961.

Perara, E., Rothman, R. E., and Lingappa, V. R., 1986, Uncoupling translocation from translation: Implications for transport of proteins across membranes, *Science* **232**:348–352.

Pfanner, N., Hartl, F. U., Guiard, B., and Neupert, W., 1987a, Mitochondrial precursor proteins are imported through a hydrophilic membrane environment, *Eur. J. Biochem.* **169:**289–293.

Pfanner, N., Tropschug, M., and Neupert, W., 1987b, Mitochondrial protein import: Nucleoside triphosphates are involved in conferring import-competence to precursors, *Cell* **49:**815–823.

Pfanner, N., Rassow, J., van der Klei, I., and Neupert, W., 1992, A dynamic model of the mitochondrial protein import machinery, *Cell* **68:**999–1002.

Phillips, G., and Silhavey, T. J., 1990, Heat-shock proteins DnaK and GroEL facilitate export of LacZ hybrid proteins in *E. coli, Nature* **344:**882–884.

Popot, J. L., and de Vitry, C., 1990, On the microassembly of integral membrane proteins, *Annu. Rev. Biophys. Chem.* **19:**369–403.

Popot, J. L., and Engelman, D. M., 1990, Membrane protein folding and oligomerization: The two-stage model, *Biochemistry* **29:**4031–4033.

Popot, J. L., Gerchman, S. E., and Engelman, D. M., 1987, Refolding of bacteriorhodopsin in lipid bilayers: A thermodynamically controlled two-stage process, *J. Mol. Biol.* **198:**655–676.

Popot, J. L., Trewhella, J., and Engelman, D. M., 1986, Reformation of crystalline purple membrane from purified bacteriorhodpsin fragments, *EMBO J.* **5:**3039–3044.

Popot, J. L., de Vitry, C., and Atteia, A., 1993, Folding and assembly of integral membrane proteins: An introduction, in *Membrane Protein Structure: Experimental Approaches* (S. H. White, ed.), Oxford University Press, in press.

Randall, L. L., Topping, T. B., and Hardy, S. J. S., 1990, No specific recognition of leader peptide by SecB, a chaperone involved in protein export, *Science* **248:**860–863.

Reading, D. S., Hallberg, R. L., and Meyers, A. M., 1989, Characterization of the yeast HSP60 gene coding for a mitochondrial assembly factor, *Nature* **337:**655–659.

Rothblatt, J. A., Deshaies, R. J., Sanders, S. L., Daum, G., and Schekman, R., 1989, Multiple genes are required for proper insertion of secretory proteins into the endoplasmic reticulum in yeast, *J. Cell Biol.* **109:**2641–2652.

Rothblatt, J. A., and Meyer, D. I., 1986, Secretion in yeast: Translocation and glycosylation of prepro-α-factor *in vitro* can occur via ATP-dependent post-translational mechanism, *EMBO J.* **5:**1031–1036.

Rothman, R. E., Andrews, D. W., Calayag, M. C., and Lingappa, V. R., 1988, Construction of defined polytopic integral transmembrane proteins: The role of signal and stop transfer sequence permutations, *J. Biol. Chem.* **263:**10470–10480.

Saier, M. H., Jr., Werner, P. K., and Muller, M., 1989, Insertion of proteins into bacterial membranes: Mechanisms, characteristics and comparisons with the eukaryotic process, *Microbiol. Rev.* **53:**333–366.

Sanders, S. L., Whitfield, K. M., Vogel, J. P., Rose, M. D., and Schekman, R. W., 1992, Sec61p and BIP directly facilitate polypeptide translocation into the ER, *Cell* **69:**353–365.

Sass, H. J., Büldt, G., Beckmann, E., Zemlin, F., van Heel, M., Zeitler, E., Rosenbusch, J. P., Dorset, D. L., and Massalski, A, 1989, Densely packed β-structure at the protein-lipid interface of porin is revealed by higher-resolution cryo-electron microscopy, *J. Mol. Biol.* **209:**171–175.

Savitz, A. J., and Meyer, D. I., 1990, Identification of a ribosome receptor in the rough endoplasmic reticulum, *Nature* **346:**540–544.

Schatz, P., Bieker, K., Otteman, K., Silhavy, T. J., and Beckwith, J., 1991, One of three transmembrane stretches is sufficient for the functioning of the SecE protein, a membrane component of the *E. coli* secretion machinery, *EMBO J.* **10:**1749–1757.

Schatz, P. J., Riggs, P. D., Jacq, A., Fath, M. J., and Beckwith, J., 1989, The SecE gene encodes an integral membrane protein required for protein export in *Escherichia coli, Genes Dev.* **3:**1035–1044.

Schatz, P. J., and Beckwith, J., 1990, Genetic analysis of protein export in *Escherichia coli, Annu. Rev. Genet.* **24:**215–248.

Scherer, P. E., Krieg, U. C., Hwang, S. T., Vestweber, D., and Schatz, G., 1990, A precursor protein partly translocated into yeast mitochondria is bound to a 70 kd mitochondrial stress protein, *EMBO J.* **9:**4310–4322.

Scherer, P. E., Manning-Krieg, U. C., Jenö, P., Schatz, G., and Horst, M., 1992, Identification of a 45-kDa protein at the protein import site of the yeast mitochondrial inner membrane, *Proc. Natl. Acad. Sci. USA* **89:**11930–11934.

Schiebel, E., Driessen, A. J. M., Hartl, F. U., and Wickner, W., 1991, $\Delta\mu H^+$ and ATP function at different steps of the catalytic cycle of preprotein translocase, *Cell* **64:**927–939.

Schiebel, E., and Wickner, W., 1992, Preprotein translocation creates a halide anion permeability in the *Escherichia coli* plasma membrane, *J. Biol. Chem.* **267:**7505–7510.

Schiffer, M., Chang, C. H., and Stevens, F. J., 1992, The functions of trytophan residues in membrane proteins, *Protein Eng.* **5:**213–214.

Schlenstedt, G., and Zimmermann, R., 1987, Import of frog prepropetide Gla into microsomes requires ATP but doe snot involve docking protein or ribosomes, *EMBO J.* **6:**699–703.

Schlenstedt, G., Gudrundsson, G. H., Boman, H. G., and Zimmermann, R., 1990, A large presecretory protein translocates both cotranslationally, using signal recognition particle and ribosome, and post-translationally, without these ribonucleoparticles, when synthesized in the presence of mammalian microsomes, *J. Biol. Chem.* **265:**13960–13968.

Schleyer, M., and Neupert, W., 1985, Transport of proteins into mitochondria: Translocation intermediates spanning contact sites between outer and inner membranes, *Cell* **43:**339–350.

Shiba, K., Ito, K., Yura, T., and Ceretti, D. P., 1984, A defined mutation in the protein export gene within the *spc* ribosomal protein operon of *Escherichia coli:* Isolation and characterization of a new temperature-sensitive *secY* mutant, *EMBO J.* 631–635.

Siefermann-Harms, D., 1985, Carotenoids in photosynthesis: location in photosynthetic membranes and light harvesting function, *Biochem. Biophys. Acta* **811:**325–355.

Siegel, V., and Walter, P., 1988, The affinity of signal recognition particle for presecretory proteins is dependent on nascent chain length, *EMBO J.* **7:**1769–1775.

Simon, S. M., and Blobel, G., 1991, A protein-conducting channel in the endoplasmic reticulum, *Cell* **65:**371–380.

Simon, S. M., and Blobel, G., 1992, Signal peptides open protein-conducting channels in *E. coli*, *Cell* **69:**677–684.

Simon, S. M., Peskin, C. S., and Oster, G. F., 1992, What drives the translocation of proteins? *Proc. Natl. Acad. Sci. USA* **89:**3770–3777.

Singer, S. J., 1971, The molecular organization of membranes, in *Structure and Function of Biological Membranes* (L. I. Rothfield, ed.), pp. 145–222, Academic Press, New York.

Singer, S. J., 1976, The fluid mosaic model of membrane structure, in *The Structure of Biological Membranes* (S. Abrahamsson and I. Passcher, eds.), pp. 443–461, Plenum Press, New York.

Singer, S. J., 1990, The structure and insertion of integral proteins in membranes, *Annu. Rev. Cell Biol.* **6:**247–296.

Skowyra, D., Geogopoulos, C., and Zyuck, M., 1990, The *E. coli* DnaK gene product, the HSP70 homolog, can reactivate heat-inactive RNA polymerase in an ATP hydrolysis-dependent manner, *Cell* **62:**939–944.

Slatin, S. L., 1988, Colicin E1 in planar lipid bilayers, *Int. J. Biochem.* **20:**737–744.

Smith, R., Thomas, D. E., Separovic, F., Atkins, A. R., and Cornell, B. A., 1989, Determination of the structure of a membrane-incorporated ion channel: Solid state nuclear magnetic resonance studies of gramicidin A, *Biophys. J.* **56:**307–314.

Söllner, T., Griffiths, G., Pfaller, R., and Neupert, W., 1989, MOM19, an import receptor for mitochondrial precursor proteins, *Cell* **59:**1061–1070.

Söllner, T., Pfaller, R., Griffiths, G., Pfanner, N., and Neupert, W., 1990, A mitochondrial import receptor for the ADP/ATP carrier, *Cell* **62:**107–115.

Stader, J., Gansheroff, L. J., and Silhavy, T. J., 1989. New suppressors of signal sequence mutations, *prl G*, are linked tightly to the *secE* gene of *Escherichia coli*, *Genes Div.* **3**:1045–1052.

Stirling, C. J., Rothblatt, J., Hosobushi, M., Deshaies, R., and Schekman, R., 1992, Protein translocation mutants defective in the insertion of integral membrane proteins into the endoplasmic reticulum, *Mol. Cell Biol.* **3**:129–142.

Strom, M. S., and Lory, S., 1986, Cloning and expression of the Pil gene of *Pseudomonas aeruginosa* PAK in *Escherichia coli*, *J. Bacteriol.* **165**:367–372.

Szczesna-Skorupa, E., Browne, N., Mead, D., and Kemper, B., 1988, Positive charges at the NH2 terminus convert the membrane-anchor signal peptide of cytochrome P-450 to a secretory signal peptide, *Proc. Natl. Acad. Sci. USA* **85**:738–742.

Szczesna-Skorupa, E., and Kemper, B., 1989, NH$_2$-terminal substitutions of basic amino acids induce translocation across the microsomal membrane and glycosylation of rabbit cytochrome P450C2, *J. Cell Biol.* **108**:1237–1243.

Tani, K., Shiozuka, K., Tokuda, H., and Mizushima, S., 1989, *In vitro* analysis of the process of translocation of OmpA across the *Escherichia coli* cytoplasmic membrane: A translocation intermediate accumulates transiently in the absence of the protonmotive force, *J. Biol. Chem.* **264**:18582–18588.

Tajima, S., Lauffer, L., Rath, V. L., and Walter, P., 1986, The signal recognition particle receptor is a complex that contains two distinct polypeptide chains, *J. Cell Biol.* **103**:1167–1178.

Tazawa, S., Umura, M., Tondokoro, J., Asano, Y., Ohsumi, T., Ichimura, T., and Sugano, H., 1991, Identification of a membrane protein responsible for ribosome binding in rough microsomal membranes, *J. Biochem.* **109**:89–98.

Thrift, R. N., Andrews, D. W., Walter, P., and Johnson, A. E., 1991, The transmembrane segment of a nascent membrane protein is located adjacent to specific ER membrane proteins until termination of protein synthesis, *J. Cell Biol.* **112**:809–821.

Tommassen, J., Filloux, A., Bally, M., Murgier, M., and Lazdunski, A., 1992, Protein secretion in *Pseudomonas aeruginosa*, *FEMS Microbiol. Rev.* **103**:73–90.

Toyn, J., Hibbs, A. R., Sanz, P., Goure, J., and Meyer, D. I., 1988, In vivo and in vitro analysis of ptl 1, a yeast is mutant with a membrane-associated defect in protein translocation, *EMBO J.* **7**:4347–4353.

Van der Goot, F. G., Gonzàlaz-Manas, J. M., Lakey, J. H., and Pattus, F., 1991, A "molten-globule" membrane-insertion intermediate of the pore-forming domain of colicin A, *Nature* **354**:408–411.

Vestweber, D., Brunner, J., Baker, A., and Schatz, G., 1989, A 42K outer-membrane protein is a component of the yeast mitochondrial protein import site, *Nature* **341**:205–209.

Vitanen, P., Garcia, M. L., and Kaback, H. R., 1984, Purified reconstituted lac carrier protein from *Escherichia coli* is fully functional, *Proc. Natl. Acad. Sci. USA* **81**:1629–1633.

Vogel, J. P., Misra, L. M., and Rose, M. D., 1990, Loss of BLP/GRP 78 function blocks translocation of secretory proteins in yeast, *J. Cell Biol.* **110**:1885–1895.

von Heijne, G., 1985, Signal sequences: The limits of variation, *J. Mol. Biol.* **184**:99–105.

von Heijne, G., 1986a, The distribution of positively charged residues in bacterial inner membrane proteins correlates with the trans-membrane topology, *EMBO J.* **5**:3021–3027.

von Heijne, G., 1986b, Mitochondrial targeting sequences may form amphiphilic helices, *EMBO J.* **5**:1335–1342.

von Heijne, G., 1988, Transcending the impenetrable: How proteins come to terms with membranes, *Biochem. Biophys. Acta* **947**:307–333.

von Heijne, G., 1989, Control of topology and mode of assembly of a polytopic membrane protein by positively charged residues, *Nature* **341**:456–458.

von Heijne, G., 1990, The signal peptide, *J. Membr. Biol.* **115**:195–201.

von Heijne, G., and Gavel, Y., 1988, Topogenic signals in integral membrane proteins, *Eur. J. Biochem.* **174**:671–678.

von Heijne, G., and Manoil, C., 1990, Membrane proteins: From sequence to structure, *Protein Eng.* **4**:109–112.

von Heijne, G., Steppahn, J., and Herrmann, R. G., 1989, Domain structure of mitochondrial and chloroplast targeting peptides, *Eur. J. Biochem.* **180**:535–545.

von Heijne, G., Wickner, W., and Dalbey, R. E., 1988, The cytoplasmic domain of *Escherichia coli* leader peptidase is a "translocation poison" sequence, *Proc. Natl. Acad. Sci. USA* **85**:3363–3366.

Wachter, C., Schatz, G., and Glick, B. S., 1992, Role of ATP in the intramitochondrial sorting of cytochrome c1 and the adenine nucleotide translocator, *EMBO J.* **11**:4787–4794.

Wada, I., Rindress, D., Cameron, P. H., Ou, W-J., Doherty, J. J., Louvard, D., Bell, A. W., Dignard, D., Thomas, D. Y., and Bergeron, J. J. M., 1991, SSR and associated calnexin are major calcium binding proteins of the endoplasmic reticulum membrane, *J. Biol. Chem.* **266**:19599–19610.

Walter, P., Ibrahimi, I., and Blobel, G., 1981, Translocation of proteins across the endoplasmic reticulum: Signal recognition protein (SRP) binds to in vitro-assembled polysomes synthesizing secretory proteins, *J. Cell Biol.* **91**:545–550.

Waters, G., and Blobel, G., 1986, Secretory protein translocation in a yeast cell-free system can occur post-translationally and requires ATP hydrolysis, *J. Cell Biol.* **12**:1543–1550.

Watson, M. E. E., 1984, Compilation of published signal sequences, *Nucleic Acid Res.* **12**:5145–5164.

Webster, R. E., 1991, The *tol* gene products and the import of macromolecules into *Escherichia coli*, *Mol. Microbiol.* **5**:1005–1011.

Weiss, M. S., Kreusch, A., Schiltz, E., Nestel, U., Welte, W., Weckesser, J., and Schulz, G. E., 1991, The structure of porin from *Rhodobacter capsulatus* at 1.8 Å resolution, *FEBS Lett.* **280**:379–382.

Wessels, H. P., and Spiess, M., 1988, Insertion of a multispanding membrane protein occurs sequentially and requires only one signal sequence, *Cell* **55**:61–70.

White, S. H., and Jacobs, R. E., 1990, Observations concerning the topology and locations of helix ends of membrane proteins of known structure, *J. Membr. Biol.* **115**:145–158.

Wickner, W., Driessen, A. J. M., and Hartl, F. U., 1991, The enzymology of protein translocation across the *Escherichia coli* plasma membrane, *Annu. Rev. Biochem.* **60**:101–121.

Wiedmann, M., Kurzchalia, T., Hartmann, E., and Rapoport, T., 1987, A signal sequence receptor in the endoplasmic reticulum membrane, *Nature* **328**:830–833.

Wiedmann, M., Goerlich, D., Hartmann, E., Kurzchalia, T. V., and Rapoport, T. A., 1989, Photo crosslinking demonstrates proximity of a 34 kDa membrane protein to different portions of preprolactin during translocation through the endoplasmic reticulum, *FEBS Lett.* **257**:263–268.

Wolfe, P. B., Rice, M., and Wickner, W., 1985, Effects of two *sec* genes on protein assembly into the plasma membrane of *Escherichia coli*, *J. Biol. Chem.* **260**:1836–1841.

Wolin, S. L., and Walter, P., 1989, Signal recognition particle mediates a transient elongation of preprolactin in reticulocyte lysate, *J. Cell Biol.* **109**:2617–2622.

Wrubel, W., Stochaj, U., Soonewald, U., Theres, C., and Ehring, R., 1990, Reconstitution of an active lactose carrier in vivo by simultaneous synthesis of two complementary protein fragments, *J. Bacteriol.* **172**:5374–5381.

Yamada, H., Tokuda, H., and Mizushima, S., 1989a, Proton motive force-dependent and -independent protein translocation revealed by an efficient *in vitro* assay system of *Escherichia coli*, *J. Biol. Chem.* **264**:1723–1728.

Yamada, H., Matsuyama, S., Tokuda, H., and Mizushima, S., 1989b, A high concentration of SecA allows proton motive force-independent translocation of a model secretory protein into *Escherichia coli* membrane vesicles, *J. Biol. Chem.* **264**:18577–18581.

Yamagushi, M., Harefi, Y., Trach, K., and Hoch, J. A., 1988, The primary structure of the mitochondrial energy-linked nicotinamide nucleotide transhydrogenase deduced from the sequence of cDNA clones, *J. Biol. Chem.* **263**:2761–2767.

Yamane, K., Ichihara, S., and Mizushima, S., 1987, In vitro translocation of protein across *Escherichia coli* membrane vesicles requires both the proton motive force and ATP, *J. Biol. Chem.* **262**:2358–2362.

Yu, Y., Sabatini, D. D., and Kreibich, G., 1990, Antiribophorin antibodies inhibit the targeting to the ER membrane of ribosomes containing secretory polypeptides, *J. Cell Biol.* **111**:1335–1342.

Zerial, M., Huylebroeck, D., and Garoff, H., 1987, Foreign transmembrane peptides replacing the internal signal sequence of transferrin receptor allow its translocation and membrane binding, *Cell* **48**:147–155.

Zimmermann, R., and Mollay, C., 1986, Import of honeybee prepromelittin into the endoplasmic reticulum: Requirements for membrane insertion, processing and sequestration, *J. Biol. Chem.* **261**:12889–12895.

Chapter 3

The Major Glycosylation Pathways of Mammalian Membranes
A Summary

Ajit Varki and Hudson H. Freeze

1. INTRODUCTION

Many types of cellular membranes have sugar chains (*oligosaccharides* or *glycans*) attached to a variety of their constituent macromolecules. Such glycosylated macromolecules are called *glycoconjugates*. This chapter will broadly scan the subject of structure and biosynthesis of oligosaccharides on glycoconjugates, primarily in higher animal cells. The classification of different types of oligosaccharides will be based on the type of *linkage* that joins them to the nonsugar component (*aglycone*), rather than on the types of sugars (*monosaccharides*) found in the chains. As shown in Table I, this classification is slightly different from the more conventional categories and nomenclature used for glycoconjugates in the past. This change is made because it is now clear that the aglycone usually determines the initiation step in creating a glycoconjugate and can have an influence upon the further buildup of a more complex glycan. Furthermore, the traditional boundaries between the different types of glycoconjugates as previously described have become blurred.

Ajit Varki and Hudson H. Freeze   Cancer Center, University of California at San Diego, and La Jolla Cancer Research Foundation, La Jolla, California 92093.
Subcellular Biochemistry, Volume 22: Membrane Biogenesis, edited by A. H. Maddy and J. R. Harris. Plenum Press, New York, 1994.

Table I
Descriptive Terms for Glycoconjugates in Relation to the Attachment Site of Sugar Chains

Historical terms	Conventional terms	Defining types of linkages and sequences
Glycoprotein	N- or O-linked glycoprotein	Scattered N-GlcNAc and/or O-GalNAc-linked sugar chains
Mucins	Mucin-like glycoproteins	Many-clustered O-GalNAc-linked chains
Mucopolysaccharides	Proteoglycans	O-Xylose-linked glycosaminoglycan chains on core proteins
		Can also carry N-GlcNAc and/or O-GalNAc-linked chains
Mucolipid, glycolipid	Glycosphingolipid	Glucose or galactose linked to ceramide
Not recognized earlier	Glycophospholipid anchored protein	Ethanolamine-phosphate in amide-linkage to proteins and glycan linked to phosphatidylinositol

The uniformity of the glycone-aglycone linkage region and of the first few sugars immediately attached to it forms a sharp dividing line between the different classes of oligosaccharides. However, this distinction becomes less prominent as more monosaccharide units are added. In fact, considerable overlap is seen in the structures of outer sequences of the different types of glycans. It is also important to remember that an individual aglycone (e.g., a protein) can carry many different types of oligosaccharide chains attached by different linkages. Most types of oligosaccharides are synthesized in the membrane-bound compartments of the ER-Golgi-plasmalemma pathway (see Figure 1). Thus, trimming and restructuring reactions in the biosynthesis of a variety of sugar chains can occur in parallel on a single glycoconjugate during its synthesis, trafficking, and maturation.

1.1. Historical Overview

The study of carbohydrate-containing molecules began with the analysis of specific classes of macromolecules. Each such class was in easily recognizable forms that could be defined and separated by their physical properties (e.g., viscous secretions, soluble blood proteins, amphipathic lipids, and gristly connective tissue molecules). The carbohydrate analytical technology of the time required large amounts of material and allowed somewhat gross analyses of these apparently different types of molecules. Thus, individual research groups tended to focus upon analysis of different types of glycosylated molecules such as glycoproteins (with N-linked oligosaccharides), mucins (with O-linked oligosaccharides), glycolipids (with lipid tails), and proteoglycans (with glycosaminoglycan chains) (see Table I). Both the diversity and complexity within each of these

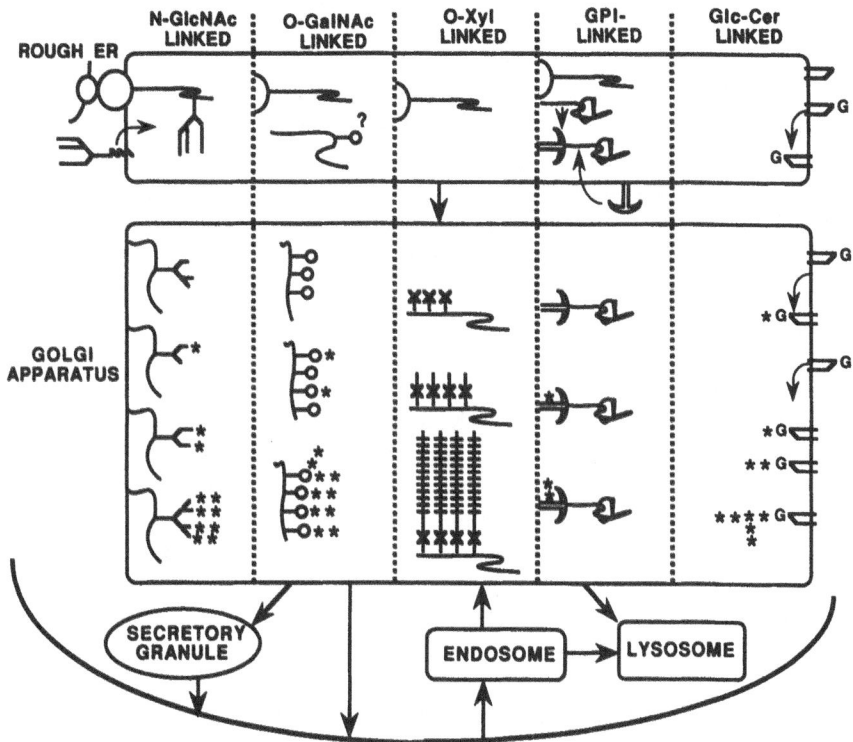

FIGURE 1. Initiation and maturation of the major types of glycoconjugates in relation to subcellular trafficking in the ER-Golgi-plasmalemma pathway. The asterisks represent the addition of outer sugars to the oligosaccharides in the Golgi apparatus. See text for a better understanding of the events depicted in the figure.

individual groups of molecules were challenging, and the major structures appeared to be clearly different in each group. Thus, research groups seldom crossed over from studying one type of molecule to another. For the same reason, prior reviews of the structure, biosynthesis, and function of glycoconjugates have tended to focus upon one or the other of such specific classes of molecules.

1.2. The Modern View

More refined and sensitive technology and the elucidation of biosynthetic pathways made it possible to analyze trace amounts of functionally important molecules (Cummings *et al.*, 1989; Varki, 1991). Such studies showed that the dividing lines between the different glycoconjugates are not as rigid as previously thought. Thus, some glycoproteins contain functionally important glycolipid portions, some glycoproteins carry glycosaminoglycan chains, and many proteoglycans have both *N*- and *O*-linked carbohydrate chains. Also, the same di- or trisaccharide unit is frequently found in the outer chains of both glycoproteins and glycolipids. Such combinations occur on many important molecules, and in some cases the glycans themselves determine function. Thus, the traditional lines of distinction have become blurred. Focus on the composition of each general class of glycoconjugate is now being replaced by analysis of individual sugar chains on physiologically important molecules. The novel term "glycobiology" describes the study of the structure, biosynthesis, distribution, regulation, and function of the entire spectrum of sugar chains found on glycoconjugates (Rademacher *et al.*, 1988).

In this chapter we will survey the major types of sugar chains and their biosynthesis. This includes the minimal requirements to initiate their synthesis, how the chains are extended, terminated, and turned over. We will also mention the subcellular localization of these biosynthetic steps in the general ER-Golgi-plasmalemma pathway, and some inhibitors and mutants useful in their study (Elbein, 1991; Winchester and Fleet, 1992). As discussed above, we will classify these sugar chains by their unique linkage regions. Because of the broad nature of this subject, the references cited consist only of review articles. For the same reason, this chapter does not deal with the function of these oligosaccharide chains, a subject deserving of a completely separate review (Cumming, 1991; Stanley, 1992; Varki, 1992b; Varki, 1993).

2. THE ER-GOLGI-PLASMALEMMA PATHWAY

The vast majority of glycoconjugates follow a common trafficking pathway involving biosynthesis in the lumen of the ER and passage through the Golgi apparatus (Farquhar, 1991; Mellman and Simons, 1992; Rothman and Orci, 1992; Chapter 4 this volume). They may then remain membrane bound, be secreted immediately, or be packaged into secretory granules or lysosomes (see Figure 1). While some bulky sugar chains are made on the cytoplasmic face of these intracellular membranes and "flipped" across to the other side, most are added to the growing chain on the *inside* of the ER or the Golgi apparatus (Hirschberg and Snider, 1987). An important topological point to remember is that whatever portion of a molecule faces the *inside* of the ER will ultimately face the *inside* of a secretory granule or lysosome, but it will face the *outside* of the cell. This is the topology of most forms of glycosylation. The biosynthetic enzymes (glycosyltrasferases) responsible for these reactions are well studied, and their location can help define various functional compartments of the ER-Golgi pathway (Paulson and Colley, 1989). A working model envisions the enzymes as physically lined up along the pathway in the sequence in which they work. This is probably an oversimplified view, and the distribution probably depends upon the cell type. The low-molecular-weight sugar nucleotides that act as donors for most of the biosynthetic steps are made in the cytosol, and specifically transported into the lumen of the organelles. A major exception to this rule is the addition of single GlcNAc residues to cytoplasmic proteins; the active site of this transferase faces the cytosol. There are also other types of cytoplasmic glycosylation, but the important point is that they tend to be very different from those found in the ER-Golgi-plasmalemma pathway.

In some cases, the oligosaccharide components of glycoconjugates (particularly the outer sugar units) appear to be turned over more rapidly than the aglycone that carries them. This implies considerable metabolic activity of these molecules and their recycling back into the Golgi for reglycosylation. The purpose of such turnover and resynthesis is currently unknown.

In the following sections, the different types of linkage regions found in the ER-Golgi-plasmalemma pathway are individually considered, with regard to their initiation signal, donor molecules, processing and termination, and specific inhibitors and mutants in their biosynthesis. The text presents limited information on the major principles, but many additional facts can be found in the accompanying figures and tables. Consult the references for more comprehensive, detailed information.

2.1. The Generation of Diversity

Animal oligosaccharides contain a limited set of monosaccharides and generate diversity by using different linkages. Of the dozens of possible monosaccharides, only a few are actually used in mammalian cell glycoconjugates (Kisailus and Allen, 1992). Figure 2 shows these sugars, along with abbrevia-

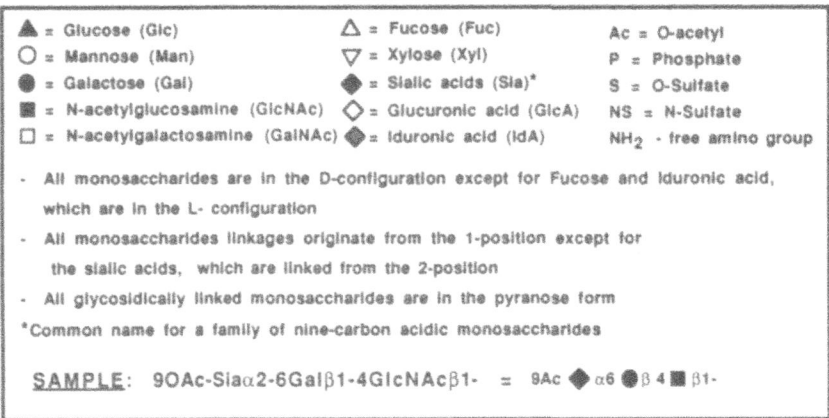

FIGURE 2. The monosaccharide units and substituents of mammalian oligosaccharide chains. Monosaccharides used in the biosynthetic pathways described in this chapter are presented, along with their standard abbreviations in parentheses. The symbol key for the monosaccharides is used throughout the remainder of the figures. Because the origin of a linkage is invariant for each monosaccharide, the remaining figures only indicate the position of attachment to an underlying sugar.

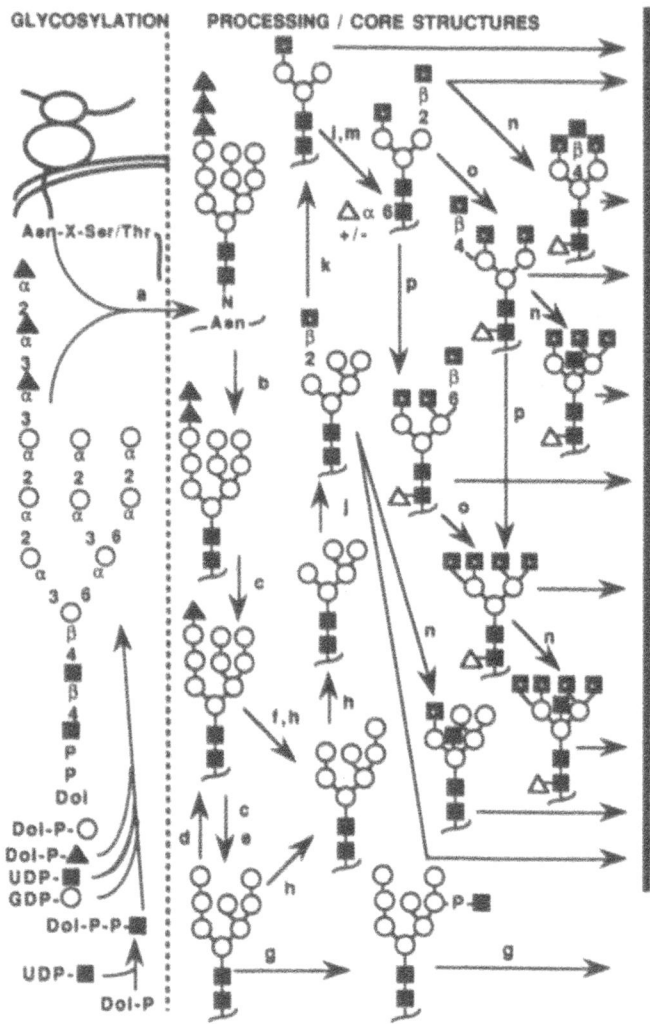

FIGURE 3. The *N*-GlcNAc-linked oligosaccharide

tions, and a symbol key that will be used for all of the figures in this chapter. Each
monosaccharide can be attached in α- or β-linkage from its 1-position (2-position
only for sialic acids) to an underlying acceptor. If that underlying acceptor is
another monosaccharide, the next number indicates the position on the acceptor
where the attachment occurs (e.g., Galβ1–4GlcNAc indicates a galactose residue
β-linked via its 1-position to the 4-position of an *N*-acetylglucosamine residue).
Because the first number is invariant for each monosaccharide, it is not indicated in
the figures. Various modifications of the monosaccharides can also occur, and

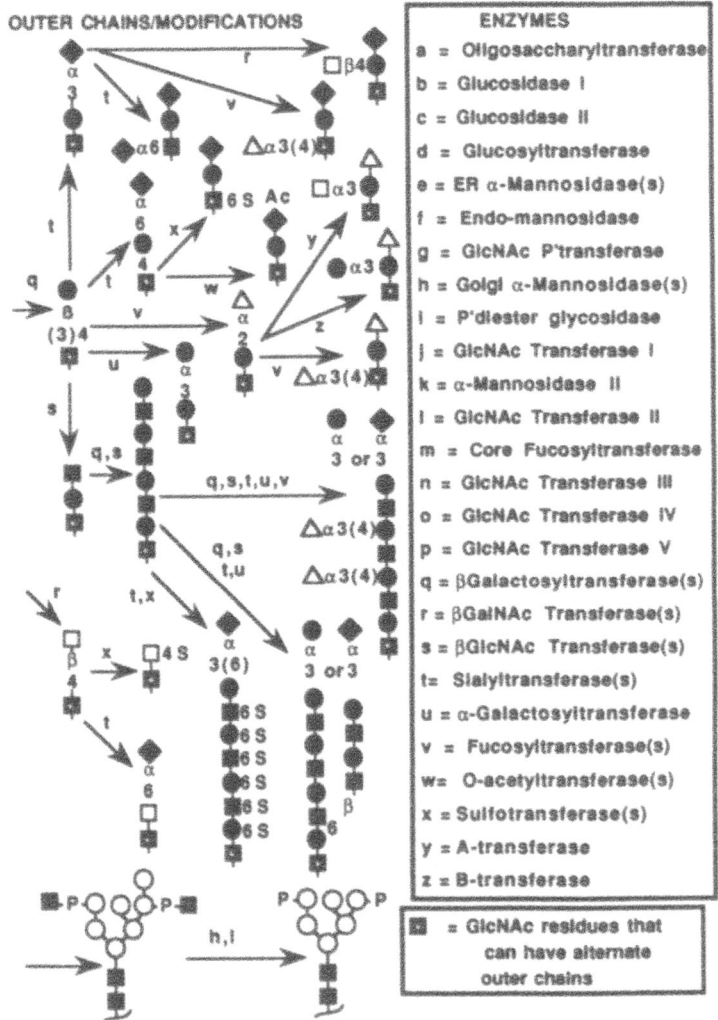

OUTER CHAINS/MODIFICATIONS

ENZYMES

a = Oligosaccharyltransferase
b = Glucosidase I
c = Glucosidase II
d = Glucosyltransferase
e = ER α-Mannosidase(s)
f = Endo-mannosidase
g = GlcNAc P'transferase
h = Golgi α-Mannosidase(s)
i = P'diester glycosidase
j = GlcNAc Transferase I
k = α-Mannosidase II
l = GlcNAc Transferase II
m = Core Fucosyltransferase
n = GlcNAc Transferase III
o = GlcNAc Transferase IV
p = GlcNAc Transferase V
q = βGalactosyltransferase(s)
r = βGalNAc Transferase(s)
s = βGlcNAc Transferase(s)
t = Sialyltransferase(s)
u = α-Galactosyltransferase
v = Fucosyltransferase(s)
w = O-acetyltransferase(s)
x = Sulfotransferase(s)
y = A-transferase
z = B-transferase

= GlcNAc residues that can have alternate outer chains

pathway. See Figure 2 for symbol key.

these are listed in Figure 2. The sialic acids designated by "Sia" actually represent a family of nine-carbon acidic sugars (Varki, 1992a; Schauer, 1991).

Literally hundreds of glycosyltransferase enzymes must be involved in the biosynthesis of the different types of oligosaccharides discussed in this chapter. No details regarding these enzymes will be presented here. Several of them have been cloned recently (Paulson and Colley, 1989). Surprisingly, while these enzymes carry out similar reactions, using similar donors and acceptors, sequence homology with one another is uncommon. However, they all show a common structural motif, being type-2 membrane-bound proteins with short cytosolic amino-terminal sequences, followed by hydrophobic membrane-anchor domains and globular carboxylterminal domains carrying the catalytic site within the Golgi lumen.

2.2. *N*-GlcNAc-Linked Oligosaccharides

These oligosaccharides are defined by the GlcNAc-*N*-Asn linkage to proteins (see Figure 3). A preformed precursor glycan is transferred from a lipid donor to the acceptor protein on the luminal side of the ER, either cotranslationally or immediately posttranslationally, to form a core structure that is subsequently processed (Kornfeld and Kornfeld, 1985; Cummings, 1992; Kobata and Takasaki, 1992; Schachter, 1991).

2.2.1. Initiation

The consensus sequence dictating the addition of these oligosaccharides is Asn-X-Ser/Thr (where X can be any amino acid except Pro). This simple tripeptide (with blocked amino and carboxy termini) can alone act as an acceptor when fed to intact cells or when mixed with isolated ER membranes. However, not all such sites are utilized in natural proteins. The structure of the protein and the rate of its synthesis can apparently influence whether a potential glycosylation site is actually used.

2.2.2. Donor

The full-sized lipid-linked oligosaccharide (LLO) is synthesized in the ER on an isoprenyl lipid called *dolichol phosphate,* by a series of sequential glycosyl transferase reactions. The first step in this synthesis is blocked by the antibiotic tunicamycin (Table II). Truncated forms of the LLO synthesized in mutant cells or those made in the presence of inhibitors can also be transferred to proteins, but usually not as efficiently as the full-sized sugar chain. Although the LLO is initiated on the outside of the ER, its completion occurs on the inner (luminal) surface of the ER where it is transferred to the acceptor. The mechanism used to flip the bulky, hydrophilic oligosaccharide through the membrane is not known. Any mutations in the synthesis of the LLO will affect *all* proteins with *N*-GlcNAc-linked chains.

2.2.3. Processing and Elongation

Figure 3 outlines the complex and variable processing of *N*-linked glycans. The actual pathway followed for an individual glycosylation site on a given protein is determined by many factors including the structure of the protein, the location (availability) of the sugar chain on the protein, the relative activities of

Table I
(Continued)

	Adhesion molecules		Receptors
Names	Regulation of surface expression	Names	Regulation of surface expression
p150-95 or CD11c/CD18	Upregulation?	ICAM-1, ELAM-1 or E-selectin GMP140 or CD62 or PADGEM or P-selectin	IL1, TNFα, slow response Thrombin, histamine, bradykinin, fast transient response
LAM 1 or L-selectin	Constitutively present on the plasma membrane and shedded by activation		

3. The β_3 integrins (cytoadhesins) involved in cell-cell aggregation

Names	Regulation of surface expression	Names	Regulation of surface expression
$\alpha_{IIb}\beta_3$ (GP$_{IIb-IIIa}$)		Fibrinogen, fibronectin, von Willebrand factor, vitronectin (platelet aggregation)	
$\alpha_v\beta_3$ (Vitronectin receptor)		Vitronectin, fibrinogen, von Willebrand factor, thrombospondin	

GMP140, originally defined as Granule Membrane protein in Platelets; ICAM-1 and -2, Intercellular Adhesion Molecule -1 and -2; ELAM-1, Endothelial Leucocyte Adhesion Molecule; MAC-1, MACrophage 1; Mo-1, MOnocyte-1; LFA-1, Lymphocyte Function associated Antigen-1; PADGEM, Platelet activation Dependent Granule External Membrane. (References: Entman et al., 1991; Wong et al., 1991; Akiyama et al., 1990; Hogg, 1989; Mantovani and Dejana, 1989; Hynes, 1987; Zimmerman et al., 1992; Ruoslahti, 1991.)

Table III
Examples of Glycosylation-Defective Tissue Culture Cell Lines

	Parental line	Mutant line	Altered step	Mature carbohydrates
Single pathway	BW5147	PHA^{R}2.7	α-Glucosidase II	Glycoproteins with Glc$_2$Man$_9$GlcNAc$_2$. Complex chains decreased.
	L cells	CL6	α-Mannosidase I	Glycoproteins with Man$_9$GlcNAc$_2$. Complex chains decreased.
	CHO	Clone 15B Lec 1	GlcNAc transferase I	Processing blocked at Man$_5$GlcNAc$_2$ processing.
	BW5147	PHA^{R}1.8	GlcNActransferase V	Decreased multiantennary N-linked oligosaccharides.
	CHO	745	Xylosyl-transferase	No glycosaminoglycan chains.
		761	Gal-transferase-I	No glycosaminoglycan chains.
	Human fibroblasts	I-cell disease	Phosphorylation of lysosomal enzymes	Lack of M-6-P residues.
Multiple pathways				
	CHO	Clone 13, Lec 8	UDP-Gal-transport into Golgi	Glycoproteins/glycolipids decreased in Gal and Sia residues. Complex glycans terminate in GlcNAc.
	CHO	Clone 1021 Lec 2	CMP-Sia transport into Golgi	Glycoproteins/glycolipids terminate in Gal and have decreeased sialic acid.
	CHO	ldl^d	UDP-Gal/GalNAc 4'-epimerase	Deficient in Gal and GalNAc-residues. Affects O- GalNAc initiation, and addition of Gal to N-GlcNAc linked and ceramide-linked molecules.

the various transferases and processing glycosidases, and the transit rate through the Golgi apparatus. The early processing steps occur in essentially all organisms, plant and animal, but continued processing leads to many diverse structures that are special for higher animals.

All of the Glc residues are removed in the ER, followed by removal of a variable number of Man residues. Metabolic inhibitors and mutant cell lines can halt processing at many of these early steps (see Tables II and III). The addition of the first GlcNAc residue represents the transition from "high-mannose" type to "complex type" oligosaccharides. However "hybrid" structures can also result. The processing of Man residues continues and it is followed by the addition of a variable number of β-GlcNAc, β-Gal, and α-sialic acid residues in the most common pathways. The α-fucose residues can be added in the core region or in the peripheral regions, and in some cases terminal α-Gal residues are substituted for sialic acids. One modification exclusive for N-linked oligosaccharides is the addition of Man6P residues to high mannose chains on lysosomal enzymes. These residues function to target the newly synthesized enzymes to the lysosome by binding to a receptor in the Golgi apparatus.

2.2.4. Termination

Certain monosaccharide units are typical of the nonreducing ends of many glycoconjugates. Chains can be terminated by α-sialic acids, α-fucose, α-Gal or blood group sugar sequences. After these transferases add their respective monosaccharides, some of the oligosaccharides can be modified further (e.g., by sulfation or acetylation) as a final step in biosynthesis. How all of these terminal reactions compete with one another is an interesting issue that is not well resolved.

2.2.5. Fate and Turnover

Proteins with N-linked chains are found in membrane-bound organelles, and as secreted molecules. On some cell-surface proteins, the terminal sugars (e.g., sialic acid) turn over faster than the peptide itself, and can also be re-added after removal. Secreted serum proteins can also lose terminal sugars. In mammals and birds, this loss may serve as a signal for the proteins to be cleared from the circulation by carbohydrate-specific receptors found on liver cells.

2.3. O-GalNAc-Linked Oligosaccharides

These sugar chains are linked to a protein at Ser or Thr residues through an α-GalNAc unit. These chains can vary greatly in size. They are often, but not always, found in clusters on the polypeptide backbone. Unlike N-linked oligosaccharides, no lipid-linked precursors or *en bloc* transfers occur (Schachter and Brockhausen, 1992; Carraway and Hull, 1991; Fukuda, 1992).

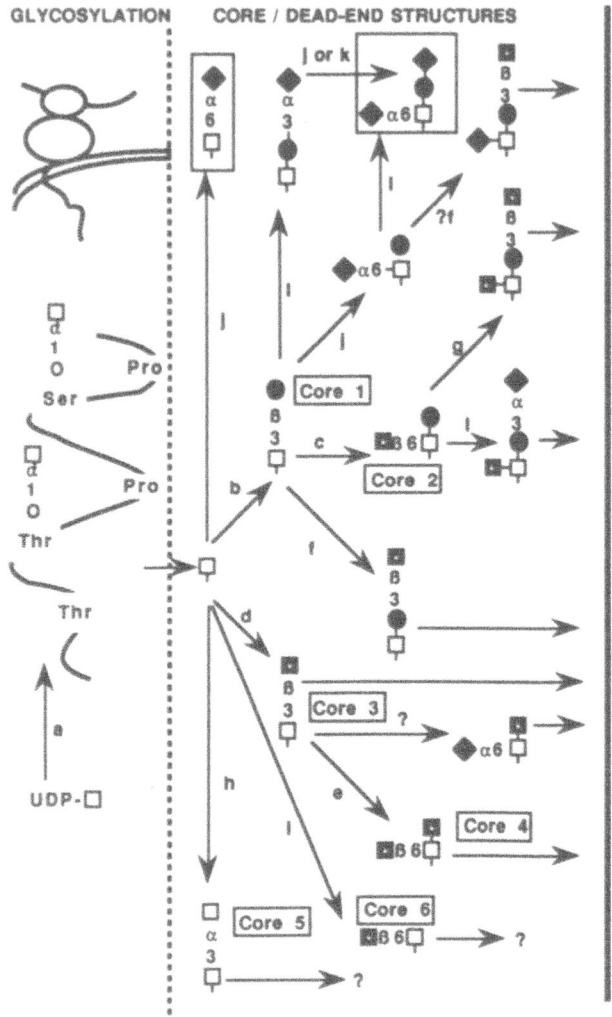

FIGURE 4. The *O*-GalNAc-linked oligosaccharide

2.3.1. Initiation

There does not appear to be a single consensus sequence for the addition of *O*-linked sugars. However, the peptide will often have a Pro-residue at positions −1 and +3 relative to the glycosylated Ser or Thr residue. Clusters of *O*-GalNAc-linked oligosaccharides are frequently found in regions of the protein with a high density of alanine, serine, threonine, and proline residues. The difference between Ser and Thr acceptors indicates that initiation requires two different transferases, for which there is now evidence. Initial *O*-glycosylation probably occurs predominantly in the Golgi apparatus.

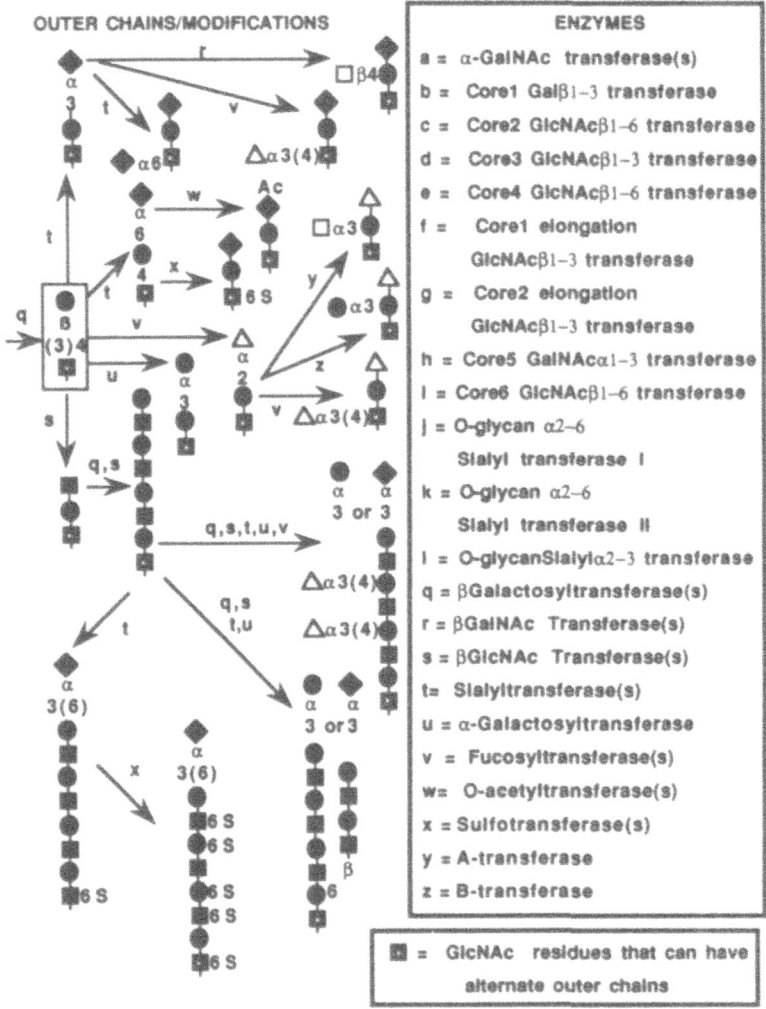

OUTER CHAINS/MODIFICATIONS

ENZYMES

a = α-GalNAc transferase(s)

b = Core1 Galβ1–3 transferase

c = Core2 GlcNAcβ1–6 transferase

d = Core3 GlcNAcβ1–3 transferase

e = Core4 GlcNAcβ1–6 transferase

f = Core1 elongation
    GlcNAcβ1–3 transferase

g = Core2 elongation
    GlcNAcβ1–3 transferase

h = Core5 GalNAcα1–3 transferase

i = Core6 GlcNAcβ1–6 transferase

j = O-glycan α2–6
    Sialyl transferase I

k = O-glycan α2–6
    Sialyl transferase II

l = O-glycanSialylα2–3 transferase

q = βGalactosyltransferase(s)

r = βGalNAc Transferase(s)

s = βGlcNAc Transferase(s)

t = Sialyltransferase(s)

u = α-Galactosyltransferase

v = Fucosyltransferase(s)

w = O-acetyltransferase(s)

x = Sulfotransferase(s)

y = A-transferase

z = B-transferase

■ = GlcNAc residues that can have
    alternate outer chains

pathway. See Figure 2 for symbol key.

2.3.2. Elongation

Several different pathways are available for the elongation of *O*-linked chains as shown in Figure 4. The early steps in these pathways generate a series of distinctive units called "core" units, numbered 1 to 6. Commitment to one pathway can block access to another pathway leading to different structures. The activity of certain inhibitors (see below) shows that the transferases acting later in this pathway do not need to recognize the protein backbone to elongate the chains.

2.3.3. Termination

The final steps are very much like those seen in the synthesis of N-linked chains discussed above.

2.3.4. Inhibitors and Mutants

This type of glycosylation can be partially blocked by adding millimolar concentrations of phenyl or benzyl α-GalNAc to cells. These hydrophobic inhibitors penetrate the cell and the Golgi and compete with the endogeneous protein acceptors. The artificial derivatives then transit through the Golgi and can act as acceptors for other sugar residues. In contrast to the N-linked oligosaccharide pathway, cellular mutants that selectively affect O-linked oligosaccharide biosynthesis are uncommon. However, several pleiotropic mutants affect both N- and O-linked oligosaccharide biosynthesis (see Table III).

2.3.5. Fate and Turnover

Many of the proteins with clusters of short O-linked chains are the traditional "mucins" that are secreted in large quantities on mucosal surfaces. On the other hand, some integral membrane proteins such as the LDL-receptor contain clusters of O-linked chains on the extracellular side of the transmembrane region. Such clustered O-linked chains tend to extend the polypeptide into a rod-like section. Thus, these clusters may help to prevent proteolysis of the proteins, to prop up the globular domains of receptors away from the membrane, and to present clustered ligands for recognition by receptors.

In at least one case, a protein with a large number of O-linked chains has been found to recycle repeatedly between the Golgi and the cell surface. On successive visits to the Golgi, new chains can be initiated and built on previously unmodified amino acids. It is not known whether this is a general property of all membrane-bound proteins with O-linked chains. This type of change does not generally happen for N-linked chains. This is because the latter are first added in the ER, which is not generally considered to be in the recycling pathway from the surface.

2.4. Ceramide-Linked Oligosaccharides (Glycosphingolipids)

These molecules are defined as the glucosylceramides and the galactosylceramides, and are predominantly, but not exclusively, found on the cell surface. They are also called glycosphingolipids because of the sphingosine chain that makes up part of the ceramide core (Hakomori, 1986, 1990; Kundu, 1992; VanEchten and Sandhoff, 1993; Shayman and Radin, 1991; Stults et al., 1989).

2.4.1. Initiation

The ceramide lipid tail is made in the ER membrane, facing the cytoplasm. The addition of glucose (the first sugar) is believed by most investigators to occur on the cytosolic face of the membranes of the ER or the Golgi. The glucosylceramide is then probably flipped into the lumen for further addition of sugars. Less is known about the initial addition of galactose to the ceramide.

2.4.2. Elongation

Figure 5 shows the major pathways for the many different classes of molecules that can arise from glucosylceramide and galactosylceramide. These are defined by the addition of specific monosaccharide units added in particular sequences. Evidence shows that the early transferases are probably specific for the glycolipids and reside in the early part of the Golgi. In contrast, many of the later-acting ones are probably shared with other types of oligosaccharides and are probably present in the late regions of the Golgi. Thus, as in the case of the N-GlcNAc-linked and O-GalNAc-linked oligosaccharides, a variety of common outer chains can be mixed and matched among the different classes to generate further diversity. The glycosphingolipids have the greatest variety of outer chains. The more unusual ones are not presented in Figure 5 owing to lack of space. The levels of the different transferases control what kinds of glycosphingolipids are made by a particular cell, but what controls the expression of the enzymes is not known. In contrast to the situation with the N-linked oligosaccharide pathway, only one of the glycosphingolipid-specific transferases has been cloned to date.

2.4.3. Termination

As in the case of the N- and O-linked sugar chains, termination occurs when specific terminal sugars (e.g., sialic acid or blood group sugars) are added. However, details of this process are not clearly understood.

2.4.4. Inhibitors and Mutants

To date, only one specific inhibitor of glycosphingolipid biosynthesis is known. PDMP was specifically synthesized as an inhibitor of glucosylceramide synthase, the enzyme that initiates the production of most major classes of glycosphingolipids. This compound has proven very useful in biosynthetic studies. However, functional studies are somewhat limited by the fact that it can also affect the related pathways for biosynthesis and turnover of sphingomyelin and sphingosine. Most of the available tissue culture mutants that affect glycosphingolipid biosynthesis are pleitropic ones affecting multiple pathways (see Table III).

GLYCOSYLATION / CORE STRUCTURES

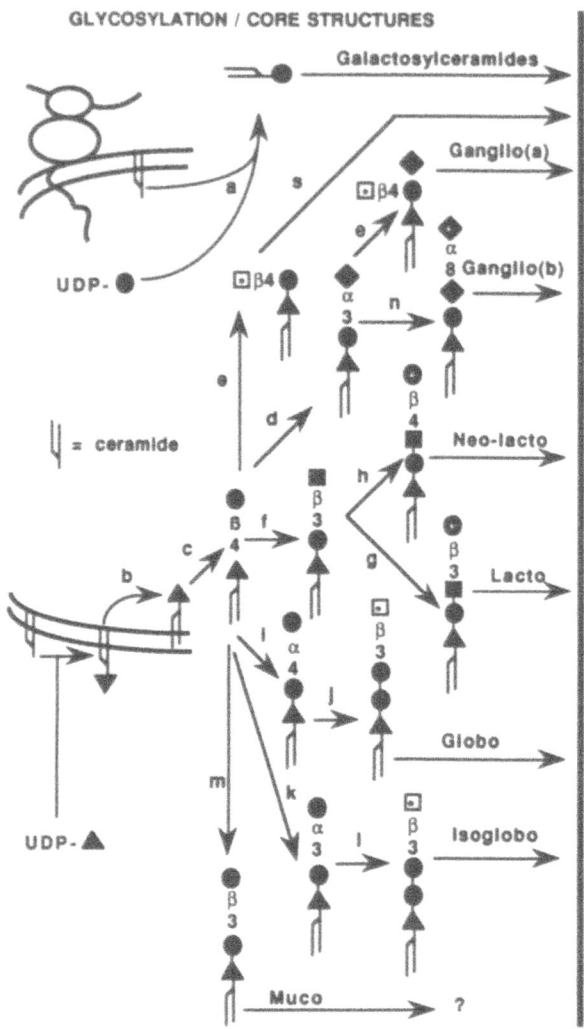

FIGURE 5. The ceramide-linked oligosaccharide

2.4.5. Turnover and Fate

The turnover of glycosphingolipids appears to be rather slow in most cells. There is some evidence that removal and replacement of terminal units can occur, similar to the *N*- and *O*-linked chains. Terminal degradation takes place in the lysosomes and involves not only hydrolases, but also specific "activator" proteins. Glycolipid transfer proteins are also well known and widespread, but their function within the cell is not well understood.

OUTER CHAINS/MODIFICATIONS

ENZYMES	
a=	Galactosylceramide synthetase
b=	Glucosylceramide synthetase
c=	Lactosylceramide synthetase
d=	Sialylα2–3 transferase
	GM3 synthetase
e=	GalNAcβ1–4 transferase
	GM2 synthetase
f=	Lacto β1–3 GlcNActransferase
g=	Lacto β1–3 Galtransferase
h=	Neoacto β1–4 Galtransferase
i=	Globo α1–4 Galtransferase
j =	Globo GalNAcβ1–3 transferase
k=	Isoglobo α1–3 Galtransferase
l =	Isoglobo GalNAcβ1–3 transferase
m =	Muco Galβ1–4 transferase
n=	Sialylα2–8 transferase
	GD3 synthetase
p =	βGlucuronyltransferase
q =	βGalactosyltransferase(s)
r =	βGalNAc Transferase(s)
s =	βGlcNAc Transferase(s)
t=	Sialyltransferase(s)
u =	α-Galactosyltransferase
v =	Fucosyltransferase(s)
w=	O-acetyltransferase(s)
x =	Sulfotransferase(s)
y =	A-transferase
z =	B-transferase

GlcNAc Gal GalNAc
residues that can have alternate outer chains

pathway. See Figure 2 for symbol key.

2.5. *O*-Xylose-Linked Chains

The transfer of β-xylose to appropriately disposed serine residues initiates these chains on a limited number of "core proteins." All of these glycans contain a shared linkage region consisting of GlcUAβ1–3Galβ1–3-Galβ1–4Xyl-*O*-Ser (see Figure 6). These cores are extended with alternating disaccharide units of either GlcUA-GlcNAc (heparin and heparan sulfate) or GlcUA-GalNAc (chondroitin sulfate and dermatan sulfate). Many of these glycosaminoglycan (GAG) chains are, in turn, extensively modified as described below. The finished product, a core protein with one or more GAG chains, is usually called a proteoglycan. Many such distinct core proteins have been cloned, and they can be transmembrane, GPI-anchored (see below), or secreted (Bhavanandan and Davidson, 1992; Kjellén and Lindahl, 1991; Ruoslahti and Yamaguchi, 1991; Yanagishita and Hascall, 1992).

2.5.1. Initiation

Xyl is transferred from UDP-Xyl to proteins that present an available Ser-Gly-X-Gly sequence. This is usually preceded by one or two acidic amino acid residues, but is not a universal signal for addition of Xyl residues. In some proteins, the Ser-Gly dipeptide occurs as a series of contiguous repeats, whereas in others it occurs singly. In the final analysis, Xyl addition is probably determined by a general three-dimensional peptide structure rather than a specific and limited sequence. Some core proteins are "part-time proteoglycans" in which only a portion of the available peptide chains are modified by GAG chains.

2.5.2. Elongation

This can be considered to occur as three distinct groups of reactions.

2.5.2a. Core Synthesis. The GlcUAβ1–3Galβ1–3-Galβ1–4Xyl-O-Ser core is added sequentially in the early part of the Golgi apparatus by a series of different sugar transferases. In some cases, this is accompanied by phosphorylation of the 2-position of Xyl and sulfation of the 4- or 6-positions of the innermost Gal residue. Both the control and the function of these additional modifications are not understood.

2.5.2b. Chain Extension. The protein itself probably determines whether GlcNAc or GalNAc is added to the GlcUA at the nonreducing end of the glycan core (Figure 6). This decision is crucial because it determines whether the chain will be a heparin/heparan sulfate type or a chondroitin/dermatan sulfate chain. Thereafter, the chains are polymerized by the sequential addition of each monosaccharide from its nucleotide donor. Some core proteins can alternately express chondroitin sulfate or heparin/heparan sulfate as the glycan, depending upon the cell in which they are synthesized.

2.5.2c. Chain Modification. The individual chains are modified during their elongation in a variety of ways shown in Figure 6. Not all of the residues are modified identically, and the control of this modification is not well understood. In the case of heparin and heparan sulfate, selected GlcNAc residues are first de-N-acetylated and then N-sulfated. At this point, some of the GlcUA can be epimerized to IdUA; O-sulfate esters can be added to GlcNAc, GlcNSO$_4$, or IdUA residues. In the case of the chondroitin sulfate chains, little GlcUA is epimerized, and GalNAc can be sulfated on the 4- and/or 6-positions. In dermatan sulfate, a portion of GlcUA can be epimerized to IdUA. All of these highly ordered and interdependent modifications happen as the proteins are rapidly moving through the late part of the Golgi apparatus.

2.5.3. Termination

Very little is known about the factors that control the size of the chain or how it is terminated.

2.5.4. Turnover and Fate

In some cell types, proteoglycans are segregated into secretory vesicles, where the glycosaminoglycan chains are depolymerized prior to controlled secretion. Cell surface or secreted proteoglycans can be taken up and degraded either by the same cell that made them, or by another cell. After uptake, some GAG chains are believed to be directed to distinct cellular compartments such as the nucleus. However, all GAG chains must eventually be broken down to their component parts in the lysosomes. Considerable evidence shows that the breakdown of GAG chains in the lysosomes occurs in a highly ordered sequence.

2.5.5. Inhibitors and Mutants

Proteoglycan synthesis can be inhibited and GAG chain synthesis greatly stimulated by feeding cells a variety of different β-xylosides. Paranitrophenyl, 4-methyl-umbelliferyl and even methyl-xyloside derivatives at submillimolar concentrations can penetrate the cells and the Golgi, where they serve as alternative acceptors for the addition of the first Gal residue. Because the xylosides are in such high concentration compared to the core proteins, they are very effective competitors. Further elongation of these chains occurs, but is heavily biased toward adding chondroitin/dermatan sulfate chains rather than heparan sulfate chains. These chains are generally smaller than those found on natural core proteins. Recently, some evidence shows that the structure of the aglycone can determine which type of GAG chain is added to the artificial core.

Several mutant CHO cell lines have been isolated that either cannot make a carbohydrate core region or do not elongate the chains. The core-deficient mutants lack xylosyl or galactosyltransferase (Figure 6). Others lack one of the chain polymerizing enzymes, or a specific sulfotransferase.

2.6. Glycophospholipid Anchors

These structures replace the carboxyl-terminal transmembrane portions of certain proteins, and thus serve as alternate ways to anchor such proteins to cell surfaces. Some proteins can exist either with transmembrane regions or with glycophospholipid (GPI) anchors. In some cases alternative splicing of mRNA

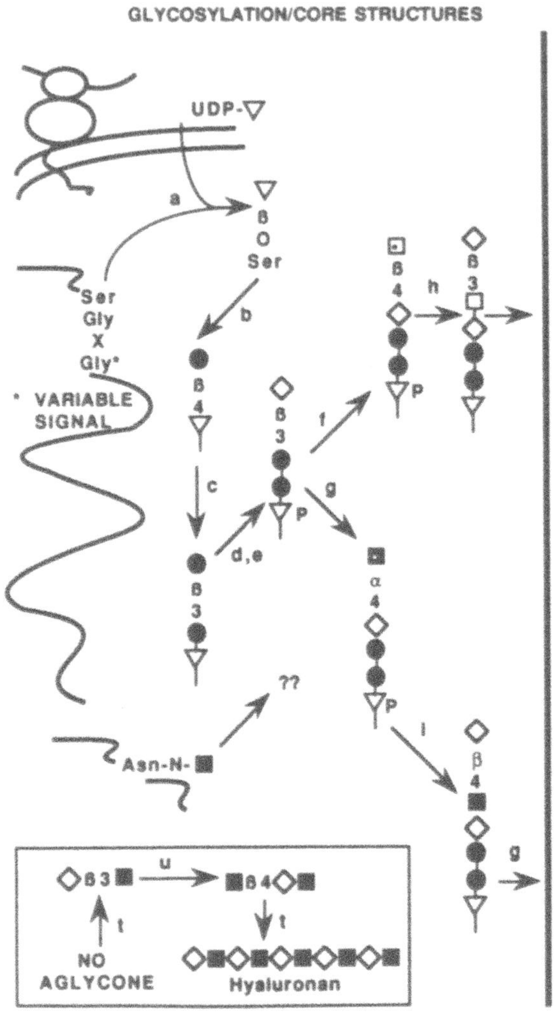

GLYCOSYLATION/CORE STRUCTURES

FIGURE 6. The *O*-xylose-linked oligosaccharide

can determine the outcome. The common core structure of these anchors can be elongated or modified in several ways (see Figure 7). These core structures can be cleaved by specific microbial and mammalian phospholipases (Doering *et al.*, 1990; Ferguson, 1992).

2.6.1.   Initiation

The core anchor structure has several unique features, including the presence of ethanolamine and a nonacetylated glucosamine. It is both performed and

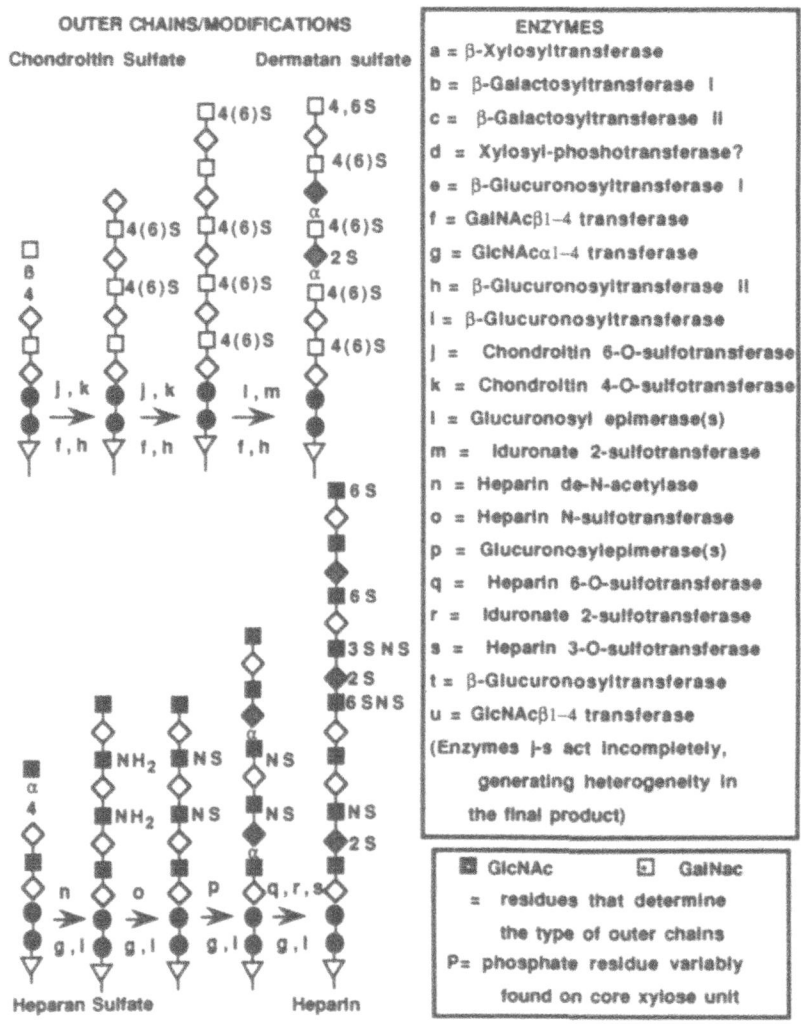

The following enzymes are shown in the figure legend:

OUTER CHAINS/MODIFICATIONS

Chondroitin Sulfate   Dermatan sulfate

ENZYMES

a = β-Xylosyltransferase
b = β-Galactosyltransferase I
c = β-Galactosyltransferase II
d = Xylosyl-phoshotransferase?
e = β-Glucuronosyltransferase I
f = GalNAcβ1–4 transferase
g = GlcNAcα1–4 transferase
h = β-Glucuronosyltransferase II
i = β-Glucuronosyltransferase
j = Chondroitin 6-O-sulfotransferase
k = Chondroitin 4-O-sulfotransferase
l = Glucuronosyl epimerase(s)
m = Iduronate 2-sulfotransferase
n = Heparin de-N-acetylase
o = Heparin N-sulfotransferase
p = Glucuronosylepimerase(s)
q = Heparin 6-O-sulfotransferase
r = Iduronate 2-sulfotransferase
s = Heparin 3-O-sulfotransferase
t = β-Glucuronosyltransferase
u = GlcNAcβ1–4 transferase
(Enzymes j–s act incompletely,
   generating heterogeneity in
   the final product)

■ GlcNAc   ☐ GalNac
= residues that determine
   the type of outer chains
P= phosphate residue variably
   found on core xylose unit

Heparan Sulfate   Heparin

pathway. See Figure 2 for symbol key.

used in the ER to replace the transmembrane region of certain proteins in a transamidation reaction. The acceptor sequence for this reaction is found at the carboxyl terminus of newly synthesized proteins and appears to include specific residues about 20 amino acids from the membrane. In some cells, a fatty acyl chain can also be added to the core inositol group, rendering the anchor resistant to phospholipases. Also, additional ethanolamine phosphate residues can be added to the core mannose residues in some cell types (not shown in Figure 7).

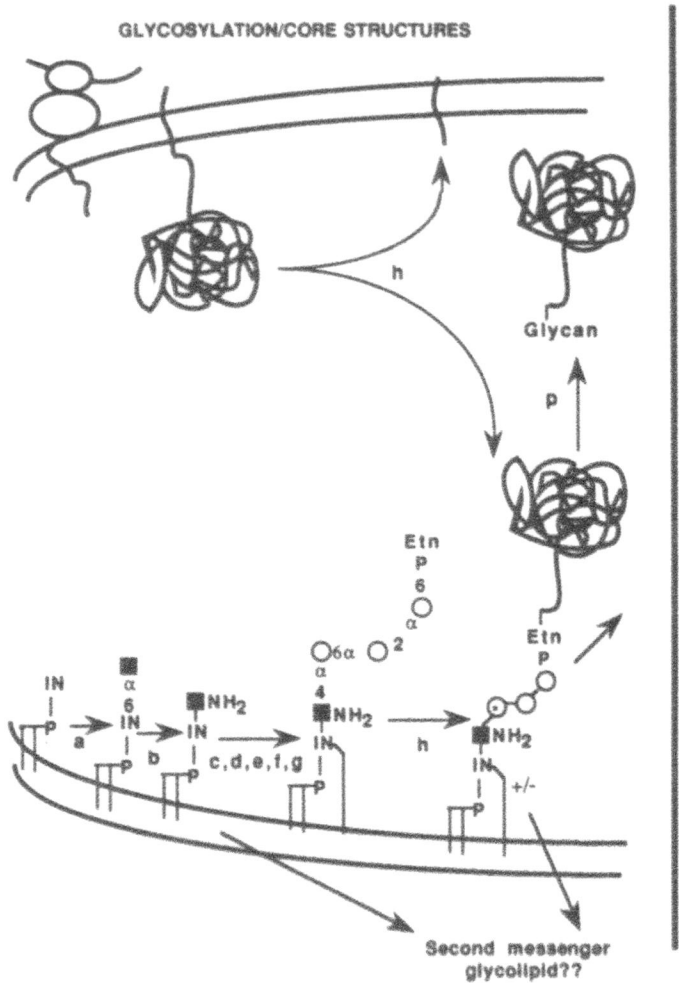

FIGURE 7. The glycophospholipid anchor

2.6.2. Extension

The fatty acyl chains of the anchor can be exchanged for different ones in some cell types after the initial biosynthesis. A variety of different side chains can be found on the GPI anchor in mature proteins, some of which are shown in Figure 7. These appear to be added later in the Golgi apparatus. They bear some similarity to the outer chains of N- and O-linked oligosaccharides, but it is not known whether they are generated by shared transferases. Their function is unknown.

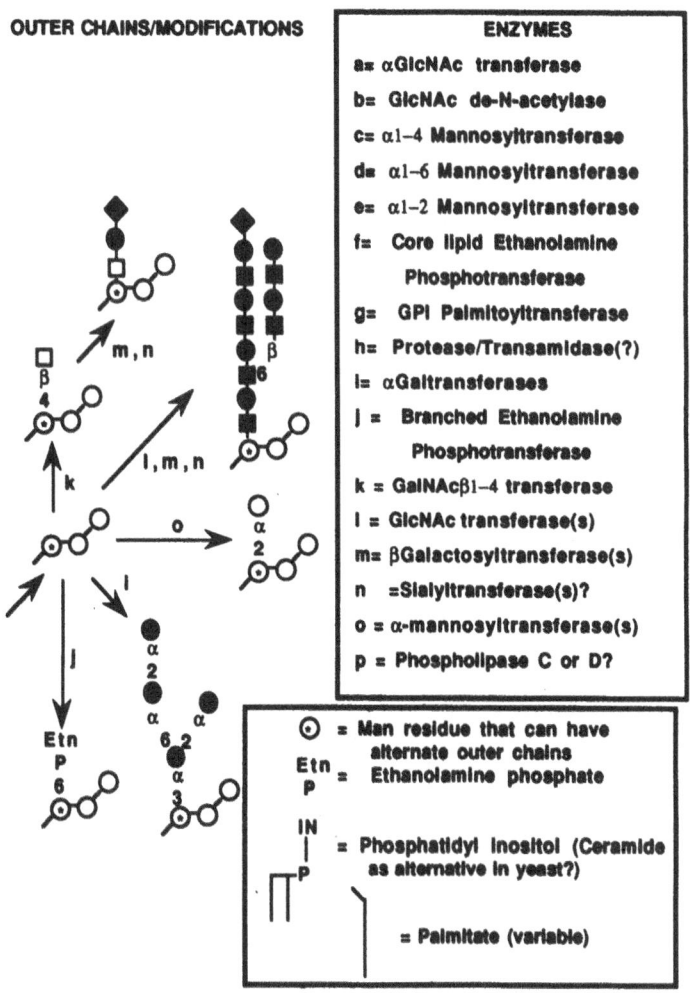

OUTER CHAINS/MODIFICATIONS

ENZYMES

a= αGlcNAc transferase

b= GlcNAc de-N-acetylase

c= α1–4 Mannosyltransferase

d= α1–6 Mannosyltransferase

e= α1–2 Mannosyltransferase

f= Core lipid Ethanolamine
   Phosphotransferase

g= GPI Palmitoyltransferase

h= Protease/Transamidase(?)

i= αGaltransferases

j = Branched Ethanolamine
   Phosphotransferase

k = GalNAcβ1–4 transferase

l = GlcNAc transferase(s)

m= βGalactosyltransferase(s)

n =Sialyltransferase(s)?

o = α-mannosyltransferase(s)

p = Phospholipase C or D?

⊙ = Man residue that can have
    alternate outer chains

Etn P = Ethanolamine phosphate

= Phosphatidyl inositol (Ceramide
  as alternative in yeast?)

= Palmitate (variable)

pathway. See Figure 2 for symbol key.

2.6.3. Termination

Apart from the side chain, no specific processing of the core structure is known. In some cases the fatty acyl chain added to the inositol ring can be removed.

2.6.4. Inhibitors and Mutants

A variety of mutants affect several steps in the assembly of the GPI anchors. These include both tissue-culture cell lines and cells from patients with an ac-

Table IV
Less Common Types of Carbohydrate–Protein Linkages

Linkage	Found in
GlcNAc-O-Serine	Cytosolic glycoproteins, transcription factors, cytokeratins, RNA polymerase, cytoskeletal proteins
GLC-P-Man-O-Serine	Parafuscin, a cytosolic glycoprotein
Fuc-O-Serine	Tissue plasminogen activator
Xyl-Xyl-Glc-O-Serine	Clotting factors
GlcNAc-1-P-Serine	Proteinase in *Dictyostelium*
Glc-O-Tyrosine	Glycogen core protein
Gal-Glc-O-Hydroxylysine	Collagen
Man-O-Ceramide	Lower invertebrate glycosphingolipids

quired disorder of blood cells called paroxysmal nocturnal hemoglobinuria (PNH). Because dolichol-phosphoryl mannose is a donor in the synthesis of the anchor, some of the mutants affect the biosynthesis of both N-linked oligosaccharides and GPI anchors.

2.6.5. Fate and Turnover

Little is known about this matter. It is suggested that, in some cases, phospholipases at the cell surface can release the protein with the oligosaccharide attached to its carboxyl terminus. Also, some controversial evidence indicates that by-products of this pathway may act as second messengers for certain hormones.

2.7. Less Common Types of Oligosaccharide Linkages

The types of glycoconjugates discussed to this point are by far the most common forms encountered, but they are not the only ones. Table IV lists some of the rarer forms of protein-carbohydrate linkages. These types are not as well studied as the commoner ones (except in the case of O-linked GlcNAc residues added in the cytoplasm, which are discussed below). It is important to keep in mind that a rarer oligosaccharide is also unique, and may be restricted to a particular protein or class of proteins. As such, it would be theoretically more likely to carry out some unique biological function(s) (Kornfeld, 1990; Troy, 1992).

2.8. Glycoconjugates Frequently Share Identical Outer Sequences

In contrast to the very distinctive features of the core regions of the different types of oligosaccharides, a large number of outer sequences are shared by a number of glycoconjugates (Kobata and Takasaki, 1992; Fukuda, 1992). These

STRUCTURE	N-GlcNAc -Linked	O-GalNAc Linked	Glc-Cer Linked	GPI- Linked	Common Name(s)
●ß4■ß-	+	+	+	+	Type 2 Lactosamine
◆α6●ß4■ß-	+	+	+		CD75, ?CD76
9(7)(4)A◇α6●ß3■ß-	+	+			
◆α3●ß4■ß-	+	+	+		
◆α8◆α3●ß4■ß-	+	+			
(◆α8◆)ₙα3●ß4■ß-	+	+			Polysialic Acid (PSA)
●α3●ß4■ß-	+		+		
△α2●ß4■ß-	+	+	+		
●ß4■ß- α3△	+	+	+		Lewisˣ (SSEA-1) CD15
△α2●ß4■ß- α3△	+	+	+		Lewisʸ
◆α3●ß4■ß- α3△	+	+	+		Sialyl Lewisˣ
◆α6●ß4■ß- 6S	+	+			
◆α3●ß4■ß- 6S	+?	+			
8S◆α8◆α3●ß4■ß-			+		
3S●ß4■ß-	+				
3S◇ß3●ß4■ß-	+?		+		HNK-1 antigen
◆α3●ß4■ß- ß4□	+	+	+		Sda/CAD antigen
●ß4■ß3(●ß4■ß3)ₙ	+	+	+	+	Polylactosamine (i antigen)
●ß4■ß3●ß4■ß3●ß4■ß3 ß6 ●ß4■ß3●	+	+	+	+?	Branched Polylactosamine (I antigen)
●ß4■ß3●ß4■ß- α3 α3 △ △	+	+	+		Dimeric Lewisˣ
△α2●ß4■ß3●ß4■ß- α3 α3 △ △	+	+	+		Dimeric Lewisʸ
◆α3●ß4■ß3●ß4■ß- α3 α3 △ △	+	+	+		Sialyl dimeric Lewisˣ
◆α3●ß4■ß3●ß4■ß- α3 △	+	+	+		VIM-2, CD65
◆α6●ß4■ß3●ß4■ß- α3 △	+	+	+		

FIGURE 8. Some outer chain sequences found on different types of glycans. See Figure 2 for symbol key.

STRUCTURE	N-GlcNAc-Linked	O-GalNAc Linked	Glc-Cer Linked	GPI-Linked	Common Name(s)
6S 6S 6S ◆α3(6)●β4■β3●β4■β3	+	+			Keratan Sulfate
●β3■β-	+	+	+		Type 1 lactosamine
◆α3●β3■β-	+	+	+		
●α3●β3■β-	+	+	+		
△α2●β3■β-	+	+	+		
●β3■β- α4△	+	+	+		Lewis[a]
△α2●β3■β- α4△	+	+	+		Lewis[b]
△α2●β3■β- α3● α4△	+	+	+		Blood Group B
△α2●β3■β- α3□ α4△	+	+	+		Blood Group A
◆α3●β3■β- α4△	+	+	+		Sialyl Lewis[a]
◆α3●β3■β- α6◆	+	+			
◆ α6 ◆α3●β3■β- α4△			+		Disialyl Lewis[a]
□β4■β-	+				
4S□β4■β-	+				
◆α6□β4■β-	+				
●β3□		+	+	+?	PNA receptor
◆α3●β3□		+	+	+?	
◆α8◆α3●β3□			+		
●α3●α3			+		
□β3●α-			+		
□α3□β3●α-			+		Forssman Antigen
○α2○α-	+			+	

FIGURE 8. (*continued*)

include sialic acids, polylactosamines, blood group sequences, etc. (see Figures 3, 5, and 8). These outer chains are probably added to the different acceptors (*O*- or *N*-linked chains, ceramide-linked chains, etc.) by overlapping and, in some cases, probably identical transferases. How this is controlled from one glycoconjugate to another is not known, but it underscores the importance of the initiation

Table V
Examples of Modifications of the Oligosaccharide Chains
on Animal Glycoconjugates

Type	Modified sugar	Found in
Phosphate	Man-6-P	Lysosomal enzymes, TGF-β precursor
	GlcNAc-6-P	Milk oligosaccharides
	Xyl-2-P	Core xylose of glycosaminoglycans
	Inositiol-1-P	Glycophospholipid anchors
	Sialic acid-9-P	Biosynthetic intermediate (free)
	GlcNAc-1-P-6 Man	Precursor of Man-6-P (see above)
	Glc-1-P-6-Man	Cytosolic parafuscin or phosphoglucomutase
	GlcNAc-1-P-Ser	Dictyostelium proteins
	GalNAc-1-P?	N-cadherin and ? other proteins
	Man-1-P-6-Man	Yeast mannans
Sulfate		
	Gal-3-S	Sulfatide (ceramide-linked) thyroglobulin (N-GlcNAc-linked)
	Gal-6-S	Keratan sulfate
	Man-6-S	Dictyostelium lysosomal enzymes
	Man-3(4)-S	Ovalbumin
	GlcNAc-6-S	Heparin/heparan sulfate, N-GlcNAc-linked chains (endothelium), keratan sulfate
	GlcNS	Heparin/haparan sulfate
	GlcNS-3-S	Haparin/heparan sulfate
	GlcNS-3,6-S	Heparin/haparan sulfate
	GalNAc-4-S	Chondroitin sulfate
	GalNAc-6-S	Chondroitin sulfate
	GlcA-3-S	Heparin sulfate
	Id-2-S	Heparin sulfate
	GlcUA-3-S	Glycosphingolipids, ? N-GlcNAc-linked oligosaccharides
	Sia-8-S	Glycosphingolipids: echinoderm, bovine
O-Acetyl	Sialic acid-4,7,8,9	O-N- and Ceramide-linked oligosaccharides
O-Methyl	Sialic acid-8	Echinoderm glycosphingolipids
O-Acyl	Inositol	Palmitoylation on GPI anchors
	Galactose	Acylated Gal-Ceramide in brain
	Sialic acid	O-lactyl groups on mucins and on free sialic acids
Ethanolamine-P	Mannose	GPI anchors

of the sugar chains. It should not be surprising that similar outer structures could be used to accomplish different functions when presented in different contexts. Tissue-specific expression of different genes whose products catalyze the same reaction but with differing substrate requirements may also be used to generate "context-specific" glycosylation.

2.9. Modifications of the Sugar Units in Oligosaccharide Chains

The different types of sugar chains described here are subject to further modification by the addition of a variety of different types of substituents (e.g., phosphate, sulfate, acetate, etc.). A list of known modifications is given in Table V. Because the addition of such substituents imparts further diversity to the common core oligosaccharide structures, it is reasonable to speculate that they may impart more specific functions to the sugar chains. In at least some cases (e.g., mannose-6-phosphate on lysosomal enzymes and specific types of sulfation in heparin), this is known to be the case (Kornfeld, 1990; Cumming, 1991).

3. NUCLEAR AND CYTOPLASMIC GLYCOSYLATION

This chapter has focused upon glycosylation that occurs in the ER-Golgi-plasmalemma pathway. Until a few years ago it was thought that glycosylation did not occur on molecules on the opposite side of the membrane (i.e., the cytosol), which is contiguous with the interior of the nucleus. However, it is now clear that the addition of single O-GlcNAc residues is very common on cytosolic and nuclear proteins (Hart *et al.*, 1989). Some proteins have many such units, sometimes closely spaced. Clues from many investigators suggest that this may only be the proverbial tip of the iceberg, and that many other types of nuclear and cytoplasmic glycosylation may occur. This topic deserves more attention than can be accorded it here.

ACKNOWLEDGMENTS. This work was supported by USPHS grants GM32373, CA38701, and GM32485, and a VA Merit Review Award. H. F. is an established investigator of the American Heart Association. We thank Michael Bevilacqua, Christoph Binkert, Daniel Donoghue, Richard Dutton, Michiko Fukuda, Chris Glass, Robert Linhardt, Elaine Muchmore and Harry Schachter for their helpful comments and suggestions.

4. REFERENCES

Bhavanandan, V. P., and Davidson, E. A., 1992, Proteoglycans: Structure, synthesis, function, in *Glycoconjugates: Composition, Structure and Function* (H. J. Allen and E. C. Kisailus, eds.), pp. 167–202, M. Dekker, Inc., New York.

Carraway, K. L., and Hull, S. R., 1991, Cell surface mucin-type glycoproteins and mucin-like domains, *Glycobiology* 1:131–138.

Cumming, D. A., 1991, Glycosylation of recombinant protein therapeutics: Control and functional implications, *Glycobiology* 1:115–130.

Cummings, R. D., Merkle, R. K., and Stults, N. L., 1989, Separation and analysis of glycoprotein oligosaccharides, *Methods Cell Biol.* 32:141–183.

Cummings, R. D., 1992, Synthesis of asparagine-linked oligosaccharides: Pathways, genetics, and metabolic regulation, in *Glycoconjugates: Composition, Structure and Function* (H. J. Allen and E. C. Kisailus, eds), pp. 333–360, M. Dekker, Inc., New York.

Doering, T. L., Masterson, W. J., Hart, G. W., and Englund, P.T., 1990, Biosynthesis of glycosyl phosphatidylinositol membrane anchors, *J. Biol. Chem.* **265**:611–614.

Elbein, A. D., 1991, Glycosidase inhibitors: Inhibitors of N-linked oligosaccharide processing, *FASEB J.* **5**:3055–3063.

Farquhar, M. G., 1991, Protein traffic through the Golgi complex, in *Intracellular Trafficking of Proteins* (C. J. Steer and J. Hanover, eds.), pp. 431–471, Cambridge University Press, New York.

Ferguson, M. A. J., 1992, Lipid anchors on membrane proteins, *Curr. Opinion Struct. Biol.* **1**:522–529.

Fukuda, M., 1992, Cell surface carbohydrates in hematopoietic cell differentiation and malignancy, in *Cell Surface Carbohydrates and Cell Development* (M. Fukuda, ed.), pp. 127–159, CRC Press, Boca Raton, Florida.

Hakomori, S., 1986, Tumor associated glycolipid antigens, their metabolism and organization, *Chem. Phys. Lipids* **42**:209–233.

Hakomori, S., 1990, Bifunctional role of glycosphingolipids: Modulators for transmembrane signaling and mediators for cellular interactions, *J. Biol. Chem.* **265**:18713–18716.

Hart, G. W., Haltiwanger, R. S., Holt, G. D., and Kelly, W. G., 1989, Glycosylation in the nucleus and cytoplasm, *Annu. Rev. Biochem.* **58**:841–874.

Hirschberg, C. B., and Snider, M. D., 1987, Topography of glycosylation in the rough endoplasmic reticulum and Golgi apparatus, *Annu. Rev. Biochem.* **56**:63–87.

Kisailus, E. C., and Allen, H. J., 1992, Nomenclature of monosaccharides, oligosaccharides, and glycoconjugates, in *Glycoconjugates: Composition, Structure and Function* (H. J. Allen and E. C. Kisailus, eds.), pp. 13–32, M. Dekker, Inc., New York.

Kjellén, L., and Lindahl, U., 1991, Proteoglycans: Structures and interactions, *Annu. Rev. Biochem.* **60**:443–475.

Kobata, A., and Takasaki, S., 1992, Structure and biosynthesis of cell surface carbohydrates, in *Cell Surface Carbohydrates and Cell Development* (M. Fukuda, ed.), pp. 1–24, CRC Press, Boca Raton, Florida.

Kornfeld, R., and Kornfeld, S., 1985, Assembly of asparagine-linked oligosaccharides, *Annu. Rev. Biochem.* **54**:631–664.

Kornfeld, S., 1990, Lysosomal enzyme targeting, *Biochem. Soc. Trans.* **18**:367–374.

Kundu, S. K., 1992, Glycolipids: Structure, synthesis, functions, in *Glycoconjugates: Composition, Structure and Function* (H. J. Allen and E. C. Kisailus, eds.), pp. 203–262, M. Dekker, Inc., New York.

Mellman, I., and Simons, K., 1992, The Golgi complex: *In vitro* veritas? *Cell* **68**:829–840.

Paulson, J. C., and Colley, K. J., 1989, Glycosyltransferases: Structure, localization, and control of cell type-specific glycosylation, *J. Biol. Chem.* **264**:17615–17618.

Rademacher, T. W., Parekh, R. B., and Dwek, R. A., 1988, Glycobiology, *Annu. Rev. Biochem.* **57**:785–838.

Rothman, J. E., and Orci, L., 1992, Molecular dissection of the secretory pathway, *Nature* **355**:409–415.

Ruoslahti, E., and Yamaguchi, Y., 1991, Proteoglycans as modulators of growth factor activities, *Cell* **64**:867–869.

Schachter, H., 1991, The "yellow brick road" to branched complex N-glycans, *Glycobiology* **1**:453–462.

Schachter, H., and Brockhausen, I., 1992, The biosynthesis of serine (threonine) N-acetylgalactosamine-linked carbohydrate moieties, in *Glycoconjugates: Composition, Structure and Function* (H. J. Allen and E. C. Kisailus, eds.), pp. 263–332, M. Dekker, Inc., New York.

Schauer, R., 1991, Biosynthesis and function of *N*- and *O*-substituted sialic acids, *Glycobiology* **1**:449–452.

Shayman, J. A., and Radin, N. S., 1991, Structure and function of renal glycosphingolipids, *Am. J. Physiol. Renal, Fluid Electrolyte Physiol.* **260**:F291–F302.

Stanley, P., 1992, Glycosylation engineering, *Glycobiology* **2**:99–107.

Stults, C. L. M., Sweeley, C. C., and Macher, B. A., 1989, Glycosphingolipids: Structure, biological source, and properties, *Methods Enzymol.* **179**:167–214.

Troy, F. A., II, 1992, Polysialylation: From bacteria to brains, *Glycobiology* **2**:5–23.

VanEchten, G., and Sandhoff, K., 1993, Ganglioside metabolism: Enzymology, topology, and regulation, *J. Biol. Chem.* **268**:5341–5344.

Varki, A., 1991, Radioactive tracer techniques in the sequencing of glycoprotein oligosaccharides, *FASEB J.* **5**:226–235.

Varki, A., 1992a, Diversity in the sialic acids, *Glycobiology* **2**:25–40.

Varki, A., 1992b, Role of oligosaccharides in the intracellular and intercellular trafficking of mammalian glycoproteins, in *Cell Surface Carbohydrates and Cell Development* (M. Fukuda, ed.), pp. 25–69, CRC Press, Boca Raton, Florida.

Varki, A., 1993, Biological roles of oligosaccharides: All of the theories are correct, *Glycobiology* **3**:97–130.

Winchester, B., and Fleet, G. W. J., Amino-sugar glycosidase inhibitors: Versatile tools for glycobiologists, *Glycobiology* **2**:199–210.

Yanagishita, M., and Hascall, V. C., 1992, Cell surface heparan sulfate proteoglycans, *J. Biol. Chem.* **267**:9451–9454.

Chapter 4

Compartments of the Early Secretory Pathway

Rob J. M. Hendriks and Stephen D. Fuller

1. INTRODUCTION

The secretory pathway consists of a set of compartments responsible for the assembly and modification of proteins destined for secretion, for transport to the plasma membrane, and to the other organelles of the cell. The classical description of the secretory pathway comprises the endoplasmic reticulum (ER) as the site of protein synthesis and first maturation steps, and the Golgi apparatus as the site of protein modification and sorting. Apart from synthesis and maturation of secreted proteins the pathway is also responsible for the production of its own components. Hence, the enzymes and factors involved in the functions of the pathway are continually renewed and are being transported to their positions within the pathway by the machinery that transports secretory proteins through it. It is this latter aspect of the pathway, its mechanism of self-renewal, that complicates both the definition and the description of the compartments of the secretory pathway. In this review we will adopt a functional approach to the description of the components of the early secretory pathway and focus on its dynamic aspects. Here the ER will be referred to as the aggregate of smooth ER (sER), rough ER

Rob J. M. Hendriks and Stephen D. Fuller   Biological Structures and Biocomputing Programme, European Molecular Biology Laboratory, D6900 Heidelberg, Germany.

Subcellular Biochemistry, Volume 22: Membrane Biogenesis, edited by A. H. Maddy and J. R. Harris. Plenum Press, New York, 1994.

(rER), and the nuclear envelope that is continuous with them. The Golgi is defined as the aggregate of cisterna and connected networks on both the *trans* (TGN, *trans*-Golgi network) and *cis* (CGN, *cis*-Golgi network) sides. We will focus on membrane traffic between the ER and Golgi compartment and will present a useful framework for understanding the membrane traffic that gives rise to these compartments.

1.1. Sorting in the Secretory Pathway

A consequence of the ability of the secretory pathway to replenish itself is that the pathway must be capable of sorting its own components. The past several years have brought a great deal of information and some consensus about the mechanisms and signals involved in sorting of proteins along the secretory pathway.

A basic premise is that of bulk flow. Early results studying ER exit of several secretory proteins (Lodish *et al.*, 1983; Scheele and Tartakoff, 1985) and transmembrane proteins (Rose and Bergmann, 1983; Williams *et al.*, 1985) suggested that a positive transport signal existed. These studies showed that transport was dramatically slowed down or stopped by mutations that changed regions of the protein. However, these results are now seen as reflections of differences in the overall structure that lead to different folding efficiencies (see for review Hurtley and Helenius, 1989; Rose and Doms, 1988). The lack of success in identifying positive signals for transport from the ER, coupled with an experiment using tripeptides to determine the bulk flow rate from the ER to the cell surface (Wieland *et al.*, 1987), informed the consensus that transport to the later stages of the exocytotic pathway is by default. Thus, proteins will be transported to the later elements of the pathway if they lack a specific signal for retention (Rothman, 1987).

Given bulk flow as the mechanism for the transport of most proteins, one can define three phenomenologically distinct sorting mechanisms utilized along the pathway (Warren, 1987). The first is retention in a compartment by direct and continuous interaction with its components. The ribophorins and components of the translocation machinery appear to exemplify this mechanism (Crimaldo *et al.*, 1987; Kreibich *et al.*, 1978). The second is selective signal mediated retrieval of components from downstream compartments. Retrieval of soluble proteins of the endoplasmic reticulum serves as the prototype for this mechanism (Dean and Pelham, 1990; Munro and Pelham, 1987; Pelham, 1989a). The third mechanism is the selective inclusion of components into a transport vesicle for transport downstream. This mechanism is used when there is a bifurcation in the pathway. For example, in the *trans*-Golgi network of a polarized cell, lysosomal, apical, and basolaterally directed proteins are separated (Griffiths and Simons, 1986;

Mellman and Simons, 1992). The first two mechanisms require that the interactions responsible for sorting continue for the life of the protein (Simons and Fuller, 1985). Hence, the signal must be a permanent part of the protein. This need not be true in the third case in which the signal can be recognized once and the protein diverted as a result.

Most of the identified sorting signals in the early secretory pathway are remarkably simple and well defined. The best-characterized is that of the carboxy-terminal four amino-acid (KDEL) signal for retention in the endoplasmic reticulum. Here, mutagenesis has clearly demarcated the functionally important sequence and demonstrated its function upon transfer to other proteins (Hardwick *et al.*, 1990; Munro and Pelham, 1987; Semenza *et al.*, 1990; Zagouras and Rose, 1989). Parsimony, the need for a limited number of receptors to recognize similar signals on a wide variety of different proteins, probably leads to a modular nature for most signals and hence their recognizability and apparent simplicity. Despite this apparent simplicity, closer examination of the evidence suggests that context is very important for the recognition and function of most signals. Further, the simplicity of the recognized signals has contributed to their identification, and other signals may be more complex. For example, three-dimensional structure is probably very important for signals that are responsible for static retention since interaction between distinct pairs of components is responsible for this phenomenon. This aspect makes these signals particularly difficult to characterize in terms of a defined sequence, and thus far none have been characterized completely.

2. COMPARTMENTS

What do we mean by the term "compartment," which we have been using so freely? Initially, compartments were defined morphologically. Electron microscopy reveals clear distinctions between smooth and rough ER, Golgi elements, and transport vesicles (Fawcett, 1981). This definition proved particularly appealing because biochemical work showed that the compartments associated with different morphologies had different components and functional activities (Palade, 1975). Unfortunately, the morphological definition or even the straightforward biochemical definition of a compartment is not useful for the description of compartments in terms of membrane traffic. For our task we need to adopt a very strong and restrictive definition that will give rise to predictable consequences in terms of membrane traffic. Such a definition will allow us to focus on the aspects of sorting and transport in the early secretory pathway. Many of the morphologically defined components of the secretory pathway, such as smooth and rough ER, will fulfill neither this stronger definition of compartment nor some of its consequences.

2.1. Definition of a Compartment

Our definition is very simple. A compartment of the secretory pathway is defined as a collection of membrane-bound structures in which the membranes are physically discontinuous from other compartments and which has a unique set of proteins characteristic for the compartment (see Figure 1). These two aspects, physical discontinuity and unique components, are obviously selected with an eye toward describing features of membrane traffic between compartments. Physical discontinuity can be assayed by morphology and by studying the distribution of freely diffusing membrane or soluble proteins in the putative compartment. These two provide a check on each other. Verifying the uniqueness of components in a given compartment is complicated by the fact that the renewal of compartments requires that residents of one compartment are transiently present in others. The question of thresholds of detection must be considered. Because our interest is in transport, we will employ a threshold of detection that is relevant for transport. As a result, very sensitive measures are not needed to demonstrate either physical discontinuity or uniqueness of components. Two membrane structures are physically continuous when their degree of continuity obviates the need for membrane transport between them. It is very difficult to demonstrate that two compartments of complex geometry have no continuities between them. However, a degree of continuity that is so small that it is difficult to detect will also be relatively insignificant in terms of transport. In none of the cases that we will examine are the elements of the definition equivocal. Continuity is always extensive and obvious, and the localization of the bulk of components is always well defined in a given condition.

An important aspect of our operational definition of a compartment is the requirement for measuring several components simultaneously. If we wish to understand the membrane traffic between compartments, then we need to know the compartment size and composition accurately. Frequently, the description of a component of a putative compartment is performed in a very qualitative way. This leads to apparent contradictions. We will discuss the compartment(s) intermediate between the ER and the Golgi at some length below. Unfortunately, many different functions have been attributed to this portion of the secretory pathway and little dual labeling and quantitation is available so that even the existence of this (these) compartment(s) remains in dispute. Faced with such confusion it is tempting to simplify the early secretory pathway to a single compartment (Mellman and Simons, 1992). However, we feel it is better to take the confusion as a measure of the state of the field and focus our attention on the need for accurate descriptions.

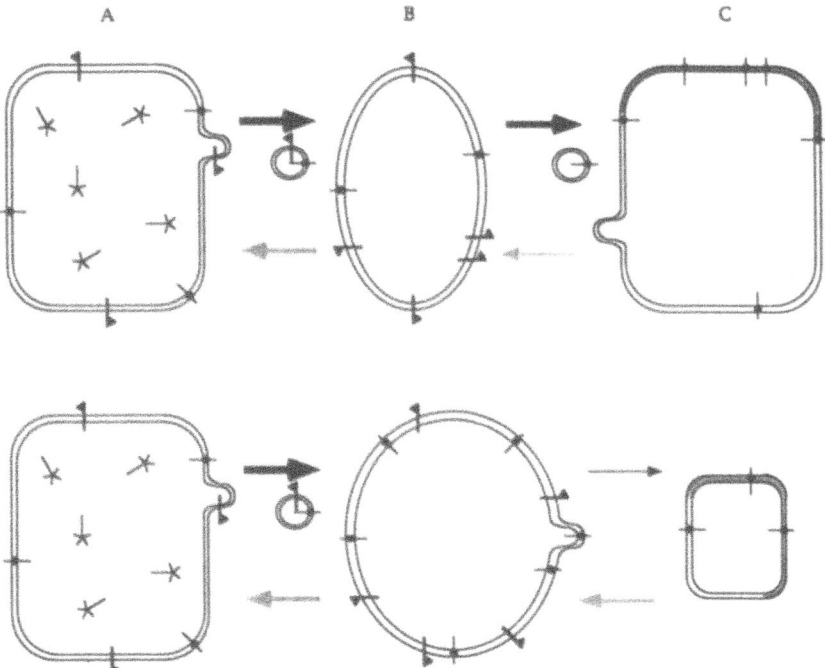

FIGURE 1. Schematic representation of our compartment definition and some of its consequences. A compartment is discontinuous from its upstream and downstream counterparts and contains unique components. Proteins ✕— and ▼— represent unique proteins of compartment (A) and (B), respectively. As depicted, compartment (B) contains none of the compartment (A) resident ✕— s; however, in addition to its own unique marker (▼—) it processes the protein —•— on transit to compartment (C). Owing to the existence of unique resident proteins, each compartment must have its sorting machinery. Traffic to and from the compartments is mediated by transport vesicles. Because a compartment has unique components, the upstream and downstream vesicles must be different. Differences in regulation of vesicular transport leading to and coming from a compartment render it dynamic. The dynamics can influence subdomains differently, as depicted in the lower panel of the figure.

2.2. Consequences of the Definition

2.2.1. Two Types of Transport Vesicles

Having established a definition, we can proceed to deduce consequences from it. The first consequence of our definition, and the one of most importance for a description of membrane traffic, is that there must be transport vesicles mediating the traffic to and from the compartment. This can be seen immediately as our definition allows no other communication between the compartment and the rest of the secretory pathway. The premise that the compartment has unique components leads to the conclusion that the vesicles mediating traffic to and from the compartment will be distinct. If this were not the case, the compartment would not be able to maintain a unique composition. If the transport vesicles are distinct then they can be regulated separately and as a result the compartment would be dynamic with respect to size and composition. This dynamism cannot be unconstrained; if otherwise, the compartment would lose its identity. Hence, there must be some components that are responsible for maintenance of size.

2.2.2. Uniqueness of Components

Three classes of components need to be distinguished. The first are the resident proteins, which are unique. These impart a unique function to the compartment (e.g., isomerization, glycosylation, etc.). As we are focusing on the functional significance of components, it is the location of the bulk of the marker that is important, not the location of all of the marker. The soluble proteins of the endoplasmic reticulum are a case in point (Dean and Pelham, 1990; Lewis and Pelham, 1992; Pelham, 1988). A second class that is characteristic but not unique are the proteins in transit to other compartments, such as the plasma membrane, endosomes, and lysosomes. Proteins involved in the vesicular transport machinery itself represent the third class of components. The rab (rat brain) proteins or the coatamer elements are examples of these (described in more detail later). These proteins can be better characterized as representing the (vesicular) connection between an upstream and its downstream compartment, rather than serving as a unique marker for an individual compartment.

2.2.3. Sorting

The presence of transport vesicles gives rise to a need for sorting mechanisms. Two cases exist. If bulk flow operates in the compartment, then there must be retention signals to remove components from the flow and establish them in the compartment. The second case is one in which the pathway bifurcates and, hence, some components must be selectively diverted from the onward bulk flow due to the presence of specific signals. (The above-mentioned features of a compartment are depicted in Figure 1.)

2.2.4. Viruses

The budding of membrane viruses serves as a useful indicator for the operation of the features of a compartment (Simons and Fuller, 1987). This follows from several key features of enveloped virus budding. Enveloped viruses bud as a result of interaction of their envelope proteins with the internal components (Simons and Fuller, 1985). The attachment of internal components can only commence when sufficient glycoproteins have clustered to form an attachment site (Fuller, 1987). Often, clustering, and budding, will only occur when the glycoproteins are properly oligomerized or processed. The clustering also excludes the membrane proteins of the host, which leads to the lack of incorporation of host proteins into the budded virion (Simons and Fuller, 1987). After the attachment site has formed and internal structures interact with and stabilize the cluster, a reorganization must occur to allow the deformation of the envelope that accompanies budding. The initial clustering must represent some tendency of the envelope proteins to interact weakly since reorganization of the envelope proteins would be hindered by strong interactions with cellular proteins or each other (Fuller, 1987). To form such clusters by weak interaction, the membrane glycoproteins must be able to diffuse in the membrane. As a result, they cannot be concentrated in a compartment unless that compartment is physically discontinuous. Further, unless transport has been blocked, the localization of the bulk of the protein reflects the sorting mechanisms of the cell. In this way, the observation that a structure is the unique site of virus budding is a strong indicator that it is functioning as a compartment.

The use of virus budding as a compartment marker needs to be accompanied by the caveat that virus infections produce cytopathic effects. In some cases these can affect the composition of the compartments of the secretory pathway. Therefore, one needs to be sure that the distribution of other markers of the compartment is not disturbed by the infection itself.

2.2.5. Unique Environment

A consequence of the unique components of a compartment is that it is able to offer a unique environment for particular functions. Among the critical elements in such an environment are its pH, its reduction potential, its divalent ion concentration, and the concentration of substrates for enzymes resident in the compartment. For example, the exchange of disulfides in the ER is possible only because the ER contains an oxidizing environment, a non-acidic pH to allow exchange, and the enzyme protein disulfide isomerase that catalyzes the exchange so it can occur at a physiological time scale (Freedman, 1987; Freedman and Hillson, 1980; Freedman, 1984).

2.2.6. Degradation

A compartment as we have defined it must also have a pathway for degradation. This is a consequence of the need for retention mechanisms and the finite lifetimes of proteins (including the ones retained in a given compartment). One can imagine several mechanisms ranging from the selective recognition and sequestration of aberrant proteins to interference with the retention signals so that bulk flow brings proteins to another compartment for degradation. There is good evidence that degradation is under fine control and mediated by a variety of mechanisms (Bonifacino and Lippincott-Schwartz, 1991; Chau *et al.*, 1989; Klausner and Sitia, 1990).

2.3. Example of the Plasma Membrane

Before proceeding to describe the early secretory pathway in terms of compartments, we would like to show how our definition and its consequences apply to the less controversial case of the plasma membrane. In an epithelial cell there are two compartments called plasma membrane: the apical membrane and the basolateral membrane. Physical discontinuity is easily shown since these compartments are separated by the tight junction. The outer leaflet of the bilayer is broken by this structure although the inner leaflet is connected (Simons and van Meer, 1988; van Meer, 1989). The consequence of this discontinuity is that membrane proteins and outer membrane lipids must enter and leave these two compartments by vesicular transport. Indeed, transcytosis is well described for this system and a start toward the characterization of the vesicles involved in transport to the plasma membrane has been made (Sztul *et al.*, 1991; Wandinger-Ness *et al.*, 1990).

Subdomains can be found in the plasma membrane compartments, and they provide a useful example of structures that fail to meet our definition. A simple one is provided by the microvillar domains. Although these have unique components, these domains are incapable of confining freely diffusible components. It is worth pointing out that these subdomains can be isolated biochemically. Hence, biochemical isolation does not serve as a reliable indicator of compartmentalization.

Viruses provide an interesting illustration of the nature of the plasma membrane compartments. It is well established that vesicular stomatitis virus (VSV) buds from the basolateral compartment while influenza virus buds from the apical surface. This budding polarity reflects the sorting of the associated membrane glycoproteins: VSV-G protein is sorted to the basolateral surface while the influenza haemagglutinin (HA) is directed to the apical surface (Hughson *et al.*, 1988; Simons and Fuller, 1985; Simons and Fuller, 1987). Hence, virus budding provides a reliable marker for compartmentalization. Both of these viruses are cytopathic and eventually cause a loss of integrity of the tight junctions separating the two compartments so that diffusible components mix. Under these conditions, virus budding loses its polarity so that the two viruses can be found budding from both surfaces (Roth and Compans, 1981).

3. ENDOPLASMIC RETICULUM

3.1. Morphology and Freely Diffusible Markers

The endoplasmic reticulum (ER) is the cell's largest compartment and typically constitutes more than half of the total cellular membrane (Weibel *et al.*, 1969). The ER provides the cell with a mechanism of segregating newly synthesized transmembrane, luminal, and secretory proteins from the cytoplasm (Palade, 1975; Walter and Lingappa, 1986). Although the ER can be divided in distinct regions (see below), morphological studies indicate that it is a continuous membrane structure (Fawcett, 1981), physically distinct from the downstream Golgi. These findings have been confirmed in experiments using the transmembrane G protein of VSV (Bergmann and Fusco, 1990) and the luminal heavy chain binding protein (BiP) (Ceriotti and Colman, 1988), which indicated that these proteins behaved as freely diffusible markers between the ER subdomains. Therefore, according to our definition of a compartment, these morphologically defined subdomains should be regarded as subcompartments, something that will be reflected in the consequences of the definition.

3.2. Subcompartments

The rough ER (rER), which is studded with ribosomes, is morphologically easily distinguishable from the smooth ER (sER), which is physically a portion of the same membrane but lacks attached ribosomes. Although many proteins are common to both subdomains, some such as ribophorin I and II (Amar-Costesec *et al.*, 1984), the docking protein (Hortsch and Meyer, 1985), or the originally designated signal sequence receptor subunits (Görlich *et al.*, 1990; Wiedmann *et al.*, 1987) are restricted to the rER, whereas one, epoxide hydrolase, has been reported to be restricted primarily to the sER in rat liver (Galteau *et al.*, 1985).

Apart from the rER and the sER, two other subdomains can be identified. The nuclear envelope (NE), which is continuous with the rER and also contains ribosomes, possesses unique structures such as the nuclear pore complexes as well as unique proteins that interact with nuclear lamins in the inner surface

(Gerace *et al.*, 1984). Another subdomain is the transitional elements (TE), which are smooth, vesicular-tubular structures facing the *cis*-Golgi and which are believed to be the ER exit sites (Palade, 1975).

The fact that there are components specific for ER subdomains indicates that some special restraining mechanisms must exist. Although the exact underlying mechanisms are still unclear, they might include complex formation between specific protein and/or lipid components in the bilayer, which exclude other proteins, and protein interactions on either side of the membrane.

3.3.   Unique Proteins

In addition to the proteins mentioned above, which are enriched in specific subdomains, the ER lumen contains a set of resident proteins that catalyze the maturation events of a nascent polypeptide chain. The most abundant members are the *glucose-regulated proteins* GRP78 (more commonly known as the heavy chain *binding protein*, BiP), GRP94, and *protein disulfide isomerase* (PDI). PDI is a multifunctional protein. Existing as a homodimer, it catalyzes the isomerization of intramolecular disulfide bridges, which is the rate-limiting step in the folding of many proteins (Freedman, 1984). But PDI is also the β-subunit of the $\alpha_2\beta_2$ tetramer of prolyl hydroxylase (Pihlajaniemi *et al.*, 1987) and has been reported to be an important component of the oligosaccharide transferase (Geetha-Habib *et al.*, 1988).

The binding protein BiP is not only involved in the translocation process of some proteins (Rose *et al.*, 1989), but also catalyzes folding by preventing aggregation of folding intermediates (Gething and Sambrook, 1990; Pelham, 1989a). Although BiP is not a heat-shock protein (hsp) in mammals, sequence homology indicates that it is a member of the hsp70 family of heat-shock proteins (Munro and Pelham, 1986), which are believed to mediate folding and stabilization of unfolded proteins in other compartments of the cell. Moreover, BiP was shown to be an ATPase that is released from its substrate by ATP hydrolysis (Flynn *et al.*, 1989; Kassenbrock and Kelly, 1989). Although the function of the calcium binding glycoprotein GRP94 has not yet been well defined, recent reports suggest that GRP94 also possesses protein-binding properties when malfolded proteins are retained in the ER (Navarro *et al.*, 1991).

3.4.   Consequences of the Definition for ER

3.4.1.   Upstream and Downstream Transport Vesicles

The most important consequence of our compartment definition is that membrane traffic between compartments must be mediated by transport vesicles. Furthermore, the upstream vesicles leading to the compartment must be different form the downstream vesicles. Based on the experiments mentioned above, which utilized BiP and VSV-G as freely diffusible markers, it is clear that this consequence does not apply to the individual ER subdomains.

The ER is an obvious exception concerning vesicular transport as means of influx of an upstream compartment. However, transmembrane, luminal, and secretory proteins are selected from the cytoplasm and targeted to the ER membrane after formation of signal recognition particle (SRP)-ribosome-nascent chain complex (Meyer and Dobberstein, 1980a; Meyer and Dobberstein, 1980b; Walter and Blobel, 1982) . The targeting step is facilitated via a specific interaction of SRP with its receptor, the docking protein (Meyer et al., 1982). After docking and release of the SRP, translocation of the polypeptide chain takes place. Although the translocation mechanism is still unknown, several proteins have been identified that are proposed to constitute a translocation complex (Görlich et al., 1992; see for review High and Dobberstein, 1992; Nunnari and Walter, 1992).

Downstream traffic from the ER is mediated by transport vesicles (Palade, 1975). Accumulating evidence suggests that the transport from a donor to an acceptor compartment can be divided into the steps of budding, targetting, and fusion. By selecting for temperature-sensitive mutants blocked early in the yeast secretory pathway, Schekman and co-workers have identified groups of synthetic lethal sec (secretory) genes (Kaiser and Schekman, 1990; Novick et al., 1980). Utilization of the powerful yeast genetics system, in combination with in vitro reconstitution of protein traffic, led to the identification of at least 15 genes whose products are involved in trafficking between the ER and the Golgi apparatus (Schekman, 1992). The product of one of the genes, sec18p, seems to be required for fusion of the budded transport vesicles with the Golgi. In mutant sec18 yeast cells, glycosylated proteins destined for secretion remain EndoH (endoglycosidase H) sensitive and accumulate in transit vesicles during incubation at a restrictive temperature. Apart from identifying genes whose products are involved in vesicle formation or fusion, this approach also led to the demonstration that these proteins function in complexes with other (sec)proteins and helped to define the localization of sorting and enzymatic events. Additional components that function in ER to Golgi transport in conjunction with these sec proteins—for example, two bet mutants (blocked early in transport; Newman and Ferro-Novick, 1987) and bos1(bet one suppresser; Shim et al., 1991)—have now been isolated.

Although the sec genes were originally identified in yeast, the importance of these proteins for secretory pathway function has led to such sequence conservation that mammalian homologues for several have been identified. For example NSF (N-ethylmaleimide-sensitive factor) is the mammalian homologue of sec18p (Wilson et al., 1989), and αSNAP (soluble NSF attachment protein) has been identified as the sec17p counterpart (Clary et al., 1990). These components of the vesicular transport machinery are used extensively to manipulate traffic between the ER and Golgi (see below).

The consensus is that non-clathrin-coated vesicles (Fine, 1989) bud from the TE of the ER and are transported to a downstream acceptor compartment to which fusion takes place. If one regards, for matter of comparison, the SRP-ribosome-nascent chain complexes as the ER's equivalent of "upstream vesicles," then they are clearly distinct form the downstream non-clathrin-coated vesicles.

3.4.2. Regulation and Dynamics

Cell-free and permeabilized cell transport systems allowed a further dissection of the biochemical and molecular basis of individual steps in vesicular traffic, thereby gaining information concerning the regulation of vesicular transport from the ER. Transport in these systems requires cytosol and an ATP energy source (Baker et al., 1988). ER to Golgi transport is also dependent upon physiological concentrations of Ca^{2+} (Beckers and Balch, 1989). These systems have also been used to investigate the role of GTP binding proteins in the regulation of transport (see below).

Indeed, the ER has been proven to be highly dynamic. For example, it can expand greatly upon an enlarged synthesis of the luminal enzyme 3-hydroxy-3-methylglutaryl-coenzyme A reductase (Bergmann and Fusco, 1990) or upon increased load of transport-incompetent protein (J. Tooze et al., 1989). Also, after exposure to drugs, such as phenobarbitol, the amount of smooth ER rapidly and reversibly increases (Jones and Fawcett, 1966). Furthermore, in response to the drug nocodazole, the ER membranes collapse, which suggests that dynamics of the ER are dependent upon intact microtubules (Terasaki, 1990).

3.4.3. Bulk Flow and Retention Signals

The consensus hypothesis of bulk flow requires that resident proteins possess a retention signal. The best-studied example of such a retention signal is that of the carboxy-terminal–KDEL signal for soluble resident ER proteins. These proteins include BiP and a whole family of glucose-regulated proteins, PDI, and a series of esterases (Pelham, 1989a; Pelham, 1990; Pelham, 1991). Many of these proteins are involved in the catalysis of protein folding and are present in the ER lumen at high concentrations. Protein disulfide isomerase, for example, is present in the ER at concentrations of nearly 1 mM (Freedman and Hillson, 1980; Freedman, 1984; Weibel et al., 1969). The current retention model for soluble resident ER proteins envisages a recycling mechanism (reviewed in Pelham, 1989a) and will be discussed below.

Although a membrane protein has been identified in yeast that seems to be sorted by the HDEL (histidyl-aspartyl-glutamyl-leucine) retention system (Sweet and Pelham, 1992) this does not seem to be the common mechanism for ER membrane protein retention. For many membrane proteins, residence appears to result from their tendency to form large aggregates (e.g., in the case of components of the translocation machinery) and/or by interactions between their membrane spanning/ectoplasmic domains and other, as yet unidentified, cellular structures (see below). Several transmembrane ER proteins contain a retention motif in their cytoplasmically exposed tails, consisting of two lysines positioned three and four or five residues from the C-terminus (Jackson et al., 1990). This motif was first recognized in the adenoviral transmembrane E3/19K glycoprotein (Nilsson et al., 1989), a protein that binds after synthesis immediately to human major histocompatibility complex class I antigens, thereby retaining them in the ER compartment. The retention does not seem to be based on simple charge interactions, since arginines or histidines cannot replace lysines. It is not yet clear how deeply proteins carrying this x-K-(x)-K-x-x signal can penetrate into the secretory pathway. The report that p53, a marker of intermediate compartment, contains this signal (Hauri and Schweizer, 1992) opens the possibility that these proteins are also retained by recycling.

Although retention signals have recently been identified in transmembrane domains, the exact signal determinants are less clear. Within the alpha chain of the multisubunit T-cell antigen receptor (TCR) complex, a transmembrane sequence containing two basic amino acid residues has been shown to act as a determinant for retention and rapid degradation in the ER. Furthermore, it was demonstrated that single basic or acidic amino acid residues can cause this targetting when placed in a context-dependent manner within the transmembrane domain of the Tac antigen, a membrane protein normally destined for the cell surface (Bonifacino et al., 1991). The phenotypic changes induced by potentially charged transmembrane residues occur without apparent alterations of the global folding or transmembrane topology of the mutant proteins. These observations suggest the ability to induce protein-protein interaction by placing charged pairs within transmembrane domains, thereby masking specific retention and degradation signals once oligomerization occurred (Cosson et al., 1991). Cloning and expression of two types of cDNAs encoding the H2 subunit of the human asialoglycoprotein receptor, differing only by the presence of a 15-bp (base pair) mini-exon encoding five amino acids (Glu-Gly-His-Arg-Gly) immediately exoplasmic to the single membrane-spanning segment, indicated the operation of a similar retention and degradation mechanism (Lederkremer and Lodish, 1991).

3.4.4. Unique Environment and Unique Functions

The ER is a highly dynamic structure possessing specific retention mechanisms to enable the preservation of its resident components. These unique components create a unique environment in the ER lumen that is intermediate between the cytosol and the extracellular space. The pH is probably close to or somewhat above neutral, and the Ca^{2+} concentration is in the millimolar range (Suzuki et al., 1991). The redox conditions in the lumen are more oxidizing than those in the cytosol (Hwang et al., 1992). This is particularly important for disulfide bond formation and therefore for the folding conditions in the ER. The unique environment provides the means for specific ER-related functions.

    3.4.4a. Protein Folding in the ER. As described before, the components of the targetting and translocation machinery make the ER a site of protein biogenesis. The folding of proteins translocated into the ER lumen usually begins on the nascent chain (Bergman and Kuehl, 1979). The number of proteins that have been characterized in detail, with respect to the *in vivo* process of protein folding within the ER, is small, but some interesting features of the process have emerged. The majority of studies concerning the role of protein conformation and ER exit rate are being performed using the viral proteins influenza virus hemagglutinin (HA) and VSV glycoprotein (VSV-G) as markers. Both HA and VSV-G are well-characterized type I transmembrane glycoproteins. Upon translocation into the ER the ectodomains become *N*-linked glycosylated. Both proteins form homotrimers, which are transported via the Golgi complex to the plasma membrane (for reviews see Burgess and Kelly, 1987; Simons and Fuller, 1985). In addition, they are stabilized by intrachain disulfide bonds.

    Using very short radioactive pulses, Braakman and colleagues (Braakman et al., 1991) analyzed the folding of the hemagglutinin precursor, HA0, in tissue culture cells. The folding rate was shown to be independent of expression level, but highly dependent on cell type and expression system, indicating that the folding kinetics were also determined by the cellular environment (Braakman et al., 1991). The efficiency of folding and subsequent trimerization was not dependent on the rate of translation, nor on temperature between 37°C and 15°C. Trimerization itself was accelerated at higher expression levels and seems to be necessary for transport to the Golgi. However, evidence from various mutant VSV-G proteins suggests that trimerization alone is not sufficient for transport (Doms et al., 1988).

    The folding of wildtype HA0 was elegantly analyzed in more detail in a further set of studies. Reduction of HA0 monomers, by addition of the reducing agent dithiothreitol to the medium of living cells infected with influenza virus, suggests that the locally folded structures of the HA0 molecule are stabilized by disulfide bonds (Braakman et al., 1992a). Proper folding of HA requires correct disulfide bond formation and glycosylation, and is energy dependent (Braakman et al., 1992b). Experiments manipulating the cellular ATP level revealed that metabolic energy was required to secure the correct folding of newly synthesized HA0, to rescue misfolded HA0 molecules from disulfide cross-linked aggregates, and to maintain already folded monomeric HA0 in a folded state. Interestingly, the rescue of a temperature-sensitive folding mutant of VSV-G, tsO45, from the aggregate after switch to the permissive temperature also depends on the presence of ATP (Doms *et al.*, 1987).

Although it is indisputable that the primary sequence determines the ultimate three-dimensional structure of a protein, it is clear that the folding process in the ER is facilitated by a set of resident proteins, the *chaperones*. Chaperones are proteins that enhance the folding or assembly efficiency but do not cause covalent modifications or become part of the final folded protein structure. The chaperones present in the ER lumen include BiP, GRP94, and various forms of *cis-trans* proline isomerase (Freedman, 1987). Association with the chaperones helps prevent aggregation of folding intermediates. In addition, association with chaperones might also contribute to the retention of improperly folded proteins (Pelham, 1989a).

3.4.4b. Glycosylation. Many proteins become glycosylated during their passage through the secretory pathway. Initial glycosylation steps start in the ER where necessary components are available. Here a lipid-linked oligosaccharide is synthesized and transferred en bloc to a nascent polypeptide chain (Kornfeld and Kornfeld, 1985). Subsequent trimming reactions of the oligosaccharide occur and may continue by further trimming reactions followed by addition of peripheral sugars in the Golgi apparatus (Roth, 1987). Various experimental observations suggest that glycosylation contributes to the folding efficiency of glycoproteins (Gallagher *et al.*, 1988; Hearing *et al.*, 1989), probably by stabilizing or shielding particular segments on the protein surface and/or by increasing the solubility properties of the folding intermediates (Marquardt and Helenius, 1992).

3.4.4c. Quality Control. Several thousand different newly synthesized proteins enter the lumen of the ER of a mammalian cell. Out of the pool of newly synthesized proteins, all functionally competent proteins have to find their appropriate localization, whereas misfolded or unassembled polypeptides need to be eliminated to prevent possible toxic effects. In the work on viral proteins described above, the unfolded or misfolded viral glycoproteins were not able to trimerize, nor to be transported to the *medial*-Golgi. This is consistent with a whole set of studies utilizing mutants of HA or VSV-G with alterations in the cytoplasmic tail (Doyle *et al.*, 1985; Gething *et al.*, 1986), in the transmembrane domain (Doyle *et al.*, 1986), or in the luminal portion (Doms *et al.*, 1988). In all cases, mutant proteins that fail to be transported to the Golgi apparatus remain in a partially unfolded state and/or are not efficiently assembled into native, trimeric structures. The misfolded proteins often form aggregates that generally contain aberrant disulfide cross-links and BiP as noncovalently associated component (Hurtley *et al.*, 1989; Marquardt and Helenius, 1992). The aggregation process itself probably reflects the poor solubility of incompletely folded polypeptide chains. A role for free thiol groups in preventing the unhindered transport of proteins through the secretory pathway has also been proposed (Alberini *et al.*, 1990). Together, these observations suggest mechanisms for quality control; however, much more needs to be learned for a complete understanding.

3.4.4d. Degradation. Because the folding and maturation processes in the ER are not completely efficient there should be a degradation mechanism that allows for the elimination of misfolded or unassembled proteins. In a variety of studies a pre-Golgi proteolytic pathway has been characterized for rapid degradation of newly synthesized T-cell receptor (TCR) subunits, which is insensitive to drugs that block lysosomal proteolysis or ER to Golgi transport (Klausner *et al.*, 1990; Lippincott-Schwartz *et al.*, 1988). The site of degradation in this pathway is either part of, or closely related to, the ER. A similar pathway has been identified for the degradation of unassembled asialoglycoprotein receptor subunits (Amara *et al.*, 1989), of transport-impaired PiZ α_1-antitrypsin variants (Le *et al.*, 1990), and of carboxy terminally truncated forms of ribophorin I (Tsao *et al.*, 1992). This "ER" degradative pathway probably plays an important role in many cells in the removal of unassembled or incompletely assembled membrane protein complexes from the secretory pathway.

3.4.5. Virus Budding and Protein Localization

Although many of the best-characterized enveloped viruses bud through the plasma membrane, some viral families bud into intracellular compartments. The issue of the precise site of budding is complicated by the cytopathic effects of virus infection as by the fact that individual virus proteins may change their localization when expressed separately.

3.4.5a. Rotavirus. The Rotaviruses are a family of nonenveloped, double-stranded RNA viruses. Their site of budding is the ER. This budding can occur through ribosome-free regions of the ER, as well as into the nuclear envelope (Estes, 1990). The crucial event for this process seems to be an interaction between the structural viral glycoprotein (VP6) and the cytoplasmic domain of the nonstructural protein NS28 (Meyer *et al.*, 1989). After budding into the ER, the immature virus is enclosed by a lipid bilayer in which the glycoproteins NS28 and VP7 are embedded. Both these proteins behave as typical ER resident membrane glycoproteins, something that is reflected in the nature of their *N*-linked oligosaccharides (Kabcenell and Atkinson, 1985; Kabcenell *et al.*, 1988). Thus, NS28 and VP7 must contain targetting information that retains them after synthesis in the ER, allowing accumulation of these proteins above a threshold level for budding. Indeed, for NS28 it was shown that the second of the three hydrophobic domains is responsible for its retention in the ER (Bergmann *et al.*, 1989). In the case of VP7, the available evidence suggests that the signal sequence and presumably the extreme N-terminus of the mature protein (Griffiths and Rottier, 1992; Stirzaker and Both, 1989) are essential for its retention in the

ER membrane. Although rotavirus particles bud into the ER, they do not enter the secretory pathway and are only released following cell lysis. After budding into the ER the lipid bilayer as well as the NS28 is lost from the virus surface, leading to mature, infectious virus particles (for reviews see Estes, 1990; Estes and Cohen, 1989).

3.4.5b. Rubella Virus.   Rubella virus (RV) is an enveloped, positive-stranded RNA virus. The RV virions contain two membrane envelope glycoproteins, E1 and E2, which form a heterodimeric complex. RV budding has been reported to occur preferentially from internal membranes at the Golgi region in baby hamster kidney (BHK) cells and at the plasma membrane in an African green monkey kidney (Vero) cell line (Bardeletti *et al.*, 1979; von Bonsdorff and Vaheri, 1969). Studies utilizing Chinese hamster ovary (CHO) cells expressing RV glycoproteins reveal that transport of E1 to the Golgi complex and plasma membrane is dependent upon interaction with the E2 glycoprotein (Hobman *et al.*, 1990; Hobman *et al.*, 1992). Unassembled subunits of RV E1 accumulate in a tubular network of smooth membranes that are in continuity with the rER. Furthermore, although this structure was shown to possess distinct properties from either the rER and the Golgi, luminal ER proteins bearing the KDEL signal and the transmembrane tsO45 VSV-G mutant accumulated at the nonpermissive temperature have access to it (Hobman *et al.*, 1992), indicating that it is a subcompartment of the ER rather than a separate entity. The site of E1 arrest appeared distal to or at the site where palmitoylation occurs and proximal to the low temperature 15°C block. These findings suggest that the site of E1 arrest corresponds to, or is located close to, the exit site from the ER, although the site of rubella budding is later in the secretory pathway.

3.4.5c. Coronavirus.   Coronaviruses are enveloped animal viruses with a single positive-stranded RNA genome, possessing two viral membrane glycoproteins, E1 and E2. Several features of coronavirus make the identification of the budding site equivocal, although some early work ascribes an ER localization for its maturation. The budding of coronavirus at early times in infection has been described to occur in a smooth-surfaced, tubulo-vesicular structure, whose membranes were associated with rough ER cisternae, including apparent transitional elements, and were also present close to the *cis* face of the Golgi stack (Tooze *et al.*, 1984; Tooze *et al.*, 1988). At later stages of infection, coronavirus buds into the rough ER and nuclear envelope. Although this process clearly can occur in the ER, the exact budding site is controversial. In part that results from the lack of available intermediate compartment markers in the early studies, a lack of consensus between results using different coronavirus strains in studies expressing individual (mutated) viral proteins, and the possible cytopathic effect of the infection on the secretory pathway. The budding process involves an interaction between the nucleocapsid and the viral glycoprotein E1 (Spaan *et al.*, 1988). This protein contains three membrane-spanning regions, each of which can individually insert and anchor the polypeptide in the membrane (Krijnse-Locker *et al.*, 1992). Most likely E1 forms a complex with a second membrane glycoprotein, the E2 protein. The restricted intracellular localization of the E1 protein is believed to play a major role of directing virus assembly at intracellular membranes (Tooze *et al.*, 1984). In the case of avian infectious bronchitis virus (IBV), this protein, expressed by itself, localizes to the tubulo-vesicular/*cis*-Golgi region (Machamer *et al.*, 1990). However, in the case of the mouse hepatitis virus (MHV) the E1 protein accumulates in the Golgi complex (Rottier and Rose, 1987). Possibly the E1 and E2 MHV proteins form a complex that is retained at the site of budding. For IBV, the domain in the E1 protein responsible for its intracellular retention was clearly shown to be contained in the first transmembrane spanning segment (Swift and Machamer, 1991). However, MHV E1 mutants that possess only the first transmembrane domain are retained in the ER (Armstrong *et al.*, 1990). Once budded into the secretory pathway, coronavirus particles are transported through the Golgi complex to the plasma membrane and secreted into the medium.

3.4.5d. Hepatitis B Virus. Hepatitis B is an enveloped, partially double-stranded DNA virus that is usually described as budding in the endoplasmic reticulum (Ganem, 1991; Ganem and Varmus, 1987). The viral surface antigen (HBsAg) is a multiple-spanning protein capable of assembling to form lipoprotein particles when expressed separately. This antigen assembly and budding process was also assigned an ER localization; however, this was based on early studies in which the existence of compartments between the ER and the Golgi was not considered. More recent work (Huovila *et al.*, 1992; Huovila and Fuller, 1993) shows that the HBsAg assembly process occurs in the intermediate compartment, as will be discussed below. It is possible that hepatitis virion formation also occurs in the intermediate compartment, but this issue needs to be re-examined.

3.4.5e. Vaccinia Virus. Vaccinia is a large DNA virus with several intracellular forms. The budding of this virus was originally believed to occur via a *de novo* membrane synthesis pathway. However, more careful recent work (Sodeik *et al.*, 1993) shows that it, like hepatitis B, utilizes an intermediate compartment (discussed below).

4. INTERMEDIATE COMPARTMENTS

The presence of compartments intermediate between the ER and the Golgi is not a novel idea. The past few years have brought a set of new tools and new markers that have begun to allow the characterization of this region of the secretory pathway. Markers such as p58 and p53 and functions such as the recapture of escaped ER proteins have been attributed to the intermediate compartments. Further, several treatments have been developed that appear capable of affecting membrane traffic through this region.

The nature of the compartments between the ER and the Golgi and even their existence as independent compartments remain extremely controversial. In light of this controversy about the existence and nature of intermediate compartments, it is worthwhile examining the importance of the following question: Does the introduction of further compartments between the ER and the Golgi simply obscure the underlying simplicity of the secretory pathway or is it necessary to understand the biology of the system? We will attempt to address this question in a functional way. There are important biological functions such as the assembly of disulfide-linked complexes that are inconstant with the environment of the ER and yet appear to occur at a pre-Golgi stage. These processes can be understood in light of an intermediate compartment, and hence the added complexity is a reasonable price to pay for an expanded understanding of the biology.

One advantage of our definition is that it is recursive; a collection of (morphologically indistinguishable) compartments is a compartment. We will consider the evidence and functions ascribed to the intermediate compartment between the ER and the Golgi first, without trying to decide whether they all occur in the same structures. This is analogous to referring to the plasma membrane of an epithelial cell as a compartment for purposes of describing transport from the Golgi, although it clearly contains two compartments.

Several reviews have been published recently that summarize work in this field (Hauri and Schweizer, 1992; Klausner *et al.*, 1992; Mellman and Simons, 1992; Saraste and Kuismanen, 1992). Each emphasizes particular characteristics of this region and advocates a simplifying view of the components and their distribution. Unfortunately, the dynamic nature of the distribution of markers that have been used make the question of localization a complex one. Furthermore, relatively few of the markers have been used for co-localization studies, so that one is left with many open questions and the possibility that different authors are describing different compartments. Our approach will be first to describe the tools and markers used to study this compartment and then to address each of the proposed descriptions of intermediate compartments separately in terms of our definition of a compartment. This will allow us to highlight some of the open questions remaining in the field and allow the reader a better perception of the limits of our knowledge in this area. A summary of the responses of the compartments of the early secretory pathway to drug and other treatments is given in Table I and Figure 2.

4.1. Recycling and Traffic between the ER and the Golgi

Evidence indicates that recycling of proteins occurs between the Golgi apparatus and the ER. The best evidence is available for the soluble proteins of

Table I
Summary of the Responses of the Compartments
of the Early Secretory Pathway to Drug Treatments[a]

	ER	IC	CGN
GTPγS	x	?	x
AIF$_{(3-5)}$/$\beta\gamma$	x	(x)	?
Mastoparan	x	?	?
α-rab1	x	?	x
α-NSF	-	?	x
CA^{2+}	-	?	x
Caffeine/20°C	-	?	x

[a]The positions of the blocks indicated in the table refer to anterograde transport. In most cases a block at an early stage obscures the effects of the same agent at a later stage. the effects of GTPγS, AIF$_{(3-5)}$, $\beta\gamma$ subunits of a trimeric G protein, mastoporan, anti-rab1, anti-NSF, and Ca^{2+} deprivation are all assayed in permeabilized cell systems. Caffeine treatments were performed in intact cells, IC, intermediate compartment; CGN, *cis*-Golgi network.

the ER that bear the KDEL retention signal (Lewis and Pelham, 1992; Pelham, 1988; Pelham, 1989b; Pelham, 1991). The lysosomal protease, cathepsin D, is modified by a series of Golgi enzymes so that it bears the mannose-6-phosphate signal required for its post-Golgi routing to the lysosomes (Kornfeld and Kornfeld, 1985). By attaching a KDEL signal to cathepsin D, Pelham and co-workers generated an elegant probe for the exposure of an ER resident protein to the enzymes of the Golgi (Dean and Pelham, 1990; Pelham, 1988). During a prolonged time course the glycans of cathepsin D–KDEL hybrid were modified by addition of GlcNAc-phosphate, although the bulk of the protein was always found in the ER. This modification indicates that KDEL-terminated proteins encounter the early Golgi, although the bulk of the protein is localized to the ER. The question of whether the modification could be accomplished by Golgi enzymes present in the ER was addressed by showing that modification did not occur in yeast that had a mutation in *sec*18, resulting in a transport block to the Golgi. Similar modifications have been demonstrated on a number of other proteins localized to the ER, indicating that this recycling is a general feature of ER proteins (Mazzarella and Green, 1987)

4.1.1. 15°C Treatment

One of the most striking and useful treatments affecting ER to Golgi transport is that of low temperature. Saraste and co-workers used immunocytochemistry to show that incubation of BHK cells at 15°C blocked the transport of viral glycoproteins to the Golgi (Saraste and Kuismanen, 1984). By taking advantage

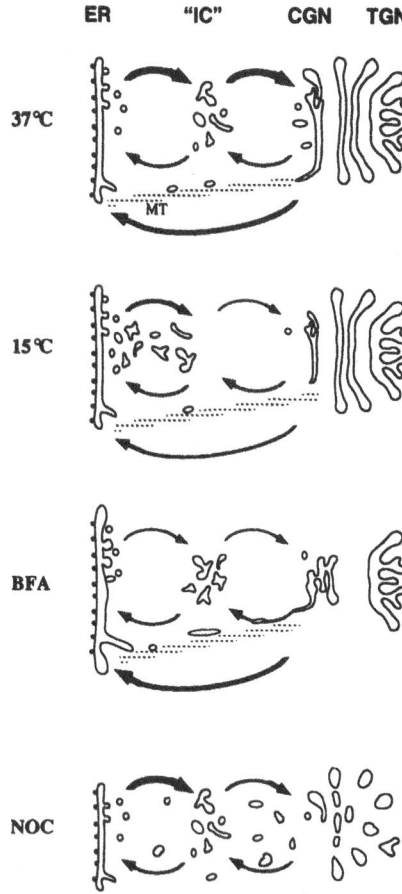

FIGURE 2. Schematic view of the responses of the compartments of the early secretory pathway. Major changes in the cellular localization and compartment size, introduced by a low temperature block and drug treatments, are depicted. IC, intermediate compartment; CGN, *cis*-Golgi network; TGN, *trans*-Golgi network; MT, microtubule.

of a temperature-sensitive mutation in the Semliki Forest virus (SFV) glycoprotein to accumulate a pulse of protein in the ER before inhibiting transport at 15°C, these researchers could characterize the morphology of the structures in which the blocked protein accumulated (Kuismanen and Saraste, 1989). They observed a striking assembly of pleimorphic vacuolar and tubulo-vesicular structures near the Golgi region as well as in the peripheral regions of the cell. The blocked proteins were present throughout both types of structures. Return of the cells to 31°C (the permissive temperature for transport of the mutant glycoprotein) reversed the block causing accumulation of protein in the tubular regions of the structures and allowed its transport to the Golgi.

The use of 15°C treatment has become quite standard for the characterization of the transport between ER and Golgi. The state of the blocked proteins can be characterized by pulse-chase analysis using labeled proteins for biochemistry or by taking advantage of the use of mutant viral glycoproteins such as the ts1 mutation of SFV or the tsO45 mutant of the G protein of VSV for immunocytochemistry (Kuismanen and Saraste, 1989). These proteins are blocked in the ER because of a folding defect at high tempreatures (39.5°C), and they exit the ER at lower temperature. Incubation at the restrictive temperature allows synthesis of protein to continue while its transport is blocked in the ER and the later stages of the secretory pathway are cleared. Switching to the permissive temperature allows the protein to fold, to exit the ER, and to define the position of the block by immunocytochemistry.

A detailed characterization of the morphology of the structures containing the VSV-G protein at 15°C allowed Bonnatti and co-workers (Lotti *et al.*, 1992) to address the question of whether this treatment generates a novel compartment or rather blocks transport in a pre-existing one. Comparison of electron microscopic immunocytochemistry and immunofluorescence microscopy allowed these authors to identify all of the structures containing G protein at 15°C. The labeled structures were localized in circular areas characterized by numerous small vesicles with an average diameter of 80 nm, which were near short tubules. The circular appearance of the areas and the close clustering of the vesicles and tubules allowed them to be distinguished from the ER and Golgi stacks. After the 1-hour incubation at 15°C, weak labeling of ER and negligible labeling of the Golgi cisternae showed that transport of VSV-G protein was slowed between these two compartments. These structures were also labeled with rab2 (Lotti *et al.*, 1992). Furthermore, the authors examined noninfected cells and cells incubated at 31°C and 37°C. Comparison with the 15°C incubated cells revealed that the same circular areas could be found in all three conditions, indicating that they were not induced either by temperature block or viral infection. The G protein was found in structures that were both peripheral and perinuclear, but it rapidly localized to the Golgi complex after release of the block by very short (<10 min) incubations of cells at 31°C.

Several important caveats should be stated when trying to interpret the results of 15°C treatment. The most important is that the block is not absolute. Hence, protein blocked for 3 hr may have advanced further along the pathway than protein blocked for only an hour. Moreover, the precise temperature of the block is quite critical, and there is no guarantee that all mammalian cell lines will behave identically with respect to 15°C incubation. These three points explain the inconsistency of the results described above with the finding (Schweizer *et al.*, 1990) that VSV-G protein could reach the *cis*-Golgi during 3-hr incubation at 15°C. Finally, one needs to remember that because some components are continually recycling, the co-localization of a component with a newly synthesized one that has blocked at 15°C may simply represent the fact that a recycling component was blocked there.

4.1.2. Brefeldin A

Brefeldin A (BFA) is a macrocyclic fungal antibiotic that blocks the transport of proteins through the secretory pathway. Takatsuki and co-workers introduced BFA as a tool when they reported (Takusuki and Tamura, 1985) that BFA treatment blocked the transport of newly synthesized VSV-G protein to the cell surface and caused it to accumulate in an endoglycosidase H-sensitive form, indicating that the glycoprotein had not encountered the *medial*-Golgi. Morphological work showed that BFA treatment caused the Golgi complex to disappear (Fujiwara *et al.*, 1988). Later work by several groups showed that established markers of the *cis*- and *medial*-Golgi return to the ER upon disassembly of the Golgi during BFA treatment (Doms *et al.*, 1989; Lippincott-Schwartz *et al.*, 1989). An exception is that of sialyl transferase, a marker of the *trans*-Golgi network (TGN). Although one group (Ivessa *et al.*, 1992) has reported that ribophorins I and II are sialylated during BFA treatment, immunocytochemistry indicates that the bulk of sialyltransferase does not return to the ER (Chege and Pfeffer, 1990). This confirms that the TGN is a functionally distinct structure from the Golgi and defines the limits of BFA-induced recycling to the ER to the early secretory pathway. The action of BFA appears to be reversible; washing out the drug causes the rapid return of the Golgi markers to their original location (Doms *et al.*, 1989; Hsu *et al.*, 1991; Klausner *et al.*, 1992; Lippincott-Schwartz *et al.*, 1990; Lippincott-Schwartz *et al.*, 1989).

Although the mechanism of action of BFA is not yet understood, the first clue has come from the identification of a peripheral Golgi protein whose redistribution to the cytoplasm is a very rapid consequence of BFA treatment. The β-coatamer protein (β-COP) has partial homology to the clathrin-coated, vesicle-associated protein β-adaptin (Duden *et al.*, 1991a; Duden *et al.*, 1991b). The β-COP has been shown to be a component of the non-clathrin-coated vesicles, which are believed to mediate transport between the successive cisternae of the Golgi. A model has been suggested in which BFA treatment prevents the formation of the nonclathrin coats of the forward (anterograde) transport vesicles while leaving unaffected the recycling (retrograde) transport that appears to be mediated by noncoated tubules along microtubules (Klausner *et al.*, 1992; Lippincott-Schwartz *et al.*, 1990). By blocking anterograde transport, BFA would allow the effect of the retrograde pathway to be manifested in the striking return of Golgi components to the ER. The effect of BFA on the endosomal system and the TGN suggests that its mechanism of action could be a quite general one at the level of inhibition of the formation of transport vesicles for several steps in intracellular traffic (Klausner *et al.*, 1992).

Two recent papers present a possible explanation for the actual mechanism of BFA action on the assembly of nonclathrin coats (Donaldson *et al.*, 1992; Helms and Rothman, 1992). The low molecular weight GTP binding protein

ADP ribosylation factor (ARF) is required for coatamer binding to membranes and probably mediates the control of guanine nucleotides of the coatomer membrane interaction (Serafini *et al.*, 1991). The new results indicate that BFA inhibits the GTP dependent interaction of ARF with the Golgi membrane by inhibiting the exchange of GTP onto ARF, which is normally catalyzed by Golgi membranes (Donaldson *et al.*, 1992; Helms and Rothman, 1992). A simple inference from these results is that BFA accomplishes its effect on the Golgi by blocking GTP-GDP exchange in ARF. This blocks the activation of ARF for membrane binding and coat assembly and hence prevents the budding of transport vesicles. Because only the GTP form of myristoylated ARF but not the GDP form can insert into membranes (Serafini *et al.*, 1991), the location of the exchange enzyme may control the localization of the ARF proteins. It is unclear whether BFA acts directly or indirectly on the exchange protein, for the protein has not yet been purified but only localized to Golgi fractions.

Whatever its mode of action, BFA has become a widely applied reagent for the dissection of the secretory pathway (Klausner *et al.*, 1992). During the redistribution of Golgi markers, tubular elements can be seen extending toward the ER, apparently showing the organization of transport along microtubules. The redistribution seen for Golgi markers does not extend to the entire early secretory pathway; several markers of intermediate compartment (see below) are unaffected by BFA treatment. Interestingly, these markers are transiently returned to the ER after BFA has been removed (Chavrier *et al.*, 1990; Hauri and Schweizer, 1992; Lippincott-Schwartz *et al.*, 1990; Vaux *et al.*, 1990). This later redistribution reflects the changes in membrane traffic as the Golgi is reassembled and provides an indication that there is no direct retrograde transport from these intermediate compartments.

4.1.3. Microtubule Agents

Microtubule disrupting agents have been useful in dissecting the retrograde and anterograde pathways of transport between the ER and the Golgi. Retrograde transport of Golgi proteins to the ER can be visualized as a series of transient tubulo-vesicular projections from the Golgi membrane (Lippincott-Schwartz *et al.*, 1990). These necklace-like Golgi processes are often seen to localize along microtubules. As treatment of cells with nocodazole results in disruption of the microtubule network, it could have an effect on this microtubule-associated retrograde transport. The Golgi is dispersed into vesicles by nocodazole treatment of normal cells. Application of BFA to nocodazole-treated cells results in no further dispersion of the Golgi and no apparent return of Golgi markers to the reticular ER. Hence, microtubules appear to be involved in the retrograde direction of the ER to Golgi transport.

4.1.4. Caffeine

Kuismanen and co-workers (Kuismanen *et al.*, 1992) have introduced caffeine as a blocker of ER to Golgi transport using a similar protocol to that employed by Saraste in the first work on the 15°C block. The effect of caffeine on SFV glycoprotein transport is temperature sensitive. At physiological temperatures, the drug affects the later stages of the secretory pathway, appearing to cause accumulation of protein in the TGN as shown by the proteolytic processing of the glycoprotein precursor p62. Treatment of cells with 10 mM caffeine at 20°C appears to block the transport of the glycoprotein prior to the acquisition of EndoH sensitivity and hence before its entrance to the *medial*-Golgi. Caffeine is known to have multiple effects on cells, including the inhibition of the breakdown of cyclic AMP, the premature triggering of events in the mitotic program, and the release of calcium from intracellular membranes (Burgoyne *et al.*, 1989; Downes *et al.*, 1990; Schlegel and Pardee, 1986). Because neither stimulation of adenylate cyclase with forskolin nor treatment with the inhibitor 8-bromo-cyclic AMP mimicked the effect of caffeine, this raises the interesting possibility that either calcium fluxes or the link between intracellular transport and mitotic events may be the basis of this effect.

4.1.5. The Use of Semiintact Cells

The use of permeabilized cell systems has allowed precise manipulation of ER to Golgi transport and demonstrated the role for a number of components. A fundamental finding was that energy is required for the exit of VSV-G protein from the ER (Balch *et al.*, 1986). Both GTPγS and Ca^{2+} depletion inhibited entrance of G protein to the *cis*-Golgi (Beckers and Balch, 1989) as defined by the acquisition of EndoD sensitivity in CHO 15B cells. The Ca^{2+} effect was shown to be fully reversible and distal to the GTPγS effect.

The observation that GTPγS inhibited transport to the *cis*-Golgi is consistent with the previous demonstration that members of both the rab and ARF families of small GTP binding proteins are necessary for ER to Golgi transport in both mammalian and yeast cells (Balch, 1992). More recent work in digitonin-permeablized cells clarified the involvement of multiple GTP binding proteins in ER to Golgi transport. Use of specific antibodies and experiments with mutant proteins (Plutner *et al.*, 1991; Schwaninger *et al.*, 1992; Tilsdale *et al.*, 1992) have confirmed that rab1A, rab1B, and rab2 are all required. The involvement of a member of the ARF family was shown by inhibition with a specific N-terminal peptide (Schwaninger *et al.*, 1992). The role of heterotrimeric G proteins was shown both by the inhibition of export from the ER by treatment with mastoporan, a G-protein activator peptide, and by the addition of purified $\beta\gamma$ subunits (Schwaninger *et al.*, 1992). The authors suggest that the role of ARF is in the formation of the coat of the transport vesicle and that heterotrimeric G proteins may function by defining the site of assembly of the vesicles (Schwaninger *et al.*, 1992). Further work from the same group defined the morphology of the compartments after these blocks (Plutner *et al.*, 1992). The effect of incubation at 15°C on the transport of G protein was similar to that seen by Bonnatti and co-workers (Lotti *et al.*, 1992) in intact cells. Morphological studies revealed that removal of calcium, or addition of antibody to the transport component NSF, blocked VSV-G protein in tubulo-vesicular structures containing the intermediate compartment marker p58 (Plutner *et al.*, 1992). Although the use of a permeablized cell system is very powerful in terms of the manipulations that are possible owing to the accessibility, it is important to note that the cytoskeleton is affected by this treatment. In particular, the microtubule network is disrupted gradually, and the authors suggest that the spread in localization of components after long times of incubation could be a reflection of this effect (Plutner *et al.*, 1992).

4.2. Markers of the Intermediate Compartment

One major problem with defining the intermediate compartment is the paucity of static markers for it. The accepted markers ascribed to the intermediate compartment are described below. These markers are usually characterized in terms of the treatments already described. However, it is important to remember that their localization can be affected by these treatments.

4.2.1. p53

Hauri and co-workers have used a monoclonal antibody directed against a protein of 53 kDa as a marker of intermediate compartment (Hauri and Schweizer, 1992; Schweizer *et al.*, 1988). The monoclonal antibody labels a collection of tubulo-vesicular membranes on the *cis* side of the Golgi. The distribution of these structures is similar to that described by Lotti *et al* (1992) after release of the 15°C block in that they are both peripheral and perinuclear (Lippincott-Schwartz *et al.*, 1990). A low level of p53 labeling is also found in the *cis*-Golgi cisterna. Biochemical characterization of the antigen shows that it is a nonglycosylated, oligomeric, transmembrane protein (Hauri and Schweizer, 1992; Schweizer *et al.*, 1988). Sequence data show that this protein contains a -x-K-(x)-K-x-x retention signal at its carboxyterminus, suggesting its function in the localization of the protein (cited in Hauri and Schweizer, 1992).

The localization of the protein overlaps with that of VSV-G protein, blocked at 15°C for 3 hours, and is found concentrated close to the Golgi (Schweizer *et al.*, 1990). Rewarming the cells causes extension of p53 positive tubules and an

ER localization of p53 for up to 1 hour before its normal distribution is recovered. BFA treatment causes a coarsening of the p53-positive structures but not the return of p53 to the ER (Lippincott-Schwartz *et al.*, 1990). Upon washing out the BFA, p53 returns transiently to the ER. Together, these results suggest that p53 is not a static marker but, rather, undergoes continual recycling between the ER and the Golgi.

Although p53 is a very useful marker, the fact that its detection has been limited to the use of a monoclonal antibody, which only reacts with the antigen in primate cells, has made it difficult to generalize to other systems. The availability of the sequence should change this situation.

4.2.2. p58

Saraste and co-workers have characterized a protein of 58 kDa that shares many properties with p53 (Lahtinen *et al.*, 1992; Saraste and Kuismanen, 1984; Saraste and Svensson, 1991). Antibodies against p58 are available that react with a variety of cell types, and so its localization to tubulo-vesicular structures adjacent to the *cis*-Golgi is well established (Lahtinen *et al.*, 1992; Saraste and Kuismanen, 1984; Saraste and Svensson, 1991). In some cell types, p58 appears to be more extensively localized to the *cis*-Golgi than p53 (Lahtinen *et al.*, 1992). The protein bears immature N- and O-linked glycans, which indicate that it is never processed by the enzymes of the *medial-* or *trans*-Golgi. The behavior of p58 upon 15°C treatment and BFA treatment is similar to that of p53, suggesting that it too recycles between the ER and the Golgi. Studies of permeablized cells indicate that treatment at 15°C, with GTPγS or with anti-rab1B antibody to block transport, causes co-localization of p58 with VSV-G protein (Plutner *et al.*, 1992) although one has to consider the possibility that this is not its normal localization but one induced by the block. An obvious question is whether p53 and p58 are homologous; however, the lack of cross-reactivity of the antibodies means that this question will only be resolved when the sequence of p58 is available and that of p53 is published.

4.2.3. p63

The fact that p53 and p58 recycle between the ER and the Golgi complicates their use as intermediate compartment markers (as discussed above). For this reason the identification of another intermediate compartment marker, p63, by a panel of monoclonal antibodies raised against a fraction enriched in p53 is very important (Hauri and Schweizer, 1992). Also, p63 is a membrane protein that overlaps in distribution with p53 and with VSV-G protein at 15°C. Distribution of p63 is unchanged by treatments with BFA or low temperature. It also has been reported to be palmitoylated even when traffic between the ER and Golgi is blocked by BFA (cited in Hauri and Schweizer, 1992). Hence, p63 represents a good candidate for a marker of the stable portion of intermediate compartment.

4.2.4. The rab Proteins

The rab (rat brain) proteins have been used as markers of intermediate compartment. Recent functional data show that rab2, rab1B, and rab1A are all required for ER to Golgi transport (Plutner et al., 1991; Schwaninger et al., 1992; Tilsdale et al., 1992) and are consistent with the localization of these rab proteins, and in particular rab2 (Chavrier et al., 1990), to the early secretory pathway. The proposed function of rab proteins in transport must be considered when they are referred to as markers of a compartment (Goud, 1992). Rab proteins are believed to function in a cyclic fashion and hence be present on both the acceptor and donor compartments involved in a transport event. Hence, the observation that rab2 is found on tubulo-vesicular structures and on the cis-Golgi can also be interpreted as demonstrating the connection of these two structures by transport vesicles rather than indicating that they represent the same compartment.

4.3. Functions Ascribed to the Intermediate Compartment

The paucity of static markers for the intermediate compartment is more than compensated by the large number of functions from salvage to protein modification that have been attributed to it. These functions have been extensively reviewed elsewhere (Hauri and Schweizer, 1992; Klausner et al., 1992; Mellman and Simons, 1992; Saraste and Kuismanen, 1992). Here we will mention them only to point out the implications of having these functions localized to an intermediate compartment.

A recurring problem in discussing the intermediate compartment functions is that we know relatively little about the functions of the cis-Golgi. The two functions usually ascribed to the cis-Golgi are the phosphorylation, which generates the mannose-6-phosphate signal on newly synthesized lysosomal proteins, and the trimming of high mannose residues (Kornfeld and Kornfeld, 1985). Unfortunately, the localization data for these activities derive from a time at which the intermediate compartment was not considered.

4.3.1. KDEL-Mediated Retention

The key observation in understanding the retention of soluble ER proteins was made by Munro and Pelham (1986, 1987), who noticed upon determining the sequence of rat BiP that the carboxy-terminal four residues matched the terminal residues of the previously established sequence of PDI (protein disulfide isomerase). They went on to show that transfer of this sequence to the carboxy-terminus of lysozyme caused this normally secreted, soluble protein to be retained in the ER. This result was interpreted as evidence that the last four residues, KDEL, comprised a necessary and sufficient signal for the retention of soluble proteins in the endoplasmic reticulum (Munro and Pelham, 1987).

Although the original work showed the modular nature of the signal, by transfer of the last six residues from BiP to lysozyme, the signal was preceded by a sequence from the *myc* protein as an upstream tag for the protein. It appears that this upstream region is necessary for KDEL function in retaining secreted proteins. Work from several groups confirms some requirement for an immediate upstream region (Buonocore and Rose, 1990; Pelham, 1989b; Pelham, 1990). The sequences of a large number of soluble ER proteins are now known, and the recognized signal varies in different organisms. The HDEL is the recognized signal in the yeast *Saccharomyces cerevisea*, whereas DDEL is recognized in the yeast *Kluyveromyces lactis* and ADEL is found to be functional in the *Schizosaccharomyces pombe*, although none of these appear to function in mammals (Hardwick *et al.*, 1990; Lewis *et al.*, 1990; Pelham, 1989a; Pelham, 1990; Pidoux and Armstrong, 1992; Semenza *et al.*, 1990). In mammalian systems, KDEL appears to be able to be substituted by RDEL and KEEL and still retain function. Our attempts at modeling these sequences suggests that the upstream region may be important in modulating the tendency of the KDEL itself to assume the recognized conformation on the protein (Fuller *et al.*, 1993).

Addition of residues after the signal ($-$KDEL $\rightarrow$ $-$KDELGL) destroyed the ability of the signal to function in retaining lysozyme (Munro and Pelham, 1987). Further, although lysozyme is secreted from the ER with a half-time of roughly 10 min, BiP without its KDEL signal is only secreted slowly with a half time of 3 hr. This suggests that the body of an ER protein as well as the signal may have a function in the retention. An elegant experiment using *Xenopus* oocytes (Ceriotti and Colman, 1988) provided clear evidence of this by showing that both BiP terminated with KDEL and with the nonfunctional signal KDELGL were free to diffuse throughout the ER but did so with a diffusion rate intermediate between that of a membrane protein and a secreted protein. Their results are consistent with a model of retention in which interactions between the walls of the ER and the body of the protein lower the effective concentration of KDEL tails, which must be recognized by the specific retention system.

4.3.1a. Retention Is Environment Dependent. The specific recognition of the KDEL signal was hypothesized to be accomplished by a receptor (Warren, 1987). The high concentration of the proteins, which must be controlled by this recognition system, and the lack of any ER membrane protein present in sufficient abundance to interact with all of them led to the hypothesis that retention was accomplished by a recycling mechanism (Munro and Pelham, 1987). Pelham and co-workers (Munro and Pelham, 1987) proposed the existence of a salvage compartment in which escaped ER proteins would be recognized on the basis of their KDEL tails and recycled to the ER. For this mechanism to function, the binding of KDEL tails by the putative receptor must be subject to modulation. Binding must be tight in the post-ER salvage compartment but weak in the ER where the proteins are released. A simple way to accomplish this modulation would be to make the interaction of the KDEL tail with the receptor dependent on the environment of the compartment. This would be analogous to the recycling of ligands during endocytosis, in which the change in pH between the endosome and the extracellular environment modifies the binding of ligand to receptor.

Although we know very little about the environment of the ER and even less about the environment of the putative salvage compartment, the available evidence suggests that the pH is higher in the ER than in the TGN, while the concentration of free calcium goes up between these two points (Suzuki et al., 1991). Koch and co-workers (Booth and Koch, 1989; Koch et al., 1988) have provided further evidence for environmental effects on retention by showing that modifying the concentration of calcium along the secretory pathway led to secretion of KDEL-terminated proteins in a particular clone of 3T3 cells, although it appears that their results are difficult to repeat in other cell lines. Either pH or Ca^{2+} concentration or both could then be candidates for the environmental switch that modulates affinity to the receptor.

4.3.1b. Identification of the Receptor for ER Retention. Identification of the KDEL receptor, the protein that recognizes the signal for retention, would be an important step toward understanding the retention system.

One approach (Vaux et al., 1990) has been the generation of anti-idiotypic antibodies that should mimic the signal structure. These antibodies identified an antigen localized between the ER and the Golgi by immunocytochemistry and Western-blotted a 72-kDa protein. This protein apparently had the appropriate properties to be the receptor. Our further work, however, now indicates that the monoclonal anti-idiotypic antibodies recognize a different protein by immunofluorescence than by Western blotting. This fact invalidates the biochemical characterization of the putative receptor and, together with other evidence that the 72-kDa protein is cytoplasmic, suggests that it is not involved in KDEL retention (Vaux et al., 1992).

Pelham and co-workers have used a fundamentally different approach taking advantage of yeast genetics to identify a protein involved in the retention of HDEL-terminated proteins in the ER (Hardwick et al., 1990; Lewis et al., 1990; Semenza et al., 1990). They have identified a protein, ERD2 (for ER retention deficient, which affects the retention of HDEL-terminated proteins. The evidence that this is the functional receptor is threefold: First, mutations in ERD2 cause proteins with an HDEL retention signal to be secreted (Semenza et al., 1990). Second, overexpression of erd2 increases the capacity of yeast cells to retain HDEL-terminated proteins (Semenza et al., 1990). Finally, transfer of the K.

lactis ERD2 into *S. cerevisea* alters the ligand specificity of retention (Lewis *et al.*, 1990).

In addition, these authors have identified a mammalian homologue of ERD2 (hERD2) and have shown that overexpression of hERD2 allows retention of DDEL terminated proteins, which are normally poorly recognized in the mammalian system (Lewis and Pelham, 1990; Lewis and Pelham, 1992). Mutation in the putative binding site of the hERD2 modifies this effect, suggesting its interaction with the ligand (Lewis and Pelham, 1992). Unfortunately, antibodies are not available against the normal ERD2 protein, so that localization and expression studies have all utilized a tagged version of the protein to allow its detection. The tagged protein is found throughout the Golgi in cells that express large amounts of ERD2 as well as in cells that stably express smaller amounts of the protein (Lewis and Pelham, 1992). By monitoring the tagged protein, Lewis and Pelham showed that overexpression of KDEL- or DDEL-terminated proteins results in the redistribution of ERD2 to the ER. A second human erd2 homologue, elp1 (*erd-like* protein), has been identified (Hsu *et al.*, 1992). It has been shown that overexpression of either erd2 or elp1 causes a profound change in the organization of the Golgi similar to that seen in BFA-treated cells (Hsu *et al.*, 1992). The relationship between this effect and the effect of ERD2 on retention remains unclear. The fact that ER retention is not essential for yeast although the ERD2 gene is essential (Semenza *et al.*, 1990) suggests that ERD2 must also have a function other than ER retention.

The localization of the ERD2 protein to the Golgi argues against an intermediate compartment function. As only the tagged protein has been localized, it is possible that the Golgi localization results from the introduction of the tag and/or reflects expression level rather than function. Until this is resolved, the connection between ER retention and the intermediate compartment will remain unclear.

4.3.2. Degradation

There is good evidence for at least two mechanisms of protein degradation in the early secretory pathway. For some proteins, such as the T-cell receptor subunits, degradation appears to occur in the ER (Lippincott-Schwartz *et al.*, 1988). Indeed, these subunits can be degraded in permeablized cells that have been deprived of the components necessary for exit from the ER (Wikstroem and Lodish, 1991). For others, such as an unassembled subunit of the asialoglycoprotein receptor (Amara *et al.*, 1989) and the truncated ribophorin I (Tsao *et al.*, 1992), degradation appears to require transport. Intermediates for degradation can be identified in the ER; however, degradation of these intermediates can be blocked with energy poisons. In the case of a truncated form of ribophorin I, the degradation can be seen to occur with biphasic kinetics. If it can be shown that this second phase occurs prior to the Golgi, this function would be associated with the intermediate compartment.

4.3.3. Complex S–S Bond Formation

Study of the assembly of hepatitis B surface antigen (HBsAg) reveals an unusual activity of the early secretory pathway. Surface antigen is a trans-membrane glycoprotein that assembles into large oligomeric lipoprotein complexes which are extensively linked by disulfide bonds. Examination of the kinetics of oligosaccharide modification indicates that the protein is blocked at a pre-Golgi stage for much of its time in the cell and then is rapidly secreted (Ganem, 1991; Ganem and Varmus, 1987). For this reason, HBsAg has been believed to be assembled in the ER. The presence of a large number of disulfide bonds in the assembled particle argues against this site of assembly, since PDI, which is present in the ER in large amounts, should resolve these cross-links. Indeed, mature HBsAg particles are not stable in the presence of PDI but are broken down to dimers (Huovila *et al.*, 1992).

Huovila *et al* (1992) showed that the formation of the disulfide links in HBsAg occurs in two stages. The first, the formation of disulfide-linked dimers, occurs rapidly and appears catalyzed by PDI. The second occurs more slowly and appears to match the rate-limiting step in assembly. This second step occurs in a compartment that excludes PDI and is insensitive to BFA treatment. The treatment with BFA also appeared to separate the two stages of disulfide bond formation since newly synthesized HBsAg only progressed as far as dimers in the presence of BFA (Huovila *et al.*, 1992). Because this function is pre-Golgi and relies on a non-ER environment, it is a strong candidate for localization to the intermediate compartment.

4.3.4. Antigen Attachment to Class I

Most antigens must be processed intracellularly before they can be presented at the cell surface in association with major histocompatibility complex molecules (MHC). It is this complex that is recognized by the antigen-specific receptor of T cells and can lead to specific killing of the presenting cell (Klausner *et al.*, 1990). Processing of antigen involves cleavage of protein antigens to smaller peptides, which then interact with MHC molecules. Class I restricted antigens include viral antigens and other cytoplasmic proteins synthesized within the presenting cell. The mechanism by which processed fragments of these antigens enter the secretory pathway and bind to the luminal domain of class I MHC molecules remains one of the important questions in understanding antigen presentation.

Two groups have shown that application of BFA to cells abolishes their ability to present endogenously synthesized antigens to class I restricted cytotoxic T cells (Nuchtern et al., 1989; Yewdell and Bennink, 1989). Both groups showed that this effect was specific for presentation of endogenous antigens by demonstrating that BFA had no effect on the presentation of exogenously added peptides at the cell surface in fixed cells. The inhibition indicates that the association of antigen with class I MHC requires transport of the molecule from the ER, and because of the previous work on the action of BFA, suggests that this interaction cannot occur later than trans-Golgi. The effect of the E19 protein on the presentation of antigens in adenovirus-infected cells has the same basis since E19 forms a complex with MHC molecules restricting them to the early portion of the secretory pathway (Nilsson et al., 1989; Paabo et al., 1987). Recent evidence that the site of interaction of processed antigen with MHC class I molecules occurs early in the secretory pathway comes from the localization of TAP1, a peptide transporter encoded in the class II region of the MHC. Specific antisera showed that the transporter was found only in the ER region and the cis-Golgi (Kleijmeer et al., 1992). Together with the evidence that MHC class I molecules circulate between the ER and the Golgi (Hsu et al., 1991), this implicates either the early Golgi or an intermediate compartment in the association of these molecules with antigen.

4.3.5. Palmitoylation

A number of viral and cell surface proteins are palmitoylated in a post-translational event, occurring before glycoproteins acquire complex-type carbohydrates (Sefton and Buss, 1987). The exact intracellular localization of palmitoylation, however, is not well defined. Combining a biochemical and kinetic study with the utilization of the 15°C block, Bonatti and co-workers (Bonatti et al., 1989) showed that quantitative palmitoylation of VSV-G and SFV E1 membrane glycoproteins seems to take place between the sites defined by low temperature and before exposure to the cis-Golgi enzyme 1,2-mannosidase I (Balch et al., 1986). Similar results were obtained previously working with VSV-infected CHO 15B cells (Dunphy et al., 1981). These results are consistent with a post-ER localization. In contrast, some workers suggest that the fatty acetyltransferase is located in the ER (Berger and Schmidt, 1985; Hobman et al., 1992); however, the extent of palmitoylation of the followed marker proteins in these cases is only low or not clear. Consistent with all these findings, work in the yeast Saccharomyces cerevisea indicates that palmitoylation can occur in a sec18-independent manner (Wen and Schlesinger, 1984) and hence prior to entrance to the Golgi. Therefore, this is a good candidate for an intermediate compartment function.

4.3.6. Virus Budding

Although virus budding may seem an unusual function for a compartment, the fact that the maturation and the assembly of viral proteins require specialized functions makes it a reasonable assay for specific compartmental functions.

Two virus-budding events have been proposed to occur in the intermediate compartment. The first is the formation of the HBsAg particle. Immunocytochemistry and morphology show that the maturation occurs in a compartment disconnected from the ER and, hence, this process has been localized to a pre-Golgi intermediate compartment (Huovila *et al.*, 1992; Huovila and Fuller, 1993). Localization of the budding of hepatitis B virus itself is not known, although evidence suggests that it occurs prior to the Golgi (Ganem, 1991; Ganem and Varmus, 1987). The second budding event now believed to be in the intermediate compartment is the budding of the intracellular infectious form of vaccinia. Originally this event was believed to occur by a process of *de novo* membrane formation; however, more recent work shows that budding occurs through the membranes of the intermediate compartment. Evidence for this localization is a combination of immunocytochemistry and morphology (Sodeik *et al.*, 1993). The fact that the initial budding event is restricted to this compartment strongly suggests that structures between the ER and the Golgi form a compartment by our restrictive definition, although continuities are seen with the ER at later times in infection.

4.4. Proposed Intermediate Compartments

In this section we will examine the evidence that various proposed intermediate compartments do in fact represent compartments in terms of intracellular transport. We will use the definitions and names provided by the original authors; thus, the names should not be considered as a description of the structure but rather as a convenient reference for it.

4.4.1. Transport Vesicles

The simplest description of the structures lying between the ER and the Golgi is that they comprise transport vesicles. This would not satisfy our definition of a compartment. Obviously, transport vesicles are physically discontinuous from either the ER or the Golgi; however, they lack unique components. This is an important functional distinction. Because the entire content of a transport vesicle is delivered to its acceptor, there can be no sorting or monitoring of the degree of completion of modification. Hence, when specific functions requiring unique components must be attributed to intermediate compartments, they cannot be accomplished in transport vesicles.

4.4.2. The CGN

There are two definitions of CGN, the *cis*-Golgi network. The more restrictive definition refers to the elaborate system of tubules that are in direct connection with the *cis* cisterna of the Golgi. This complex network is reminiscent of endosomal tubules believed to function in the separation of ligand and receptor during endocytosis (Tooze and Hollinshead, 1991). This definition of CGN is an apt description of morphology; however, it is not a compartment as it is not disconnected from the *cis*-Golgi itself. The CGN and the *cis*-Golgi are analogous to the smooth ER and rough ER.

A more liberal definition combines this elaborate network with the *cis*-Golgi itself (Mellman and Simons, 1992). Defined in this way the CGN is clearly a compartment, and it is this definition that we will use below. The various features we attribute to compartments all apply. There are clearly transport vesicles, and there is evidence that the compartment can change in size upon alteration of the amount of membrane traffic (Griffiths *et al.*, 1989). Functions that have been assigned to this compartment include all of the early functions of the Golgi such as the attachment of the lysosomal sorting signal, irreversible palmitoylation, and initial *O*-glycosylation and the trimming of high mannose residues (Hauri and Schweizer, 1992; Kornfeld and Kornfeld, 1985; Mellman and Simons, 1992; Schweizer *et al.*, 1991).

4.4.3. ERGIC

Schweitzer and Hauri (Hauri and Schweizer, 1992) have defined ERGIC (ER–Golgi intermediate compartment) by use of the marker p53. ERGIC refers to the disconnected tubulo-vesicular elements between the ER and the Golgi. Immunocytochemistry shows that the bulk of p53 is in this compartment under normal conditions. The p53 is not a unique marker, because it can also be found in the CGN; however, this can easily be explained by its recycling. A more useful marker is p63, which does not appear to recycle (cited in Hauri and Schweizer, 1992).

A cell fractionation procedure has been developed that enriches for p53 (Schweizer *et al.*, 1991). Unfortunately, this procedure is not a purification and cannot be used in many other cell types. It shows clearly that some components are not enriched with p53 and that the functions described above for the CGN are not found in ERGIC. Reversible palmitoylation appears to be localized here since both p62 (which co-localizes with p58) and p63 are palmitoylated (Hauri and Schweizer, 1992; Mundy and Warren, 1992).

ERGIC remains the best characterized of the putative intermediate compartments; however, much remains to be done. Although a large amount of ER proteins including both PDI and the membrane protein ribophorin is found in the p53-enriched fractions, we assume that this reflects contamination. Because we have assumed that the functionally significant continuities will be extensive, the lack of observed continuities fulfills our definition's requirement. The size of ERGIC appears to fluctuate with incubations at 15°C and during BFA treatment, although the marker, p53, being used to follow these events can relocalize. A full characterization of ERGIC as a compartment will require exploiting markers such as p63 to define the boundaries and components more precisely.

4.4.4. The Pre-Golgi Compartment

Saraste and Kuismanen (Saraste and Kuismanen, 1992) have used the pre-Golgi as marked by p58 to describe the tubulo-vesicular compartments between the ER and the Golgi. It is not clear whether this protein marks the same structures as p53; indeed, the distribution of p58 is skewed toward the CGN. In describing the pre-Golgi, the authors draw the analogy with a maturation model of the elements of the endocytotic pathway (Saraste and Kuismanen, 1992). Hence, traffic between the ER and the Golgi pass through a series of intermediates whose properties more and more closely approximate that of the CGN until they reach the CGN. One aspect of this model is that the localization of the elements of this portion of the secretory pathway changes from peripheral to perinuclear during transport. This is, indeed, observed by several groups. If we follow the authors' suggestion that the pre-Golgi represents a continuum of structures that mature from an ER character to that of the *cis*-Golgi (Saraste and Kuismanen, 1992) then the pre-Golgi does not fit our definition of a compartment. We would not expect to be able to attribute control of membrane traffic and the other features we suggest for compartments to such structures. Indeed, transport vesicles would be unnecessary. It is also very difficult to test their maturation model with the markers presently available.

4.4.5. The 15°C Compartment

The careful work of Bonnatti and co-workers (Lotti *et al.*, 1992) have established that the collection of vesicles and tubules that they see marked with VSV-G protein during short incubations at 15°C is a separate and discontinuous structure from the ER. The morphology of this structure seems different from those labeled with p53; hence, it is presumably distinct from ERGIC, although double labeling has not been performed. Unfortunately, we are lacking any unique components of this compartment, although these structures are labeled with rab2 (Lotti *et al.*, 1992). The *in vitro* studies of Balch and co-workers suggest that p58 accumulates in similar structures at 15°C. Their work also indicates control by GTP-binding proteins over the exit and entrance. We do not know, however, whether the structures identified by Lotti *et al* (1992) are identical with those for which functional data are available. An interesting, and still unresolved, question is whether the KDEL-terminated proteins have been excluded from this very early portion of the intermediate compartment. Until unique components of this compartment can be identified, the 15°C compartment cannot be considered a compartment by our definition.

4.4.6. The Salvage Compartment

As more is learned about the mechanism of ER protein retention, evidence for the existence of an intermediate compartment involved in salvage of ER proteins becomes less convincing (Pelham, 1991; Warren, 1987). The only remaining candidates for the receptor that marks this compartment do not localize to the pre-Golgi region of the cell and, in fact, appear to be present in the late Golgi (Lewis and Pelham, 1992; Hsu *et al.*, 1992). This could reflect the fact that these proteins cannot be detected unless they are expressed with a tag allowing their immunolocalization. Hence, they may be mis-localized owing to overexpression or the addition of the tag although they still remain functional. Other evidence still supports the existence of such a compartment. In most cells, one cannot detect KDEL proteins in the Golgi or indeed in most of the post-ER region by immunocytochemistry (Huovila *et al.*, 1992; Huovila and Fuller, 1993; Tooze *et al.*, 1989). Together with the fact that KDEL-terminated proteins contain only early Golgi modifications, this suggests that salvage occurs at an early stage in the secretory pathway. The need for modulation of the interaction of KDEL tail with receptor argues for the existence of a separate compartment. However, until comprehensive morphological studies can be combined with the biochemistry and genetics, this prototype of the intermediate compartment will remain enigmatic.

4.4.7. The Rubella E1 Compartment

Recent work (Hobman *et al.*, 1992) has suggested that the E1 of rubella marks a unique intermediate compartment when it is expressed in the absence of the other viral proteins. This structure does not meet our definition of a compartment because it is both continuous with the ER and freely accessible to ER markers. Consistent with our definition, this structure is not the site of rubella virus budding. The other envelope glycoprotein of the virion, E2, forms a complex with E1, which allows it to exit the ER during normal infection (Oker-Blom *et al.*, 1983).

4.4.8. The HBs Assembly Compartment

The HBsAg matures in a structure that is discontinuous from the ER and the Golgi (Huovila *et al.*, 1992). This structure appears to fulfill the requirements of a compartment. Morphometry shows that it can expand to become almost twice the size of the ER and has a tubulo-vesicular structure (Huovila *et al.*, 1992; Huovila and Fuller, 1993). The only ascribed function of this compartment is that of formation of complex disulfide links. For this purpose it is necessary that it maintains a distinct environment from the ER. Further, exit from this compartment awaits the completion of assembly of the particle so that sorting must occur here. Rab2 has been found to co-localize with this compartment (Huovila and Fuller, 1993). A great deal more work needs to be done to define this structure adequately. In particular, we need to show its relationship to the other markers of this region.

5. CONCLUSIONS

This chapter addressed a basic question concerning the early compartments of the secretory pathway. The first phase of the characterization of these compartments is drawing to a close; a number of markers have been defined and a series of manipulations have been established modulating traffic in this region. This would have been adequate as a characterization of the more classical and less dynamic compartments of the cell, but is clearly not sufficient for an understanding of ER to Golgi traffic. The next phase, in which the separate identities of the compartments in this region will be established, is just beginning. It is hoped that such functions as reversible palmitoylation and complex S–S bond formation will become better characterized as a result. A complete understanding must rest on a careful and quantitative synthesis of morphological, immunocytochemical, and biochemical work on this region. It will also require a combined effort to compare compartments identified in different ways using different markers. We have tried to show here that the biology of this region of the cell is still obscured by the lack of such comparisons. Until we can develop coherent quantitative models for these compartments, their interesting biology will remain hidden.

ACKNOWLEDGMENTS. We are pleased to acknowledge the many discussions with our colleagues that have led to the ideas in this chapter. We particularly wish to thank Prof. Kathryn Howell (University of Colorado Medical School, Denver) and Drs. Beate Sodeik and Neil Emans (EMBL) for careful critical readings of the manuscript and for many useful comments.

NOTE ADDED IN PROOF: Hong and coworkers (Tang *et al.*, 1993) have recently localized the endogenous bovine ERD2 homologue to the *cis*-side of the Golgi apparatus and to a spotty intermediate compound.

ABBREVIATIONS USED IN THIS MANUSCRIPT

-x-K-(x)-K-x-x:	retention motif identified in several transmembrane ER proteins, consisting of two lysines positioned 3 and 4 or 5 residues from the C-terminus
3T3:	a mouse fibroblast cell line
ADEL:	alanyl-aspartyl-glutamyl-leucine; retention signal used for *S. pombe* BiP
ARF:	ADP-ribosylating factor; a small GTP-binding protein
β-COP:	β-coatamer protein
bet:	blocked early in transport
BFA:	Brefeldin A; a macrocyclic fungal antibiotic
BHK:	cell line derived from baby hamster kidney
BiP:	heavy chain binding protein or GRP78
bos:	bet one suppresser
CGN:	*cis*-Golgi network
CHO:	cell line derived from Chinese hamster ovary
DDEL:	aspartyl-aspartyl-glutamyl-leucine; retention signal functional in *K. lactis*
E1/E2:	viral envelope transmembrane glycoprotein 1 or 2
EndoD:	endoglycosidase D; sensitivity to this enzyme indicates that the glycans of a protein have been exposed to the *cis*-Golgi, but were not processed by *medial*-Golgi glycosidases.
EndoH:	endoglycosidase H; acquisition of resistance to this enzyme indicates that the glycans of a protein have been exposed to the *medial*-Golgi.
ER:	endoplasmic reticulum
ERD:	ER-retention defective
ERGIC:	ER-Golgi intermediate compartment
GDP:	guanosine diphosphate
GRP:	glucose-regulated protein
GTP:	guanosine triphosphate
HA:	influenza virus hemagglutinin
HBsAg:	hepatitis B surface antigen
HDEL:	histidyl-aspartyl-glutamyl-leucine; retention signal used in *S. cerevisea* and *K. lactis*
hsp:	heat-shock protein; members belonging to this group are stress-inducible proteins
IBV:	avian infectious bronchitis virus, a member of the coronavirus family
KDEL:	lysyl-aspartyl-glutamyl-leucine; retention signal used in mammalian cells

KDELGL: lysyl-aspartyl-glutamyl-leucyl-glycyl-leucine;   C-terminal
                  extension to retention signal rendering it nonfunctional
KEEL: lysyl-glutamyl-glutamyl-leucine; retention signal functional
                  in mammalian cells
MHC: major histocompatibility complex
MHV: mouse hepatitis virus, a member of the coronavirus family
NE: nuclear envelope, a subcompartment of the ER
NS: nonstructural; NS proteins are not part of the mature virus
NSF: N-ethylmaleimide-sensitive factor
PDI: protein disulfide isomerase
PiZ: naturally occurring variant of the Z type α_1-antitrypsin gene
rab: rat brain; ras-like small GTP binding protein, first cloned as
                  mammalian copy from rat brain
RDEL: arginyl-aspartyl-glutamyl-leucine; retention signal function-
                  al in mammalian cells
rER: rough ER, a subcompartment of the ER
RV: rubella virus
sec: secretory
sER: smooth ER, a subcompartment of the ER
SFV: Semliki Forest virus
SNAP: soluble NSF attachment protein, the mammalian homo-
                  logue of sec17p
SRP: signal recognition particle
Tac: interleukin-2 receptor α-chain
TAP: translocator of antigenic peptides, also called RING4
TCR: T-cell receptor
TE: transitional elements, a subcompartment of the ER
TGN: $trans$-Golgi network
tsO45/ts1:   temperature-sensitive mutants of VSV (tsO45), or SFV (ts1)
Vero: cell line derived from African green monkey kidney
VP: viral protein that is eventually integrated in the mature virus
VSV: vesicular stomatitis virus
VSV-G: vesicular stomatitis virus transmembrane G-protein

6. REFERENCES

Alberini, C. M., Bet, P., Milstein, C., and Sitia, R., 1990, Secretion of immunoglobulin M
    assembly intermediates in the presence of reducing agents, *Nature* **347**:485–487.
Amar-Costesec, A., Todd, J. A., and Kreibich, G., 1984, Segregation of the polypeptide transloca-
    tion apparatus to regions of the endoplasmic reticulum containing ribophorins and ribosomes. I:
    Functional tests on rat liver microsomal subfractions, *J. Cell Biol.* **99**:2247–2253.

Amara, J. F., Lederkremer, G., and Lodish, H. F., 1989, Intracellular degradation of unassembled asialoglycoprotein receptor subunits: A pre-Golgi nonlysosomal endoproteolytic cleavage, *J. Cell Biol.* **110**:3315–3324.

Armstrong, J., Patel, S., and Riddle, P., 1990, Lysosomal sorting mutants of coronavirus E1 protein, a Golgi membrane protein, *J. Cell Sci.* **95**:191–197.

Baker, D., Hicke, L., Rexach, M., Schleyer, M., and Schekman, R., 1988, Reconstitution of SEC gene product-dependent intercompartmental protein transport, *Cell* **54**:335–344.

Balch, W., 1992, From G minor to G major, *Current Biology* **2**:157–169.

Balch, W., Elliot, M., and Keller, D., 1986, ATP coupled transport of vesicular stomatitis virus G protein between the endoplasmic reticulum and the Golgi apparatus, *J. Biol. Chem.* **261**:14681–14689.

Bardeletti, G., Tektoff, J., and Gautheron, D., 1979, Rubella virus maturation and production in two host cell systems, *Intervirology* **11**:97–103.

Beckers, C. J. M., and Balch, W. E., 1989, Calcium and GTP: Essential components in vesicular trafficking between the endoplasmic reticulum and Golgi apparatus, *J. Cell Biol.* **108**:1245–1256.

Berger, M., and Schmidt, M. F. 1985, Protein fatty acyltransferase is located in the rough endoplasmic reticulum, *FEBS Lett.* **187**:289–294.

Bergman, L. W., and Kuehl, W. M., 1979, Formation of an intrachain disulfide bond on nascent immunoglobulin light chains, *J. Biol. Chem.* **254**:8869–8876.

Bergmann, C. C., Maass, D., Poruchynsky, M. S., Atkinson, P. H., and Bellamy, A. R., 1989, Topology of the non-structural rotavirus receptor glycoprotein NS28 in the rough endoplasmic reticulum, *EMBO J.* **8**:1695-1703.

Bergmann, J. E., and Fusco, P. J., 1990, The G protein of vesicular stomatitis virus has free access into and egress from the smooth endoplasmic reticulum of UT-1 cells, *J. Cell Biol.* **110**:625–635.

Bonatti, S., Migliaccio, G., and Simons, K., 1989, Palmitoylation of viral membrane glycoproteins takes place after exit from the endoplasmic reticulum, *J. Biol. Chem.* **264**:12590–12595.

Bonifacino, J., and Lippincott-Schwartz, J., 1991, Degradation of proteins within the endoplasmic reticulum, *Curr. Opin. Cell Biol.* **3**:592–600.

Bonifacino, J. S., Cosson, P., Shah, N., and Klausner, R. D., 1991, Role of potentially charged transmembrane residues in targeting proteins for retention and degradation within the endoplasmic reticulum, *EMBO J.* **10**:2783–2793.

Booth, C., and Koch, G. L., 1989, Perturbation of cellular calcium induces secretion of luminal ER proteins, *Cell* **59**:729–737.

Braakman, I., Helenius, J., and Helenius, A., 1992a, Manipulating disulfide bond formation and protein folding in the endoplasmic reticulum, *EMBO J.* **11**:1717–1722.

Braakman, I., Helenius, J., and Helenius, A., 1992b, Role of ATP and disulphide bonds during protein folding in the endoplasmic reticulum, *Nature* **356**:260–262.

Braakman, I., Hoover-Litty, H., Wagner, K. R., and Helenius, A., 1991, Folding of influenza hemagglutinin in the endoplasmic reticulum, *J. Cell Biol.* **114**:401–411.

Buonocore, L., and Rose, J. K., 1990, Prevention of HIV-1 glycoprotein transport by soluble CD4 retained in the endoplasmic reticulum, *Nature* **345**:625–628.

Burgess, T. L., and Kelly, R. B., 1987, Constitutive and regulated secretion of proteins, *Annu. Rev. Cell Biol.* **3**:243–293.

Burgoyne, R. D., Cheek, T. R., Morgan, A., O'Sullivan, A. J., Moreton, R. B., Berridge, M. J., Mata, A. M., Colyer, J., Lee, A. G., and East, J. M., 1989, Distribution of two distinct Ca-ATPase likie proteins and their relationship to the agonist-sensitive calcium store in adrenal chromaffin cells, *Nature* **342**:72–74.

Ceriotti, A., and Colman, A., 1988, Binding to membrane proteins within the endoplasmic reticulum cannot explain the retention of the glucose-regulated protein GRP78 in Xenopus oocytes, *EMBO J.* **7**:633–638.

Chau, V., Tobias, J., Bachmair, A., Marriot, D., Echer, D., Gonda, D., and Varshavsky, A., 1989, A multiubiquitin chain is confined to specific lysine in a targeted short-lived protein, *Science* **243**:1576–1583.

Chavrier, P., Parton, R. G., Hauri, H. P., Simons, K., and Zerial, M., 1990, Localization of low molecular weight GTP binding proteins to exocytic and endocytic compartments, *Cell* **62**:317–329.

Chege, N., and Pfeffer, S., 1990, Compartmentalization of the Golgi complex: Brefeldin A distinguishes *trans*-Golgi cisternae from the *trans*-Golgi network, *J. Cell Biol.* **111**:893–899.

Clary, D. O., Griff, I. C., and Rothman, J. E., 1990, SNAPs, a family of NSF attachment proteins involved in intracellular membrane fusion in animals and yeast, *Cell* **61**:709–721.

Cosson, P., Lankford, S. P., Bonifacino, J. S., and Klausner, R. D., 1991, Membrane protein association by potential intramembrane charge pairs, *Nature* **351**:414–416.

Crimaldo, C., Hortsch, M., Gausepohl, H., and Meyer, D., 1987, Human ribophorins I and II: The primary structure and membrane topology of two highly conserved rough endoplasmic reticulum-specific glycoproteins, *EMBO J.* **6**:75–82.

Dean, N., and Pelham, H. R., 1990, Recycling of proteins from the Golgi compartment to the ER in yeast, *J. Cell Biol.* **111**:369–377.

Doms, R. W., Keller, D. S., Helenius, A., and Balch, W. E., 1987, Role for adenosine tri-phosphate in regulating the assembly and transport of vesicular stomatitis virus G protein trimers, *J. Cell Biol.* **105**:1957–1969.

Doms, R. W., Russ, W. G., and Yewdell, J. W., 1989, Brefeldin A redistributes resident and itinerant Golgi proteins to the endoplasmic reticulum, *J. Cell Biol.* **109**:61-72.

Doms, R. W., Ruusala, A., Machamer, C., Helenius, J., Helenius, A., and Rose, J. K., 1988, Differential effects of mutations in three domains on folding, quaternary structure, and intracellular transport of vesicular stomatitis virus G protein, *J. Cell Biol.* **107**:89–99.

Donaldson, J., Finazzi, D., and Klausner, R., 1992, Brefeldin A inhibits Golgi membrane-catalysed exchange of guanine nucleotide onto ARF protein, *Nature* **360**:350–352.

Downes, C. S., Musk, S. R. R., Watson, J. V., and Johnson, R. T., 1990, Caffeine overcomes a restriction point associated with DNA replication but does not accelerate mitosis, *J. Cell Biol.* **110**:1855–1859.

Doyle, C., Roth, M. G., Sambrook, J., and Gething, M.-J., 1985, Mutations in the cytoplasmic domain of the influenza virus hemagglutinin affect different stages of intracellular transport, *J. Cell Biol.* **100**:704–714.

Doyle, C., Sambrook, J., and Gething, M.-J., 1986, Analysis of progressive deletions of the transmembrane and cytoplasmic domains of influenza hemagglutinin, *J. Cell Biol.* **103**:1193–1204.

Duden, R., Allan, V., and Kreis, T., 1991a, Involvement of β-COP in membrane traffic through the Golgi complex, *Trends Cell Biol.* **1**:14–19.

Duden, R., Griffiths, G., Frank, R., Argos, P., and Kreis, T., 1991b, β-COP, 110kD protein associated with non-clathrin-coated vesicles and the Golgi complex, shows homology to β-adaptin, *Cell* **64**:649–665.

Dunphy, W. G., Fries, E., Urbani, L. J., and Rothman, J. E., 1981, Early and late functions associated with the Golgi apparatus reside in distinct compartments, *Proc. Natl. Acad. Sci. USA* **78**:7453–7457.

Estes, M. K., 1990, Rotaviruses and their replication, in *Virology,* pp. 1329–1352, Raven Press, New York.

Estes, M. K., and Cohen, J., 1989, Rotavirus gene structure and function, *Microbiol Rev* **53**:410–449.

Fawcett, D. W., 1981, *The Cell,* 2nd ed., pp. 332–333, W. B. Saunders, Philadelphia.

Fine, R. E., 1989, Vesicles without clathrin: Intermediates in bulk flow exocytosis, *Cell* **58**:609–610.

Flynn, G. C., Chappell, T. G., and Rothman, J. E., 1989, Peptide binding and release by proteins implicated as catalysts of protein assembly, *Science* **245**:385–390.

Freedman, R., 1987, Folding into the right shape, *Nature* **329**:196–197.

Freedman, R., and Hillson, D., 1980, Formation of disulphide bonds, in *The Enzymology of Post-translational modification* (R. Freedman and H. Hawkins, eds.), Academic Press, London.

Freedman, R. B., 1984, Native disulphide bond formation in protein biosynthesis: Evidence for the role of protein disulphide isomerase, *Trends Biochem. Sci.* **9**:438–441.

Fujiwara, T., Oda, K., Yokota, A., Takatsuki, A., and Ikehara, Y., 1988, Brefeldin A causes disassembly of the Golgi complex and accumulation of secretory proteins in the endoplasmic reticulum, *J. Biol. Chem.* **265**:18545–18552.

Fuller, S. D., 1987, The T-4 envelope of Sindbis virus is organized by interactions with a complementary T-3 capsid, *Cell* **48**:923–934.

Fuller, S. D., Vriend, G., Pastore, A., and Eder, A., 1993, A model for the structure of the KDEL ER retention sequence leads to a model of environment dependent recognition, in preparation.

Gallagher, P., Henneberry, J., Wilson, I., Sambrook, J., and Gething, M.-J., 1988, Addition of carbohydrate side chains at novel sites on influenza virus hemagglutinin can modulate the folding, transport, and activity of the molecule, *J. Cell Biol.* **107**:2059–2073.

Galteau, M. M., Antoine, B., and Reggio, H., 1985, Epoxide hydrolase is a marker for the smooth endoplasmic reticulum in rat liver, *EMBO J.* **4**:2793–2800.

Ganem, D., 1991, Assembly of hepadnaviral virions and subviral particles, *Curr. Top. Microbiol. Immunol.* **168**:61–83.

Ganem, D., and Varmus, H., 1987, The molecular biology of hepatitis B viruses, *Annu. Rev. Biochem.* **56**:561–693.

Geetha-Habib, M., Noiva, R., Kaplan, H. A., and Lennarz, W. J., 1988, Glycosylation site binding protein, a component of oligosaccharyl transferase, is highly similar to three other 57-kd luminal proteins of the ER, *Cell* **54**:1053–1060.

Gerace, L., Comeau, C., and Benson, M., 1984, Organization and modulation of nuclear lamina structure, *J. Cell Sci.* (Suppl.) **1**:137–160.

Gething, M.-J., McCammon, K., and Sambrook, J., 1986, Expression of wild-type and mutant forms of influenza hemagglutinin: The role of folding in intracellular transport, *Cell* **46**:939–950.

Gething, M.-J., and Sambrook, J., 1990, Transport and assembly processes in the endoplasmic reticulum, *Semin. Cell Biol.* **1**:65–72.

Görlich, D., Hartmann, E., Prehn, S., and Rapoport, T. A., 1992, A protein of the endoplasmic reticulum involved early in polypeptide translocation, *Nature* **357**:47–52.

Görlich, D., Prehn, S., Hartmann, E., Herz, J., Otto, A., Kraft, R., Wiedman, M., Knepsel, S., Dobberstein, B., and Rapoport, T. A., 1990, The signal sequence receptor has a second subunit and is part of a translocation complex in the endoplasmic reticulum as probed by bifunctional reagents, *J. Cell Biol.* **111**:2283–2294.

Goud, B., 1992, Small GTP binding proteins as compartmental markers, *Semin. Cell Biol.* **3**:301–307.

Griffiths, G., Fuller, S. D., Back, R., Hollinshead, M., Pfeiffer, S., and Simons, K., 1989, The dynamic nature of the Golgi complex, *J. Cell Biol.* **108**:277–297.

Griffiths, G., and Rottier, P., 1992, Cell biology of viruses that assemble along the biosynthetic pathway, *Semin. Cell Biol.* **3**:367–381.

Griffiths, G., and Simons, K., 1986, The *trans*-Golgi network: Sorting at the site of exit from the Golgi complex, *Science* **234**:438–443.

Hardwick, K. G., Lewis, M. J., Semenza, J., Dean, N., and Pelham, H. R., 1990, ERD1, a yeast gene required for the retention of luminal endoplasmic reticulum proteins, affects glycoprotein processing in the Golgi apparatus, *EMBO J.* **9**:623–630.

Hauri, H.-P., and Schweizer, A., 1992, The endoplasmic reticulum-Golgi intermediate compartment, *Curr. Opin. Cell Biol.* **4**:600–608.

Hearing, J., Gething, M.-J., and Sambrook, J., 1989, Addition of truncated oligosaccharides to influenza virus hemagglutinin results in its temperature-conditional cell-surface expression, *J. Cell Biol.* **108**:355–365.

Helms, J., and Rothman, J., 1992, Inhibition by Brefeldin A of a Golgi membrane enzyme that catalyses exchange of guanine nucleotide bound to ARF, *Nature* **360**:352-354.

High, S., and Dobberstein, B., 1992, Mechanisms that determine the transmembrane disposition of proteins, *Curr. Opin. Cell Biol.* **4**:581–586.

Hobman, T. C., Lundstrom, M. E., and Gillam, S., 1990, Processing and transport of rubella virus structural proteins in COS cells, *Virology* **178**:122-133.

Hobman, T. C., Woodward, L., and Farquhar, M. G., 1992, The rubella virus E1 glycoprotein is arrested in a novel post-ER, pre-Golgi compartment, *J. Cell Biol.* **118**:795–811.

Hortsch, M., and Meyer, D. I., 1985, Immunochemical analysis of rough and smooth microsomes from rat liver: Segregation of docking protein in rough membranes, *Eur. J. Biochem.* **150**:559–564.

Hsu, V. W., Shah, N., and Klausner, R. D., 1992, A Brefeldin A-like phenotype is induced by the overexpression of a human ERD-2-like protein, ELP-1, *Cell* **69**:625-635.

Hsu, V. W., Yuan, L. C., Nuchtern, J. G., Lippincott-Schwarz, J., Hammerling, G. J., and Klausner, R. D., 1991, A recycling pathway between the endoplasmic reticulum and the Golgi apparatus for retention of unassembled MHC class I molecules, *Nature* **352**:441-444.

Hughson, E., Wandinger-Ness, A., Gausepohl, H., Griffiths, G., and Simons, K., 1988, The cell biology of enveloped virus infection of epithelial tissues, in *Molecular Biology and Infectious Diseases*, pp. 75–89, Elsevier, Paris.

Huovila, A.-P., Eder, A., and Fuller, S., 1992, Hepatitis B surface antigen assembles in a post-ER, pre-Golgi compartment, *J. Cell Biol.* **118**:1305–1320.

Huovila, A.-P., and Fuller, S., 1993, An ER–Golgi intermediate compartment is not continuous with endoplasmic reticulum, submitted.

Hurtley, S. M., Bole, D. G., Hoover, L. H., Helenius, A., and Copeland, C. S., 1989, Interactions of misfolded influenza virus hemagglutinin with binding protein (BiP), *J. Cell Biol.* **108**:2117–2126.

Hurtley, S. M., and Helenius A., 1989, Protein oligomerization in the endoplasmic reticulum, *Annu. Rev. Cell Biol.* **5**:277-307.

Hwang, C., Sinskey, A., and Lodish, H., 1992, The oxidized redox potential in the endoplasmic reticulum: Glutathione as the principal redox buffer, *Science* **257**:1496-1502.

Ivessa, N., De Lemos-Chiarandini, C., Tsao, Y., Takatsuki, A., Adesnik, M., Savatini, D., and Kreibich, G., 1992, O-glycosylation of intact and truncated ribophorins in Brefeldin A-treated cells: Newly synthesized ribophorins are only transiently accessible to the relocated glycosyltransferases, *J. Cell Biol.* **117**:949–958.

Jackson, M. R., Nilsson, T., and Peterson, P. A., 1990, Identification of a consensus motif for retention of transmembrane proteins in the endoplasmic reticulum, *EMBO J.* **9**:3153–3162.

Jones, A. L., and Fawcett, D. W., 1966, Hypertrophy of the agranular endoplasmic reticulum in hamster liver induced by phenobarbital, *J. Histochem. Cytochem.* **14**:215–232.

Kabcenell, A. K., and Atkinson, P. H., 1985, Processing of the rough endoplasmic reticulum membrane glycoprotein of rotavirus SA11, *J. Cell Biol.* **101**:1270-1280.

Kabcenell, A. K., Poruchynsky, M. S., Bellamy, A. R., Greenberg, H. B., and Atkinson, P. H., 1988, Two forms of VP7 are involved in the assembly of SA11 rotavirus in the endoplasmic reticulum, *J. Virol.* **62**:2929-2941.

Kaiser, C. A., and Schekman, R., 1990, Distinct sets of SEC genes govern transport vesicle formation and fusion early in the secretory pathway, *Cell* **61**:723-733.

Kassenbrock, C. K., and Kelly, R. B., 1989, Interaction of heavy chain binding protein (BiP/GRP78) with adenine nucleotides, *EMBO J.* **8**:1461-1467.

Klausner, R., Donaldson, J., and Lippincott-Schwartz, J., 1992, Brefeldin A: Insights into the control of membrane traffic and organelle structure, *J. Cell Biol.* **116**:1071-1080.

Klausner, R., and Sitia, R., 1990, Protein degredation in the endoplasmic reticulum, *Cell* **62**:611-614.

Klausner, R. D., Lippincott-Schwartz, J., and Bonifacino, J. S., 1990, The T-cell antigen receptor: Insights into organelle biology, *Annu. Rev. Cell Biol.* **6**:403-431.

Kleijmeer, M. J., Kelly, A., Geuze, H. J., Slot, J. W., Townsend, A., and Trowsdale, J., 1992, Location of MHC-encoded transporters in the endoplasmic reticulum and cis-Golgi, *Nature* **357**:342-344.

Koch, G. L. E., Booth, C., and Wooding, F. B. P., 1988, Dissociation and reassembly of the endoplasmic reticulum in live cells, *J. Cell Sci.* **91**:511-522.

Kornfeld, R., and Kornfeld, S., 1985, Assembly of asparagine linked oligosaccharides, *Annu. Rev. Biochem.* **54**:631-664.

Kreibich, G. B., Ulrich, L., and Sabatini, D., 1978, Proteins of the rough endosomal membrane related to ribosome binding. I: Identification of ribophorins I and II, membranes characteristic of rough microsomes, *J. Cell Biol.* **77**:464-487.

Krijnse-Locker, J., Rose, J. K., Horzinek, M. C., and Rottier, P. J. M., 1992, Membrane assembly of the triple-spanning coronavirus M protein: Individual transmembrane domains show preferred orientation, *J. Biol. Chem.* **267**:21911-21918.

Kuismanen, E., Jäntti, J., Mäkiranti, V., and Sariola, M., 1992, Effect of caffeine on intracellular transport of Semliki Forest virus membrane glycoproteins, *J. Cell Sci.* **102**:505-513.

Kuismanen, E., and Saraste, J., 1989, Low temperature-induced transport blocks as tools to manipulate membrane traffic, *Meth. Cell Biol.* **32**:257-274.

Lahtinen, U., Dahllöf, B., and Saraste, J., 1992, Characterization of a 58 kD *cis*-Golgi protein in pancreatic exocrine cells, *J. Cell Sci.* **103**:321-333.

Le, A., Graham, K. S., and Sifers, R. N., 1990, Intracellular degradation of the transport-impaired human PiZ a_1-antitrypsin variant, *J. Biol. Chem.* **265**:14001-14007.

Lederkremer, G. Z., and Lodish, H. F., 1991, An alternatively spliced miniexon alters the subcellular fate of the human asialoglycoprotein receptor H2 subunit: Endoplasmic reticulum retention and degradation or cell surface expression, *J. Biol. Chem.* **266**:1237-1244.

Lewis, M. J., and Pelham, H. R. B., 1990, A human homologue of the yeast HDEL receptor, *Nature* **348**:162-163.

Lewis, M. J., and Pelham, H. R. B., 1992, Ligand-induced redistribution of a human KDEL receptor from the Golgi complex to the endoplasmic reticulum, *Cell* **68**:353-364.

Lewis, M. J., Sweet, D. J., and Pelham, H. R., 1990, The ERD2 gene determines the specificity of the luminal ER protein retention system, *Cell* **61**:1359-1363.

Lippincott-Schwartz, J., Bonifacino, J. C., Yuan, L. C., and Klausner, R. D., 1988, Degradation from endoplasmic reticulum: Disposing of newly synthesized proteins, *Cell* **54**:209-220.

Lippincott-Schwartz, J., Donaldson, J. G., Schweizer, A., Berger, E. G., Hauri, H.-P., Yuan, L. C., and Klausner, R. D., 1990, Microtubule-dependent retrograde transport of proteins into the ER in the presence of Brefeldin A suggests an ER recycling pathway, *Cell* **60**:821-836.

Lippincott-Schwartz, J., Yuan, L. C., Bonifacino, J. S., and Klausner, R. D., 1989, Rapid redistribution of Golgi proteins into the ER in cells treated with Brefeldin A: Evidence for membrane cycling from Golgi to ER, *Cell* **56**:801-813.

Lodish, H. F., Kong, N., Snider, M., and Strous, G. J., 1983, Hepatoma secretory proteins migrate from rough endoplasmic reticulum to Golgi at characteristic rates, *Nature* **304**:80-83.

Lotti, L. V., Torrisi, M.-R., Pascale, M. C., and Bonatti, S., 1992, Immunocytochemical analysis of the transfer of vesicular stomatitis virus G glycoprotein from the intermediate compartment to the Golgi complex, *J. Cell Biol.* **118**:43-50.

Machamer, C. E., Mentone, S. A., Rose, J. K., and Farquhar, M. G., 1990, The E1 glycoprotein of an avian coronavirus is targeted to the *cis* Golgi complex, *Proc. Natl. Acad. Sci. USA* **87**:6944–6948.

Marquardt, T., and Helenius, A., 1992, Misfolding and aggregation of newly synthesized proteins in the endoplasmic reticulum, *J. Cell Biol.* **117**:505–513.

Mazzarella, R. A., and Green, M., 1987, ERp99, an abundant, conserved glycoprotein of the endoplasmic reticulum, is homologous to the 90-kDa heat shock protein (hsp90) and the 94-kDa glucose-regulated protein (GRP94), *J. Biol. Chem.* **260**:6926–6931.

Mellman, I., and Simons, K., 1992, The Golgi complex: In vitro veritas, *Cell* **68**:829–840.

Meyer, D. I., and Dobberstein, B., 1980a, Identification and characterization of a membrane component essential for the translocation of nascent proteins across the membrane of the endoplasmic reticulum. *J. Cell Biol.* **87**:503–508.

Meyer, D. I., and Dobberstein, B., 1980b, A membrane component essential for vectorial translocation of nascent proteins across the endoplasmic reticulum: Requirements for its extraction and reassociation with the membrane, *J. Cell Biol.* **87**:498–502.

Meyer, D. I., Krause, E., and Dobberstein, B., 1982, Secretory protein translocation across membranes—the role of the "docking protein," *Nature* **297**:647–650.

Meyer, J. C., Bergmann, C. C., and Bellamy, A. R., 1989, Interaction of rotavirus cores with the nonstructural glycoprotein NS28, *Virology* **171**:98–107.

Mundy, D., and Warren, G., 1992, Mitosis and inhibition of intracellular transport stimulate palmitoylation of a 62kD protein, *J. Cell Biol.* **116**:135–146.

Munro, S., and Pelham, H. R., 1987, A C-terminal signal prevents secretion of luminal ER proteins, *Cell* **48**:899–907.

Munro, S., and Pelham, H. R. B., 1986, An hsp70-like protein in the ER: Identity with the 78 kD glucose regulated protein and immunoglobulin heavy chain binding protein, *Cell* **46**:291–300.

Navarro, D., Qadri, I., and Pereira, L., 1991, A mutation in the ectodomain of herpes simplex virus 1 glycoprotein B causes defective processing and retention in the endoplasmic reticulum, *Virology* **184**:253–264.

Newman, A. P., and Ferro-Novick, S., 1987, Characterization of new mutants in the early part of the yeast secretory pathway isolated by a [3H]mannose suicide selection, *J. Cell Biol.* **105**:1587–1594.

Nilsson, T., Jackson, M., and Peterson, P. A., 1989, Short cytoplasmic sequences serve as retention signals for transmembrane proteins in the endoplasmic reticulum, *Cell* **58**:707–718.

Novick, P., Field, C., and Schekman, R., 1980, Identification of 23 complementation groups required for post-translational events in the yeat's secretory pathway, *Cell* **21**:205–215.

Nuchtern, J. G., Bonifacino, J. S., Biddison, W. E., and Klausner, R. D., 1989, Brefeldin A implicates egress from endoplasmic reticulum in class I restricted antigen presentation, *Nature* **339**:223–226.

Nunnari, J., and Walter, P., 1992, Protein targeting to and translocation across the membrane of the endoplasmic reticulum, *Curr. Opin. Cell Biol.* **4**:573–580.

Oker-Blom, C., Kakkinen, N., Kaarianen, L., and Petterson, R.-F., 1983, Rubella virus contains one capsid protein and three envelope glycoproteins, E1, E2a, and E2b, *J. Virol.* **46**:964–973.

Paabo, S., Bhat, B. M., Wold, W. S., and Peterson, P. A., 1987, A short sequence in the COOH-terminus makes an adenovirus membrane glycoprotein a resident of the endoplasmic reticulum, *Cell* **50**:311–317.

Palade, G., 1975, Intracellular aspects of the process of protein secretion, *Science* **89**:347–358.

Pelham, H. R. B., 1988, Evidence that luminal ER proteins are sorted from secreted proteins in a post-ER compartment, *EMBO J.* **7**:913–918.

Pelham, H. R. B., 1989a, Control of protein exit from endoplasmic reticulum, *Annu. Rev. Cell Biol.* **5**:1–23.

Pelham, H. R. B., 1989b, The selectivity of secretion: Protein sorting in the endoplasmic reticulum, *Biochem. Soc. Trans.* **17**:795–802.

Pelham, H. R. B., 1990, The retention signal for soluble proteins of the endoplasmic reticulum, *Trends Biochem. Sci.* **15**:483–486.

Pelham, H. R. B., 1991, Recycling of proteins between the endoplasmic reticulum and the Golgi complex, *Curr. Opin. Cell Biol.* **3**:585–591.

Pidoux, A. L., and Armstrong, J., 1992, Analysis of the BiP gene and identification of an ER retention signal in *Schizosaccharomyces pombe, EMBO J.* **11**:1583–1591.

Pihlajaniemi, T., Helaakoski, T., Tasanen, K., Myllyla, R., Huhtala, M. L., Koivu, J., and Kivirik-ko, K. I., 1987, Molecular cloning of the beta-subunit of human prolyl 4-hydroxylase: This subunit and protein disulphide isomerase are products of the same gene, *EMBO J.* **6**:643–649.

Plutner, H., Cox, A. D., Pind, S., Khosravi, F. R., Bourne, J. R., Schwaninger, R., Der, C. J., and Balch, W. E., 1991, Rab1b regulates vesicular transport between the endoplasmic reticulum and successive Golgi compartments, *J. Cell Biol.* **115**:31–43.

Plutner, H., Davidson, H., Saraste, J., and Balch, W., 1992, Morphological analysis of protein transport from the ER to Golgi membranes in Digitonin-permeablized cells: Role of p58-containing compartment, *J. Cell Biol.* **119**:1097–1116.

Rose, J. K., and Bergmann, J. E., 1983, Altered cytoplasmic domains affect intracellular transport of the vesicular stomatitis virus glycoprotein, *Cell* **34**:513–524.

Rose, J. K., and Doms, R. W., 1988, Regulation of protein export from the endoplasmic reticulum, *Annu. Rev. Cell Biol.* **4**:257–288.

Rose, M. D., Misra, L. M., and Vogel, J. P., 1989, KAR2, a karyogamy gene, is the yeast homolog of the mammalian BiP/GRP78 gene, *Cell* **57**:1211–1221.

Roth, J., 1987, Subcellular organization of glycosylation in mammalian cells, *Biochim. Biophys. Acta* **906**:405–436.

Roth, M. G., and Compans, R. W., 1981, Delayed appearance of pseudotypes between vesicular stomatitis and influenza viruses during mixed infection of MDCK cells, *J. Virol.* **40**:848-860.

Rothman, J. E., 1987, Protein sorting by selective retention in the endoplasmic reticulum and Golgi stack, *Cell* **50**:521–522.

Rottier, P. J. M., and Rose, J. K., 1987, Coronavirus E1 glycoprotein expressed from cloned cDNA localizes in the Golgi region, *J. Virol.* **61**:2042-2045.

Saraste, J., and Kuismanen, E., 1984, Pre- and post-Golgi vacuoles operate in the transport of Semliki Forest virus membrane glycoproteins to the cell surface, *Cell* **38**:535–539.

Saraste, J., and Kuismanen, E., 1992, Pathways of protein sorting and membrane traffic between the rough endoplasmic reticulum and the golgi complex, *Semin. Cell Biol.* **3**:343–355.

Saraste, J., and Svensson, K., 1991, Distribution of the intermediate elements operating in ER to Golgi transport, *J. Cell Sci.* **100**:415–430.

Scheele, G., and Tartakoff, A., 1985, Exit of nonglycosylated secretory proteins from the rough endoplasmic reticulum is asynchronous in the exocrine pancreas, *J. Biol. Chem.* **260**:926-931.

Schekman, R., 1992, Genetic and biochemical analysis of vesicular traffic in yeast, *Curr. Opin. Cell Biol.* **4**:587–592.

Schlegel, R., and Pardee, A. B., 1986, Caffeine-induced uncoupling of mitosis from the completion of DNA replication in mammalian cells, *Science* **232**:1264-1266.

Schwaninger, R., Plutner, H., Bokoch, G., and Balch, W., 1992, Multiple GTP binding proteins regulate vesicular traffic from the ER to Golgi membranes, *J. Cell Biol.* **119**:1077–1096.

Schweizer, A., Fransen, J. A. M., Baechi, T., Hauri, H. P., and Ginsel, G., 1988, Identification, by a monoclonal antibody, of a 53-kD protein associated with a tubulo-vesicular compartment at the *cis*-side of the Golgi apparatus, *J. Cell Biol.* **107**:1643–1653.

Schweizer, A., Matter, K., Ketcham, C. M., and Hauri, H. P., 1991, The isolated ER-Golgi intermediate compartment exhibits properties that are different from ER and cis-Golgi, J. Cell Biol. 113:45-54.

Schweizer, J., Fransen, J. A. M., Matter, K., Kries, T. E., Ginsel, G., and Hauri, H.-P., 1990, Identification of an intermediate compartment involved in protein transport from endoplasmic reticulum to Golgi apparatus, Eur. J. Cell Biol. 53:185-196.

Sefton, B. M., and Buss, J. E., 1987, The covalent modification of eukaryotic proteins with lipid, J. Cell Biol. 104:1449-1453.

Semenza, J. C., Hardwick, K. G., Dean, N., and Pelham, H. R., 1990, ERD2, a yeast gene required for the receptor-mediated retrieval of luminal ER proteins from the secretory pathway, Cell 61:1349-1357.

Serafini, T., Orci, L., Amherd, M., Brunner, M., Kahn, R., and Rothman, J., 1991, ADP-ribosylation factor is a subunit of the coat of Golgi-derived COP-coated vesicles: A novel role for a GTP-binding protein, Cell 67:239-253.

Shim, J., Newman, A. P., and Ferro-Novick, S., 1991, The BOS1 gene encodes an essential 27-kD putative membrane protein that is required for vesicular transport from the ER to the Golgi complex in yeast, J. Cell Biol. 113:55-64.

Simons, K., and Fuller, S. D., 1985, Cell surface polarity in epithelia, Ann. Rev. Cell Biol. 1:243-288.

Simons, K., and Fuller, S. D., 1987, The budding of enveloped viruses: A paradigm for membrane sorting?, in Biological Organization: Molecular Interactions at High Resolution (R. M. Burnett and H. J. Vogel, eds.), pp. 139-150, Academic Press, New York.

Simons, K., and van Meer, G., 1988, Lipid sorting in epithelial cells, J. Biochem. 27:6197-6202.

Sodeik, B., Doms, R. W., Ericsson, M., Hiller, G., Machamer, C. E., van't Hof, W., van Meer, G., Moss, B., and Griffiths, G., 1993, Assembly of vaccinia virus: Role of the intermediate compartment between the endoplasmic reticulum and the Golgi stacks, J. Cell Biol. 121:521-541.

Spaan, W., Cavanagh, D., and Horzinek, M. C., 1988, Coronavirus: Structure and genome expression, J. Gen. Virol. 69:2939-2952.

Stirzaker, S. C., and Both, G. W., 1989, The signal peptide of the rotavirus glycoprotein VP7 is essential for its retention in the ER as an integral membrane protein, Cell 56:741-747.

Suzuki, C. K., Bonifacino, J. S., Lin, A. Y., Davis, M. M., and Klausner, R. D., 1991, Regulating the retention of T-cell receptor α chain variants within the endoplasmic reticulum: Ca²⁺-dependent association with BiP, J. Cell Biol. 114:189-205.

Sweet, D. J., and Pelham, H. R., 1992, The Saccharomyces cerevisiae SEC20 gene encodes a membrane glycoprotein which is sorted by the HDEL retrieval system, EMBO J. 11:423-432.

Swift, A. M., and Machamer, C. E., 1991, A Golgi retention signal in a membrane-spanning domain of coronavirus E1 protein, J. Cell Biol. 115:19-30.

Sztul, E., Kaplin, A., Saucan, L., and Palade, G., 1991, Protein traffic between distinct plasma membrane domains: Isolation and Characterization of vesicular carriers involved in transcytosis, Cell 64:81-89.

Takusuki, A., and Tamura, G., 1985, Brefeldin A, a specific inhibitor of intracellular translocation of vesicular stomatitis G protein: Intracellular accumulation of high mannose-type sugars and inhibition of its cell surface expression, Agric. Biol. Chem. 45:899-902.

Tang, B. L., Wong, S. M., Qi, X. L., Low, S. H., and Hong, W., 1993, Molecular cloning characterization, subcellular localization and dynamics of p23, the mammalian KDEL receptor; J. Cell Biol. 120: 325-328.

Terasaki, M., 1990, Recent progress on structural interactions of the endoplasmic reticulum, Cell Motil. 15:71-75.

Tilsdale, E., Bourne, J., Khosravi-Far, R., Der, C., and Balch, W., 1992, GTP binding mutants of rab1 and rab2 are potent inhibitors of endoplasmic reticulum (ER) to Golgi transport, J. Cell Biol. 119:749-761.

Tooze, J., and Hollinshead, M., 1991, Tubular early endosomal networks in AtT20 and other cells, *J. Cell Biol.* **115**:635–654.

Tooze, J., Kern, H. F., Fuller, S. D., and Howell, K. E., 1989, Condensation-sorting events in the rough endoplasmic reticulum of exocrine pancreatic cells, *J. Cell Biol.* **109**:35–50.

Tooze, J., Tooze, S., and Warren, G., 1984, Replication of coronavirus MHV-A59 in sac-cells: Determination of the first site of budding of progeny virions, *Eur. J. Cell Biol.* **33**:281–293.

Tooze, S. A., Tooze, J., and Warren, G., 1988, Site of addition of *N*-acetylgalactosamine to the E1 glycoprotein of mouse hepatitus virus-A59, *J. Cell Biol.* **106**:1475–1487.

Tsao, Y. S., Ivessa, N. E., Adesnik, M., Sabatini, D. D., and Kreibich, G., 1992, Carboxy terminally truncated forms of ribophorin I are degraded in pre-Golgi compartments by a calcium-dependent process, *J. Cell Biol.* **116**:57–67.

van Meer, G., 1989, Lipid traffic in animal cells, *Annu. Rev. Cell Biol.* **5**:247–275.

Vaux, D., Tooze, J., and Fuller, S., 1990, Identification by anti-idiotype antibodies of an intracellular membrane protein that recognizes a mammalian endoplasmic reticulum retention signal, *Nature* **345**:495–502.

Vaux, D., Tooze, J., and Fuller, S., 1992, Identification by anti-idiotype antibodies of an intracellular membrane protein that recognizes a mammalian endoplasmic reticulum retention signal (retraction), *Nature* **360**:372.

von Bonsdorff, C., and Vaheri, A., 1969, Growth of rubella virus BHK21 cells: Electron microscopy of morphogenesis, *J. Gen. Virol.* **5**:47–51.

Walter, P., and Blobel, G., 1982, Signal recognition particle contains a 7S RNA essential for protein translocation across the endoplasmic reticulum, *Nature* **299**:691–698.

Walter, P., and Lingappa, V., 1986, Mechanisms of protein translocation across the endoplasmic reticulum membrane, *Annu. Rev. Cell Biol.* **2**:499–516.

Wandinger-Ness, A., Bennett, M., Antony, C., and Simons, K., 1990, Distinct transport vesicles mediate the delivery of plasma membrane proteins to the apical and basolateral domains of epithelial cells, *J. Cell Biol.* **111**:987–1000.

Warren, G., 1987, Signals and salvage sequences, *Nature* **327**:17–18.

Weibel, E. R., Staübli, W., Gnägi, H. R., and Hess, F. A., 1969, Correlated morphometric and biochemical studies on the liver cell. I: Morphometric model, stereological methods and normal morphometric data for rat liver, *J. Cell Biol.* **42**:68–91.

Wen, D., and Schlesinger, M. J., 1984, Fatty acid-acylated proteins in secretory mutants of *Saccharomyces cerevisiae*, *Mol. Cell Biol.* **4**:688–694.

Wiedmann, M., Kurzchalia, T. V., Hartmann, E., and Rapoport, T. A., 1987, A signal sequence receptor in the endoplasmic reticulum membrane, *Nature* **328**:830–833.

Wieland, F. T., Gleason, M. L., Serafini, T. A., and Rothman, J. E., 1987, The rate of bulk flow from the endoplasmic reticulum to the cell surface, *Cell* **50**:289–300.

Wikstroem, L., and Lodish, H., 1991, Non-lysosomal pre-Golgi degredation of unassembled asialoglycoprotein receptor subunits: A TLCK-and TPCK-sensitive cleavage within the ER, *J. Cell Biol.* **113**:997–1007.

Williams, D. B., Swiedler, S. J., and Hart, G. W., 1985, Intracellular transport of membrane glycoproteins: Two closely related histocompatibility antigens differ in their rates of transit to the cell surface, *J. Cell Biol.* **101**:725–734.

Wilson, D. W., Wilcox, C. A., Flynn, G. C., Chen, E., Kuang, W.-J., Henzel, W. J., Block, M. R., Ullrich, A., and Rothman, J. E., 1989, A fusion protein required for vesicle-mediated transport in both mammalian cells and yeast, *Nature* **339**:355–359.

Yewdell, J. W., and Bennink, J. R., 1989, Brefeldin A specifically inhibits presentation of protein antigens to cytotoxic T lymphocytes, *Science* **244**:715–718.

Zagouras, P., and Rose, J. K., 1989, Carboxy-terminal SEKDEL sequences retard but do not retain two secretory proteins in the endoplasmic reticulum, *J. Cell Biol.* **109**:2633–2640.

Chapter 5

Assembly of Mitochondrial Membranes

Elizabeth M. Ellis and Graeme A. Reid

1. INTRODUCTION

1.1. Structure and Function of the Mitochondrial Membranes

The mitochondria of eukaryotes are highly specialized organelles that catalyze oxidative phosphorylation, providing ATP that is used by the cell to drive biosynthetic reactions and other energy-requiring processes. Many biosynthetic processes are carried out within the organelle, but most of the ATP produced by oxidative phosphorylation is exported and utilized in the cytosol. Mitochondria differ from most other organelles in two particular ways that are relevant to a discussion of their biogenesis. Within the mitochondrial matrix is a DNA genome that encodes a small number of mitochondrial polypeptides. Most of the organelle's proteins are, however, imported from the cytosol. Second, the mitochondrion is enclosed by two distinct membranes, the inner and the outer membrane. Imported proteins must be targeted and assembled specifically into one or the other of these membranes, or to one of the two aqueous compartments delineated by them, namely the intermembrane space and the matrix (Figure 1).

The mitochondrial outer membrane contains relatively few polypeptides,

Elizabeth M. Ellis   Biomedical Research Centre, University of Dundee, Ninewells Hospital and Medical School, Dundee, DD1 9SY. Scotland.   **Graeme A. Reid**   Institute of Cell and Molecular Biology, University of Edinburgh, Edinburgh EH9 3JR, Scotland.
Subcellular Biochemistry, Volume 22: Membrane Biogenesis, edited by A. H. Maddy and J. R. Harris. Plenum Press, New York, 1994.

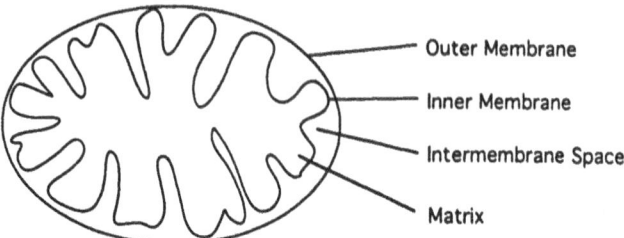

FIGURE 1. Mitochondrial compartmentalization. Mitochondria are composed of two structurally and functionally distinct membranes, the outer and the inner membrane, which delineate the aqueous compartments of the matrix and the intermembrane space. The inner membrane is, in mitochondria isolated from most sources, extensively invaginated to form "cristae" that are enriched in respiratory chain enzymes.

one of which is a porin that acts as a nonspecific transporter of small molecules (Table I). Several other outer-membrane proteins have recently been shown to be required for protein transport into the mitochondrion as described below (Sections 2.3 and 2.4). The inner membrane, in contrast, is metabolically very active; it contains ATP synthase and most of the components of the respiratory chain (reviewed by Hatefi, 1985). The matrix is the site of the TCA cycle, β-oxidation of fatty acids, mitochondrial DNA replication, transcription and protein synthesis, and many other biochemical processes. The intermembrane space contains a

Table I
Some Mitochondrial Membrane Proteins and Their Sites of Synthesis

Outer membrane		Mammals	Yeast
Porin			Cytoplasm[a]
Amine oxidase		Cytoplasm[a]	
Inner membrane			
QH$_2$-cytochrome c reductase	Cytochrome b	Mitochondria	Mitochondria
	Other subunits	Cytoplasm	Cytoplasm
NADH dehydrogenase	Subunits 1–6	Mitochondria	Cytoplasm
	Other subunits	Cytoplasm	Cytoplasm
Cytochrome oxidase	Subunits I,II,III	Mitochondria	Mitochondria
	Subunits IV–VIII	Cytoplasm	Cytoplasm
F$_1$ATPase	Subunits α,β,γ,δ,ε	Cytoplasm	Cytoplasm
F$_0$ATPase	Subunits 6,8	Mitochondria	Mitochondria
	Subunit 9	Cytoplasm	Mitochondria
ADP/ATP translocator		Cytoplasm	Cytoplasm

[a]Although not demonstrated directly, these components can be assumed to be encoded in the nucleus since their sequences are not found in mitochondrial DNA.

few enzymes that are components of the respiratory chain plus adenylate kinase. The biochemistry of mitochondrial function is eloquently described, in detail that cannot be matched here, in the monograph by Tzagoloff (1982).

1.2. Mitochondrial Biogenesis: Contribution of the Mitochondrial and Nuclear Genomes

Mitochondria are generally believed to have evolved from prokaryotes that developed an endosymbiotic relationship with the ancestors of eukaryotes. During the period of over 1 billion years since this relationship began, genetic information has been transferred from the mitochondrion to the nucleus. What we see today as mitochondrial DNA is thus a remnant of an entire prokaryotic genome. The organization of the mitochondrial genome is rather variable among different groups of organisms. In animals, it consists of a circular molecule of only 16 kbp in which functional sequences are extremely packed together. In yeast, the mitochondrial genome is about fivefold larger, not because of an increased coding capacity but because of differences in structure, including the presence of intergenic "spacers" and intragenic introns. In higher plants the mitochondrial genome varies from 300–3000 kbp.

Mitochondrial DNA encodes a small number of polypeptides, each of which remains within the organelle, and the rRNA and tRNA molecules that are involved in the translation of mitochondrial mRNA. The mitochondrial protein synthesizing system depends on these and many cytoplasmically synthesized components to produce some of the subunits of cytochrome c oxidase, ATP synthase, cytochrome c reductase, and NADH dehydrogenase (Table I). Each of these complexes is composed of several subunits, some of which are encoded in the mitochondrion and others in the nucleus. This situation presents an interesting problem for the coordination of gene expression from both genomes.

Both the maintenance and the expression of this organellar genome are entirely dependent on functions encoded by nuclear genes. The nuclear genome affects the transcription, splicing, and translation of mitochondrial genes (Lustig et al., 1982; McGraw and Tzagaloff, 1983; Fox, 1986). The molecular mechanisms of these regulatory processes are now being established. For example, Fox and colleagues have identified nuclear-encoded proteins that regulate translation of specific mitochondrial gene products (Costanzo and Fox, 1990). In addition, the work of Butow and colleagues (Parikh et al., 1987) has shown that the expression of several nuclear genes depends on the nature of the mitochondrial genome, demonstrating the existence of an uncharacterized signaling pathway. The coupling between nuclear and mitochondrial gene expression is not absolute as inhibition of protein synthesis in the mitochondria does not stop cytoplasmic synthesis of mitochondrial proteins, nor their import and partial assembly (van Loon et al., 1983; Weiss and Kolb, 1979). In the yeast Saccharomyces cere-

visiae, the most important controls of mitochondrial assembly and biogenesis appear to be carbon source and oxygen regulation, and, in all cells, growth and cell division have effects on mitochondrial biogenesis. These interactions are described in detail by Grivell (1989).

The great majority of mitochondrial proteins are encoded by nuclear DNA, synthesized in the cytosol, and transported into or across the mitochondrial membranes. This chapter will focus largely on the targeting and assembly of these proteins into membranes, not only because of the importance of this process for mitochondrial biogenesis, but also because that is where current knowledge is most detailed. Recent interest in mitochondrial protein targeting has led to rapid advances in our understanding of this process at the molecular level. We also discuss the synthesis and insertion of lipids into the mitochondrial membranes, about which relatively little is known but which has obvious importance in membrane biogenesis.

2. SYNTHESIS AND TARGETING OF NUCLEAR-ENCODED MITOCHONDRIAL PROTEINS: THE GENERAL TARGETING PATHWAY

2.1. Targeting Sequences: Structure and Function

Most cytoplasmically synthesized mitochondrial proteins are made as larger precursors with an amino terminal targeting sequence that is cleaved off during or immediately after translocation into the mitochondrial matrix. The importance of the amino terminal sequence for the correct localization of the protein has been demonstrated using fusion proteins consisting of the targeting sequence attached to a protein that normally remains in the cytosol (Douglas *et al.*, 1984; Hurt *et al.*, 1984). Mouse dihydrofolate reductase (DHFR) has been found particularly suitable for the construction of fusion proteins in that it is small and monomeric and seems to be readily transported across biological membranes when endowed with a targeting sequence. The use of fusion proteins has demonstrated that as few as nine amino acid residues are required for targeting to the mitochondrial matrix (Keng *et al.*, 1986) and that targeting sequence activity is remarkably insensitive to sequence modifications (von Heijne, 1985; Walker *et al.*, 1990).

Comparison of the N-terminal amino acid sequences of many mitochondrial precursor proteins shows little sequence similarity in common with the situation that exists with secretory signal sequences. Mitochondrial targeting sequences do share some important features, particularly in that they are enriched for the positively charged amino acids lysine and arginine, and are normally lacking in acidic residues (Hurt and van Loon, 1986; von Heijne, 1985). The amino acids are arranged in such a way that they are predicted to form amphiphilic structures, most commonly α-helices (von Heijne, 1986), and experimental evidence has demonstrated that the amphiphilicity of natural and modified targeting sequences correlates with their ability to support targeting (Allison and Schatz, 1986; Roise *et al.*, 1988). Chemically synthesized peptides corresponding to mitochondrial targeting sequences have been shown to be surface-active (Tamm, 1986), to inhibit targeting of precursor polypeptides in a competitive manner (Gillespie *et al.*, 1985), and to assume a helical conformation in a membrane-like environment (Roise *et al.*, 1986). It has been suggested that the amphiphilicity of mitochondrial targeting sequences may account for much of their behavior and that the proton electrochemical gradient across the inner membrane may in some way be sensed by the targeting sequences such that they insert only into mitochondrial membranes (Roise *et al.*, 1986; Pfanner and Neupert, 1985). This hypothesis would account for the specificity of targeting to mitochondrial membranes and obviate the need for receptors. However, such a model now seems at best incomplete since the involvement of specific receptor proteins in the outer membranes is well documented (see Section 2.3). It has also been shown that specific binding of precursor proteins to the outer membrane, prior to their insertion into the inner membrane, does not require an energized membrane. Nevertheless, the $\Delta\psi$-dependent* insertion of proteins into the inner membrane may well involve a mechanism similar to that suggested by Roise *et al.* (1986) but with the supplementary involvement of protein components of the targeting machinery (reviewed in Pfanner and Neupert, 1990).

In general, the mitochondrial targeting sequence must promote an ordered set of events, some of which are intimately linked both temporally and spatially, others of which are clearly separate. These events include the binding of the precursor protein to the mitochondrial outer membrane, the insertion and subsequent translocation of the protein through one or both membranes, cleavage at a specific site in those cases where the precursor is processed, and, where the final destination is not the matrix, the targeting sequence must contain information for intramitochondrial sorting. The information for targeting to the matrix is in general located at the extreme N-terminus, with processing and sorting signals C-terminal to this (Hurt and van Loon, 1986). Targeting to the matrix is apparently a default pathway once mitochondrial import is initiated. Precursors destined for other mitochondrial subcompartments possess additional signals that specify their ultimate location (Figure 2). For example, a nonpolar membrane anchor sequence immediately after the matrix-targeting domain targets proteins to the mitochondrial outer membrane (Freitag *et al.*, 1982; Hase *et al.*, 1984). Precursors destined for the intermembrane space and some inner membrane proteins possess presequences that contain a long stretch of mainly uncharged

---

*$\Delta\psi$ is the electrical components of the proton electrochemical gradient ($\Delta\mu H^+$) across the mitochondrial inner membrane.

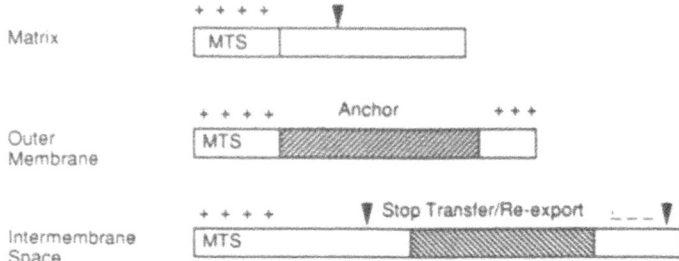

FIGURE 2. Domain structure of mitochondrial targeting and sorting sequences. The amino terminal presequences of mitochondrial proteins possess regions that specify intramitochondrial destinations. MTS, matrix targeting signal. The shaded regions indicate uncharged, putative membrane-spanning regions. Positively (+) and negatively (−) charged regions are indicated, and the arrowheads represent proteolytic cleavage sites.

amino acids adjacent to the matrix-targeting region (Gasser *et al.*, 1982b; Kaput *et al.*, 1982).

2.2. Chaperones and the Maintenance of a Translocation-Competent State

A fusion protein consisting of the N-terminal targeting sequence from subunit IV of yeast cytochrome *c* oxidase attached to DHFR (CoxIV-DHFR) is efficiently transported across both mitochondrial membranes. Eilers and Schatz (1986) demonstrated that transport of this polypeptide into isolated mitochondria could be blocked by addition of methotrexate, a ligand that binds tightly to DHFR, stabilizing the folded structure. This finding indicated that proteins must be at least partially unfolded in order to cross the mitochondrial membranes, and this conclusion has been borne out by a great deal of subsequent work. The *in vitro* transport of precursor polypeptides into mitochondria is very much faster if they are denatured in urea immediately before dilution into a mitochondrial suspension and this denaturation overcomes the observed requirement for extra-mitochondrial ATP (Verner and Schatz, 1987). It thus appears that ATP is required either to maintain precursor polypeptides in a translocation-competent state or to unfold proteins to enable translocation where folding has already progressed too far (Pfanner *et al.*, 1987c).

The maintenance of proteins in an incompletely folded state has been attributed to a class of proteins termed "molecular chaperones" (Ellis, 1987). One class of chaperones includes heat-shock proteins of the Hsp70 family, and these have been implicated on targeting not only to mitochondria but also to the endoplasmic reticulum. Depletion of these chaperones in *S. cerevisiae* results in the accumulation of precursor polypeptides in the cytosol (Deshaies *et al.*, 1988).

A yeast homologue of the bacterial heat-shock protein DnaJ, Mas5p, has been implicated in mitochondrial protein targeting, and is located either in the cytosol or at the mitochondrial surface (Atencio and Yaffe, 1992). It is thought that Mas5p may, like its bacterial homologue, interact with Hsp70s. It is not clear from these experiments whether heat-shock proteins are directly involved in targeting and, although they may catalyze ATP-dependent unfolding, it may be that this role is normally undertaken by other proteins more specific to the mitochondrial targeting pathway, in a manner analogous to the involvement of Signal Recognition Particle (SRP) in the targeting of proteins to the endoplasmic reticulum.

Cytosolic factors required for efficient import have been identified that may, indeed, have a more specific role in mitochondrial protein targeting. Precursor Binding Factor (PBF) has been purified from rabbit reticulocyte lysate, and this protein prevents aggregation or inappropriate folding of the precursor of ornithine carbamoyltransferase by binding to its presequence (Murakami and Mori, 1990; Murakami et al., 1990). Ono and Tuboi (1990a) have similarly purified a 28-kDa protein that stimulates import in vitro. Using a genetic approach in yeast, Ellis and Reid (1992) have suggested that a ribonucleoprotein may be involved at this early step in the targeting pathway. These cytosolic factors may have a role in promoting translocation competence, or may be required for the recognition of precursors preceding their binding to receptors on the mitochondrial surface (Figure 3).

2.3. Receptors for Targeted Proteins

The specificity of targeting to the mitochondrion indicates that there are receptor molecules associated with the organelle that recognize mitochondrial targeting sequences. Several lines of evidence suggest that these receptors are proteinaceous. First, pretreatment of isolated mitochondria with proteinase results in an abolition of import in vitro. (Riezman et al., 1983; Zwizinski et al., 1984). Second, antibodies raised against outer-membrane proteins inhibit in vitro import (Ohba and Schatz, 1987). More recently, several proteins, identified by different approaches, have the properties expected of receptor molecules (Table II), and their range of precursor substrates has been determined. Antibodies raised against outer-membrane proteins of Neurospora crassa mitochondria have been screened for their ability to block transport and the binding of radiolabeled precursor proteins to the mitochondrial surface (Söllner et al., 1989; Söllner et al., 1990). A 19-kDa polypeptide, MOM19, was thus implicated in the binding and transport of many precursors with amino terminal targeting sequences including the β-subunit of ATP synthase (Söllner et al., 1989) and a second receptor, MOM72, is apparently required for recognition of the adenine nucleotide translocator (ANT). The S. cerevisiae equivalent of MOM72 is known as

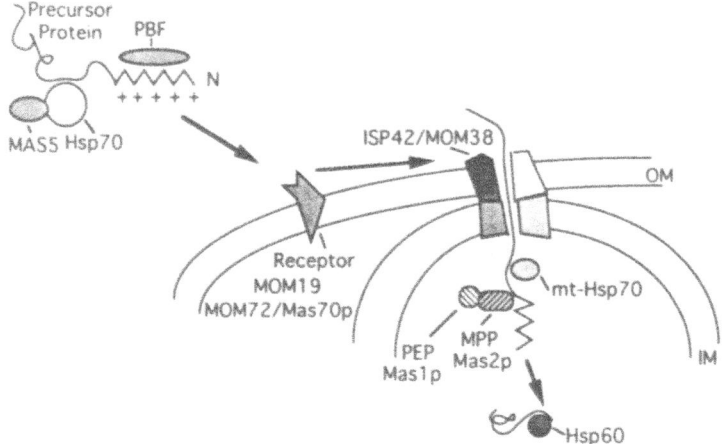

FIGURE 3. The general targeting pathway. Proteins synthesized in the cytosol interact with heat-shock proteins Hsp70 and Mas5p as well as with other polypeptides such as Precursor Binding Factor (PBF). After migration to the mitochondrial surface the precursor proteins bind to receptors such as MOM19 or MOM72/Mas70p. Interaction with MOM38/ISP42 leads to the initiation of translocation into and across the mitochondrial membranes. Proteins can be transported across both membranes simultaneously at translocation contact sites. On emerging into the matrix, precursor proteins interact with the mitochondrial Hsp70 immediately prior to interaction with Hsp60, which is required for proper folding and assembly. Most precursor polypeptides undergo proteolytic processing, mediated by the matrix processing peptidase (MPP) and the processing enhancing protein (PEP).

Mas70p, but from antibody and proteinase studies the receptor in yeast seems to be involved in the import of β-subunit as well as ANT, although not some of the precursors recognized by MOM72. Disruption of the yeast Mas70 gene results in the slower import of many precursors, but is not lethal (Hines *et al.*, 1990). These findings suggest that there are several receptors, each with different but perhaps overlapping sets of precursor substrates.

MOM19 and MOM72/Mas70p are both integral proteins of the mito-chondrial outer membrane and are exposed to the cytosol, enabling interaction with precursor polypeptides. In *N. crassa*, immunoprecipitation experiments have shown that MOM19 and MOM72 are associated with two other outer-membrane proteins, MOM38 and MOM22. MOM38 is thought to be a second-ary receptor or General Insertion Protein (GIP) (Kiebler *et al.*, 1990) and is 67% similar to the yeast protein ISP42, which has been implicated as a component of the translocation site (see below).

Another approach using anti-idiotypic antibodies to a mitochondrial prese-quence resulted in the identification of a 32-kDa outer-membrane protein, p32 (Pain *et al.*, 1990). Again, disruption of the gene encoding this protein in yeast is

Table II
Components of the Mitochondrial Protein Targeting Pathway

Location	Component	Function	Reference
Cytosol	Hsp70	Maintenance of precursor in import-competent state	a,b
	MAS5	Interaction with Hsp70?	c
	PBF	Binds to presequences	d
	MTS1	?	e
Outer membrane	Mas70p/MOM72	Receptor	f,g
	MOM19	Receptor	h
	MOM22	Component of receptor complex	i
	p32	Receptor?	j,k
	ISP42/MOM38 (GIP)	Insertion of precursors into the outer membrane	i,l
Matrix	mt-Hsp70	Translocation and folding	m,n,o
	MPP/Mas2p	Component of matrix proteinase Cleavage of targeting sequences	p,q,r
	PEP/Mas1p	Component of matrix proteinase Cleavage of targeting sequences	p,q,r,s
	Hsp60	Folding, assembly (and re-export?) of proteins	t
Inner membrane	IMP1	Cleavage of cytochrome b_2 and cytochrome c oxidase subunit II	u,v
Intermembrane space	Cytochrome c heme lyase	Import of cytochrome c by attachmemt of heme	w

References: a, Deshaies et al., 1988; b, Murakami et al., 1988; c, Atencio and Yaffe, 1992; d, Murakami and Mori, 1990; e, Ellis and Reid, 1992; f, Hines et al., 1990; g. Söllner et al., 1990; h, Söllner et al., 1989; i, Kiebler et al., 1990; j, Pain et al., 1990; k, Murakami et al., 1990; l, Baker et al., 1990; m, Kang et al., 1990; n, Ostermann et al., 1990; o, Scherer et al., 1991; p, Böhni et al., 1983; q, Pollock et al., 1988; r, Hawlitschek et al., 1988; s, Witte et al., 1988; t, Cheng et al., 1989; u, Behrens et al., 1992; v, Schneider et al., 1991; w, Nicholson et al., 1989.

not lethal, suggesting that it too may be one of several overlapping receptors. Conversely, this protein has also been identified as the mitochondrial phosphate transporter (Phelps et al., 1991), and its role in targeting is unclear. A further putative receptor has been purified from rat liver by Ono and Tuboi (1990b).

2.4. Contact Sites and the Translocation Event

The mitochondrial inner and outer membranes have been observed under the electron microscope to converge at various regions, and it has been proposed that these contact sites may be the site of translocation of proteins across the membranes (Kellems et al., 1975). Mitochondrial precursor polypeptides have been artificially trapped during transport and found to accumulate at contact sites under certain conditions, showing that these proteins can indeed cross both mitochondrial membranes together and directly implicating contact sites in the translocation process. Preincubation of a mitochondrial precursor with antibodies that recognize the carboxyl terminus of the protein resulted in the blocking of translocation across the membranes (Schleyer and Neupert, 1985). In addition, low temperatures and low levels of nucleoside triphosphates (NTPs) cause precursors to accumulate at contact sites (Pfanner et al., 1987a), and fusion proteins with a tightly folded carboxy terminus can be trapped at these sites (Rassow et al., 1989; Vestweber and Schatz, 1988). In most cases, the trapped proteins can be seen, using electron microscopy, to accumulate at contact sites. They are accessible to the matrix proteinase at the amino terminus, yet at the same time are sensitive to externally added proteinases at regions of the carboxy terminus.

The number of contact sites has been estimated at between 100 and 1000 per mitochondrion, and these sites can be saturated by the artificially trapped intermediates, thus preventing translocation of several different wild-type mitochondrial precursors (Rassow et al., 1989; Vestweber and Schatz, 1988). This and other lines of evidence demonstrate that different precursors use the same route of translocation into the mitochondria (Wienhues et al., 1991).

The structure of contact sites is not fully determined, but at least a proportion appears to be relatively stable on isolation (Schwaiger et al., 1987). Precursors trapped at contact sites are accessible to aqueous perturbants, and are therefore likely to be passing through a proteinaceous apparatus rather than through the lipid bilayer (Pfanner et al., 1987a). These protein components may be directly or indirectly involved in the translocation process (Pfanner et al., 1988).

Protein components required for the translocation of mitochondrial precursor proteins have been identified by a variety of techniques (Table II). Photochemical cross-linking has led to the identification in yeast of a protein, ISP42, that associates with a precursor fusion protein trapped at contact sites (Vestweber et al., 1989). ISP42 is essential for viability in yeast cells, and antibodies against it block import in vitro (Baker et al., 1990). The protein has been shown to be located in the mitochondrial outer membrane, but surprisingly is not enriched at contact sites. To explain this phenomenon, it has been proposed that there is a dynamic interaction between ISP42 and other components of the translocation site during the translocation process. In N. crassa, the homologue of ISP42 is MOM38, and this has been shown to associate with outer-membrane proteins including the putative receptors MOM19 and MOM72 (Kiebler et al., 1990). These results suggest that there is a considerable degree of interaction between the receptors in the outer membrane and the site of translocation (Figure 3).

Recent work has led to the discovery that the mitochondrial inner membrane has a complete protein transporting machinery that can operate independently of the outer membrane (Glick et al., 1991; Hwang et al., 1991). Thus, mitoplasts (mitochondria in which the outer membrane has been disrupted) are capable of recognizing and transporting precursor polypeptides. Cytochrome c heme lyase is transported to its destination in the intermembrane space by a process that is independent of the inner membrane but involves the same receptor and general insertion protein components required for transport to the matrix. At the same time, those proteins that are transported across both membranes are not released into the intermembrane space but are apparently always transported directly through translocation contact sites. These contact sites may therefore be considered as sites where the transport machinery of each of the two membranes is brought together to form an active complex. Pfanner et al. (1992) have proposed a dynamic model of assembly and disassembly of contact sites to account for the available data.

The process of translocation has been studied in some detail, particularly with reference to the energy required for the precursor to cross the membrane. In vitro import is clearly energy-dependent (Gasser et al., 1982a), and a requirement has been observed both for nucleoside triphosphates and a proton electrochemical potential gradient (Pfanner and Neupert, 1986). The hydrolysis of a phosphodiester bond in either ATP or guanosine triphosphate (GTP) occurs outside the mitochondria (Chen and Douglas, 1987), and it is thought that this is required for maintaining the precursor protein in a translocation-competent state (Pfanner et al., 1987c). ATP hydrolysis also occurs in the mitochondrial matrix, coupled to protein translocation (Section 2.5).

The addition of an uncoupler of oxidative phosphorylation such as carbonylcyanide-m-chlorophenylhydrazone (CCCP) prevents the processing of some mitochondrial precursor polypeptides by blocking translocation (Nelson and Schatz, 1979; Reid and Schatz, 1982). The effects of a range of mitochondrial poisons on protein transport indicate that the electrical component ($\Delta\psi$) of the proton electrochemical gradient is required. Dissection of the transport pathway into a sequence of events with distinct intermediates has shown that $\Delta\psi$ is required for the insertion of the N-terminal region of the transported protein into the inner membrane and not for subsequent translocation of the remainder of the polypeptide chain (Neupert et al., 1990). The molecular basis for the involvement of $\Delta\psi$ is not clear, but it may involve an electrophoretic effect on the positively charged amino terminus of the precursor protein.

2.5. Posttranslocational Folding and Assembly

As a precursor protein emerges from the site of translocation into the matrix, it is bound by the mitochondrial heat-shock protein, mt-Hsp70 (Figure 3) (Table II). Precursors that have accumulated at contact sites can be cross-linked to or immunoprecipitated as a complex with mt-Hsp70 (Ostermann et al., 1990; Scherer et al., 1990). Mitochondria from a yeast strain carrying a mutant mt-Hsp70 gene accumulate precursors at translocation sites in vitro, and it is thought that the mt-Hsp70 is required for the unfolding of the precursor after it has inserted into the translocation site (Kang et al., 1990). This suggests a close relationship between the translocation event and posttranslocation processes. It has been proposed that the binding of a chaperone such as mt-Hsp70 to the unfolded precursor may result in the vectorial movement of the protein through the translocation site (Neupert et al., 1990; Simon et al., 1992). Once in the matrix, the precursor is released in an ATP-dependent manner (Scherer et al., 1990) and Hsp60, a chaperone that is homologous to GroEL, binds to the precursor and catalyzes its ATP-dependent folding and assembly (Cheng et al., 1989; Ostermann et al., 1989). Both these chaperones are essential for viability in yeast (Craig et al., 1987; Cheng et al., 1989).

The posttranslocational assembly of proteins into multisubunit complexes has been examined in some detail for individual complexes, both in the matrix and the inner membrane and is described in Section 4.3.

2.6. Proteolytic Processing

Many N-terminal targeting sequences are cleaved by a specific metallo-endoproteinase upon reaching the matrix (Böhni et al., 1983; McAda and Douglas, 1982; Yang et al., 1991). This step most likely takes place while the precursor is still bound to one of the chaperones, Hsp70 or Hsp60 (Figure 3). The proteinase consists of two structurally related components, encoded by the Mas1 and Mas2 genes in yeast. The processing event is essential for the correct assembly or sorting of many proteins. In yeast a temperature-sensitive mutation in either component of the proteinase is lethal (Yaffe et al., 1985; Yaffe and Schatz, 1984) and leads to the accumulation of precursors outside the matrix, again highlighting the relationship between translocation and later events. The catalytic component of the proteinase is known as MPP (matrix processing peptidase) and its partner as PEP (processing enhancing protein). Not only do these two polypeptides interact to form an optimally active enzyme but their sequences show that they are closely related to each other (Pollock et al., 1988). In Neurospora, PEP is identical to one of the subunits of cytochrome c reductase (cytochrome bc_1 complex), whereas two different but related proteins are found in yeast (Schulte et al., 1989).

Many inner-membrane and intermembrane-space proteins undergo a second processing event. The proteinase responsible for processing cytochrome b_2 and subunit 2 of cytochrome c oxidase (COXII) has been identified as Inner-Membrane Proteinase I (IMP1; Schneider et al., 1991). Other proteins including the inner-membrane protein cytochrome c_1 are processed by yet another proteinase (Behrens et al., 1991).

3. ASSEMBLY OF PROTEINS INTO THE MITOCHONDRIAL OUTER MEMBRANE

Some of the proteins targeted to the outer membrane appear to follow many of the initial steps of the general targeting route as outlined above (Figure 4). For

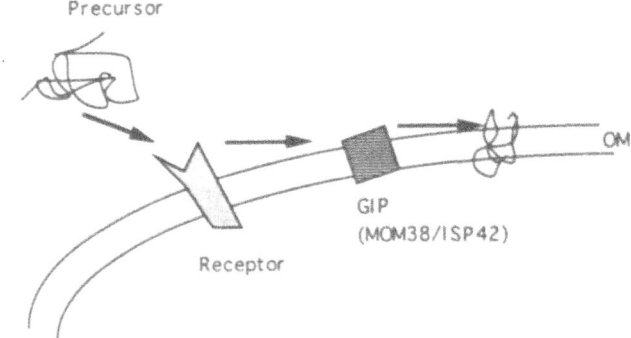

FIGURE 4. Insertion of proteins into the mitochondrial outer membrane. As with proteins transported across both mitochondrial membranes, at least some of the proteins of the outer membrane are
transported by a pathway that depends on the MOM19 and MOM72/Mas70p receptors and the
General Insertion Protein, GIP (MOM38/ISP42). The pathway to the outer membrane is therefore a
branch from the general pathway described in Figure 2 with both pathways sharing common early
steps. Targeting to the outer membrane does not require an energized inner membrane and does not
involve proteolytic processing.

example, porin has been shown to bind to receptors with high affinity (Pfaller
and Neupert, 1987) and in doing so to compete for binding with the inner-
membrane protein ANT (adenine nucleotide translocator; Pfanner *et al.*, 1988).
As porin competes for binding with other precursors, this suggests that the
initial part of the import pathway is shared by proteins with different destinations. This is exemplified by the fact that the 70-kDa outer-membrane protein
Mas70p can be mistargeted to the matrix by the introduction of a point mutation
in the protein (Hase *et al.*, 1984) and that the extreme N-terminal segment of this
protein can target a fusion protein to the mitochondrial matrix (Hase *et al.*,
1986).

   Under normal circumstances, however, outer-membrane proteins do not
pass through the matrix, and most, such as Mas70p and porin, are not cleaved
(Hase *et al.*, 1984; Freitag *et al.*, 1982). In some cases, including Mas70p, the
protein possesses a typical N-terminal matrix-targeting domain, but C-terminal
to this is a region that acts as a membrane anchor or stop-transfer domain. In
other cases, such as porin, no typical targeting sequence can be seen; instead,
regions of both the C-terminus and N-terminus are required for targeting (Hamajima *et al.*, 1988). Other proteins, such as the 45-kDa outer-membrane protein,
possess a membrane-spanning domain that anchors it in the outer membrane
(Yaffe *et al.*, 1989). However, ISP42/MOM38 has no membrane-spanning
domain or matrix targeting sequence (Baker *et al.*, 1990; Kiebler *et al.*, 1990).
It is possible that rather different pathways exist for the targeting of these proteins.

4. ASSEMBLY OF PROTEINS IN THE MITOCHONDRIAL INNER MEMBRANE

4.1. Targeting of Nuclear-Encoded Inner-Membrane Proteins

The mechanism by which inner membrane proteins are localized depends to some extent on their position with respect to the membrane. Three classes exist: peripheral proteins that are located on the matrix face of the inner membrane; peripheral proteins that are on the intermembrane space side of the inner membrane; and intrinsic inner membrane proteins (Figure 5).

4.1.1. Matrix Face Proteins

Most proteins that are associated with the inner membrane on the matrix side are targeted in a similar way to soluble matrix proteins. These include well-studied proteins such as β-subunit of F_1ATPase. Most, if not all, have typical

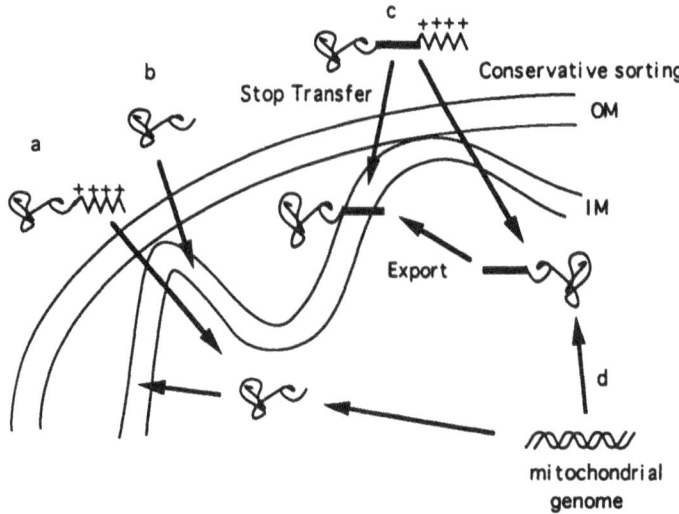

FIGURE 5. Sorting pathways to the mitochondrial inner membrane. (A) Many proteins with an amino-terminal presequence are targeted via the matrix and are assembled or integrated into the inner membrane after removal of the presequence; (B) proteins without a presequence are inserted directly into the inner membrane without passing through the matrix; (C) proteins that face the intermembrane space are thought either to pass through the matrix before being re-exported via an ancestral secretory pathway (conservative sorting) or to become anchored in the inner membrane during transport through contact sites due to the presence of a stop-transfer sequence; (D) proteins encoded by the mitochondrial genome are inserted into the membrane and assembled together with nuclear-encoded proteins.

matrix targeting sequences, they are generally processed by the matrix proteinase, and their assembly is dependent on the mitochondrial Hsp60 protein
(Cheng et al., 1989).

4.1.2. Proteins Facing the Intermembrane Space

Some peripheral inner-membrane proteins facing the intermembrane space,
such as cytochrome c_1 and the Rieske Fe/S protein of ubiquinol:cytochrome c
reductase, are thought to be targeted in a similar manner to intermembrane-space
proteins such as cytochrome b_2 and cytochrome c peroxidase. Like these
intermembrane-space proteins, cytochrome c_1 has a long targeting sequence with
two cleavage sites (Ohashi et al., 1982; Sadler et al., 1984). Originally it was
proposed that the N-terminal region of cytochrome c_1 precursor was targeted to
the matrix, with the remainder of the protein failing to cross the inner membrane
because of a hydrophobic stretch in the C-terminal region of the targeting sequence, which could act as a "stop-transfer" sequence (Gasser et al., 1982b;
Ohashi et al., 1982; van Loon and Schatz, 1987). Cleavage of the N-terminus by
the matrix proteinase results in an intermediate protein, which is then cleaved by
a proteinase located on the intermembrane-space side of the inner membrane to
release the mature polypeptide. Removal or mutation of the hydrophobic stretch
results in cytochrome c_1 being targeted to the matrix (van Loon et al., 1987).
However, at low temperatures and using matrix proteinase inhibitors, the precursor form of cytochrome c_1 could be detected in the matrix (Nicholson et al.,
1989), thus suggesting that the entire precursor is transferred to the matrix before
being retranslocated back across the inner membrane, anchored by the hydrophobic region before being processed. In this "conservative sorting" hypothesis,
the second portion of the targeting sequence is proposed to act as a mitochondrial
export signal (Hartl et al., 1987), in a way similar to a bacterial export signal
from which it is proposed to have evolved. Indeed, the processing site for
cytochrome c_1 resembled the processing site for bacterial leader peptidase (Romisch et al., 1987; Sadler et al., 1984).

Similarly, cytochrome b_2 is proposed to follow a re-export route to the
intermembrane space, and the conservative sorting hypothesis is further supported by evidence that the second processing peptidase for cytochrome b_2
(IMP1) in mitochondria is similar to E. coli signal peptidase (Schneider et al.,
1991; Behrens et al., 1991). In addition, the E. coli chaperone GroEL, which is
involved in the export of bacterial proteins, has its homologue in mitochondria
Hsp60. The folding, assembly, and re-export of proteins that pass through the
matrix would therefore be predicted to be dependent on Hsp60. A yeast Hsp60
mutant accumulates the intermediate form of cytochrome b_2 (Cheng et al.,
1989). However, contradictory evidence suggests that the intermediate form of
cytochrome c_1 does not accumulate in this mutant (Glick and Schatz, 1991).

Also, since Hsp60 is an ATPase, proteins whose assembly or re-export is Hsp60-dependent should require ATP. It seems that the correct localization of a cyto-chrome c_1-DHFR fusion can take place when ATP levels are reduced, suggesting no Hsp60 involvement (Glick and Schatz, 1991).

Finally, import and processing of a cytochrome c_1-DHFR fusion by mito-plasts can take place when the DHFR moiety is bound to methotrexate, thus preventing it from entering the matrix. It therefore remains to be established when cytochrome c_1 follows the re-export pathway in its entirety or whether the intermediate form is transferred from the translocation site while still bound to the inner membrane.

The processing of cytochrome c_1 requires NADH. This may be required for the covalent attachment of heme by cytochrome c_1-heme lyase, which is a prerequisite for the second processing step (Ohashi et al., 1982).

The Rieske Fe/S is also processed in two steps (Hartl et al., 1986), but both cleavage steps appear to take place within the matrix. The protein is then re-exported through the inner membrane, and it is possible that this takes place in a similar manner to the transport of intermembrane-space proteins.

4.1.3. Integral Proteins of the Inner Membrane

Some integral inner-membrane proteins are known to pass through the ma-trix and are processed before being assembled into complexes. These include cytochrome c oxidase subunit IV (Hurt et al., 1984) and the ATP synthase subunit 9 from N. crassa. Subunit 9 has a 66-amino acid presequence, which is processed by the matrix proteinase. It also possesses a hydrophobic stretch within it, yet the precursor is able to pass through both membranes without hindrance (Mahlke et al., 1990). Cytochrome p450 SSC, an integral inner-membrane protein that faces the matrix, also passes through the matrix before being integrated (Ou et al., 1986).

In contrast to these proteins, the adenine nucleotide translocator (ANT) is synthesized without a cleavable presequence, and instead possesses internal tar-geting information (Pfanner et al., 1987b; Smagula and Douglas, 1988). The protein is anchored in the inner membrane by three pairs of amphiphilic helices. However, ANT does appear to require at least part of the general targeting pathway for its transport, interacting as it does with the Mas70p/MOM72 re-ceptor and the General Insertion Protein GIP. ANT does not pass through the matrix and is therefore not dependent on Hsp60 for folding and assembly (Ma-hlke et al., 1990); interestingly, it does require mt-Hsp70. It is possible that there is a direct transfer of the protein from the translocation site to the inner mem-brane.

Inner-membrane proteins that are structurally similar to adenine nucleo-tide translocator include rat Uncoupling Protein (UCP) and the phosphate carrier. Like ANT, UCP has no extreme N-terminal targeting sequence (Aquila *et al.*, 1987) and targeting information is internal (Liu *et al.*, 1990). Phosphate carrier, although a very similar protein, possesses a typical N-terminal presequence in animals (Runswick *et al.*, 1987) though not in yeast (Phelps *et al.*, 1991).

4.2. Targeting of Mitochondrially Encoded Inner-Membrane Proteins

Most mitochondrially encoded polypeptides are membrane associated and assemble into complexes with nuclear-encoded polypeptides to form components of the respiratory pathway. These must have signals that target the proteins to the inner membrane. Given the endosymbiotic origin of mitochondria, this process would be expected to be akin to the export of proteins from bacterial cells (Hartl and Neupert, 1990). In addition to mitochondrially encoded polypeptides following this pathway, it is thought that some nuclear-encoded polypeptides may take this route after being targeted to the mitochondrial matrix (e.g., cytochromes c_1 and b_2; see Section 4.1.2).

4.2.1. Mitochondrial Export Signals

In *S. cerevisiae*, the mitochondrially encoded cytochrome c oxidase subunit II (COXII) is known to be synthesized with a cleavable amino-terminal presequence (Pratje *et al.*, 1983) and, in *N. crassa*, cytochrome c oxidase subunits I and II have presequences (Werner and Bertrand, 1979; Machleidt and Werner, 1979; van den Boogart *et al.*, 1982). COXII is assembled with other subunits of cytochrome c oxidase in the mitochondrial inner membrane, and it is likely that the information contained within its presequence is responsible for targeting it to the inner membrane. In yeast the COXII presequence is composed of 15 amino acids, six of which are polar. The nature of targeting information in other mitochondrially encoded polypeptides is currently unknown.

4.2.2. Requirement for a Membrane Potential

The maturation of COXII can be blocked in an *in vitro* system using mitochondrial energy poisons. The precursor form accumulates under conditions where formation of the transmembrane proton electrochemical gradient ($\Delta\mu H^+$) is blocked, but the protein is nevertheless inserted into the inner membrane. This has led to the proposal that $\Delta\mu H^+$ is required for processing (but not insertion) of COXII (Clarkson and Poyton, 1989). Although $\Delta\mu H^+$ is essential for the biogenesis of polypeptides inserted into the inner membrane both from within the matrix and from outside the organelle (Section 2.4), its mode of action appears quite different for these two groups of proteins.

4.2.3.   Processing of Mitochondrially Encoded Peptides

The precursor to COXII is processed to the mature form by Inner Membrane Proteinase I (IMP1; Schneider *et al.*, 1991). This enzyme of 21.4 kDa is an integral protein of the inner membrane. A yeast mutant lacking the enzyme cannot process COXII, nor the nuclear-encoded cytochrome b_2 (Pratje and Guiard, 1986). The gene for IMPI (PET2528) is nuclear, and the predicted amino acid sequence shows some similarity to *E. coli* leader peptidase, supporting the idea that the mitochondrial export pathway may have evolved from a bacterial protein export pathway (Behrens *et al.*, 1991).

4.3.   Assembly of Proteins into Enzyme Complexes

Assembly of mitochondrial proteins into complexes has recently received a great deal of attention. This has been made possible by the availability of techniques to introduce point mutations or deletions into yeast genes, and to express them *in vivo*, thereby facilitating the identification of assembly defects. In addition, yeast mutants that are defective in the assembly of one or more complexes have been isolated. For example, two yeast genes required for the function of cytochrome *c* oxidase have been identified (Nobrega *et al.*, 1990; Tzagaloff *et al.*, 1990). Their products are mitochondrially located, but are not part of the cytochrome *c* oxidase complex, and may be required for assembly, posttranslational modifications or the addition of the heme and copper redox centers.

Assembly of the yeast ubiquinol-cytochrome *c* reductase complex has been studied in some detail (Grivell, 1989). By mutating or deleting one or more of the eight nuclear-encoded subunits, it has been possible to determine the role of each subunit in the stability and assembly of the complex. In addition, the contribution of components to the enzyme's catalytic activity has been studied.

The assembly of both ATP synthase and cytochrome *c* oxidase has been shown to occur in a sequential order (Pfanner and Neupert, 1987). *In vitro*, the assembly of F_1 ATPase occurs from an intramitochondrial pool of subunits (Lewin and Norman, 1983).

5.   LIPID SYNTHESIS AND LOCALIZATION

5.1.   Composition

Mitochondrial membranes contain a specific mix of lipids required for membrane function (Daum, 1985). The composition of mitochondrial membranes varies slightly in different cell types, but in all cases the major lipid constituents are phosphatidylcholine (PC), phosphatidylethanolamine (PE), cardiolipin (CL), and phosphatidylinositol (PI), with lesser amounts of cholesterol (or ergosterol in yeast) and phosphatidylserine (PS). (See Table III.) The phos-

Table III
Lipid Composition of Mitochondrial Membranes

Lipid	Yeast[a]	Rat Liver[b]
Phosphatidylcholine	40.2%	40.3%
Phosphatidylethanolamine	26.5%	34.6%
Phosphatidylinositol	14.6%	4.6%
Phosphatidylserine	3.0%	0.7%
Cardiolipin	13.3%	17.8%
Ergosterol (mol/mol phospholipid)	0.2	—
Cholestorol (mol/mol phospholipid)	—	0.03

Data expressed as percent of total phospholipid.
[a]Data from Zinser et al., 1991.
[b]Data from van Meer, 1989.

pholipids in mitochondrial membranes contain slightly lower amounts of saturated fatty acids than those in the plasma membrane (Colbeau et al., 1971). Overall, the composition of mitochondrial membranes means that they are more fluid than membranes such as the plasma membrane (Zinser et al., 1991). The relationship between composition and function is not fully understood, but altered membrane composition can affect respiration and ion movement (Daum, 1985). Acidic phospholipids such as CL may be involved in the translocation of mitochondrial precursor proteins (Eilers et al., 1989). Differences in percentage lipid composition are also found between the outer and inner membranes of the mitochondria, most notably that the inner membrane has higher levels of CL and the outer membrane contains more PI (Sperka-Gottlieb et al., 1988). In addition, factors such as senescence affect the composition of lipids in mitochondria and may be the cause of loss of activity of some enzymes (Paradies et al., 1992; Vorbeck et al., 1982).

5.2. Site of Synthesis of Lipids

The major steps in the biosynthesis of most of the lipid constituents of mitochondrial membranes have been shown by fractionation of cell membranes to be localized in either the endoplasmic reticulum (ER) or the mitochondria (Dennis and Kennedy, 1972; Cobon et al., 1974). For some lipids, there seems to be more than one site of synthesis, and the situation appears to vary between different species and different cell types.

5.2.1. Lipids Synthesized in the Mitochondria

Phospholipids required specifically for mitochondrial function (PE and CL) are synthesized within the organelle (Daum, 1985). PE is synthesized in the mitochondrial inner membrane by decarboxylation of PS (Dennis and Kennedy,

1972; Zborowski *et al.*, 1983; Kuchler *et al.*, 1986). CL is also synthesized in the inner membrane, as is its precursor phosphatidylglycerol (Kuchler *et al.*, 1986; van Meer, 1989).

5.2.2. Lipids Synthesized in the Endoplasmic Reticulum

The endoplasmic reticulum is the main site of synthesis for PC, PS, PI, and cholesterol, and PI is synthesized here as well as in the mitochondria. PC is synthesized in the ER of animal cells from CDP-choline and is also synthesized from PE by phosphatidylethanolamine *N*-methyltransferase (Vance and Vance, 1985; Kuchler *et al.*, 1986; Zinser *et al.*, 1991). PS is synthesized in the ER of mammalian cells from other phospholipids by head-group exchange (Bishop and Bell, 1988) and, in yeast, from serine and CDP-diacylglycerol (Kuchler *et al.*, 1986). PI is synthesized in the ER in many animal cells but, in yeast, PI synthase has been detected in the mitochondrial outer membrane (Kuchler *et al.*, 1986). More recently, however, PI synthase has been identified not in the mitochondria but in the ER and secretory vesicles (Zinser *et al.*, 1991).

5.2.3. Other Extramitochondrial Sites of Synthesis

A separate membrane fraction involved in phospholipid biosynthesis has recently been identified from mammalian and yeast cells using improved membrane fractionation techniques (Vance, 1990; Zinser *et al.*, 1991). A mitochondria-associated membrane fraction that is distinct from both the ER and the mitochondrial outer membrane in rat liver cells has been shown to contain PS synthase (Vance, 1990). PE *N*-methyltransferase, an enzyme of phosphatidylcholine biosynthesis, is found in this fraction as well as in the ER. In *S. cerevisiae*, PS synthase activity was originally detected in the mitochondrial outer membrane as well as in the ER (Kuchler *et al.*, 1986), but more recently it has been shown to occur in a nonmitochondrial membrane fraction (Zinser *et al.*, 1991).

5.3. Transport of Lipids to Mitochondria

Lipids that are synthesized extramitochondrially need to be transported to the organelle in a specific manner, and a mechanism must exist for this transport to take place (for review see Voelker, 1991b). By using radioactive lipids and fluorescent markers, it has been possible to study the extent of this lipid traffic *in vivo* and *in vitro*.

5.3.1. Transport of Lipids in the Cell

Three mechanisms have been proposed for the transfer of phospholipids from their site of synthesis to other membranes (e.g., the plasma membrane).

First, specific or nonspecific transfer proteins, several of which have been isolated, may be involved in the *in vivo* movement of lipids (Wirtz, 1991). These
carriers need to promote not only exchange of lipid but also a net transfer to the
acceptor membrane. The PI transfer protein (PIT) has been isolated from *S.
cerevisiae*, the gene identified and shown to be identical to Sec14 (Bankaitis *et
al.*, 1990). Mutations in this gene block protein secretion at the level of the Golgi
apparatus. Second, lipids could be transferred laterally via zones of transient
continuity between donor and acceptor membranes. Third, the transfer could be
achieved by the budding of vesicles from the donor membrane followed by
fusion to the recipient membrane, a phenomenon observed extensively in the
secretory pathway in eukaryotes. These types of transfer may be specific for
different lipids or different acceptor organelles.

5.3.2. Transport to the Mitochondria

To investigate the mechanism of mitochondrial lipid transport, their movement from the ER to mitochondria has been studied. After synthesis, PC is
rapidly transported to the mitochondrial outer membrane into which it is fully
equilibrated in 5 min (Yaffe and Kennedy, 1983). As this process is so rapid, the
mechanism is thought to involve phospholipid transfer proteins that have specificity for PC (reviewed in Helmkamp, 1986). This rapid transport appears to be
particularly important as mitochondria cannot synthesize their own phosphatidylcholine. In comparison, PE transport to the mitochondria is slower and takes
up to 2 hr. Mitochondria can synthesize PE from PS in the inner membrane, and
it seems that the main flow of PE is away from the organelle. Conversely, PS
must be transferred from the ER to the mitochondrial inner membrane in order to
be decarboxylated. This transport process is much slower, taking up to 7 hr, and
can be blocked by ATP-depletion (Voelker, 1989a). A membrane-bound intermediate may be involved in the transport of PS in permeabilized cells (Voelker,
1991a).

Import of PS has also been studied with isolated mitochondria from rat liver
(Vance, 1990; Voelker, 1989b), and it appears that the rate-limiting step is the
transfer of lipid from the donor membrane to the mitochondrial outer membrane.
Uncouplers do not affect this process, nor is ATP required. In a two-step model,
the ATP-dependent step brings the donor membrane into close proximity with the
mitochondrial outer membrane. The transfer of lipid into the mitochondrial outer
membrane is ATP-independent, and is thought to occur by collision of the donor
and acceptor membranes. It has been proposed that the donor membrane could be
the mitochondrially associated membranes described by Vance (1990). Their
association with the mitochondria of liver cells and the fact that they are able to
transfer PS to the mitochondria *in vitro* makes them likely candidates.

Cholesterol has been shown to be transported to the mitochondrial outer membrane in adrenal tissue, and from there to the inner membrane. This process is thought to involve the sterol carrier protein 2 (SCP-2) at both stages of transport (Scallen et al., 1985). This nonspecific lipid transfer protein can enhance cholesterol uptake by isolated mitochondria (Chanderbhan et al., 1982).

In summary, no single mechanism explains the movement of lipids to the mitochondrial membranes. Several aspects remain to be established: whether vesicles play a role in this transport; the nature of the requirement for ATP; and the number and nature of lipid transfer proteins that are involved.

5.4. Transfer between Outer and Inner Membranes

In yeast, PS is transported from the outer to the inner membrane of mitochondria where it is decarboxylated to PE. The PE is then transported back to the outer membrane. This process requires neither ATP (Voelker, 1991b) nor a membrane potential (Simbeni et al., 1990). It has been suggested that PS transfer may occur at contact sites formed where the outer and inner membranes come together (Simbeni et al., 1990, 1991; Ardail et al., 1991). Contact sites have been isolated and have been shown to contain a lipid composition distinct from either the outer or inner membrane (Simbeni et al., 1991). The role of contact sites in mitochondrial protein import is well established and described above (Section 2.4). In addition, adriamycin inhibits mitochondrial uptake of PS, as well as inhibiting mitochondrial protein import (Voelker, 1991b; Eilers et al., 1989). The site of action of this drug is thought to be between the inner and outer membranes (Voelker, 1991b).

PI is also transported between the outer and inner membrane. In yeast, PI synthase is thought to be associated with the outer membrane, and the transfer of PI to the inner membrane appears to be partially coupled to its synthesis (Simbeni et al., 1990).

Cholesterol is transported from the outer membrane to the inner side of the inner membrane in adrenal tissue where it is used in the biosynthesis of steroids (reviewed in van Amerongen et al., 1989). However, in these cells transport can be blocked by cholesterol sulfate, thus adding to the evidence for a regulated tissue-specific cholesterol translocator.

5.5. Regulation of Lipid Synthesis and Incorporation

As the function of mitochondrial membranes is dependent on the correct lipid composition, it is vital that regulation of synthesis and incorporation is achieved. The addition of the correct ratios of each lipid to the mitochondrial membranes is essential for growth of mitochondria, and its regulation plays an important role in mitochondrial division. In HeLa cells, the total amount of mitochondrial membrane and the number of mitochondria vary very little throughout the cell cycle (Posakony *et al.*, 1977), suggesting some coordinate regulation with other cell membranes. Moreover, regulation of lipid biosynthesis and incorporation should be able to accommodate changes that mitochondria undergo during cell differentiation or metabolic adaptation, where the growth of mitochondrial membranes exceeds that of other membranes (Lang and Herbener, 1972). In yeast, where changes in mitochondrial morphology can be extreme (Pon and Schatz, 1991), phospholipid synthesis is regulated both by control of transcription of the genes encoding biosynthetic enzymes and by control of enzyme activity (Carman and Henry, 1989).

Pools of lipids are known to exist within cells—for example, PE, PC (Vance and Vance, 1986), and cholesterol (Khan *et al.*, 1989). These reservoirs may be required for the transport of lipids within the cell, and may also serve some regulatory role.

6. CONCLUSIONS

The biogenesis of mitochondrial membranes depends upon the close cooperation between two genomes. Most components of mitochondria are synthesized outside the organelle and specifically transported not only to the mitochondrion but to a particular subcompartment. Recent advances have greatly improved our understanding of both the nature of the targeting signals that direct proteins to mitochondria and of the pathway for their recognition and translocation across or into the outer and inner membranes. The picture that emerges is of a highly dynamic and flexible pathway with several options open to individual polypeptides during their transport: The options taken at each branch point in the general pathway determine the ultimate location of the polypeptide within the mitochondrion.

As yet we know little of the molecular basis for the recognition of targeting sequences that are poorly conserved with no simple sequence motif. Recognition must be very different from other well-characterized protein:protein interactions such as those between antibodies and antigens. Furthermore, our understanding of the molecular mechanism of transmembrane movement of proteins is extremely vague. With the intensity of current research in these and other areas of mitochondrial biogenesis, the situation will clarify over the next few years.

REFERENCES

Allison, D. S., and Schatz, G., 1986, Artificial mitochondrial presequences, *Proc. Natl. Acad. Sci. USA* **83**:9011–9015.

Aquila, H., Link, T. A., and Klingenberg, M., 1987, Solute carriers in energy transfer of mitochondria form a homologous protein family, *FEBS Letts.* **212**:1–9.

Ardail, D., Lerme, F., and Louisot, P., 1991, Involvement of contact sites in phosphatidylserine import into liver mitochondria, *J. Biol. Chem.* **266**:7978–7981.

Atencio, D. P., and Yaffe, M. P., 1992, MAS5, a yeast homolog of DnaJ involved in mitochondrial protein import, *Mol. Cell Biol.*, **12**:283–291.

Baker, K. P., Schaniel, A., Vestweber, D., and Schatz, G., 1990, A yeast mitochondrial outer membrane protein essential for protein import and cell viability, *Nature* **348**:605–609.

Bankaitis, V. A., Aitken, J. R., Cleves, A. E., and Dowhan, W., 1990, An essential role for a phospholipid transfer protein in yeast Golgi fraction, *Nature* **347**:561–562.

Behrens, M., Michaelis, G., and Pratje, E., 1991, Mitochondrial inner membrane protease I of *Saccharomyces cerevisiae* shows sequence similarity to the *Escherichia coli* leader peptidase, *Mol. Gen. Genet.* **228**:167–176.

Bishop, W. R., and Bell, R. M., 1988, Assembly of phospholipids into cellular membranes: Biosynthesis, transmembrane movement and intracellular translocation, *Annu. Rev. Cell Biol.* **4**:579–610.

Böhni, P. C., Daum, G., and Schatz, G., 1983, Import of proteins into mitochondria: Partial purification of a matrix-located protease involved in cleavage of mitochondrial precursor polypeptides, *J. Biol. Chem.* **258**:4937–4943.

Carman, G. M., and Henry, S. A., 1989, Phospholipid biosynthesis in yeast, *Annu. Rev. Biochem.* **58**:635–669.

Chanderbhan, R., Noland, B. J., Scallen, T. J., and Vahouny, G. V., 1982, Sterol carrier protein 2, *J. Biol. Chem.* **257**:8928–8934.

Chen, W.-J., and Douglas, M. G., 1987, Phosphodiester bond cleavage outside the mitochondria is required for the completion of protein import into the mitochondrial matrix, *Cell* **49**:651–658.

Cheng, M. Y., Hartl, F.-U., Martin, J., Pollock, R. A., Kalousek, F., *et al.*, 1989, Mitochondrial heat-shock protein hsp60 is essential for assembly of proteins into mitochondria, *Nature* **337**:620–625.

Clarkson, G. H. D., and Poyton, R. O., 1989, A role for membrane potential in the biogenesis of cytochrome *c* oxidase subunit II, a mitochondrial gene product, *J. Biol. Chem.* **264**:10114–10118.

Cobon, G. S., Crowfoot, P. D., and Linnane, A. W., 1974, Biogenesis of mitochondria: Phospholipid synthesis in vitro by yeast mitochondrial and microsomal fractions, *Biochem. J.* **144**:265–275.

Colbeau, A., Nachbaur, J., and Vignais, P. M., 1971, Enzymic characterization and lipid composition of rat liver subcellular membranes, *Biochim. Biophys. Acta* **249**:462–492.

Costanzo, M. C., and Fox, T. D., 1990, Control of mitochondrial gene expression in *Saccharomyces cerevisiae*, *Annu. Rev. Genet.* **24**:91–113.

Craig, E. A., Kramer, J., and Kosic-Smith, J., 1987, SSC1, a member of the 70-kDa heat shock protein multigene family of *Saccharomyces cerevisiae* is essential for growth, *Proc. Natl. Acad. Sci. USA* **84**:4156–4160.

Daum, G., 1985, Lipids of mitochondria, *Biochim. Biophys. Acta* **822**:1–42.

Dennis, E. A., and Kennedy, E. P., 1972, Intracellular sites of lipid synthesis and the biogenesis of mitochondria, *J. Lipid Res.* **13**:263–267.

Deshaies, R. J., Koch, B. D., Werner-Washburne, M., and Craig, E. A., 1988, A sub-family of stress proteins facilitates translocation of secretory and mitochondrial precursor polypeptides, *Nature* **332**:800–805.

Douglas, M. G., Geller, B. L., and Emr, S. D., 1984, Intracellular targeting and import of and F_1ATPase β-subunit-β-galactosidase hybrid protein into yeast mitochondria, *Proc. Natl. Acad. Sci. USA* **81**:3983–3987.

Eilers, M., and Schatz, G., 1986, Binding of a specific ligand inhibits import of a purified precursor during import into isolated mitochondria, *Nature* **322**:228–232.

Eilers, M., Endo, T., and Schatz, G., 1989, Adriamycin, a drug interacting with acid phospholipids, blocks import of precursor proteins by isolated yeast mitochondria, *J. Biol. Chem.* **264**:2945–2950.

Ellis, E. M., and Reid, G. A., 1993, The *MTS1* gene from *Saccharomyces cerevisiae* encodes a putative RNA-binding protein involved in mitochondrial protein targeting, *Gene* **132**:175–183.

Ellis, R. J., 1987, Proteins as molecular chaperones, *Nature* **328**:378–379.

Fox, T. D., 1986, Nuclear gene-products required for translation of specific mitochondrially coded messenger RNAs in yeast, *Trends Genet.* **2**:97–100.

Freitag, H., Janes, M., and Neupert, W., 1982, Biosynthesis of mitochondrial porin and insertion into the outer mitochondrial membrane of *Neurospora crassa, Eur. J. Biochem.* **126**:197–202.

Gasser, S. M., Daum, G., and Schatz, G., 1982a, Import of proteins into mitochondria: Energy-dependent uptake of precursors by isolated mitochondria, *J. Biol. Chem.* **257**:13034–13041.

Gasser, S. M., Ohashi, A., Daum, G., Böhni, P. C., Gibson, J., Reid, G. A., Yonetani, T., and Schatz, G., 1982b, Imported mitochondrial proteins cytochrome b_2 and cytochrome c_1 are processed in two steps, *Proc. Natl. Acad. Sci. USA* **79**:267–271.

Gillespie, L. L., Argan, C., Taneja, A. T., Hodges, R., Freeman, K. B., and Shore, G. C., 1985, A synthetic signal peptide blocks import of precursors destined for the mitochondrial inner membrane or matrix, *J. Biol. Chem.* **260**:16045–16048.

Glick, B., and Schatz, G., 1991, Import of proteins into mitochondria, *Annu. Rev. Genet.* **25**:21–44.

Glick, B., Wachter, C., and Schatz, G., 1991, Protein import into mitochondria: Two systems acting in tandem? *Trends Cell Biol.* **1**:99–103.

Grivell, L. A., 1989, Nucleo-mitochondrial interactions in yeast mitochondrial biogenesis, *Eur. J. Biochem.* **182**:477–493.

Hamajima, S., Sakaguchi, N., Mihara, K., Ono, S., and Sato, R., 1988, Both amino- and carboxy-terminal portions are required for insertion of yeast porin into the outer mitochondrial membrane, *J. Biochem.* **104**:362–367.

Hartl, F.-U., and Neupert, W., 1990, Protein sorting to mitochondria: Evolutionary conservation of folding and assembly, *Science* **247**:930–938.

Hartl, F.-U., Schmidt, B., Wachter, E., Weiss, H., and Neupert, W., 1986, Transport into mitochondria and intramitochondrial sorting of the Fe/S protein of ubiquinol-cytochrome c reductase, *Cell* **47**:939–951.

Hartl, F.-U., Ostermann, J., Guiard, B., and Neupert, W., 1987, Successive translocation into and out of the mitochondrial matrix: Targeting of proteins to the intermembrane space by a bipartite signal peptide, *Cell* **51**:1027–1037.

Hase, T., Muller, U., Riezman, H., and Schatz, G., 1984, A 70-kd protein of the yeast mitochondrial outer membrane is targeted and anchored via its extreme amino terminus, *EMBO J.* **3**:3157–3164.

Hase, T., Nakai, M., and Matsubara, H., 1986, The N-terminal 21 amino acids of a 70 kDa protein of the yeast mitochondrial outer membrane direct *Escherichia coli* beta-galactosidase into the mitochondrial matrix in yeast cells, *FEBS Lett.* **197**:199–203.

Hatefi, Y., 1985, The mitochondrial electron transport and oxidative phosphorylation system, *Annu. Rev. Biochem.* **54**:1015–1069.

Hawlitschek, G., Schneider, H., Schmidt, B., Tropschug, M., Hartl, F.-U., and Neupert, W., 1988, Mitochondrial protein import: Identification of processing peptidase and of PEP, a processing enhancing protein, *Cell* **53**:795–806.

Helmkamp, G. M. Jr., 1986, Phospholipid transfer proteins: Mechanism of action, *J. Bioenerg. Biomembr.* **18**:71–91.

Hines, V., Brandt, A., Griffiths, G., Horstmann, H., Brutsch, H., and Schatz, G., 1990, Protein import into yeast mitochondria is accelerated by the outer membrane protein MAS70, *EMBO J.* 9:3191–3200.

Hurt, E. C., and van Loon, A. P. G. M., 1976, How proteins find mitochondria and intramitochondrial compartments, *Trends Biochem. Sci.* 11:204–207.

Hurt, E. C., Pesold-Hurt, B., and Schatz, G., 1984, The amino terminal region of an imported mitochondrial precursor polypeptide can direct cytoplasmic dihydrofolate reductase into the mitochondrial matrix, *EMBO J.* 3:3149–3156.

Hwang, S. T., Wachter, C., and Schatz, G., 1991, Protein import into the yeast mitochondrial matrix—a new translocation intermediate between the two mitochondrial membranes, *J. Biol. Chem.* 266:21083–21089.

Kang, P. J., Ostermann, J., Schilling, J., Neupert, W., Craig, E. A., and Pfanner, N., 1990, Hsp70 in the mitochondrial matrix is required for translocation and folding of precursor proteins, *Nature* 348:137–143.

Kaput, J., Goltz, S., and Blobel, G., 1982, Nucleotide sequence of the yeast nuclear gene for cytochrome *c* peroxidase precursor: Functional implications of the presequence for protein transport into mitochondria, *J. Biol. Chem.* 257:15054–15058.

Kellems, R. E., Allison, V. F., and Butow, R. A., 1975, Cytoplasmic type 80S ribosomes associated with yeast mitochondria. IV: Attachment of ribosomes to the outer membrane of isolated mitochondria, *J. Cell Biol.* 65:1–14.

Keng, T., Alami, E., and Guarente, L., 1986, The nine aminoterminal residues of δ-aminolevulinate synthase direct β-galactosidase into the mitochondrial matrix, *Mol. Cell. Biol.* 6:355–364.

Khan, B., Wilcox, H. G., and Heimberg, M., 1989, Cholesterol is required for secretion of very low density lipoprotein by rat liver, *Biochem. J.* 259:807–816.

Kiebler, M., Pfaller, R., Söllner, T., Griffiths, G., Horstmann, H., et al., 1990, Identification of a mitochondrial receptor complex required for recognition and membrane insertion of precursor proteins, *Nature* 348:610–616.

Kuchler, K., Daum, G., and Paltauf, F., 1986, Subcellular and submitochondrial localization of phospholipid synthesizing enzyme in *Saccharomyces cerevisiae*, *J. Bacteriol.* 165:901–910.

Lang, C. A., and Herbener, G. H., 1972, Quantitative comparison of the mitochondrial populations in the liver of newborn and weanling rats, *Dev. Biol.* 29:176–182.

Lewin, A. S., and Norman, D. K., 1983, Assembly of F_1-ATPase in isolated mitochondria, *J. Biol. Chem.*, 258:6750–6755.

Liu, X., Freeman, K. B., and Shore, G. C., 1990, An amino terminal signal sequence abrogates the intrinsic membrane-targeting information of mitochondrial uncoupling protein, *J. Biol. Chem.* 265:9–12.

Lustig, A., Levens, D., and Rabinowitz, M., 1982, The biogenesis and regulation of yeast mitochondrial RNA polymerase, *J. Biol. Chem.* 257:5800–5808.

Machleidt, W., and Werner, S., 1979, Is the mitochondrially made subunit 2 of cytochrome oxidase synthesized as a larger precursor in *Neurospora crassa*? *FEBS Lett.* 107:327–330.

Mahlke, K., Pfanner, N., Martin, J., Horwich, A. L., Hartl, F.-U., and Neupert, W., 1990, Sorting pathways of mitochondrial inner membrane proteins, *Eur. J. Biochem.* 192:551–555.

McAda, P., and Douglas, M. G., 1982, A neutral metallo-endoprotease involved in the processing of an F_1 ATPase subunit precursor in mitochondria, *J. Biol. Chem.* 257:3177–3182.

McGraw, P., and Tzagaloff, A., 1983, Assembly of the mitochondrial membrane system: Characterization of a yeast nuclear gene involved in the processing of the cytochrome *b* pre-messenger RNA, *J. Biol. Chem.* 258:9459–9468.

Murakami, K., and Móri, M., 1990, Purified presequence binding factor (PBF) forms an import-competent complex with a mitochondrial precursor protein, *EMBO J.* 9:3201–3208.

Murakami, K., Tokunaga, F., Iwanaga, S., and Mori, M., 1990, Presequence does not prevent folding of a purified mitochondrial precursor protein and is essential for association with a reticulocyte lysate cytosolic factor(s), *J. Biochem.*, **108**:207–214.

Nelson, N., and Schatz, G., 1979, Energy-dependent processing of cytoplasmically made precursors to mitochondrial proteins, *Proc. Natl. Acad. Sci. USA* **76**:4365–4369.

Neupert, W., Hartl, F.-U., Craig, E. A., and Pfanner, N., 1990, How do polypeptides cross the mitochondrial membranes? *Cell* **63**:447–450.

Nicholson, D. W., Stuart, R. A., and Neupert, W., 1989, Biogenesis of cytochrome c_1: Role of cytochrome c_1 heme lyase and of the two proteolytic processing steps during import into mitochondria, *J. Biol. Chem.* **264**:10156–10168.

Nobrega, M. P., Nobrega, F. C., and Tzagaloff, A., 1990, *COX10* codes for a protein homologous to the *ORF1* product of *Paracoccus denitrificans* and is required for the synthesis of yeast cytochrome oxidase, *J. Biol. Chem.* **265**:14220–14226.

Ohashi, A., Gibson, J., Gregor, I., and Schatz, G., 1982, Import of proteins into mitochondria: The precursor of cytochrome c_1 is processed in two steps, one of them heme-dependent, *J. Biol. Chem.* **257**:13042–13047.

Ohba, M., and Schatz, G., 1987, Protein import into yeast mitochondria is inhibited by antibodies raised against 45-kd proteins of the outer membrane, *EMBO J.* **6**:2109–2115.

Ono, H., and Tuboi, S., 1990a, Purification and identification of a cytosolic factor required for import of precursors of mitochondrial proteins into mitochondria, *Arch. Biochem. Biophys.* **280**:299–304.

Ono, H., and Tuboi, S., 1990b, Purification of the putative import-receptor for the precursor of the mitochondrial protein, *J. Biochem.* **107**:840–847.

Ostermann, J., Horwich, A. L., Neupert, W., and Hartl, F.-U., 1989, Protein folding in the mitochondria requires complex formation with hsp60 and ATP hydrolysis, *Nature* **341**:125–130.

Ostermann, J., Voos, W., Kang, P. J., Craig, E. A., Neupert, W., and Pfanner, N., 1990, Precursor proteins in transit through mitochondrial contact sites interact with hsp70 in the matrix, *FEBS Lett.* **277**:281–284.

Ou, W., Ito, A., Morohashi, K., Fujikuriyama, Y., and Omura, T., 1986, Processing-independent in vitro translocation of cytochrome P-450(SSC) precursor across mitochondrial membranes, *J. Biochem.* **100**:1287–1296.

Pain, D., Murakami, H., and Blobel, G., 1990, Identification of a receptor for protein import into mitochondria, *Nature* **347**:444–449.

Paradies, G., Ruggico, F. M., and Dinoi, P., 1992, Decreased activity of the phosphate carrier and modification of lipids in cardiac mitochondria from senescent rats, *Int. J. Biochem.* **24**:783–787.

Parikh, V. S., Morgan, M. M., Scott, R., Clements, L. S., and Butow, R. A., 1987, The mitochondrial genotype can influence nuclear gene expression in yeast, *Science* **235**:576–580.

Pfaller, R., and Neupert, W., 1987, High-affinity binding sites involved in the import of porin into mitochondria, *EMBO J.* **6**:2635–2642.

Pfanner, N., Hartl, F.-U., Guiard, B., and Neupert, W., 1987a, Mitochondrial precursor proteins are imported through a hydrophilic membrane environment, *Eur. J. Biochem.* **169**:289–293.

Pfanner, N., Hartl, F.-U., and Neupert, W., 1988, Import of proteins into mitochondria: A multi-step process, *Eur. J. Biochem.* **175**:205–212.

Pfanner, N., Hoeben, P., Tropschug, M., and Neupert, W., 1987b, The carboxy-terminal two-thirds of the ADP/ATP carrier polypeptide contain sufficient information to direct translocation into mitochondria, *J. Biol. Chem.* **262**:14851–14854.

Pfanner, N., and Neupert, W., 1985, Transport of proteins into mitochondria: A potassium diffusion potential is able to drive the import of ADP/ATP carrier, *EMBO J.* **4**:2819–2825.

Pfanner, N., and Neupert, W., 1986, Transport of F_1ATPase subunit β into mitochondria depends on both a membrane potential and nucleoside triphosphates, *FEBS Lett.* **209**:152–156.

Pfanner, N., and Neupert W., 1987, Biogenesis of mitochondrial energy transducing complexes, *Curr. Top. Bioenerg.* **15:**177–219.

Pfanner, N., and Neupert, W., 1990, The mitochondrial protein import apparatus, *Annu. Rev. Biochem.* **59:**331–353.

Pfanner, N., Rassow, J., van der Klei, I. J., and Neupert, W., 1992, A dynamic model of the mitochondrial import machinery, *Cell* **68:**999–1002.

Pfanner, N., Tropschug, M., and Neupert, W., 1987c, Mitochondrial protein import: Nucleoside triphosphates are involved in conferring import competence to precursors, *Cell* **49:**815–823.

Phelps, A., Schobert, C. T., and Wohlrab, H., 1991, Cloning and characterization of the mitochondrial phosphate transport protein gene from the yeast *Saccharomyces cerevisiae, Biochemistry* **30:**248–252.

Pollock, R. A., Hartl, F.-U., Cheng, M. Y., Ostermann, J., Horwich, A., and Neupert, W., 1988, The processing peptidase of yeast mitochondria: The two co-operating components MPP and PEP are structurally related, *EMBO J.* **7:**3493–3500.

Pon, L., and Schatz, G., 1991, Biogenesis of yeast mitochondria, in *The Molecular and Cellular Biology of the Yeast Saccharomyces: Genome Dynamics, Protein Synthesis and Energetics,* pp. 333–406, Cold Spring Harbor Laboratory Press, Cold Spring Harbor, N.Y.

Posakony, J. W., England, J. M., and Attardi, G., 1977, Mitochondrial growth and division during the cell cycle in HeLa cells, *J. Cell Biol.* **74:**468–491.

Pratje, E., and Guiard, B., 1986, One nuclear gene controls the removal of transient presequences from two yeast proteins: One encoded by the nuclear, the other the mitochondrial genome, *EMBO J.* **5:**1313–1317.

Pratje, E., Mannhaupt, G., Michaelis, G., and Beyreuther, K., 1983, A nuclear mutation prevents processing of a mitochondrially encoded membrane protein in *Saccharomyces cerevisiae, EMBO J.* **2:**1049–1054.

Rassow, J., Guiard, B., Wienhues, U., Herzog, V., Hartl, F.-U., and Neupert, W., 1989, Translocation arrest by reversible folding of a precursor protein imported into mitochondria: A means to quantitate translocation contact sites, *J. Cell Biol.* **109:**1421–1428.

Reid, G. A., and Schatz, G., 1982, Import of protein into mitochondria: Yeast cells grown in the presence of carbonylcyanide *m*-chlorophenylhydrazone accumulate massive amounts of some mitochondrial precursor polypeptides, *J. Biol. Chem.* **257:**13056–13061.

Riezman, H., Hay, R., Witte, C., Nelson, N., and Schatz, G., 1983, Yeast mitochondrial outer membrane specifically binds cytoplasmically-synthesized precursors of mitochondrial proteins, *EMBO J.* **2:**1113–1118.

Roise, D., Horvath, S. J., Tomich, J., Richards, J. H., and Schatz, G., 1986, A chemically synthesized presequence of an imported mitochondrial protein can form an amphiphilic helix and perturb artificial phospholipid bilayers, *EMBO J.* **5:**1327–1334.

Roise, D., Theiler, F., Horvath, S. J., Tomich, J. M., Richards, J. H., Allison, D. S., and Schatz, G., 1988, Amphiphilicity is essential for mitochondrial presequence function, *EMBO J.* **7:**649–653.

Romisch, J., Tropschug, M., Sebald, W., and Weiss, H., 1987, The primary structure of cytochrome c_1 from *Neurospora crassa, Eur. J. Biochem.* **164:**111–115.

Runswick, M. J., Powell, S. J., Nyren, P., and Walker, J. E., 1987, Sequence of the bovine mitochondrial phosphate carrier: Structural relationship to ADP/ATP translocase and the brown fat mitochondria uncoupling protein, *EMBO J.* **6:**1367–1374.

Sadler, I., Suda, K., Schatz, G., Kaudewitz, F., and Haid, A., 1984, Sequencing of the nuclear gene for the yeast cytochrome c_1 precursor reveals an unusually complex amino-terminal presequence, *EMBO J.* **3:**2137–2143.

Scallen, T. J., Pastuszyn, A., Noland, B. J., Chanderbhan, R., Kharroubi, A., and Vahouny, G. V., 1985, Sterol carrier and lipid transfer proteins, *Chem. Phys. Lipids* **38:**239–261.

Scherer, P. E., Krieg, U. C., Hwang, S. T., Vestweber, D., and Schatz, G., 1990, A precursor protein partly translocated into yeast mitochondria is bound to a 70-kd mitochondrial stress protein, *EMBO J.* 9:4315–4322.

Schleyer, M., and Neupert, W., 1985, Transport of proteins into mitochondria: Translocation intermediates spanning contact sites between outer and inner membranes, *Cell* 43:339–350.

Schneider, A., Behrens, M., Scherer, P., Pratje, E., Michaelis, G., and Schatz, G., 1991, Inner membrane protease I, an enzyme mediating intramitochondrial protein sorting in yeast, *EMBO J.* 10:247–254.

Schulte, U., Arretz, M., Schneider, H., Tropschug, M., Wachter, E., Neupert, W., and Weiss, H., 1989, A family of mitochondrial proteins involved in bioenergetics and biogenesis, *Nature* 339:147–149.

Schwaiger, M., Herzog, V., and Neupert, W., 1987, Characterization of translocation contact sites involved in the import of mitochondrial proteins, *J. Cell Biol.* 105:235–246.

Simbeni, R., Paltauf, F., and Daum, G., 1990, Intramitochondrial transfer of phospholipids in the yeast *Saccharomyces cerevisiae*, *J. Biol. Chem.* 265:281–285.

Simbeni, R., Pon, L., Zinser, E., Paltauf, F., and Daum, G., 1991, Mitochondrial membrane contact sites of yeast: Characterization of lipid components and possible involvement in intramitochondrial translocation of phospholipids, *J. Biol. Chem.* 266:10047–10049.

Simon, S. M., Peskin, C. S., and Oster, G. F., 1992, What drives the translocation of proteins? *Proc. Natl. Acad. Sci. USA* 89:3770–3774.

Smagula, C., and Douglas, M. G., 1988, Mitochondrial import of the ADP/ATP carrier protein in *Saccharomyces cerevisiae:* Sequences required for receptor-binding and membrane translocation, *J. Biol. Chem.* 263:6783–6790.

Söllner, T., Griffiths, G., Pfaller, R., and Neupert, W., 1989, MOM19, an import receptor for mitochondrial precursor proteins, *Cell* 59:1061–1070.

Söllner, T., Pfaller, R., Griffiths, G., Pfaller, R., and Neupert, W., 1990, A mitochondrial import receptor for the ADP/ATP carrier, *Cell* 62:107–115.

Sperka-Gottlieb, C. D. M., Hermetter, A., Paltauf, F., and Daum, G., 1988, Lipid topology and physical properties of the outer mitochondrial membrane of the yeast *Saccharomyces cerevisiae*, *Biochim. Biophys. Acta* 946:227–234.

Tamm, L. K., 1986, Incorporation of a synthetic mitochondrial signal peptide into charged and uncharged phospholipid monolayers, *Biochemistry* 25:7470–7476.

Tzagaloff, A., 1982, *Mitochondria,* Plenum Press, New York.

Tzagaloff, A., Capitanio, N., Nobrega, M. P., and Gatti, D., 1990, Cytochrome oxidase assembly in yeast requires the product of *COX11*, a homolog of the *P. denitrificans* protein encoded by *ORF3*, *EMBO J.* 9:2759–2764.

van Amerongen, A., van Noort, M., van Beckhoven, J. R. C. M., Rommerts, F. F. G., Orly, J., and Wirtz, K. W. A., 1989, The subcellular distribution of the non-specific lipid transfer protein (sterol carrier protein 2) in rat liver and adrenal gland, *Biochim. Biophys. Acta* 1001:243–248.

van den Boogart, P., van Dijk, S., and Agsteribbe, E., 1982, The mitochondrially made subunit 2 of *Neurospora crassa* cytochrome aa_3 is synthesized as a precursor protein, *FEBS Lett.* 147:97–100.

van Loon, A. P. G. M., Brändli, A. W., Pesold-Hurt, B., Blank, D., and Schatz, G., 1987, Transport of proteins to the mitochondrial intermembrane space: The "matrix-targeting" and "sorting" domains in the cytochrome c_1 presequence, *EMBO J.* 6:2433–2439.

van Loon, A. P. G. M., and Schatz, G., 1987, Transport of proteins to the mitochondrial intermembrane space: The "sorting" domain if the cytochrome c_1 presequence is a stop-transfer sequence specific for the mitochondrial inner membrane, *EMBO J.* 6:2441–2448.

van Loon, A. P. G. M., van Eijk, E., and Grivell, L. A., 1983, Biosynthesis of the ubiquinol:Cytochrome c reductase complex in yeast: Disco-ordinate synthesis of the 11-kd subunit in response to increased gene copy number, EMBO J. 2:1765–1770.

van Meer, G., 1989, Lipid traffic in animal cells, Annu. Rev. Cell Biol. 5:247–275.

Vance, D. E., and Vance, J. E., 1986, Specific pools of phospholipids are used for lipoprotein secretion by cultured rat hepatocytes, J. Biol. Chem. 261:4486–4491.

Vance, J. E., and Vance, D. E., 1985, The role of phosphatidylcholine biosynthesis in the secretion of lipoproteins from hepatocytes, Can. J. Biochem. Cell Biol. 63:870–881.

Vance, J. E., 1990, Phospholipid synthesis in a membrane fraction associated with mitochondria, J. Biol. Chem. 265:7248–7256.

Verner, K., and Schatz, G., 1987, Import of an incompletely folded precursor protein into isolated mitochondria requires an energized inner membrane, but no added ATP, EMBO J. 6:2449–2456.

Vestweber, D., and Schatz, G., 1988, A chimeric mitochondrial precursor protein with internal disulfide bridges blocks import of authentic precursors into mitochondria and allows quantitation of import sites, J. Cell Biol. 107:2037–2043.

Vestweber, D., Brunner, J., Baker, A., and Schatz, G., 1989, A 42K outer-membrane protein is a component of the yeast mitochondrial protein import site, Nature 341:205–209.

Voelker, D. R., 1989a, Phosphatidylserine translocation to the mitochondrion is an ATP dependent process in permeabilized animal cells, Proc. Natl. Acad. Sci. USA 86:9921–9925.

Voelker, D. R., 1989b, Reconstitution of phosphatidylserine import into rat liver mitochondria, J. Biol. Chem. 264:8019–8025.

Voelker, D. R., 1991a, Adriamycin disrupts phosphatidylserine synthesis and translocation in permeabilized animal cells, J. Biol. Chem. 266:12185–12188.

Voelker, D. R., 1991b, Organelle biogenesis and intracellular lipid transport in eukaryotes, Microbiol. Rev. 55:543–560.

von Heijne, G., 1985, Signal sequences: The limits of variation, J. Mol. Biol. 184:99–105.

von Heijne, G., 1986, Mitochondrial targeting signals may form amphiphilic helices, EMBO J. 5:1335–1342.

Vorbeck, M. L., Martin, A. P., Long, J. W., Smith, J. M., and Orr, P. R., 1982, Aging-dependent modification of lipid composition and lipid structural order parameters of hepatic mitochondria, Arch. Biochem. Biophys. 217:351–361.

Walker, M. E., Valentin, E., and Reid, G. A., 1990, Transport of the yeast ATP synthase β-subunit into mitochondria: Effects of amino acid substitutions on targeting, Biochem. J. 266:227–234.

Weiss, H., and Kolb, H. J., 1979, Isolation of mitochondrial succinate:ubiquinol reductase, cytochrome c reductase and cytochrome c oxidase from Neurospora crassa using nonionic detergent, Eur. J. Biochem. 99:139–149.

Werner, S., and Bertrand, H., 1979, Conversion of a mitochondrial precursor polypeptide into subunit 1 of cytochrome oxidase of the mi-3 mutant of Neurospora crassa, Eur. J. Biochem. 99:463–470.

Wienhues, U., Becker, K., Schleyer, M., Guiard, B., Tropschug, M., Horwich, A. L., Pfanner, N., and Neupert, W., 1991, Protein folding causes an arrest of preprotein translocation into mitochondria in vivo, J. Cell Biol. 115:1601–1609.

Wirtz, K. W. A., 1991, Phospholipid transfer proteins, Annu. Rev. Biochem. 60:73–99.

Yaffe, M. P., and Kennedy, E. P., 1983, Intracellular phospholipid movement and the role of phospholipid transfer proteins in animal cells, Biochemistry 22:1497–1507.

Yaffe, M. P., Ohta, S., and Schatz, G., 1985, A yeast mutant temperature sensitive for mitochondrial assembly is deficient in a mitochondrial protease activity that cleaves imported precursor polypeptides, EMBO J. 4:2069–2074.

Yaffe, M. P., and Schatz, G., 1984, Two nuclear mutations that block mitochondrial protein import in yeast, Proc. Natl. Acad. Sci. USA 81:4819–4823.

Yaffe, M. P., Jensen, R. E., Guido, E. C., 1989, The major 45-kDa protein of the yeast mitochondrial outer membrane is not essential for cell growth or mitochondrial function, *J. Biol. Chem.* **264**:21091–21096.

Yang, M., Geli, V., Oppliger, W., Suda, K., James, P., and Schatz, G., 1991, The MAS-encoded processing protease of yeast mitochondria: Interaction of the purified enzyme with signal peptides and a purified precursor protein, *J. Biol. Chem.* **266**:6416–6423.

Zborowski, J., Dygas, A., and Wojtczak, L., 1983, Phosphatidylserine decarboxylase is located on the external side of the inner mitochondrial membrane, *FEBS Lett.* **157**:179–182.

Zinser, E., Sperka-Gottlieb, C. D. M., Fasch, E.-V., Kohlwein, S. P., Paltauf, F., and Daum, G., 1991, Phospholipid synthesis and lipid composition of subcellular membranes in the unicellular eukaryote *Saccharomyces cerevisiae*, *J. Bacteriol.* **173**:2026–2034.

Zwizinski, C., Schleyer, M., and Neupert, W., 1984, Proteinaceous receptors for the import of mitochondrial precursor proteins, *J. Biol. Chem.* **259**:7850–7856.

Chapter 6

The Assembly of Chloroplast Membranes

Colin Robinson

1. INTRODUCTION

A hallmark of most plant cells is the complexity of the subcellular architecture. An intricate system of compartmentation is maintained by a wide variety of membrane-bound organelles, each of which possesses a characteristic complement of proteins. Because the vast majority of these proteins are synthesized in a common location (i.e., the cytosol), it follows that this system of compartmentation is totally dependent on the operation of efficient, specific mechanisms of protein targeting into and across the appropriate membranes. As a result of research carried out over the last 20 years, it is now evident that these targeting mechanisms can be divided into two broad categories. Proteins destined for one of the organelles of the endomembrane system (the endoplasmic reticulum, Golgi apparatus, vacuole, and protein bodies) are almost always initially transported into the endoplasmic reticulum by a co-translational process (reviewed by Bennett and Osteryoung, 1991). In contrast, proteins are targeted directly into higher plant chloroplasts and mitochondria (and probably also glyoxysomes) by distinct, posttranslational mechanisms (Highfield and Ellis, 1978; Chua and Schmidt, 1978).

Colin Robinson   Department of Biological Sciences, University of Warwick, Coventry CV4 7AL, England.

Subcellular Biochemistry, Volume 22: Membrane Biogenesis, edited by A. H. Maddy and J. R. Harris. Plenum Press, New York, 1994.

The biogenesis of the chloroplast and mitochondrion is also complicated by two other factors. First, each of these organelles synthesizes a small but significant proportion of the resident proteins. Second, both organelles are composed of more than one type of membrane, which means that proteins must not only be specifically targeted *into* each these organelles, but also accurately "sorted" *within* the organelle to the correct membrane or soluble phase. Intensive efforts have been made to determine the mechanisms involved in these processes, and the aim of this chapter is to review recent advances in the chloroplast field, particularly with respect to intraorganellar protein sorting. For reviews on mitochondrial protein import, the reader is referred to Nicholson and Neupert (1988), Glick and Schatz (1991), and Chapter 5 this volume.

2. CHLOROPLAST STRUCTURE AND BIOGENESIS

2.1. The Structure and Functions of Higher Plant Chloroplasts

The chloroplast is the most complex organelle known, in structural terms, for it contains three distinct types of membrane (the outer and inner envelope membranes and the thylakoid membrane) enclosing three soluble phases (the interenvelope membrane space, the stroma, and the thylakoid lumen). Each of these subcompartments is responsible for one or more metabolic processes. The outer envelope membrane is relatively porous to small molecules, and it functions mainly in biosynthetic reactions, (particularly lipid synthesis) and protein import (Douce and Joyard, 1979). In contrast, the inner membrane is a permeability barrier that mediates the exchange of metabolites between the cytosol and the chloroplast interior (Heber and Heldt, 1981). At present, essentially nothing is known about either the functions or the occupants of the interenvelope membrane space. The critical light and dark reactions of photosynthesis take place in the thylakoid network and stroma, respectively and the stroma is also the site of other essential processes, including protein synthesis and amino acid synthesis.

The six-compartment structure-function model described above has been developed primarily from numerous studies on chloroplasts from a variety of higher plant species, and also from the unicellular green alga *Chlamydomonas reinhardtii*. It is, however, worth noting that in some species of alga, the chloroplast is bounded by an envelope containing one or more additional membranes. For example, *Euglena gracilus* chloroplasts contain a three-membrane envelope, and emerging evidence suggests that protein import into these organelles may differ mechanistically from protein import into higher plant chloroplasts (Shashidhara *et al.*, 1992).

2.2. The Sites of Synthesis of Chloroplast Proteins

The chloroplast contains its own DNA and protein-synthesizing machinery, and synthesizes about 100 different proteins (approximately 20% of the total number in the chloroplast). All of these proteins appear at present to be located in the stroma, thylakoid membrane, or thylakoid lumen; there is no evidence of any "export" of proteins either into or across the inner envelope membrane (Shimada and Sugiura, 1991). The remainder of the chloroplast proteins are imported after synthesis in the cytosol. Studies on the origins of a variety of chloroplast proteins have shown that imported proteins are targeted into all of the organellar subcompartments (with the possible exception of the interenvelope membrane space), emphasizing the need for efficient intraorganellar sorting mechanisms. Figure 1 illustrates the complexity of both the chloroplast structure and the biogenesis of the resident proteins.

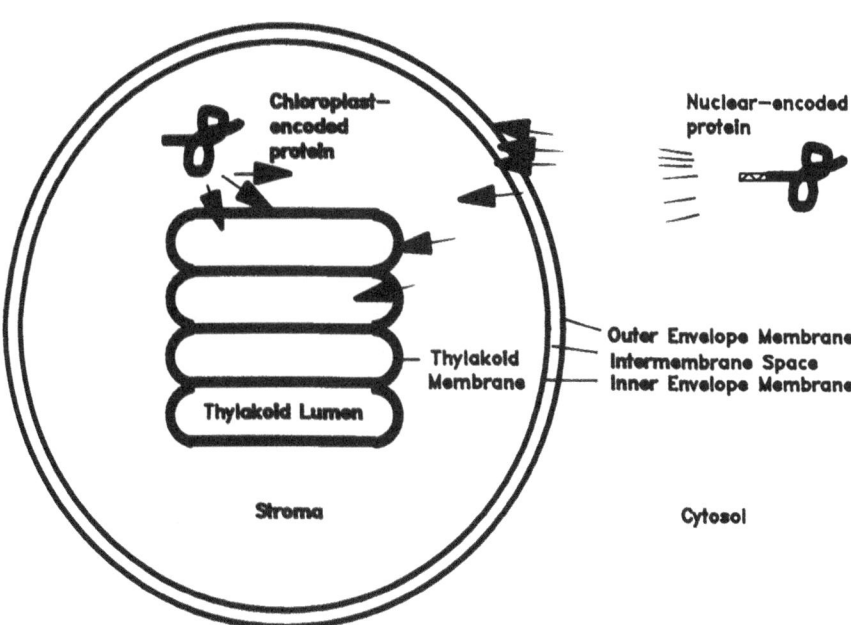

FIGURE 1. Chloroplast structure and biogenesis. The chloroplast genome encodes about 20% of the organellar proteins; these either remain in the soluble stromal phase or are targeted into, or across, the thylakoid membrane. The remainder are synthesized in the cytosol with aminoterminal presequences (with the exception of outer envelope membrane proteins), which direct transport into each of the six chloroplast subcompartments.

3. TRANSPORT OF PROTEINS INTO THE STROMA

Early studies on the import of proteins into chloroplasts focused almost exclusively on a single stromal protein: Rubisco (ribulose bisphosphate carboxylase-oxygenase) small subunit. The sheer abundance of this protein (Rubisco accounts for up to 50% of total soluble leaf protein) made the small subunit an especially useful tool for protein transport studies. In 1978, two groups showed that the precursor to this protein could be imported into isolated chloroplasts (Highfield and Ellis, 1978; Chua and Schmidt, 1978) and that this process takes place *posttranslationally*. Further studies using this protein, and later others, have identified a number of key steps in the early stages of the import mechanism.

3.1. Synthesis of Imported Proteins as Larger Precursors

Cytosolically synthesized proteins destined for the inner envelope, the stroma, the thylakoid membrane, or the thylakoid lumen have one feature in common: every protein studied to date is initially synthesized as a larger precursor containing an amino-terminal presequence. Outer-envelope membrane proteins, which are discussed later in this chapter, appear not to be synthesized with cleavable presequences. The presequences of different imported proteins vary considerably in size—from about 2 kDa to 15 kDa—but most are about 5–8 kDa.

Following the initial observations of precursor forms of chloroplast proteins (Dobberstein *et al.*, 1977) it was naturally suggested that the presequences may contain essential targeting information. This was convincingly confirmed for stromal protein precursors by Van den Broeck *et al.* (1985) and Schreier *et al.* (1985), who generated chimeric proteins consisting of the presequence of Rubisco small subunit linked to bacterial neomycin phosphotransferase. Import of this passenger protein was found to take place both *in vitro* and *in vivo*, demonstrating that the Rubisco small subunit presequence contains all of the information required to direct the import of a foreign protein into the chloroplast stroma.

Subsequent to these initial demonstrations, a wide variety of foreign proteins have been successfully targeted into the chloroplast stroma (Gatenby *et al.*, 1988; Kavanagh *et al.*, 1988; Lubben *et al.*, 1989). Indeed, it appears likely that almost any soluble protein can be targeted to the stroma using this approach. However, it is notable that different constructs can be imported with markedly differing efficiencies. In several studies, it was found that efficient import was observed only if the passenger protein was preceded by both the presequence *and* a section of mature protein from a stromal protein precursor (Wasmann *et al.*, 1986; Kavanagh *et al.*, 1988). These findings raise the possibility that chloroplast-targeting signals are not in fact located solely in the presequences of imported stromal proteins, and that the targeting signal extends into the mature protein in some cases. Alternatively, it may be that in these chimeric proteins, the sections of mature chloroplast protein serve as useful linker sequences, projecting the presequences away from the passenger proteins and making them more accessible to the chloroplast import machinery.

3.2. Structures of Stroma-Targeting Presequences

Numerous cDNAs encoding chloroplast protein precursors have now been isolated and sequenced, and it is thus possible to compare the primary structures of a large number of stroma-targeting presequences. It was originally anticipated that this type of information would yield valuable insights into the properties of the signals that specifically direct the precursors into chloroplasts. Unfortunately, this has proved not to be the case; the presequences of stromal proteins display essentially no primary sequence similarity, and no common secondary structures can be deduced with any reliability. The presequences do share some common features, in that the amino terminal half (approximately) is usually uncharged and rich in hydroxylated residues, whereas the carboxyterminal section is basic. However, it is difficult to predict from this information how the presequence is recognized so specifically by the chloroplast import machinery.

3.3. Possible Role of Cytosolic Factor(s) in the Import of Some Precursor Proteins

For many years it was unclear whether cytosolic proteins were involved in the chloroplast protein import mechanism. Although precursors synthesized in a cell-free translation system (for example, wheat-germ lysate or reticulocyte lysate) are almost always "import-competent," such systems obviously contain cytosolic proteins that may assist in the *in vitro* import process. More recently, however, several precursor proteins have been purified in significant quantities after expression in *Escherichia coli*. In the case of pre-ferredoxin, Pilon *et al.* (1992) demonstrated convincingly that no additional factors were required for the import of purified precursor protein into isolated chloroplasts. In contrast, Waegemann *et al.* (1990) found that the purified precursor of the light-harvesting chlorophyll-binding protein of photosystem 2 (pre-LHC2), a major thylakoid membrane protein, could not be imported. Import-competence could, however, be restored if the precursor was first unfolded in the presence of 8M urea and then dialyzed in the presence of leaf extract, prior to incubation with chloroplasts. These findings suggest that one or more cytosolic factors are involved in maintaining newly synthesized pre-LHC2 in an import-competent conformation, perhaps by preventing aggregation of this hydrophobic protein. Clearly, it will be interesting to test whether other imported proteins interact with cytosolic factors, and to determine the mode(s) of action of the proteins involved.

3.4. Binding of Precursors to Receptors on the Chloroplast Surface

After synthesis in the cytosol, precursor proteins are believed to bind to specific receptors on the chloroplast surface. Direct evidence for the existence of such receptors was provided by Cline *et al.* (1985), who showed that protease treatment of intact chloroplasts drastically reduced their ability to both bind and import *in vitro*-synthesized precursors of stromal and thylakoid proteins. Very little, however, is currently known about either the mechanisms by which these receptors recognize precursor proteins or the identities of the receptors themselves. Pain *et al.* (1988), and Schnell *et al.* (1990), using an indirect, anti-idiotypic antibody approach, claimed to have identified the pre-small subunit receptor as a 36 kDa envelope protein. However, Flügge *et al.* (1991) have recently provided evidence that this protein is in fact the phosphate translocator. Clearly, further studies are required to resolve this issue, and to determine whether one or more types of receptor operate in the import of chloroplast proteins.

3.5. Protein Translocation across the Chloroplast Envelope

Following the binding to import receptors, a critical event in the import of all stromal (and thylakoidal) proteins is the translocation across the envelope membranes. This event is of the utmost interest, since hundreds of different proteins of widely varying size, charge, and hydrophobicity are imported into the chloroplast across two membrane bilayers, one of which (the inner) is a highly selective permeability barrier. Unfortunately, and perhaps surprisingly, this element of the import process is, in mechanistic terms, one of the most obscure. It is generally assumed that proteins are transported at "contact sites" between the inner and outer envelope membranes, and that the proteins are unfolded prior to or during translocation. However, these ideas are largely inspired by elegant studies on mitochondrial protein import, in which these processes have been convincingly demonstrated (Schleyer and Neupert, 1985). As yet, we have relatively little information on the mechanisms by which proteins are transported across the chloroplast envelope, although the energy requirements for this process have been analyzed in some detail. There is good evidence that protein transport into the stroma requires ATP at two stages. Low levels of ATP (outside the stroma) are required for the binding of precursors to the chloroplast envelope, and higher levels of stromal ATP are required for the translocation process. There is no evidence for any role of the thylakoidal proton-motive force other than to provide ATP (Olsen *et al.*, 1989; Theg *et al.*, 1989; Grossman *et al.*, 1980).

3.6.   Proteolytic Processing of Stromal Protein Precursors

During or shortly after import into the stroma, precursors of stromal proteins are cleaved to the mature sizes by a stromal processing peptidase (SPP).
This enzyme has been partially purified from pea chloroplasts, and *in vitro*
processing assays have shown SPP to be a metal-dependent peptidase that is
highly specific for chloroplast protein precursors (Robinson and Ellis, 1984).
However, as with the signals specifying targeting to the chloroplast, very little is
known about the signal within the presequence that specifies cleavage by SPP.
The cleavage sites within different precursors of stromal proteins display essentially no primary sequence similarity, and it is therefore difficult to explain the
basis for the highly specific SPP cleavage reaction.

4.   THE IMPORT AND SORTING OF THYLAKOID PROTEINS

4.1.   Two-Phase Pathway for the Import of Thylakoid Lumen Proteins

Considerable attention has centered on the biogenesis of cytosolically synthesized thylakoid proteins in order to determine how the correct "sorting" of
these proteins is achieved after import into the chloroplast. The biogenesis of
hydrophilic thylakoid lumen proteins is especially intriguing, since these proteins
must cross all three chloroplast membranes to reach their site of function. Such
proteins include plastocyanin, a small, soluble electron carrier, and the 33, 23,
and 16 kDa proteins (33K, 23K, 16K) of the photosystem II oxygen-evolving
complex. All four are initially synthesized as precursor proteins in the cytosol
(Smeekens *et al.*, 1985; Westhoff *et al.*, 1985), and import into the thylakoid
lumen is believed to occur by a two-phase mechanism. In the proposed mechanism, each of these proteins is initially synthesized with a bipartite presequence
consisting of two targeting signals in tandem. The first "envelope transit" signal
functions to direct the protein into the stroma, after which it is removed by SPP.
The second, "thylakoid transfer" signal subsequently directs transport of the
intermediate form across the thylakoid membrane, after which it is removed by a
second, thylakoidal processing peptidase (TPP).

This two-phase import pathway, which is depicted in Figure 2, was initially
prompted by the finding that the two processing steps can be reconstructed *in
vitro*. Precursors of these lumenal proteins were specifically processed only to
intermediate forms by partially purified SPP, and complete maturation required
the presence of detergent-solublized thylakoids (Hageman *et al.*, 1986; James *et
al.*, 1989; Ko and Cashmore, 1989). Further analysis of TPP has shown that this
enzyme is highly hydrophobic, with the active site on the lumenal face on the
thylakoid membrane (Kirwin *et al.*, 1987, 1988).

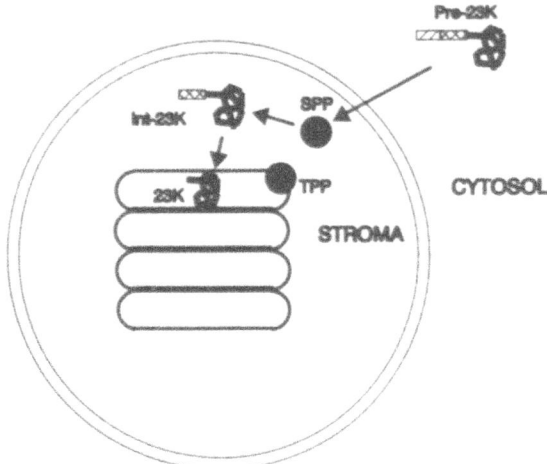

FIGURE 2. Two-phase pathway for the import of proteins into the thylakoid lumen. Lumenal proteins such as the 23 kDa protein of the oxygen-evolving complex (23K) are synthesized with a bipartite presequence consisting of an envelope transit signal and a thylakoid transfer signal in tandem. After synthesis in the cytosol, pre-23K is imported into the stroma and processed to an intermediate form (i23K) by a specific stromal processing peptidase (SPP); i23K is then transported across the thylakoid membrane and processed to the mature size by a thylakoidal processing peptidase (TPP).

Further evidence in favor of the model came from two other observations. First, transient intermediate-size polypeptides have been observed in import experiments using lumenal protein precursors (Smeekens *et al.*, 1986; James *et al.*, 1989). Second, it is clear that the presequences of lumenal proteins contain two domains that are structurally and functionally distinct. The SPP cleavage sites have been determined within *Sileane pratensis* pre-plastocyanin, wheat pre-23K, and wheat pre-33K (Bassham *et al.*, 1991), and some of the salient features of the envelope transit and thylakoid transfer domains are illustrated in Figure 3. It is apparent that the envelope transit domains are similar, in structural terms, to the presequences of stromal proteins, in that they contain an uncharged, serine/threonine rich amino-terminal domain followed by a domain containing a series of basic residues. In contrast, the thylakoid transfer domains contain a hydrophilic amino-terminal section, with both basic and acidic residues (acidic residues are very rarely found in stroma-targeting presequences) followed by a hydrophobic stretch of residues. Immediately prior to the terminal cleavage site, the -3 and -1 residues are always short-chain amino acids, usually alanine (von Heijne *et al.*, 1989; Halpin *et al.*, 1989). These common features are particularly interesting because the hydrophobic core region and -3, -1 motif are also critical

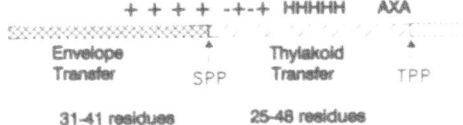

FIGURE 3. Structures of lumenal protein presequences. The first "envelope transfer" signal within the presequences of thylakoid lumen proteins is generally rich in hydroxylated residues and residues with small, hydrophobic side-chains (for example, Ala and Leu) but contains no obvious structural motif. The aminoterminal section is usually uncharged, whereas the carboxyterminal section contains a number of basic residues. The thylakoid transfer signal has three distinct domains: an aminoterminal, charged domain (with both basic and acidic residues); a central, hydrophobic core domain (HHH); and a terminal domain in which the -1 and -3 residues are short-chain, usually Ala (AXA). The latter two features are also found in "signal peptides" that direct transport across the bacterial plasma membrane. The two signals are removed by SPP and TPP, respectively.

elements of "signal" sequences that direct the export of proteins in bacteria (von Heijne, 1986). Two lines of evidence have confirmed that the two types of peptide do indeed have similar properties. First, it is clear from in vitro processing assays that the thylakoidal processing peptidase, TPP, and E. coli signal peptidase have very similar reaction specificities (Halpin et al., 1989; Shackleton and Robinson, 1991). Second, Seidler and Michel (1990) demonstrated that the thylakoid transfer domain of spinach 33K was able to direct export of the protein into the periplasm when expressed in E. coli.

Major insights into the likely course of evolution of this two-phase pathway have emerged from studies on cyanobacteria, which are widely believed to be the progenitors of chloroplasts. Many species of cyanobacteria contain internal thylakoid membranes, and in several such species it has been found that lumenal proteins (such as 33K) are synthesized with presequences that closely resemble the thylakoid transfer signals of their higher plant chloroplast counterparts (Kuwabara et al., 1987).

It is therefore believed that the evolution of the chloroplast from an endosymbiotic cyanobacterium involved the transfer of genes for both stromal and lumenal proteins into the nucleus, and the acquisition of envelope transit sequences to target the proteins back into the stroma. Further transport of lumenal proteins across the thylakoid membrane is then mediated by the ancestral thylakoidal translocation system.

Although many of the studies described above have pointed to the existence of a thylakoidal protein translocation system, it is only recently that the operation of such a system has been demonstrated biochemically. Kirwin et al. (1989) initially demonstrated moderately efficient ATP-stimulated import of 33K by isolated thylakoids, and more recently a light-driven thylakoid import assay has been developed in which 33K, 23K, and 16K are all imported with high efficiency (Mould and Robinson, 1991; Mould et al., 1991; Cline et al., 1992; Klösgen et al., 1992).

4.2. Energy Requirements for the Transport of Lumenal Proteins across the Thylakoid Membrane

Following the observation that the import of 33K, 23K, and 16K into thylakoids requires light *in vitro*, it was established that the transport of all three proteins across the thylakoid membrane requires the transthylakoidal proton motive force. Specifically, it was found that translocation was driven by the proton gradient component, ΔpH, of the proton motive force; the electrical potential $\Delta\psi$ is not a requirement (Mould and Robinson, 1991; Cline *et al.*, 1992; Klösgen *et al.*, 1992). As yet, however, little is known about the mechanism by which the ΔpH drives the translocation of these proteins. The situation is made considerably more complex by the demonstration that the transport of plasto-cyanin into the thylakoid lumen is completely unaffected by the dissipation of the ΔpH (Theg *et al.*, 1989). These findings raise the interesting possibility that more than one type of protein translocation system operates in the thylakoid membrane.

Alternatively, it may be that a single type of system is particular adaptable in its mode of action, depending on the protein being translocated. Interestingly, it has been found that ATP is not required for the transport of 23K or 16K across the thylakoid membrane (Cline *et al.*, 1992). The energy requirements for pro-tein translocation across envelope membranes and for transport of (at least) 23K and 16K across the thylakoid membrane are thus completely different, since the former process requires ATP but not a membrane potential (Theg *et al.*, 1989). The overall import pathway and the varying requirements of lumenal proteins are illustrated in Figure 4, together with the import pathway for a major thylakoid membrane protein (see below).

4.3. Import and Integration of Thylakoid Membrane Proteins

Situated between the hydrophilic proteins of the stroma and thylakoid lumen are a number of hydrophobic integral thylakoid membrane proteins. Many of these proteins are synthesized in the cytosol, and the biogenesis of one such protein, LHC2, has been studied in considerable detail. LHC2 is synthesized as a larger precursor but, unlike the presequences of thylakoid lumen proteins, the LHC2 presequence functions only to target the protein as far as the chloroplast stroma. If the presequence is replaced by that of Rubisco small subunit, the hybrid protein is still imported into the chloroplast and LHC2 is efficiently integrated into the thylakoid membrane (Lamppa, 1988). This finding clearly indicates that the mature LHC2 protein contains the information specifying inte-gration into the thylakoid membrane. Integration of LHC2 into isolated thy-lakoids was demonstrated by Cline (1986), and subsequent work has shown that this process requires both ATP and the activity of one or more stromal proteins

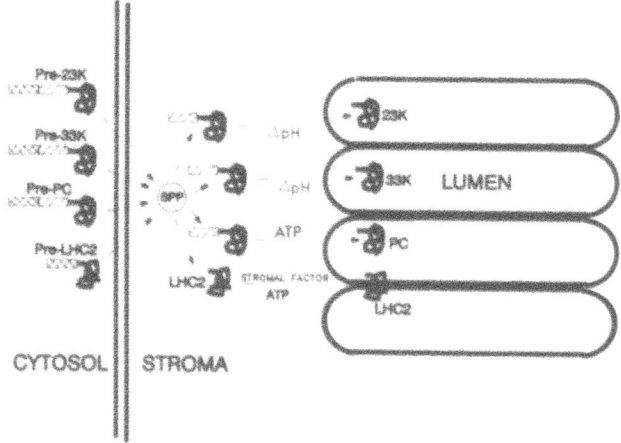

FIGURE 4. Protein targeting into the thylakoid membrane and lumen. Although the 23 kDa and 33 kDa oxygen-evolving complex proteins (23K, 33K) and plastocyanin (PC) are all targeted into the thylakoid lumen by a two-phase mechanism, the requirements for translocation across the thylakoid membrane can differ significantly. The transport of plastocyanin into the lumen requires only ATP, whereas a proton gradient across the thylakoid membrane is essential for translocation of 23K and 33K. The integration of the light-harvesting chlorophyll binding protein (LHC2) into the thylakoid membrane is mediated by information in the mature protein, rather than a cleavable signal; this process also requires ATP and a stromal factor.

(Fulson and Cline, 1988). Possibly, the stromal factor operates as an "unfoldase" or "antifoldase," preventing LHC2 from aggregating and maintaining the protein in an integration-competent conformation.

5. TRANSPORT OF PROTEINS INTO THE CHLOROPLAST ENVELOPE

Despite the critical roles of envelope proteins in regulating metabolite exchange between the stroma and the cytosol, and in a variety of key biosynthetic reactions, relatively little is known about the biogenesis of these proteins. The main reason for this is that chloroplast envelope proteins are of very low abundance (the chloroplast envelope accounts for only 1–2% of total chloroplast membrane). To date, only four cDNA clones encoding envelope proteins have been isolated, and in some respects import analyses are at an early stage. Nevertheless, interesting data have already emerged, especially concerning outer-membrane proteins. Salomon *et al.* (1990) and Li *et al.* (1991) isolated cDNA clones encoding outer-envelope membrane proteins of 6.7 kDa and 14 kDa, respectively. In each case, it was demonstrated that the import mechanism differed from those of other chloroplast proteins in several critical respects. Neither protein is synthesized with a cleavable presequence, and both proteins are imported into the envelope in the absence of ATP hydrolysis. Furthermore, protease treatment of intact chloroplasts, which drastically inhibits the subsequent import of stromal or thylakoid proteins, has no apparent effect on the *in vitro* import of these outer-envelope membrane proteins, providing convincing evidence that these proteins are imported by a completely different mechanism. It remains to be determined whether import is mediated by a distinct, protease-resistant receptor, or whether these proteins interact directly, and specifically, with chloroplast envelope lipids.

A very different mechanism is involved in the import of the two inner envelope membrane proteins cloned to date: the putative phosphate translocator, and a 37 kDa polypeptide. In each case, the proteins are initially synthesized as larger precursors, import is ATP-dependent, and prior protease treatment of chloroplast markedly inhibits import *in vitro*. In the initial stages, at least, these proteins thus appear to be imported by the mechanism elucidated for stromal and thylakoid proteins. The only unusual feature apparent from these studies concerns the presequences of these proteins. In each case, an amphiphilic α-helix is predicted within the presequence (a feature completely absent from other chloroplast presequences, but characteristic of mitochondrial protein presequences). Whether this motif plays a role in targeting the proteins to the inner envelope membrane is presently unclear. Indeed, there is currently little or no available information on the mechanism by which this aspect of chloroplast protein sorting is accomplished, and it will be of particular interest to determine how hydrophobic proteins are specifically directed to either the inner envelope membrane or the thylakoid membrane. In addition, there is still uncertainty concerning the precise import pathway for inner envelope membrane proteins. These proteins might be arrested en route to the stromal phase by an envelope-specific signal (perhaps a hydrophobic "stop-transfer" signal that has an affinity for proteins or lipids in the inner envelope membrane). Alternatively, the proteins may be fully translocated into the stroma, and subsequently insert into the membrane by virtue of an envelope-specific integration signal.

6. SUMMARY

During the last five or six years there has been a shift in focus in the field of chloroplast protein transport, with greater emphasis being placed on understanding the sorting of proteins to the thylakoids and the envelope membranes. As a result, we have a much-improved understanding of the variety of import pathways that function during chloroplast biogenesis. It is now clear that a considerable number of distinct intraorganellar sorting mechanisms operate to direct imported proteins to their correct destinations. Some of the underlying mechanisms are also beginning to emerge, although it is accurate to say that we are still a long way from understanding in genuine detail how proteins are translocated across chloroplast membranes. However, the availability of such a range of efficient *in vitro* import assays should ensure that rapid progress is made in coming years.

The major gaps in this field now concern the identities and roles of the elements of the import apparatus: Although at least two distinct translocation systems operate during chloroplast biogenesis, none of these components has been identified, purified, or cloned. This is primarily because these proteins are often difficult to assay individually, and they are usually of very low abundance. Nevertheless, it is essential that progress is made in this area soon in order to maintain the present momentum.

7. REFERENCES

Bassham, D. C., Bartling, D., Mould, R. M., Dunbar, B., Weisbeek, P., Herrmann, R. G., and Robinson, C., 1991, Transport of proteins into chloroplasts: Delineation of envelope transit and thylakoid transfer signals within the presequences of three imported thylakoid lumen proteins, *J. Biol. Chem.* **266**:23606–23610.

Bennett, A. B., and Osteryoung, K. W., 1991, Protein transport and targeting within the endomembrane system of plants, in "*Plant Genetic Engineering*" (D. Grierson, ed.) pp. 199–230. Blackie Press, Glasgow.

Chua, N-H., and Schmidt, G. W., 1978, Post-translational transport into intact chloroplasts of a precursor to the small subunit of ribulose 1,5-bisphosphate carboxylase, *Proc. Natl. Acad. Sci USA* **75**:6110–6117.

Cline, K., Werner-Washburne, M., Lubben, T. H., and Keegstra, K., 1985, Precursors to two nuclear-encoded chloroplast proteins bind to the outer envelope membrane before being imported into chloroplasts, *J. Biol. Chem.* **260**:3691–3696.

Cline, K., 1986, Import of proteins into chloroplasts: Membrane integration of a thylakoid precursor protein reconstituted in chloroplast lysates, *J. Biol. Chem.* **261**:14804–14809.

Cline, K., Ettinger, W., and Theg, S. M., 1992, Protein-specific energy requirements for protein transport across or into thylakoid membranes, *J. Biol. Chem.* **267**:2688–2696.

Dobberstein, B., Blobel, G., and Chua, N-H., 1977, In vitro synthesis and processing of a putative precursor for the small subunit of ribulose bisphosphate carboxylase of *Chlamydomonas reinhardtii, Proc. Natl. Acad. Sci. USA* **74**:1082–1087.

Douce, R., and Joyard, J., 1979, Structure and function of the plastid envelope. *Adv. Bot. Res.* **7**:1–14.

Flügge, U-I., Weber, A., Fischer, K., Lottspeich, F., Eckershorn, C., Waegmann, K., and Soll, J., 1991, The major chloroplast envelope polypeptide is the phosphate translocator and not the protein import receptor, *Nature* **353**:364–367.

Fulson, D. R., and Cline, K., 1988, A soluble protein factor is required *in vitro* for membrane insertion of the thylakoid precursor protein pLHCP, *Plant Physiol.* **88**:1146–1151.

Gatenby, A. A., Lubben, T. H., Ahlquist, P., and Keegstra, K., 1988, Imported large subunits of ribulose bisphosphate carboxylase, but not imported β-ATP synthase subunits, are assembled into holoenzyme in isolated chloroplasts, *EMBO J.* **7**:1307–1314.

Glick, B. S., and Schatz, G., 1991, Import of proteins into mitochondria, *Annu. Rev. Genet.* **25**:21–44.

Grossman, A. R., Bartlett, S. G., and Chua, N-H., 1980, Energy-dependent uptake of cytoplasmically synthesised polypeptides by chloroplasts, *Nature* **285**:625–628.

Hageman, J., Robinson, C., Smeekens, S., and Weisbeek, P., 1986, A thylakoid processing peptidase is required for complete maturation of the lumen protein plastocyanin, *Nature* **324**:567–569.

Halpin, C., Elderfield, P. D., James, H. E., Zimmermann, R., Dunbar, B., and Robinson, C., 1989, The reaction specificities of the thylakoidal processing peptidase and *Escherichia coli* leader peptidase are identical, *EMBO J.* **8**:3917–3922.

Heber, U., and Heldt, H. W., 1981, The chloroplast envelope: Structure, function and role in leaf metabolism, *Annu. Rev. Plant Physiol.* **32**:139–152.

Highfield, P. E., and Ellis, R. J., 1978, Synthesis and transport of the small subunit of chloroplast ribulose bisphosphate carboxylase, *Nature* **271**:420–424.

James, H. E., Bartling, D., Musgrove, J. E., Kirwin, P. M., Herrmann, R. G., and Robinson, C., 1989, Transport of proteins into chloroplasts: Import and maturation of precursors to the 33, 23 and 16 KDa proteins of the oxygen-evolving complex, *J. Biol. Chem.* **264**:19573–19576.

Kavanagh, T. A., Jefferson, R. A., and Bevan, M. W., 1988, Targeting a foreign protein to chloroplasts using fusions to the transit peptide of a chlorophyll a/b protein, *Mol. Gen. Genet.* **215**:38–45.

Kirwin, P. M., Elderfield, P. D., and Robinson, C., 1987, Transport of proteins into chloroplasts: Partial purification of a thylakoidal processing peptidase involved in plastocyanin biogenesis, *J. Biol. Chem.* **262**:16386–16390.

Kirwin, P. M., Elderfield, P. D., Williams, R. S., and Robinson, C., 1988, Transport of proteins into chloroplasts: Organisation, orientation and lateral distribution of the plastocyanin processing peptidase in the thylakoid network, *J. Biol. Chem.* **263**:18128–18132.

Kirwin, P. M., Meadows, J. W., Shackleton, J. B., Musgrove, J. E., Elderfield, P. D., Hay, N. A., and Robinson, C., 1989, ATP-dependent import of a lumenal protein by isolated thylakoid vesicles, *EMBO J.* **8**:3917–3921.

Klösgen, R. B., Brock, I. W., Herrmann, R. G., and Robinson, C., 1992, Proton gradient-driven import of the 16KDa oxygen-evolving complex protein as the full precursor protein by isolated thylakoids, *Plant Mol. Biol.* **18**:1031–1034.

Ko, K., and Cashmore, A. R., 1989, Targeting of proteins to the thylakoid lumen by the bipartite transit peptide of the 33 kd oxygen-evolving protein, *EMBO J.* **8**:3187–3194.

Kuwabara, R., Reddy, K. J., and Sherman, L. A., 1987, Nucleotide sequence of the gene from the cyanobacterium *Anacystis nidulans* R2 encoding the Mn-stabilising protein involved in photosystem II water oxidation, *Proc. Natl. Acad. Sci. USA* **84**:8230–8236.

Lamppa, G. K., 1988, The chlorophyll a/b binding protein inserts into the thylakoids independent of its cognate transit peptide, *J. Biol. Chem.* **263**:14996–14999.

Li, H., Moore, T., and Keegstra, K., 1991, Targeting of proteins to the outer envelope membrane uses a different pathway than transport into chloroplasts, *Plant Cell* **3**:709–717.

Lubben, T., Gatenby, A. A., Ahlquist, P., and Keegstra, K., 1989, Chloroplast import characteristics of chimeric proteins, *Plant Mol. Biol.* **12**:13–18.

Mould, R. M., and Robinson, C., 1991, A proton gradient is required for the transport of two lumenal oxygen-evolving proteins across the thylakoid membrane, *J. Biol. Chem.* **266**:12189–12193.

Mould, R. M., Shackleton, J. B., and Robinson, C., 1991, Protein transport into chloroplasts: Requirements for the efficient import of two lumenal oxygen-evolving complex proteins into isolated thylakoids, *J. Biol. Chem.* **266**:17286–17290.

Nicholson, D. W., and Neupert, W., 1988, Synthesis and assembly of mitochondrial proteins, in *Protein Transfer and Organelle Biogenesis* (R. C. Ras and P. W. Robbins, eds.), pp. 677–746, Academic Press, San Diego.

Olsen, L., Theg, S., Selman, B., and Keegstra, K., 1989, ATP is required for the binding of precursor proteins to chloroplasts, *J. Biol. Chem.* **264:**6724–6729.

Pain, D., Kanwar, Y. S., and Blobel, G., 1988, Identification of a receptor for protein import into chloroplasts and its localisation to envelope contact zones, *Nature* **331:**232–235.

Pilon, M., de Krwijff, B., and Weisbeek, P., 1992, New insights into the import mechanism of the ferredoxin precursor into chloroplasts, *J. Biol. Chem.* **267:**2548–2556.

Robinson, C., and Ellis, R., 1984, Transport of proteins into chloroplasts: Partial purification of a chloroplast protease involved in the processing of imported precursor polypeptides, *Eur. J. Biochem.* **142:**337–342.

Salomon, M., Fischer, K., Flügge, U-I., and Soll, J., 1990, Sequence analysis and protein import studies of an outer chloroplast envelope polypeptide, *Proc. Natl. Acad. Sci. USA* **87:**5778–5782.

Schleyer, M., and Neupert, W., 1985, Transport of proteins into mitochondria: Translocation intermediates spanning contact sites between outer and inner membranes, *Cell* **43:**339–346.

Schnell, D. J., Blobel, G., and Pain, D., 1990, The chloroplast import receptor is an integral membrane protein of chloroplast envelope contact sites, *J. Cell. Biol.* **111:**1825–1838.

Schreier, P. H., Seftor, E. A., Schell, J., and Bohnert, H. J., 1985, The use of nuclear-encoded sequences to direct the light-regulated synthesis and transport of a foreign protein into plant chloroplasts, *EMBO J.* **4:**25–31.

Seidler, A., and Michel, H., 1990, Expression in *Escherichia coli* of the *psb0* gene encoding the 33KD protein of the oxygen-evolving complex from spinach, *EMBO J.* **8:**1743–1748.

Shackleton, J. B., and Robinson, C., 1991, Transport of proteins into chloroplasts: The thylakoidal processing peptidase is a signal-type peptidase with stringent substrate requirements at the -3 and -1 positions, *J. Biol. Chem.* **266:**12152–12156.

Shashidhara, L. S., Saw Hoon, L., Shackleton, J. B., Robinson, C., and Smith, A., 1992, Targeting of porphobilinogen deaminase into isolated *Euglena* chloroplasts, *J. Biol. Chem.* **267:**12885–12891.

Shimada, H., and Sugiura, M., 1991, Fine structural features of the chloroplast genome: Comparison of the sequenced chloroplast genomes, *Nucleic Acids Res.* **19:**983–994.

Smeekens, S., Bouerle, C., Hageman, J., Keegstra, K., and Weisbeek, P., 1986, The role of the transit peptide in the routing of precursors toward different chloroplast compartments, *Cell* **46:**365–375.

Smeekens, S., de Groot, M., von Binsbergen, J., and Weisbeek, P., 1985, Sequence of the precursor of the chloroplast thylakoid lumen protein plastocyanin, *Nature* **317:**456–458.

Theg, S., Bauerle, C., Olsen, L., Selman, B., and Keegstra, K., 1989, Internal ATP is the only energy requirement for the translocation of precursor proteins across chloroplastic membranes, *J. Biol. Chem.* **264:**6730–6736.

Van den Broeck, G., Timko, M. P., Kausch, A. P., Cashmore, A. R., Van Montagu, M., and Herrera-Estrella, L., 1985, Targeting of foreign protein to chloroplasts by fusion to the transit peptide from the small subunit of ribulose 1,5-bisphosphate carboxylase, *Nature* **313:**358–363.

von Heijne, G., 1986, A new method for predicting signal sequence cleavage sites, *Nucleic Acids Res.* **14:**4683–4690.

von Heijne, G., Steppuhn, J., and Herrmann, R. G., 1989, Domain structure of mitochondrial and chloroplast targeting peptides, *Eur. J. Biochem.* **180:**535–541.

Waegemann, K., Paulsen, H., and Soll, J., 1990, Translocation of proteins into isolated chloroplasts requires cytosolic factors to obtain import competence, *FEBS Lett.* **261:**89–92.

Wasmann, C. C., Reiss, B., Bartlett, S. G., and Bohnert, H. J., 1986, The importance of the transit peptide and the transported protein for protein import into chloroplasts, *Mol. Gen. Genet.* **205:**446–453.

Westhoff, P., Jansson, C., Klein-Hitpass, K., Berzborn, R., Latsson, C., and Bartlett, S. G., 1985, Intracellular coding sites of polypeptiodes associated with photosystem II, *Plant Mol. Biol.* **4:**137–146.

Chapter 7

Biogenesis of the Lysosomal Membrane

Minoru Fukuda

1. INTRODUCTION

Lysosomes serve as a major digestive compartment of eukaryotic cells. They are responsible for the degradation of both foreign material internalized by endocytosis and intracellular material delivered to lysosomes during autophagocytosis (de Duve, 1983). Lysosomes are defined as dense vacuoles containing a variety of acid-dependent hydrolases allowing them to perform their functions. However, lysosomes fuse with endosomes, phagosomes, and a plasma membrane, and they can play a central role in a membrane's flow and dynamics. Because such membrane flow can come from both the endocytic and biosynthetic pathways, it is difficult to define lysosomes. However, it appears that the definition of de Duve, one of the discoverers of lysosomes, is still appropriate and was recently elaborated upon by Kornfeld and Mellman (1989). Thus, lysosomes are defined as the *final* repository of macromolecules for degradation products derived from the extracellular space (via endocytosis) or from within the cell (via autophagocytosis). Lysosomes are enriched with various hydrolases that separate them from other cytoplasmic components. The lysosomal membrane that forms such a unique vacuole must have a distinct structure and function. Thus, the lysosomal membrane is characterized by its resistance to degradation by lysoso-

Minoru Fukuda   La Jolla Cancer Research Foundation, Cancer Research Center, La Jolla, California 92037.
Subcellular Biochemistry, Volume 22: Membrane Biogenesis, edited by A. H. Maddy and J. R. Harris. Plenum Press, New York, 1994.

mal hydrolases, its maintainence and generation of an acidic intralysosomal environment, and its ability to selectively transport the products of hydrolysis by lysosomal hydrolases (Cohn and Ehrenreich, 1969; Ohkuma and Poole, 1978; Neufeld and Ashwell, 1980; Helenius et al., 1983). It is also likely that the lysosomal membrane has a specificity with which it interacts and fuses with other membrane organelles such as endosomes, phagosomes, and the plasma membrane (Mellman, 1984).

Another major component of lysosomes is hydrolases present within their vacuoles. The enzymes are essential for degradation of macromolecules delivered to lysosomes. The majority of these enzymes are soluble and are delivered by mechanisms different from those used to transport lysosomal membrane proteins. In fact, the understanding of such mechanisms of lysosomal enzymes has advanced greatly and can be summarized as follows (Kornfeld, 1986; von Figura and Hasilik, 1986).

Lysosomal enzymes are glycoproteins and are synthesized initially in the same biosynthetic pathways as secretory and membrane glycoproteins. Thus, newly synthesized lysosomal enzymes are inserted into the lumen of the rough endoplasmic reticulum (RER), their signal sequences are cleaved, and Asn-linked oligosaccharide is transferred to a nascent polypeptide through the transfer of a lipid-linked, glucose-containing, high-mannose oligosaccharide. The latter is composed of three glucoses, nine mannoses, and two N-acetylglucosamines. While the polypeptide is still in the RER, the glucoses and two α-1 $\rightarrow$ 2-linked mannose residues are removed from the nine-mannose oligosaccharide. The processing that follows takes place in secretory and membrane glycoproteins. Such processing trims down the nine-mannose, by Golgi α-mannosidase I, into an oligosaccharide with five mannose residues. This then triggers the next reactions, which eventually form sialylated complex Asn-linked oligosaccharides.

Shortly after exit from the RER, the first reaction, which is unique to the synthesis of lysosomal enzymes, is the acquisition of Man-6-phosphate. First, lysosomal enzyme N-acetylglucosamine-1-phosphotransferase (phosphotransferase) transfers N-acetylglucosamine-1-phosphate from UDP-GlcNAc to mannose residues in lysosomal enzymes. The resulting N-acetylglucosamine, attached to mannose by a phosphodiester, is then removed by N-acetylglucosamine-1-phospho-diesterase, α-N-acetylglucosamidase, forming a phosphate attached to C-6 of mannose residues. It was demonstrated that phosphotransferase recognizes the protein domain of lysosomal enzymes, which is distinct from those in other proteins or glycoproteins that are not transported to lysosomes (Baranski et al., 1990). Because such a protein domain is unique to lysosomal enzymes, it is this phosphotransferase reaction that assures the phosphorylation of only lysosomal enzymes.

The newly formed Man-6-P marker is now recognized by specific mannose-6-phosphate receptors, MPRs. The receptor-lysosomal enzyme forms a complex, and this complex is eventually transported out at trans-Golgi to pre-lysosomal compartments. The complex is dissociated, leaving lysosomal en-

zymes in prelysosomal vacuoles. These enzymes are then transported to lysosomes as prelysosomal compartments that mature to lysosomes. MPR, on the
other hand, recycles back to the Golgi complex, again ready to complex with
lysosomal enzymes. This recycling is economical and eliminates the need for a
continuous supply of MPR through biosynthesis. The newly synthesized MPR is
only necessary to supplement the MPR degraded during these processes.

It is evident from the above description that lysosomal enzymes are transported from the Golgi to prelysosomes by virtue of a carrier, MPR. Man-6-P is
essential for these lysosomal enzymes to be carried, since its recognition by MPR
is necessary for complex formation. However, this does not tell us how such a
complex is then transported to prelysosomes. Before discussing this, I shall first
summarize our current knowledge of MPRs and then discuss the molecular
signals for the transport of these molecules.

2. MANNOSE PHOSPHATE RECEPTORS

2.1. Two Receptors for Mannose-6-Phosphate

Mannose-6-phosphate receptors, MPRs, are the carriers that transport
lysosomal enzymes to prelysosomes. There are two different kinds of MPRs
(Dahms et al., 1989). One is cation-dependent MPR, CD-MPR; and the other is
cation-independent MPR, CI-MPR. The bovine CD-MPR contains a 159-residue
extraplasmic domain, a single 25-residue transmembrane region, and a 67-
residue cytoplasmic domain. The CD-MPR has one binding site for mannose-6-
phosphate and appears to be functional as a dimer. Bovine CI-MPR, on the other
hand, contains a large 2219-residue extracytoplasmic domain, a single 23-
residue transmembrane domain, and a 163-residue cytoplasmic domain. The
extracytoplasmic domain of CI-MPR is composed of 15 contiguous repeating
segments, and these repeats appear to be the minimum size for binding of
mannose-6-phosphate. Of these repeating segments, two are involved in actual
binding to mannose-6-phosphate receptors. Furthermore, each repeating segment
is homologous to the extracellular domain of CD-MPR (Dahms et al., 1989;
Kornfeld and Mellman, 1989). These results strongly suggest that CD-MPR
binds to mannose-6-phosphate as a dimer, while two different repeating segments
of MPR bind to two mannose-6-phosphates. Such intracellular binding may be
the reason why CI-MPR binds to a diphosphorylated oligosaccharide with a
much stronger affinity than does the CD-MPR (Tong and Kornfeld, 1989).

2.2. Intracellular Trafficking of Mannose-6-Phosphate Receptors

Once lysosomal enzymes bind to MPR, the complex is then transported to
prelysosomes. Because mannose-6-phosphate residues are synthesized at the *cis-*

Golgi, the exit of the complex of the MPR and lysosomal enzymes can be anywhere between the *cis* and the *trans*-Golgi network (TGN). Several data suggest, however, such a complex exits at the *trans*-Golgi network. In particular, some of the lysosomal enzymes were found to have fully processed *N*-glycans, indicating that the enzymes are passed through the *trans*-Golgi or the *trans*-Golgi network (Fedde and Sly, 1985). Second, several studies indicate that MPR routinely recycles to the Golgi compartments that contain sialyltransferase, while recycling to the earlier compartments of the Golgi is observed much less frequently (Duncan and Kornfeld, 1988; Goda and Pfeffer, 1988).

Once the complex of lysosomal enzymes and MPR exits from the *trans*-Golgi, it is then transported to endosomes and late endosomes (prelysosomes). When the complex reaches late endosomes, the lysosomal enzymes dissociate from the receptors, owing to the acidic pH in the endosomes. It was shown that such dissociation takes place easily at a pH less than 5.5, and thus the dissociation of enzymes and the receptor occurs most often in a late endosomal compartment of pH 5.5 or less. In this compartment, enzymes transported through endocytosis likely meet enzymes transported from the Golgi and together they are transported to lysosomes (Griffith *et al.*, 1988).

2.3. Signals for Lysosomal Delivery

In addition to the direct routing of lysosomal enzymes from the Golgi complex, MPRs transport lysosomal enzymes from the cell surface by rapid endocytosis. In this situation, lysosomal enzymes transported to the plasma membrane by a default pathway or by secretion can be taken up by the endocytic pathway to endosomes and eventually to lysosomes. Structural determinants of the MPR traffic signal were identified in the cytoplasmic tail of the receptor.

First, Lobel *et al.* (1989) demonstrated that the deletion of the 89-amino acid residue from the COOH-terminus to the cytoplasmic tail did not impair the endocytosis capability, whereas the same deletion impaired the direct sorting of newly synthesized lysosomal enzymes to prelysosomes. When the cytoplasmic tail was further truncated and contained only 20-amino acid residues from the transmembrane, it was no longer functional in either direct sorting or endocytosis. Moreover, mutation of the tyrosines at 24 or 26 positions rendered the MPR incapable of endocytosis. In these tyrosine mutations, direct sorting at the Golgi was minimally impaired. The tyrosine-motif recognized in the MPR turns out to represent a generic signal that is used in a wide variety of proteins that are subject to endocytosis. It was shown that endocytosis of the LDL receptor is critically dependent on the cytoplasmic tyrosine residue (Davis *et al.*, 1986, 1987). Similarly, the tyrosine-containing motif was found to be crucial for endocytosis of the transferrin receptor (Jing *et al.*, 1990) and for the rapid internalization of mutated influenza hemagglutinin (Lazarovits and Roth, 1988). The data obtained after detailed mutation studies, however, point toward two different structural motifs for the tyrosine-containing signal. One group can be represented by F-X-X-X-V-Y, where the critical tyrosine residue is at the carboxyl-terminal end. The other group can be expressed as Y-X-X-R, where R is bulky hydrophobic amino acids (Table I). At this point, it appears that different motifs are used in different proteins. However, more recent studies suggest that Y-X-X-R may be the most critical structural element (Jadot *et al.*, 1992).

The novel motif was recognized recently as a signal for rapid endocytosis. This discovery demonstrated that the mannose-6-phosphate receptor utilized another motif for direct sorting to lysosomes. The previous studies already showed that the most terminal carboxyl sequence is critical for the direct sorting of MPR (Lobel *et al.*, 1989). When this initial finding was extended and the amino acid residues close to the carboxyl-terminus were examined in detail, a novel motif consisting of Leu-Leu was identified (Johnson and Kornfeld, 1992a,b). In both CD-MPR and CI-MPR, the studies revealed that L-L-X-X or H-L-L-X-X at the carboxyl-terminal end act as lysosomal delivery signals (Table II). Furthermore, it was shown that this lysosomal delivery signal works for the direct sorting at the *trans*-Golgi network.

These results suggest that the MPR utilizes two different transport signals. The tyrosine-containing motif directs the MPR at the cell surface to the endocytic pathway. In this transport, lysosomal enzymes at the plasma membrane or at the extracellular medium are taken up to lysosomes through the endocytic pathway. The Leu-Leu motif directs the MPR at the Golgi complex to prelysosomes. In this transport, newly synthesized lysosomal enzymes are sorted at the *trans* Golgi

Table I
Tyrosine-Containing Motif for Internalization Signals

Proteins	Amino acid sequence[a]
Human lamp-1[b]	RSHAG YQTI
Human lamp-2[c]	HHHAG YEQF
Rat/mouse lamp-2[c]	RHHTG YEQF
Rat/human Limp I[c]	SIRSG YEVM
Human acid phosphatase	AEPPG YRHV ADGQD . . .
Human transferrin receptor	GEPLS YTRF SLARQ . . .
Human LDL receptor	SIN FDNPVY QKTTE . . .
Bovine CI-MPR	NVS YKYSKV NKEEE . . .
Bovine CD-MPR	GME FPHLAF WQDLG . . .
Rabbit poly(Ig) receptor	EADLA YSAF LLQSN . . .

[a]Critical amino acid residues are typed in boldface.
[b]Rat, mouse, and chicken lamp-1 have the identical sequence.
[c]The critical aino acid residues in these sequences have not been identified. The sequences are also in Y-X-X-R motif.

Table II
Comparison of Leu-Ile Motif Sequences

	18
Limp II	. . . T A D E R A P **L I** R T
(rat/human)	
	127
Man-6-P/IGF-II receptor	. . . D R V G **L** V R G . . .
(human)	162
	. . . S D E D **L L** H I
	132
Man-6-P/IGF-II receptor	. . . D R V G **L** V R G . . .
(bovine)	161
	. . . S D E D **L L** H V
	65
CD-M-MPR	. . . S E E R D D **H L L** P M
(bovine)	
	131
T-cell antigen receptor	. . . S D K Q T **L L** Q N . . .
γ chain	

Amino acid sequences that are critical for lysosomal delivery are denoted in boldface. The internal sequences in bovine Man-6-P/IGF-II receptor were not tested yet.

network and delivered to lysosomes. However, it is not known yet whether these motifs are utilized only in one direction.

3. ISOLATION OF LYSOSOMAL MEMBRANE GLYCOPROTEINS

As described above, synthesized lysosomal enzymes are transported to pre-lysosomes or late endosomes by mannose-6-phosphate receptors that complex with these enzymes. Once the complex reaches prelysosomes, the mannose-6-phosphate receptors dissociate from the complex and only lysosomal enzymes are transported to lysosomes. Thus, it is certain that in prelysosomes, lysosomal enzymes are trapped in vesicles or vacuoles that eventually become lysosomes. In fact, it was shown that MPR and lysosomal membrane glycoproteins co-exist in these prelysosomes (Griffith *et al.*, 1988).

The other critical question is how such vacuoles are formed and what kind of protein components are present in the membrane of these lysosomal vacuoles. Such components must satisfy several requirements. First, lysosomal membrane proteins must be resistant to acid hydrolases present in the lumen of lysosomes. As individual proteins, or as a collection of those proteins, the lysosomal membrane proteins must provide a barrier to lysosomal enzymes. In doing so, the lysosomal membrane separates acid hydrolases from the rest of the cytoplasm, protecting them from intracellular degradation.

Second, it is apparent that some of the membrane components are involved in specific interaction and fusion with other membrane organelles, such as endosomes, phagosomes, and the plasma membrane. It is not known at this point which lysosomal membrane proteins have such functions.

Third, some of the lysosomal membrane proteins must be transporters. One group of transporters engages in the transport of amino acids or carbohydrates that are degraded by acid hydrolases. Another group is a protein group that maintains the acidity of the intralumenal portion of lysosomes (see Reggio et al., 1984; Nezu et al., 1992).

Of these proteins, characterization of the major glycoproteins that constitute the lysosomal membrane has been advanced so far. Initially, the work was carried out to identify membrane proteins or glycoproteins that are uniquely present in lysosomes but absent in the plasma membrane. The analysis of isolated lysosomal membrane revealed two glycoproteins, with Mr ~120,000 and ~60,000 that are specifically present in lysosomes (Yamamoto et al., 1980; Burnside and Schneider, 1982). These studies were difficult because it is not easy to isolate a large amount of lysosomes that are not contaminated by other membrane components. Thus, later studies were directed to raise monoclonal antibodies that specifically react with isolated, purified lysosomal membrane. Those studies identified four different lysosomal membrane proteins that are the major components of lysosomal membrane. The first and the second members, termed lamp-1 and lamp-2, are related to each other and were identified in various laboratories. Lamp-1 was isolated from human (Carlsson et al., 1988), mouse (Chen et al., 1985), rat (Lewis et al., 1985; Barriocanal et al., 1986), and chicken (Lippincott-Schwartz and Fambrough, 1987) cells. In chicken cells, this glycoprotein is called LEP-100, which traverses the plasma membrane, endosomes to lysosomes. Mouse lamp-2 was also isolated as one of the macrophage antigens, Mac-3 (Ho and Springer, 1983; Chen et al., 1985). These results already tell us that lamp-1 and lamp-2 can be on the cell surface. Lamp-1 and lamp-2, isolated from human cells, are the major glycoproteins containing poly-N-acetyllactosamines (Carlsson et al., 1988; Fukuda et al., 1988; Viitala et al., 1988). Poly-N-acetyllactosamines are very complex carbohydrates with various ligand and antigenic structures (Fukuda, 1985). It is important to recognize that lamp-1 and lamp-2 contain unique carbohydrates (see below). Most recently, more extensive studies have been done on lamp-1 and lamp-2 molecules than on any other lysosomal membrane glycoprotein.

3.1. Structure of Lamp-1 and Lamp-2

Lamp-1 and lamp-2 were isolated from various species, and the cDNAs encoding their polypeptide chains were isolated from human (Carlsson et al.,

1988; Viitala *et al.*, 1988), mouse (Chen *et al.*, 1988), rat (Howe *et al.*, 1988; Himeno *et al.*, 1989; Noguchi *et al.*, 1989), and chicken (Fambrough *et al.*, 1988) cells. Lamp-1 is also called Limp III (Barriocanal *et al.*, 1986), lgp120 (Howe *et al.*, 1988), and LEP-100 (Lippincott-Schwartz and Fambrough, 1987), whereas lamp-2 is also called Limp IV and lgp110 (see Table III). The deduced amino acid sequences and the protein chemistry of lamp-1 and lamp-2 isolated form human (Carlsson and Fukuda, 1989, 1990) and mouse (Arternburn *et al.*, 1990) cells revealed the following characteristics for these molecules.

1. The glycoproteins consist of a polypeptide core of ~40 kDa. A large part of the molecule is located in the lumenal side of lysosomes, and the domain is connected to the transmembrane domain, which is extended to a short cytoplasmic tail. Lamp-1 and lamp-2 are homologous to each other, and they are similar in their domain structures (Figure 1).
2. The intralumenal domain contains 349- (human) to 361- (chicken) amino acid residues. The intralumenal portion can be divided into an internally homologous domain that is separated by a region rich in proline residues.
3. The intralumenal domain contains eight half-cysteine residues. If the second and third, and the sixth and seventh half-cysteine residues were connected, the size of the disulfide loops would be close to that observed

Table III
Nomenclature and Properties of the Major Lysosomal Membrane Glycoproteins

Properties	Lamp-1	Lamp-2	Limp I	Limp II
Synonyms				
Human	hlamp-1	hlamp-2	CD63 ME491	
Rat	Limp III lgp120	Limp IV lgp110	LIMP I	lgp85
Mouse	mlamp-1	mlamp-2		
Chicken	LEP100			
Apparent molecular mass (kDa)	90–120	95–120	35–55	60–85
Copies/cell (x10⁻⁴)	30–60	20–60	<20	<20
Polypeptide chain (residues)	382–396	381–389	238	478
Number of *N*-glycans attached	17–20	16–17	3	up to 11
O-glycans	5–6	7–8	?	?
Carbohydrate content (%)	55–65	55–65	30–55	20–45
Tyr motif	Yes	Yes	Yes	No
Leu-Ile motif	No	No	No	Yes

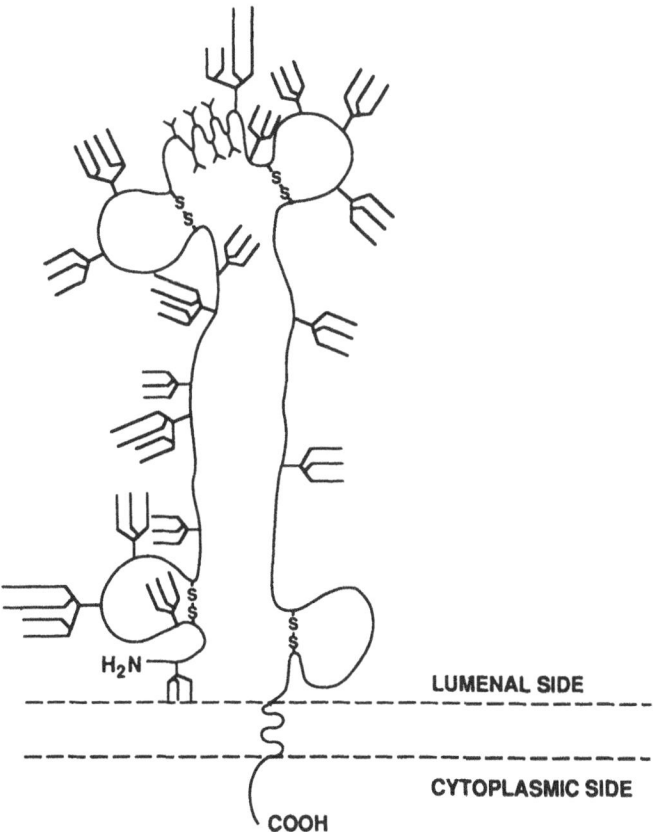

FIGURE 1. Depicted structure of lamp-1. *N*-glycans and *O*-glycans are indicated by ψ and Y, respectively. Some of the *N*-glycans are bulky poly-*N*-acetyllactosamine, while *O*-glycans are attached to the hinge region. Because the hinge region can move freely, lamp-1 probably has more than one configuration. Lamp-2 has an almost identical configuration (From Fukuda, 1991, and Arterburn, *et al.*, 1990).

in the immunoglobulin superfamily. In addition, the COOH-terminal ends of these putative loops could contain a consensus sequence, Tyr-X-Cys, found in immunoglobulin-related domains (Williams and Barclay, 1988). The protein chemical analysis demonstrated, however, that the neighboring half-cysteine residues are connected to each other, forming four disulfide loops in the lumenal portion (Carlsson and Fukuda, 1989; Arterburn, 1990). The results indicate that lamp molecules represent a new family of membrane glycoproteins different from the immunoglobulin superfamily.

4. The proline-rich region is homologous to the immunoglobulin hinge-region. This hinge-region is flexible in structure, and lamp molecules

can be bent at this region (Carlsson and Fukuda, 1989). Throughout species, the hinge-region of lamp-1 molecules is enriched with proline and serine residues, whereas that of lamp-2 is enriched with proline and threonine residues (Sawada *et al.*, 1993a).

5. Comparison of the amino acid sequences of lamp-1 and lamp-2 revealed the following interesting features. Human lamp-1 is more homologous to chicken lamp-1 (51.5% identity) than it is to human lamp-2 (36.7%). Similarly, lamp-2 from one species is more homologous to lamp-2 from another species than to lamp-1 of the same species.

6. The recently elucidated structure of the genomic DNA that encodes human lamp-1, lamp-2, and chicken lamp-1 has the following characteristics (Zot and Fambrough, 1990; Sawada *et al.*, 1993a). Human lamp-1 and chicken lamp-1 have similar gene organization in terms of intron size, whereas that of human lamp-2 differs from the above two genes. However, human and chicken lamp-1 and human lamp-2 have identical exon usage. The coding triplet of introns 1, 3, 4, 7, and 8 is interrupted at the first and second nucleotides. Introns 2, 5, and 6, on the other hand, have no interruption in their respective coding triplets. Moreover, each exon of the lamp-1 and lamp-2 genes encodes almost identical portions of these genes (Figure 2). These results strongly suggest that lamp-1 and lamp-2 were derived from the same primordial gene. The results, however, also strongly suggest that such duplication took place early in evolution, but both lamp-1 and lamp-2 have conserved their structures relatively well during evolution. Thus, it is likely that lamp-1 and lamp-2 have distinct functions. Consistent with this conclusion is data indicating that human lamp-1 and lamp-2 are encoded by genes located on different chromosomes, 13q34 and Xq24, respectively (Mattei *et al.*, 1990; Manoni *et al.*, 1991).

7. It was shown that the two intralumenal domains separated by a hinge-like structure are homologous to each other (Chen *et al.*, 1988; Carlsson *et al.*, 1988). It was thus suggested that a primordial gene encoding the NH_2- or the COOH-terminal half of the intralumenal portion was duplicated, and then a gene segment encoding the hinge-region was inserted during and after this duplication. However, the genomic structure does not support such a hypothesis. This is because exons II, III, and IV do not correspond to exons VI, VII, and VIII. This is particularly evident since, in contrast to other exons, exon VII does not code for cysteine (Figure 2). The results suggest instead that the NH_2-terminal half and the COOH-terminal half of the intralumenal domain, the hinge-like domain, and the transmembrane and cytoplasmic domains are assembled to form this gene (Sawada *et al.*, 1993a).

8. The intralumenal domain is extensively decorated by a significant num-

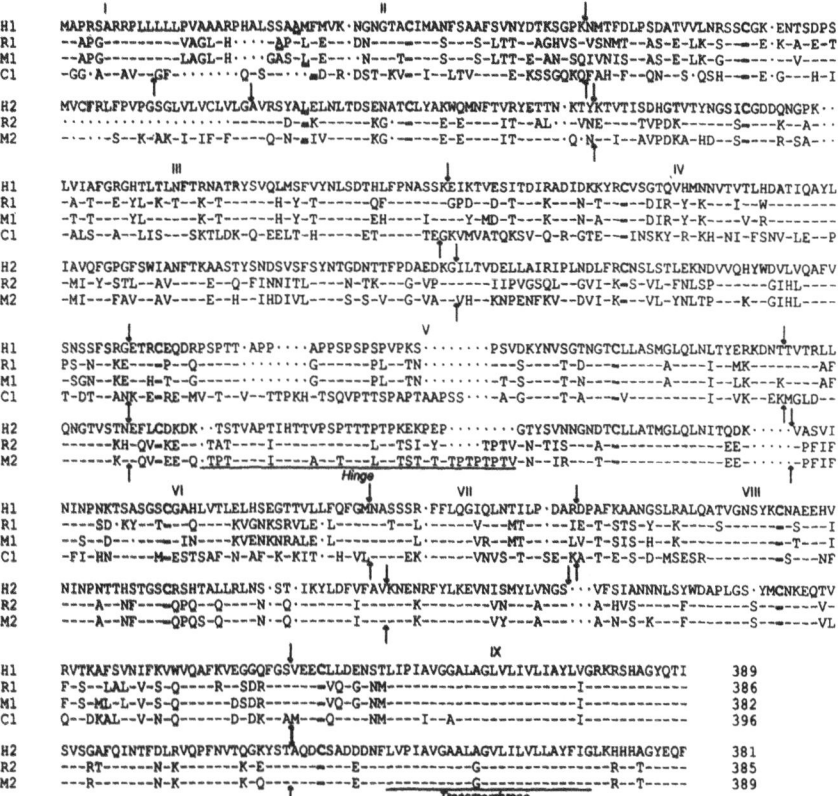

FIGURE 2. Comparison of the amino acid sequences of lamp-1 and lamp-2 from various species. H1, R1, M1, and C1 denote human, rat, mouse, and chicken lamp-1, respectively. H2, R2, and M2 denote human, rat, and mouse lamp-2, respectively. The amino acid residues identical to human lamp-1 or lamp-2 are denoted by hyphens. The dots denote a gap and the vertical arrows indicate the position of the introns. The number of the exon encoding each domain is shown by a roman numeral. The amino terminus of mature proteins is indicated by bold underlines. The intron 1 in human lamp-1 and introns 1 and 7 in mouse lamp-2 have not been determined yet. The genomic analysis of RI, MI, and R2 have not yet been carried out (From Sawada *et al.*, 1993a).

ber (16–20) of *N*-glycans, some of which are very complex poly-*N*-acetyllactosamines (see below). Lamp-1 and lamp-2 also contain O-glycans that are exclusively attached to hinge-like regions (Carlsson *et al.*, 1993). It is critical to have O-glycans in these molecules in order to protect the hinge-region from proteolytic digestion. It is expected that molecules lacking O-glycans would be cleaved into two halves in lysosomes. The carbohydrase constitutes 55–65% of the total mass of the lamp molecules.

3.2. Structure of Limp I and Limp II

In addition to the lamp-1 and lamp-2 described above, Barriocanal *et al.* (1986) found there are two other major lysosomal membrane glycoproteins that are smaller than lamp-1 and lamp-2. They are designated Limp I and Limp II, according to the size of the molecules. Although Limp I and Limp II are not related to lamp-1 and lamp-2, they share some properties with the lamp molecules. Limp I is identical to CD63, from which cDNA was isolated (Metzelaar *et al.*, 1991). CD63 was originally identified as the molecule present in activated platelets (Nieuwenhuis *et al.*, 1987). It was also shown that this molecule is induced on the cell surface of basophilic granulocytes after activation of formyl-methionyl-leucyl-phenylalanine. Limp I consists of 238 amino acid residues, but mature Limp I behaves as a heterogenous molecule with Mr ~30,000 to ~53,000 (Metzelaar *et al.*, 1991).

Judging by the heterogeneity and the number of bands in SDS-gel electrophoresis, it is likely that its *N*-glycosylation sites are variably utilized. The extreme heterogeneity in apparent molecular weight strongly suggests that Limp I also contains a large amount of poly-*N*-acetyllactosamine. The molecule does not have a typical cleavable leader peptide at its NH$_2$-terminus. Instead, the hydrophobic domains separately traverse the lipid bilayer three times, as depicted in Figure 3. There are several cysteine residues, but it is not known whether they form disulfide loops. The COOH-terminal cytoplasmic tail, which presumably extends toward the cytoplasm, has an amino acid sequence containing a tyrosine residue, which is the fourth residue from the COOH-terminal end. This sequence, which includes the bulky hydrophobic amino acid (methionine) at the COOH-terminal end, has the same motif as the tyrosine-containing motif that was shown to serve as a transport signal to lysosomes for lamp-1 and lamp-2 (see Table I). In this sense, Limp I belongs to a member of lamp-1 and lamp-2. Although no specific function of Limp I was identified, it is noteworthy that this molecule is identical to ME491 (Hotta *et al.*, 1988). ME491 was identified as a

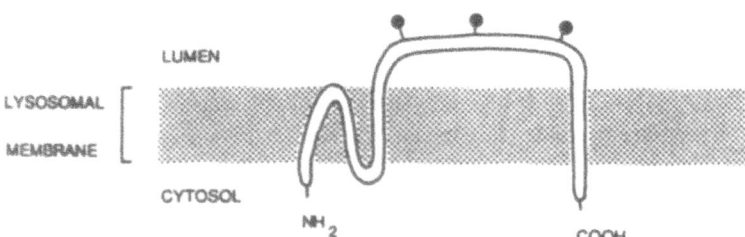

FIGURE 3. Model for the orientation of Limp I in the lysosomal membrane. The intralumenal region contains three *N*-glycosylation sites that are utilized. The protein transverses the lipid bilayer three times. The intralumenal domain contains six half-cysteine residues.

tumor-associated antigen, abundant in human malignant melanomas, neuroen-docrine tumors, and adrenocarciomas (Atkinson *et al.*, 1984; Ernst *et al.*, 1986). Thus, Limp I was also shown to be present at the plasma membrane.

cDNA from Limp II was isolated from rat (Vega *et al.*, 1991b) and human (Fujita *et al.*, 1992) cells. This protein has 477 or 478 amino acid residues, and the mature protein shows an apparent molecular weight of 74,000~85,000. The protein has two transmembrane residues—one is an uncleavable signal/anchor peptide, and the other is the transmembrane domain near the COOH-terminus (Figure 4). The presumed intralumenal portion contains 11 potential *N*-gly-cosylation sites, many of which must be utilized since the mature protein is much larger than the molecular weight of the polypeptide backbone. It has five cysteine residues, some of which participate in disulfide loop formation. Limp II is very homologous to CD36, which apparently participates in the adhesion of various cells (Tandon *et al.*, 1989; Greenwalt *et al.*, 1992). Both proteins contain cys-teine residues at identical positions in seven out of ten residues found in CD36. Moreover, CD36 contains both *N*- and O-linked oligosaccharides, suggesting that Limp II is likely to contain O-glycans as well. It was shown that *N*-linked oligosaccharides of CD36 are responsible for the resistance of membrane-bound CD36 to proteolysis (Greenwalt *et al.*, 1991). Because CD36 is exposed to endogenous protease in blood and in milk, the *N*-linked oligosaccharides and most likely O-linked oligosaccharides appear to be important for protecting them against proteolysis. Despite these similarities, the COOH-terminal sequences of Limp II and CD36 are different. The cytoplasmic tail of CD36 is short, consist-

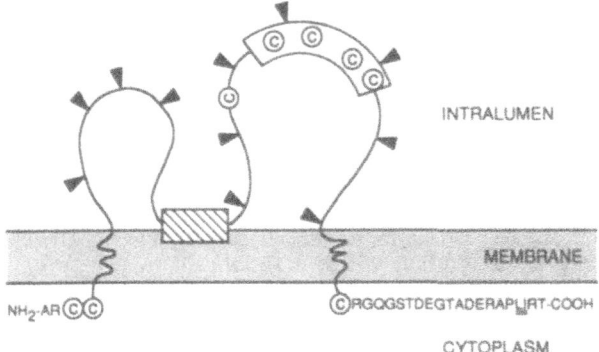

FIGURE 4. Schematic representation of Limp II. The protein transverses the lipid bilayer twice, and one segment in the intralumenal domain may be associated with the membrane, depicted by the hatched box. Cysteine residues are indicated by circles, and the open box denotes a proline-rich region. Solid triangles show potential *N*-glycosylation sites. The amino acid sequences in the presumed cytoplasmic domains are shown.

ing of six amino acids, among which there are three basic amino acids (Oquendo *et al.*, 1989).

Limp II, on the other hand, has 20 amino acid residues that are distinct from those present in CD36. Considering that CD36 is a plasma membrane glycoprotein and Limp II is mainly located in lysosomes, it is expected that the unique cytoplasmic tail of Limp II is responsible for lysosomal transport (see below). It may be noteworthy that both Limp I and Limp II have a molecular architecture similar to type-III membrane glycoproteins (Wickner and Lodish, 1985). Erythrocyte band 3 is one of the first proteins to be identified as a type-III membrane glycoprotein, and it is the first glycoprotein shown to contain poly-*N*-acetyllactosamines. It is likely that such hydrophobic membrane proteins tend to be modified by poly-*N*-acetyllactosamines, as discussed previously (Fukuda *et al.*, 1984). In any event, it is apparent that these lysosomal membrane glycoproteins are characterized by extensive glycosylation in the lumenal portion of the molecule.

4. ROLES OF CYTOPLASMIC TAILS IN INTRACELLULAR TRAFFIC

4.1. Roles of the Cytoplasmic Tail of Lamp Molecules

Several observations suggest that traffic of lysosomal membrane glycoproteins to lysosomes is mediated by mechanisms different from those of lysosomal enzymes. First, it was shown that the lysosomal membrane glycoproteins reach the lysosomes even when *N*-glycan synthesis was inhibited by tunicamycin (Barriocanal *et al.*, 1986). Second, subcellular localization examined by immunoelectron microscopy indicates that the lysosomal membrane glycoproteins reside predominantly in lysosomes, whereas mannose-6-phosphate receptors reside mainly in the Golgi complex and prelysosomal compartments (Brown *et al.*, 1986; Griffith *et al.*, 1988). These results indicate that the molecular signal for targeting of lysosomal membrane glycoproteins differs from that of lysosomal enzymes and most likely is present in the peptide moiety of lysosomal membrane glycoproteins. It was also predicted that the peptide signal essential for lysosomal trafficking is the portion where a significant homology exists between lamp-1 and lamp-2 molecules. In addition, the extracytoplasmic domain or lumenal domain of both lamp-1 and lamp-2 are heavily decorated by glycans so that the polypeptide backbone of that domain is unlikely to be recognized. Thus, the cytoplasmic segment is a prime candidate for the peptide signal. This prediction is supported by the fact that the His-Ala-Gly-Tyr sequence is present in both lamp-1 and lamp-2 molecules. By expressing lamp-1 cDNA in monkey COS-1 cells, it was first determined that the tyrosine residue in the cytoplasmic tail is essential for the delivery of lamp molecules to lysosomes. Furthermore, it was demonstrated that the cytoplasmic tail of lamp-1 enables a reporter molecule to be delivered to lysosomes (Williams and Fukuda, 1990). It was also noteworthy that by changing the relative position of a tyrosine residue in the cytoplasmic tail the delivery signal was abolished. These results demonstrated that the tyrosine residue, in conjunction with the whole cytoplasmic tail, provides a sufficient signal for trafficking of lamp molecules to lysosomes (Williams and Fukuda, 1990).

Other studies revealed that the cytoplasmic tyrosine is essential for trafficking during receptor-mediated endocytosis involving low-density lipoproteins (Davis *et al.*, 1987), transferrin (Jing *et al.*, 1990), and mannose-6-phosphate (Jadot *et al.*, 1992) receptors. The amino acid sequence surrounding this tyrosine consists of a motif expressed as Tyr-X-X-R, where R is a bulky hydrophobic amino acid (Table I). As shown in Figure 5, the tyrosine-containing sequence present in lamp-1, lamp-2, and Limp I also contains this motif. It is most likely that changing the position of the tyrosine abolished the lysosomal signal because tyrosine in such positions was not in the form of Tyr-X-X-R. If it were in this form, a tyrosine introduced into an otherwise dormant sequence would give an endocytosis signal, as shown in the influenza virus hemagglutin (Lazarovitz and Roth, 1988). When a tyrosine was introduced into a position that did not satisfy the requirement for the tyrosine motif, such an amino acid sequence did not work as a signal. Almost identical results were obtained on trafficking of lysosomal acid phosphatase, LAP, to lysosomes. Although LAP is eventually processed to a soluble enzyme in lysosomes, it is initially delivered to lysosomes as a membrane protein. This delivery was also found to be dependent on its cytoplasmic tyrosine (Peters *et al.*, 1990). The sequence surrounding this tyrosine residue also conforms to the requirement for the tyrosine motif.

Why then does such a requirement exist for the tyrosine-containing motif? Molecular modeling and nuclear magnetic resonance (NMR) analysis showed that this tyrosine motif is critical to the presentation of the tyrosine residue in a β-turn structure. Such a turn structure is found in the peptide sequence synthesized according to the amino acid sequences of LDL (Bansal and Gierasch, 1991), LAP (Eberle *et al.*, 1991; Lehmann *et al.*, 1992) and transferrin receptors (Collawn *et al.*, 1990). Such conformation is apparently important to present efficiently the tyrosine motif, Tyr-X-X-R, to a carrier molecule that recognizes a cytoplasmic tyrosine (see below).

4.2. Roles of the Cytoplasmic Tail of the Limp II Molecule

As described previously, Limp II is different from Limp I, lamp-1, and lamp-2 because it lacks a cytoplasmic tyrosine. The amino acid sequence of its cytoplasmic tail is entirely different from those proteins with a tyrosine motif. It

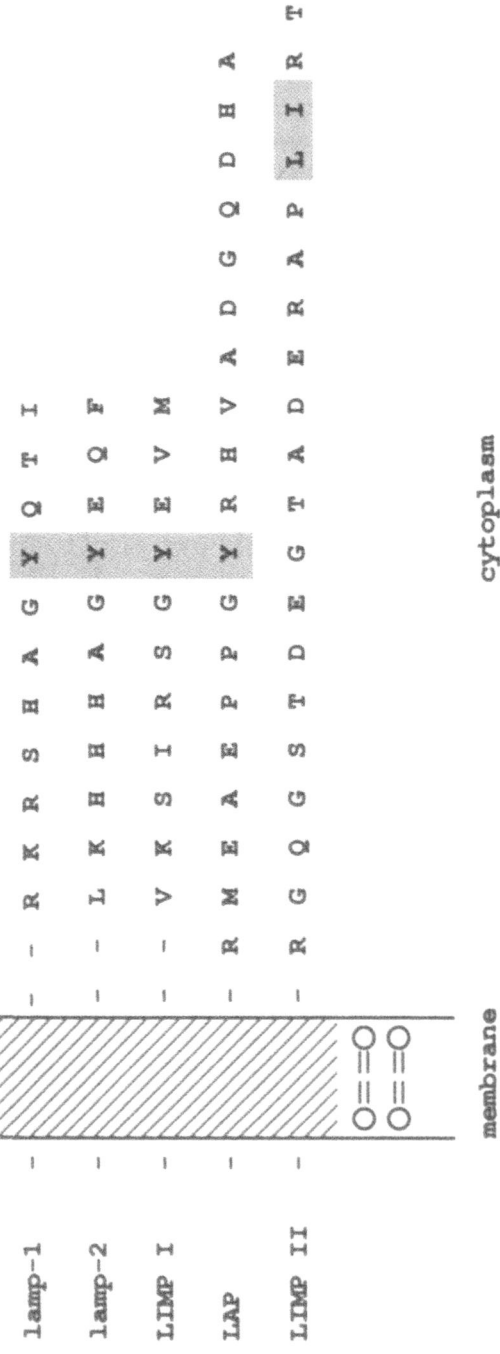

FIGURE 5. Comparison of cytoplasmic tail sequences of membrane proteins transported to lysosomes. The sequences shown are for human lamp-1, lamp-2, Limp I, and LAP (lysosomal acid phosphatase), and rat Limp II. Lamp-1, lamp-2, Limp I, and LAP share the tyrosine motif (see Table II), while Limp II has the Leu-Ile motif (see Table III).

was then expected that Limp II must utilize a distinct amino acid sequence for its trafficking to lysosomes.

First, Vega *et al.* (1991a) made a chimeric protein in which the cytoplasmic tail had the Limp II sequence. Once this protein was expressed, it was transported to lysosomes. When the cytoplasmic tail contained the CD8 sequence, however, the chimeric protein was not transported to lysosomes. These results established that the 20-amino acid cytoplasmic tail of Limp II contains a necessary and sufficient signal for lysosomal targeting. To identify the amino acid residues critical for this targeting, Ogata and Fukuda (1993) performed the following experiments. First, amino acid residues were gradually deleted from the COOH-terminal end of the cytoplasmic tail. The results indicated that the deletion of the last three or four residues abolished the delivery signal. Thus, Leu-Ile-Arg-Thr were identified to be critical (see Figure 5). When Leu or Ile were each replaced with alanine, sorting capacity was impaired. Alanine replacement of the other individual amino acids, including the proline residue next to the Leu-Ile sequence, however, had minimal effect. It was also noted that the replacement of Arg had minimal effect, although the deletion of this amino acid did somewhat impair the sorting capacity. These results established that the lysosomal targeting of Limp II requires a novel Leu-Ile-X motif.

It was also noted that a sequence similar to the Leu-Ile motif is found in both CI-MPR and CD-MPR at their COOH-terminal end. In addition, there is an internal homologous sequence in CI-MPR that resides 35 residues from the COOH-terminus. When the last nine amino acid sequence in the Limp II tail were replaced with these CI-MPR homologous sequences, both sequences worked well as a lysosomal targeting signal (Ogata and Fukuda, 1993). These results established that Limp II and MPR share the same signal for their transport to lysosomes. The next question asked was whether such a Leu-Ile motif can be transported next to the transmembrane. However, in this series of experiments deletion of the ten amino acid residues from the transmembrane domain impaired its sorting function. Moreover, the replacement of serine five residues from the transmembrane domain impaired the sorting capacity, while replacement of the threonine ten residues from the transmembrane domain had no effect (see Figure 5). These two hydroxyl amino acids are potential sites for casein kinase II phosphorylation, although no evidence of phosphorylation was obtained, while these same experiments did provide evidence for phosphorylation of the MPR tail. This latter observation is consistent with the recent report (Méresse and Hoflack, 1993). However, no impairment of the lysosomal targeting signal was observed when all potential phosphorylation sites of the CI or CD-MPR tail were mutated (Johnson and Kornfeld, 1992b). These results established that a novel Leu-Ile (or Leu-Leu) motif must be present in a specific configuration to signal the transport of Limp II to lysosomes. The spacer sequence between the transmembrane and the Leu-Ile motif is apparently critical for lysosomal targeting of

Limp II, while such function was not apparent when the 64-residue or the 163-residue tail of MPR was mutated. This is probably because the Limp II tail is much shorter than that of MPR. The short Limp II tail is much less likely to tolerate change(s) in amino acid sequences.

These combined results predict that the Leu-Ile motif in Limp II and the MPR utilizes the same mechanisms for sorting of those proteins to prelyso-some/lysosome compartments. It was shown that the CD-MPR is directly sorted at the Golgi network by recognition of the Leu-Leu motif (Johnson and Kornfeld, 1992a). These results strongly suggest that the proteins utilized the Leu-Leu or Leu-Ile motif are directly sorted at the *trans*-Golgi or the *trans*-Golgi network. Related to this finding, it was shown that lamp-1 and lamp-2 apparently utilize the same carrier, since overexpression of the one protein resulted in cell surface expression of the other (Harter and Mellman, 1992). However, lamp-1, lamp-2, and Limp II were not apparently recognized by the same carrier since overexpression of lamp-1 did not lead to cell surface expression of Limp II (Ogata and Fukuda, 1993). Considering that the majority of lamp molecules are also directly sorted at the *trans*-Golgi (see below), it is important to identify the molecule(s) that recognize the tyrosine-containing motif and the Leu-Ile motif, respectively.

4.3. Possible Roles of Adaptors for Recognition of the Tyrosine-Containing Motif and the Leu-Ile Motif

To understand the mechanisms of lysosomal targeting, it is necessary to understand how the tyrosine-containing motif or the Leu-Ile motif is utilized for their transport. The proteins with the tyrosine-containing motif are endocytosed through clathrin-coated pits. Pearse and Robinson (1990) isolated a class of molecules, termed *adaptors,* that constitute the coated pits. Together with clathrin, adaptors are found in coated pits that are present both at the plasma membrane and in the Golgi region. When adaptors were isolated from these two different sources, it was found that each adaptor consists of four distinct subunits. These two different adaptors are separated by hydroxyl-apatite column chromatography, and are termed the HA-1 and HA-2 adaptors, according to the elution positions from the column. Thus, HA-1 consists of β'-adaptin, γ-adaptin, a 47 kDa subunit, and a 20 kDa subunit; HA-2 consists of β-adaptin, α-adaptin, a 50 kDa subunit, and a 17 kDa subunit. Among them, the β-subunit has some homology with the other β-subunit, but no other subunits are related to each other.

By immunofluorescence, the HA-1 adaptor is found in Golgi coated pits and the HA-2 adaptor is found in the plasma membrane (Robinson and Pearse, 1986). Because endocytosis from the plasma membrane has been studied more extensively than sorting at the Golgi, studies on the HA-2 adaptor have provided some critical information. First, the HA-2 was shown to interact with tyrosine residues of the CI-MPR cytoplasmic tail, which are critical for the rapid endocytosis of the CI-MPR. If those tyrosine residues are mutated, the HA-2 no longer binds to the CI-MPR.

In contrast, the Golgi adaptor HA-1 binds to a segment of the cytoplasmic tail of MPR, which is different from the tyrosine-containing segment. The binding of HA-1 was not affected at all by mutation at the tyrosine residues (Glickman *et al.*, 1989). As discussed above, it was shown that the Leu-Leu motif in MPR is required for the direct sorting at the Golgi region (Johnson and Kornfeld, 1992a). The combined results thus suggest that the Leu-Leu or the Leu-Ile motif is required for the binding of the Golgi adaptor to the cytoplasmic tails of Limp II or MPR. Further studies are of significance in order to understand the roles of the Leu-Leu or the Leu-Ile motif in intracellular sorting and endocytosis. In relation to these studies, it is worth mentioning that the intracellular degradation of CD3 is mediated by the same Leu-Leu motif. By analyzing the roles of CD3 tails in transport to lysosomes, Letourneur and Klausner (1992) obtained almost identical results. In their studies, the signal was functional only when certain deletions in the cytoplasmic tail took place. Similarly, such a signal was found in one of the human invariant chains, Iip31 (Lotteau *et al.*, 1990). This protein is transported to lysosomes if it is expressed, whereas a similar protein lacking this motif stays in the ER. These results indicate that Leu-Ile motif is utilized in a wide variety of proteins that are transported to lysosomes.

As described above, it is likely that the tyrosine-containing motif is also utilized for direct sorting at the *trans*-Golgi. It is thus possible that the Golgi adaptor also binds to this motif or the plasma membrane adaptor is utilized. If the latter is the case, the distribution of the plasma membrane adaptor, HA-2, is less restricted, so a small amount of HA-2 should also be present around the Golgi complex. In fact, data obtained thus far do not exclude this possibility. Alternatively, it is possible that there is another adaptor molecule at the Golgi complex that utilizes the tyrosine-containing motif. Further studies are necessary to resolve the issue.

5. CELL SURFACE EXPRESSION OF LYSOSOME MEMBRANE GLYCOPROTEINS

Although lysosomal membrane glycoproteins are predominantly found in lysosomes and endosomes, they can also be present at the plasma membrane. For instance, Mac-3, which is identical to mouse lamp-2, was originally identified as a macrophage cell surface molecule (Ho and Springer, 1983). Limp I is identical to CD63 and ME491, and CD63 was originally determined to be a cell surface antigen on activated platelets. ME491, on the other hand, was found on human malignant melanoma cells (Hotta *et al.*, 1988). Surface expression of Limp I on activated platelets and basophils correlates with the exocytosis of lysosomal contents.

Additionally, increased amounts of lamp-1 and lamp-2 are on the cell surface of activated platelets (Febbraio and Silverstein, 1990; Silverstein and Febbraio, 1992). These results are consistent with the notion that surface expression of lysosomal membrane proteins is achieved by fusion of the lysosomal membrane with the plasma membrane. However, considering the following case, the presence of lysosomal membrane proteins may not simply be a by-product of lysosome-plasma membrane fusion. Cytotoxic T-lymphocytes exocytose granules during specific interaction with target cells. Such granules contain a family of serine esterases and lethal protein, perforin. Immunocytochemical examinations revealed that lamp-1, lamp-2, and CD63 are present on the limiting outer membrane in mature granules, while mannose-6-phosphate can also be detected in immature granules (Peters *et al.*, 1991a). After granules fuse with the plasma membrane, lysosomal membrane glycoproteins apparently protect the plasma membrane from the contents of granules. These results indicate that lysosomal membrane glycoproteins are utilized when the cytoplasm and the cell membrane need protection from hydrolases and other lethal proteins by sequestering those proteins in separate vesicles.

In relation to these findings, it is worth mentioning the results obtained on MHC class II molecules. MHC class II molecules preferentially bind peptides derived from exogenous antigens degraded in the exocytic pathway. Immunoelectron microscopic examination revealed that MHC class II molecules are transported to structures with lysosomal characteristics. These molecules are segregated from other molecules, including the MHC I molecules, most likely at the Golgi complex (Peters *et al.*, 1991b). It is now evident that the invariant chain directs the MHC class II complex to lysosomal compartments, because the invariant chain contains a Leu-Leu motif (Lotteau *et al.*, 1990). These results suggest that lysosomal membrane glycoproteins and the lysosomal targeting signal play critical roles in the many occasions where membranes are fused with endosomal/lysosomal compartments.

In certain instances, cell surface expression of lamp-2 may be indicative of some of the immunological responses. For instance, it was found that there is an increased amount of lamp-2 expressed on the cell surface of lymphocytes derived from scleroderma patients, compared to those of normal individuals (Holcombe *et al.*, 1993). This increased expression is probably caused by the chronic activation of B-lymphocytes in such autoimmune disease. Moreover, it was recently discovered that the amount of cell surface lamp-2 is likely to reflect platelet activation in cancer patients with prethromboitic conditions (Kannan *et al.*, 1993). This research reinforces the importance of understanding how lamp-2 is transported to the plasma membrane, rather than to lysosomes, in normal and pathologic conditions.

With respect to how the proteins are transported to lysosomes, the cell surface expression of lysosomal membrane glycoproteins poses another question. Because the tyrosine-motif can be utilized for rapid endocytosis, there is a possibility that lysosomal membrane proteins are first transported to the plasma membrane and then directed to lysosomes through the endocytic pathway. In this case, lysosomal membrane proteins are not sorted at the Golgi complex, but instead are transported to the plasma membrane simply by a default pathway. The lysosomal membrane proteins are then sorted at the plasma membrane and directed into the endocytic pathway. Once the proteins reach prelysosomal compartments, other proteins, such as receptors, have a signal that directs the proteins back to the plasma membrane or to the *trans*-Golgi (in the case of MPR). The lysosomal membrane proteins, on the other hand, lack such a signal; thus, they are transported to lysosomes as prelysosomes mature to lysosomes.

An alternative pathway is that the lysosomal membrane proteins are directly sorted at the Golgi complex and then transported to lysosomes via prelysosome compartments. In this case, other receptors such as MPRs are not transported to lysosomes since they have a signal that redirects the receptors to the *trans*-Golgi. If the receptor is transported to lysosomes, those receptors are susceptible to intralumenal acid hydrolases and will not be detected.

The data accumulated thus far indicate the latter case (Green *et al.*, 1987; Carlsson and Fukuda, 1992; Harter and Mellman, 1992). In particular, the appearance of newly synthesized lamp in lysosomes is much sooner than would be expected if the proteins had reached the plasma membrane and then were transported to lysosomes. Moreover, it was recently shown that mannose-6-phosphate receptors are directly sorted at the *trans*-Golgi (Johnson and Kornfeld, 1992a). However, it is still possible that the routing via the plasma membrane could be major traffic for some of the lysosomal membrane proteins.

For example, it was shown that LAP (lysosomal acid phosphatase) is first transported to the plasma membrane and then enters the endocytic pathway. After repeated recycling between the plasma membrane and endosome compartments, the protein is eventually transported to lysosomes (Braun *et al.*, 1989). This was also suggested for trafficking of LEP-100, chicken lamp-1 (Mathews *et al.*, 1992). This intracellular movement is exactly the same as that exhibited by receptors for transferrin and low density lipoprotein. The weakness in this work is that LAP was overexpressed in these experiments so that sorting at the Golgi complex could be saturated and the majority of LAP molecules thus needed to be sorted by an endocytic pathway, although the endocytic pathway is a secondary pathway to retrieve LAP. However, it is also possible that LAP utilizes this endocytic pathway as a major route. Lamp-1, lamp-2, and Limp I have a tyrosine-motif at the COOH-terminal end. LAP, on the other hand, has this motif in the internal sequence and additional amino acids are present at the COOH-terminal site of the tyrosine motif (see Figure 5). Almost all of the other receptors involved in rapid endocytosis also contain the tyrosine-motif at the internal sequences (Table I). If the adaptor in the Golgi complex only recognizes the tyrosine-motif at the COOH-terminus, those receptors and LAP are prevented from binding to the adaptors at the *trans*-Golgi and can only be recognized by the plasma membrane adaptor. On the other hand, the tyrosine motif in lysosomal membrane glycoproteins is recognized by the Golgi adaptor because it is present at the COOH-terminus. This is a plausible hypothesis because the receptors need to be at the plasma membrane. Also, if these proteins must be sorted directly at the Golgi complex, they must have another motif, Leu-Leu (or Leu-Ile), as shown for MPR. Future studies are necessary to determine whether this hypothesis is correct (see also Figure 6).

6. ROLES OF CARBOHYDRATES IN LAMP MOLECULES

As mentioned above, lamp molecules were identified as having a significant amount of poly-N-acetyllactosamine, which is composed of Galβ1 → 4GlcNAcβ1 → 3, N-acetyllactosamine repeats in side-chains (Carlsson *et al.*, 1988; Viitala *et al.*, 1988). These long side-chains appear to be better acceptors for various glycosyltransferases, including fucosyltransferases (Do *et al.*, 1990). This is the reason why the terminus of poly-N-acetyllactosamine often contains unique structures such as sialyl Le^x, NeuNAc2 → 3Galβ1 → 4 (Fucα1 → 3) GlcNAc (for review see Fukuda, 1985).

Lamp molecules were identified as the major carriers for poly-N-acetyllactosamines in many cells. The half-life of these lysosomal membrane proteins was appreciably decreased when they were synthesized in the presence of tunicamycin, which inhibits N-glycosylation (Barriocanal *et al.*, 1986). For instance, the half-life of Limp II in tunicamycin-treated normal rat kidney (NRK) cells exhibited only 5% of that of control cells without tunicamycin treatment. The half-lives of lamp-1 and lamp-2 were dramatically increased when HL-60 cells were induced to differentiate into granulocytic cells. During this differentiation, more of N-glycosylation sites acquired poly-N-acetyllactosamines (Lee *et al.*, 1990). Thus, a high content of carbohydrates, in particular bulky carbohydrates such as poly-N-acetyllactosamine, apparently contributes to the stability of lamp and Limp molecules. It is most likely that the peptide moiety containing disulfide-loops also contributes to resistance to intralumenal lysosomal enzymes since these proteins are resistant to proteases unless they are reduced. Future studies will be of significance in order to address these points.

Nonetheless, the abundance of lamp and Limp molecules is so great that these molecules may form a nearly continuous coat on the inner surface of the lysosomal membrane, serving as a barrier to soluble hydrolases (Granger *et al.*, 1990). These results indicate that poly-N-acetyllactosamines attached to lysoso-

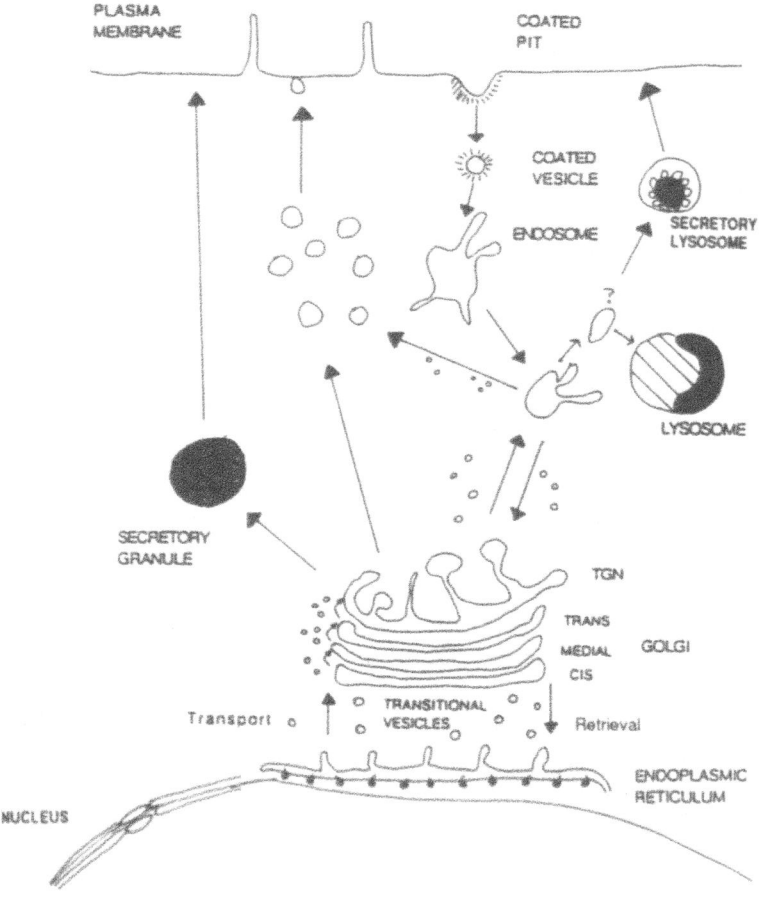

FIGURE 6. Possible trafficking of newly synthesized proteins. Newly synthesized lysosomal membrane proteins exit from the *trans*-Golgi and enter the endosome compartment. The majority of lysosomal membrane proteins are transported to prelysosomes together with lysosomal enzymes. Once these proteins reach the prelysosome compartment, the majority of them are transported to lysosomes, but some of them are transported to the plasma membrane and exocytosed. The latter movement is more prominent in special cases, such as in cytotoxic T-lymphocytes. A small portion of newly synthesized lysosomal membrane proteins and lysosomal enzymes are transported to the plasma membrane, probably through a default pathway. These proteins then return to endosomes, and then to prelysosomes through the endocytic pathway, and eventually they are transported to lysosomes.

mal membrane glycoproteins play a critical role in maintaining the stability of lysosomal membranes. It is thus conceivable that differentiated HL-60 cells contain more active lysosomes because they have more stable lysosomes. This is reasonable since these cells are more phagocytic and require more functional lysosomes. It is possible that differentiated CaCO-2 colonic tumor cells require less lysosomes while differentiated F9 embryocarcinoma cells require more lysosomes, since the former decreases its content of poly-N-acetyllactosamines after differentiation, whereas the latter increases its content (Youakim et $al.$, 1989; Amos and Lotan, 1990).

The carbohydrates of lysosomal membrane glycoproteins are enriched with poly-N-acetyllactosamines. It was reported that tumor cells, including highly metastatic tumor cells, express more poly-N-acetyllactosamines than their normal counterparts or poorly metastatic counterparts (Yamashita et $al.$, 1984; Pierce and Arango, 1986; Yousefi et $al.$, 1991). Conversely, several glycosylation inhibitors, such as tunicamycin, castanospermine, and swainsonine, dramatically reduce tumor formation when they are administered into animals (Irimura et $al.$, 1981; Humphries et $al.$, 1986; Dennis, 1986). Those inhibitors either completely block N-glycan synthesis or reduce poly-N-acetyllactosamine since the latter two block the processing of N-glycans from high-mannose to complex-type oligosaccharides (Elbein, 1987). These results strongly suggest that the amount of poly-N-acetyllactosamine is directly related to the tumorigenicity or metastatic capability of tumor cells. Furthermore, it was shown in one of the studies that the major carrier for poly-N-acetyllactosamines is lamp-1 (Laferte and Dennis, 1989).

How then does the increased amount of poly-N-acetyllactosamines in lamp-1, and possibly other lysosomal membrane glycoproteins, lead to a higher incidence of tumor metastasis? Recently, it was discovered that sialyl Le^x, and its isomer sialyl Le^a, NeuNAcα2 $\rightarrow$ 3Galβ1 $\rightarrow$ 3 (Fucα1 $\rightarrow$ 4) GlcNAc, are ligands for adhesion molecules, E- and P-selectin, on endothelial cells and platelets (Lowe et $al.$, 1990; Phillips et $al.$, 1990; Walz et $al.$, 1990; Larsen et $al.$, 1990; Berg et $al.$, 1991). It also was repeatedly observed that tumor cells, in particular carcinoma cells, are enriched with sialyl Le^x and sialyl Le^a structures (Magnani et $al.$, 1982; Fukushima et $al.$, 1984; Hoff et $al.$, 1989) and that some tumor cells adhere to endothelial cells or platelets through E-selectin (Rice and Bevilacqua, 1989; Hession et $al.$, 1989). These results suggest that highly metastatic tumor cells bind more efficiently to endothelial cells at metastatic sites, because these cells are enriched with sialyl Le^x and sialyl Le^a structures.

It was also discovered that highly metastatic tumor cells express more lamp molecules on the cell surface than their low metastatic counterparts (Saitoh et $al.$, 1992). The most recent studies indicate that such highly metastatic tumor cells adhere more efficiently to E-selectin-expressing cells than their low metastatic counterparts (Sawada et $al.$, 1993b). The level of these surface lamp molecules can be increased by genetic manipulation. This is possible since either an overexpression of the lamp molecule or expression of mutated lamp increases the amount of cell surface lamp. The latter is feasible since lamp molecules with a mutated cytoplasmic tyrosine are transported to the plasma membrane by a default pathway instead of to lysosomes. When such a genetic manipulation was made, the increased level of cell surface lamp was proportional to the increased efficiency in adhesion to E-selectin-expressing cells (Sawada *et al.*, 1993b).

These results suggest that lamp molecules on the cell surface are critical because they present ligands for adhesion molecules. In tumor metastasis, it is known that tumor cells aggregated with platelets in blood circulation (Nicolson, 1989). Such aggregated tumor cells then could be trapped in capillary tubes, which could then trigger the activation of endothelial cells leading to expression of E-selectin. Tumor cells then slowed down in the bloodstream, and such "rolling" likely leads to stronger adhesion to endothelial cells through integrins and counter-receptor interactions. It was shown that a similar process takes place at inflammatory sites (Lawrence and Springer, 1991; Ley *et al.*, 1991). It is possible that these events result in the lodging of tumor cells in capillary beds at junctions between endothelials, establishing metastasis.

Recently, it was shown that lung inflammation induced by IgG immunocomplexes is mediated by E-selectin (Mulligan *et al.*, 1991). Moreover, it was demonstrated that such inflammation can be inhibited by administration of sialyl Le^x-glycopeptides (Mulligan *et al.*, 1993). Similarly, it was shown that tumor cell adhesion is inhibited by a genetically engineered soluble lamp-1 that contains sialyl Le^x structures (Sawada *et al.*, 1993b). It will be important to test whether the increased expression of lamp-1 at the cell surface leads to a higher incidence of metastasis, and if such a process can be inhibited by soluble lamp-1 expressing sialyl Le^x structures.

As shown in this chapter, lysosomal membrane glycoproteins play a critical role in the dynamics of membranes. Understanding the trafficking of these proteins, in addition to their roles in subcellular compartments and at the plasma membrane, is expected to reveal the significance of these molecules in nucleated cells.

ACKNOWLEDGMENTS. I wish to thank Drs. Sven Carlsson, Randall Holcombe, Ni Lee, Kentaro Maemura, Shigenori Ogata, Osamu Saitoh, Ritsuko Sawada, Juha Viitala, Wei-Chun Wang, and Mark Williams for providing data summarized in this chapter. I also thank Drs. Michiko Fukuda, John Lowe, Marie-Geneviève Mattei, and Jürgen Roth for collaborative efforts with us and Ms. Melissa Moore for preparing and editing the manuscript. The work in our laboratory was supported by RO1 CA48737 from the National Cancer Institute and in part by PO1 AI33189 from the National Institute of Allergy and Infectious Diseases.

7. REFERENCES

Amos, B., and Lotan, R., 1990, Modulation of lysosomal-associated membrane glycoproteins during retinoic acid-induced embryonal carcinoma cell differentiation, *J. Biol. Chem.* **265**:19192–19198.

Arterburn, L. M., Earles, B. J., and August, J. T., 1990, The disulfide structure of mouse lysosomes-associated membrane protein 1, *J. Biol. Chem.* **265**:7419–7423.

Atkinson, B., Ernst, C. S., Ghrist, B. F. D., Herlyn, M., Blaszczyk, M., Ross, A. H., Herlyn, D., Steplewski, Z., and Koprowski, H., 1984, Identification of melanoma-associated antigens using fixed tissue screening of antibodies, *Cancer Res.* **44**:2577–2581.

Bansal, A., and Gierasch, L. M., 1991, The NPXY internalization signal of the LDL receptor adopts a reverse-turn conformation, *Cell* **67**:1195–1201.

Baranski, T. J., Faust, P. I., and Kornfeld, S., 1990, Generation of a lysosomal enzyme targeting signal in the secretory protein pepsinogen, *Cell* **63**:281–291.

Barriocanal, J. G., Bonifacino, J. S., Yuan, L., and Sandoval, I. V., 1986, Biosynthesis, glycosylation, movement through the Golgi system, and transport to lysosomes by an *N*-linked carbohydrate-independent mechanism of three lysosomal integral membrane proteins, *J. Biol. Chem.* **261**:16755–16763.

Berg, E. L., Robinson, M. K., Mansson, O., Butcher, E. C., and Magnani, J. L., 1991, A carbohydrate domain common to both sialyl Lea and sialyl Le^x is recognized by the endothelial cell leukocyte adhesion molecule ELAM-1, *J. Biol. Chem.* **266**:14869–14872.

Braun, M., Waheed, A., and von Figura, K., 1989, Lysosomal acid phosphatase is transported to lysosomes via the cell surface, *EMBO J.* **8**:3633–3640.

Brown, W. J., Goodhouse, J., and Farquhar, M. G., 1986, Mannose 6-phosphate receptors for lysosomal enzymes cycle between the Golgi complex and endosomes, *J. Cell Biol.* **103**:1235–1247.

Burnside, J., and Schneider, D. J., 1982, Characterization of rat liver lysosomes: Composition, enzyme activities and turnover, *Biochem. J.* **204**:525–534.

Carlsson, S. R., and Fukuda, M., 1992, The lysosomal membrane glycoprotein lamp-1 is transported to lysosomes by two alternative pathways, *Arch. Biochem. Biophys.* **296**:630–639.

Carlsson, S. R., Roth, J., Piller, F., and Fukuda, M., 1988, Isolation and characterization of human lysosomal membrane glycoproteins, h-lamp-1 and h-lamp-2, *J. Biol. Chem.* **263**:18911–18919.

Carlsson, S. R., and Fukuda, M., 1989, Structure of human lysosomal membrane glycoprotein 1: Assignment of disulfide bonds and visualization of its domain arrangement, *J. Biol. Chem.* **264**:20526–20531.

Carlsson, S. R., and Fukuda, M., 1990, The polylactosaminoglycans of human lysosomal membrane glycoproteins, lamp-1 and lamp-2: Localization on the peptide backbones, *J. Biol. Chem.* **265**:20488–20495.

Carlsson, S. R., Lyckell, P.-O., and Fukuda, M., 1993, Assignment of O-glycan attachment sites to the hinge-like regions of human lysosmal membrane glycoproteins lamp-1 and lamp-2, *Arch. Biochem. Biophys.*, **304**:65–73.

Cha, Y., Holland, S. M., and August, J. T., 1990, The cDNA sequence of mouse LAMP-2, *J. Biol. Chem.* **265**:5008–5013.

Chen, J. W., Murphy, T. L., Willingham, M. C., Pastan, I., and August, J. T., 1985, Identification of two lysosomal membrane glycoproteins, *J. Biol. Chem.* **101**:85–95.

Chen, J. W., Cha, Y., Yudsel, K. U., Gracy, R. W., and August, J. T., 1988, Isolation and sequencing of a cDNA clone encoding lysosomal membrane glycoprotein mouse LAMP-1, *J. Biol. Chem.* **263**:8754–8758.

Cohn, Z. A., and Ehrenreich, B. A., 1969, The uptake, storage, and intracellular hydrolysis of carbohydrates by macrophages, *J. Exp. Med.* **129**:201–225.

Collawn, J. F., Stangel, M., Kuhn, L. A., Esekogwu, V., Jing, S., Trowbridge, I. S., and Tainer, J. A., 1990, Transferrin receptor internalization sequence YXRF implicates a tight turn as the structural recognition motif for endocytosis, *Cell* **63**:1061–1072.

Dahms, N. M., Lobel, P., and Kornfeld, S., 1989, Mannose 6-phosphate receptors and lysosomal enzyme targeting, *J. Biol. Chem.* **264**:12115–12118.

Davis, C. G., Lehrman, M. A., Russell, D. W., Anderson, R. G. W., Brown, M. S., and Goldstein, J. L., 1986, The J.D. mutation in familial hypercholesterolemia: Amino acid substitution in cytoplasmic domain impedes internalization of LDL receptors, *Cell* **45**:15–24.

Davis, C. G., van Driel, I. R., Russell, D. W., Brown, M. S., and Goldstein, J. L., 1987, The low density lipoprotein receptor, *J. Biol. Chem.* **262**:4075–4082.

de Duve, C., 1983, Lysosomes revisited, *Eur. J. Biochem.* **137**:391–397.

Dennis, J. W., 1986, Effects of swainsonine and polyinosinic: Polycytidylic acid on murine tumor cell growth and metastasis, *Cancer Res.* **46**:5131–5136.

Do, K.-Y., Smith, D. F., and Cummings, R. D., 1990, Lamp-1 in CHO cells is a primary carrier of poly-*N*-acetyllactosamine chains and is bound preferentially by a mammalian S-type lectin, *Biochem. Biophys. Res. Commun.* **173**:1123–1128.

Duncan, J. R., and Kornfeld, S., 1988, Intracellular movement of two mannose 6-phosphate receptors return to the Golgi apparatus, *J. Cell Biol.* **106**:617–628.

Eberle, W., Sander, C., Klaus, W., Schmidt, B., von Figura, K., and Peters, C., 1991, The essential tyrosine of the internalization signal in lysosomal acid phosphatase is part of a β turn, *Cell* **67**:1203–1209.

Elbein, A. D., 1987, Inhibitors of the biosynthesis and processing of *N*-linked oligosaccharide chains, *Annu. Rev. Biochem.* **56**:497–534.

Ernst, C. S., Shen, J. W., Litwen, S., Herlyn, M., Koprowski, H., and Sears, H. F., 1986, Multiparameter evaluation of the expression in situ of normal and tumor-associated antigens in human colorectal carcinoma, *J. Natl. Cancer Inst.* **77**:387–395.

Fambrough, D. M., Takeyasu, K., Lippincott-Schwartz, J., Siegel, N., and Somerville, D., 1988, Structure of LEP100, a glycoprotein that shuttles between lysosomes and the plasma membrane, deduced from the nucleotide sequence of the encoding cDNA, *J. Cell Biol.* **96**:61–67.

Febbraio, M., and Silverstein, R. L., 1990, Identification and characterization of LAMP-1 as an activation-dependent platelet surface glycoprotein, *J. Biol. Chem.* **265**:18531–18537.

Fedde, K. N., and Sly, W. S., 1985, Ricin-binding properties of acid hydrolases from isolated lysosomes implies prior processing by terminal transferases of the *trans*-Golgi apparatus, *Biochem. Biophys. Res. Commun.* **133**:614–620.

Fujita, H., Takata, Y., Kobata, A., Tanaka, A., Takahashi, T., Himeno, M., and Kato, K., 1992, Isolation and sequencing of a cDNA clone encoding the 85 kDa human lysosomal sialoglycoprotein (hL GP85) in human metastatic pancreas islet tumor cells, *Biochem. Biophys. Res. Comm.* **164**:604–611.

Fukuda, M., 1985, Cell surface glycoconjugates as onco-differentiation markers in hematopoietic cells, *Biochim. Biophys. Acta* **780**:119–150.

Fukuda, M., Dell, A., and Fukuda, M. N., 1985, Structure of fetal lactosaminoglycan: The carbohydrate moiety of band 3 isolated from human umbilical cord erythrocytes, *J. Biol. Chem.* **259**:4782–4791.

Fukuda, M., Viitala, J., Matteson, J., and Carlsson, S. R., 1988, Cloning of cDNAs encoding human lysosomal membrane glycoproteins, h-lamp-1 and h-lamp-2, *J. Biol. Chem.* **263**:18920–18928.

Fukuda, M., 1991, Lysosomal membrane glycoproteins: Structure, biosynthesis and intracellular trafficking, *J. Biol. Chem.* **266**:21327–21330.

Fukushima, K., Hirota, M., Terasaki, P. I., Wakisaka, A., Togashi, H., Chia, O., Suyama, N., Fukushi, Y., Nudelman, E., and Hakomori, S-I., 1984, Characterization of sialosylated Lewis[x] as a new tumor-associated antigen, *Cancer Res.* **44:**5279–5285.

Glickman, J. N., Conibear, E., and Pearse, B. M. F., 1989, Specificity of binding of clathrin adaptors to signals on the mannose-6-phosphate/insulin-like growth factor II receptor, *EMBO J.* **8:**1041–1047.

Goda, Y., and Pfeffer, S. R., 1988, Selective recycling of the mannose-6-phosphate, insulin-like growth factor II receptor to the *trans* Golgi network in vitro, *Cell* **55:**309–320.

Granger, B. L., Green, S. A., Gabel, C. A., Howe, C. L., Mellman, I., and Helenius, A., 1990, Characterization and cloning of lgp 110, a lysosomal membrane glycoprotein from mouse and rat cells, *J. Biol. Chem.* **265:**12036–12043.

Green, S. A., Zimmer, K.-P., Griffiths, G., and Mellman, I., 1987, Kinetics of intracellular transport and sorting of lysosomal membrane and plasma membrane proteins, *J. Cell Biol.* **105:**1227–1240.

Greenwalt, D. E., Tandon, N. N., and Jamieson, G. A., 1991, Role of carbohydrate on the conformation of human and bovine GPIV (CD36), *Thromb. Haemost.* **65:**1153 (abstract).

Greenwalt, D. E., Lipsky, R. H., Ockenhouse, C. F., Ikeda, H., Tandon, N. N., and Jamieson, G. A., 1992, Membrane glycoprotein CD36: A review of its role in adherence, signal transduction, and transfusion medicine, *Blood* **80:**1105–1115.

Griffith, G., Hoflack, B., Simons, K., Mellman, I., and Kornfeld, S., 1988, The mannose-6-phosphate receptor and the biogenesis of lysosomes, *Cell* **52:**329–341.

Harter, C., and Mellman, I., 1992, Transport of the lysosomal membrane glycoprotein lgp120 (lgp-A) to lysosomes does not require appearance on the plasma membrane, *J. Cell Biol.* **117:**311–325.

Helenius, A., Mellman, I., Wall, D., and Hubbard, A., 1983, Endosomes, *Trends Biochem. Sci.* **8:**245–250.

Hession, C., Osborn, L., Goff, D., Chi-Rosso, G., Vassallo, C., Pasek, M., Pittack, C., Tizard, R., Goelz, S., McCarthy, K., Hopple, S., and Lobb, R., 1989, Endothelial leukocyte adhesion molecule 1: Direct expression cloning and functional interactions, *Proc. Natl. Acad. Sci. USA* **87:**1673–1677.

Himeno, M., Noguchi, Y., Sasaki, H., Tanaka, Y., Furuno, K., Kono, A., Sakaki, Y., and Kato, K., 1989, Isolation and sequencing of a cDNA clone encoding 107 kDa sialglycoprotein in rat liver lysosomal membranes, *FEBS Lett.* **244:**351–356.

Ho, M.-K., and Springer, T. A., 1983, Tissue distribution, structural characterization, and biosynthesis of Mac-3, a macrophage surface glycoprotein exhibiting molecular weight heterogeneity, *J. Biol. Chem.* **258:**636–642.

Hoff, S., Matsushita, Y., Ota, D. M., Cleary, K. R., Yamori, T., Hakomori, S., Irimura, T., 1989, Increased expression of sialyl-dimeric Lex antigen in advanced primary colorectal carcinomas and liver metastases, *Cancer Res.* **49:**6883–6888.

Holcombe, R. F., Baethge, B. A., Stewart, R. M., Betzing, K., Hall, V. C., Fukuda, M., and Wolf, R. E., 1993, Cell surface expression of lysosome-associated membrane proteins (LAMPs) in scleroderma: Relationship of lamp2 to disease duration, anti-Sc170 antibodies, serum interleukin-8 and soluble interleukin-2 receptor levels, *Clin. Immunol. Immunopath.*, **67:**31–39.

Hotta, H., Ross, A. H., Huebner, K., Isobe, M., Wendeborn, S., Chao, M. V., Ricciardi, R. P., Tsujimoto, Y., Croce, C. M., and Koprowski, H., 1988, Molecular cloning of lysosomal membrane antigen CD63, *Cancer Res.* **48:**2955–2962.

Howe, C. L., Granger, B. L., Hull, M., Green, S. A., Gabel, C. A., Helenius, A., and Mellman, I., 1988, Derived protein sequence, oligosaccharides, and membrane insertion of the 120-kD lysosomal membrane glycoprotein (lgp120): Identification of a highly conserved family of lysosomal membrane glycoproteins, *Proc. Natl. Acad. Sci. USA* **85:**7577–7581.

Humphries, M. J., Matsumoto, K., White, S. L., and Olden, K., 1986, Oligosaccharide modification by swainsonine treatment inhibits pulmonary colonization by B16-F10 murine melanoma cells, *Proc. Natl. Acad. Sci. USA* **83:**1752–1756.

Irimura, T., Gonzalez, R., and Nicolson, G. L., 1981, Effects of tunicamycin on B16 metastatic melanoma cell surface glycoproteins and blood-borne arrest and survival properties, *Cancer Res.* **51:**3411–3418.

Jadot, M., Canfield, W. M., Gregory, W., and Kornfeld, S., 1992, Characterization of the signal for rapid internalization of the bovine mannose-6-phosphate/insulin-like growth factor-II receptor, *J. Biol. Chem.* **267:**11069–11077.

Jing, S., Spencer, T., Miller, K., Hopkins, C., and Trowbridge, I. S., 1990, Role of the human transferring receptor cytoplasmic domain in endocytosis: Localization of a specific signal sequence for internalization, *J. Cell Biol.* **110:**283–294.

Johnson, K. F., and Kornfeld, S., 1992a, A His-Leu-Leu sequence near the carboxyl terminus of the cytoplasmic domain of the cation-dependent mannose-6-phosphate receptor is necessary for the lysosomal enzyme sorting function, *J. Biol. Chem.* **267:**17110–17115.

Johnson, K. F., and Kornfeld, S., 1992b, The cytoplasmic tail of the mannose-6-phosphate/insulin-like growth factor-II receptor has two signals for lysosomal enzyme sorting in the Golgi, *J. Cell Biol.* **119:**249–257.

Kannan, K., Divers, S. G., Lurie, A., Chervenak, R., Fukuda, M., and Holcombe, R. F., 1993, Lysosome-associated membrane protein-2 (lamp-2) and CD63 as markers for platelet activation, *J. Immunol.* **150:**76a.

Kornfeld, S., 1986, Trafficking of lysosomal enzymes in normal and disease states, *J. Clin. Invest.* **77:**1–6.

Kornfeld, S., and Mellman I., 1989, The biogenesis of lysosomes, *Annu. Rev. Cell Biol.* **5:**483–525.

Laferte, S., and Dennis, J. W., 1989, Purification of two glycoproteins expressing β1-6 branched Asn-linked oligosaccharides from metastatic tumor cells, *Biochem. J.* **259:**569–576.

Larsen, E., Palabrica, T., Sajer, S., Gilbert, G. E., Wagner, D. D., Furle, B. C., and Furle, B., 1990, PADGEM-dependent adhesion of platelets to monocytes and neutrophils is mediated by a lineage-specific carbohydrate, LNF III (CD15), *Cell* **63:**467–474.

Lawrence, M. B., and Springer, T. A., 1991, Leukocytes roll on a selectin at physiologic flow rates: Distinction from and prerequisite for adhesion through integrins, *Cell* **65:**859–873.

Lazarovits, J., and Roth, M., 1988, A single amino acid change in the cytoplasmic domain allows the influenza virus hemagglutinin to be endocytosed through coated pits, *Cell* **53:**743–752.

Lee, N., Wang, W.-C., and Fukuda, M., 1990, Granulocytic differentiation of HL-60 cells is associated with increase of poly-N-acetyllactosamine in Asn-linked oligosaccharides attached to human lysosomal membrane glycoproteins, *J. Biol. Chem.* **265:**20476–20487.

Lehmann, L. E., Eberle, W., Krull, S., Prill, V., Schmidt, B., Sander, C., von Figura, K., and Peters, C., 1992, The internalization signal in the cytoplasmic tail of lysosomal acid phosphatase consists of the hexapeptide PGYRHV, *EMBO J.* **11:**4391–4399.

Letourneur, F., and Klausner, R. D., 1992, A novel di-leucine motif and a tyrosine-based motif independently mediate lysosomal targeting and endocytosis of CD3 chains, *Cell* **69:**1143–1157.

Lewis, V., Green, S. A., Marsh, M., Vihko, P., Helenius, A., and Mellman, I., 1985, Glycoproteins of the lysosomal membrane, *J. Cell Biol.* **100:**1839–1847.

Ley, K., Gaehtgens, P., Fennie, C., Singer, M. S., Lasky, L. A., and Rosen, S. D., 1991, Lectin-like cell adhesion molecule 1 mediates leukocyte rolling in mesenteric venules in vivo, *Blood* **77:**2553–2555.

Lippincott-Schwartz, J., and Fambrough, D. M., 1987, Cycling of the integral membrane glycoprotein, LEP100, between plasma membrane and lysosomes: Kinetic and morphological analysis, *Cell* **49:**669–677.

Lobel, P., Fujimoto, K., Ye, R. D., Griffiths, G., and Kornfeld, S., 1989, Mutations in the cytoplasmic domain of the 275 kd mannose-6-phosphate receptor differentially alter lysosomal enzyme sorting and endocytosis, *Cell* **57**:787–796.

Lotteau, V., Teyton, L., Peleraux, A., Nilsson, T., Karlsson, L., Schmid, S. L., Quaranta, V., and Peterson, P. A., 1990, Intracellular transport of class II MHC molecules directed by invariant chain, *Nature* **348**:600–605.

Lowe, J. B., Stoolman, L. M., Nair, R. P., Larsen, R. D., Berhend, T. L., and Marks, R. M., 1990, ELAM-1-dependent cell adhesion to vascular endothelium determined by a transfected human fucosyltransferase cDNA, *Cell* **63**:475–484.

Magnani, J. L., Nilsson, B., Brockhaus, M., Zopf, D., Steplewski, Z., Koprowski, H., and Ginsburg, V., 1982, A monoclonal antibody-defined antigen associated with gastrointestinal cancer is a gangliosidase containing sialylated lacto-*N*-fucopentaose II, *J. Biol. Chem.* **257**:14365–14369.

Manoni, M., Tribioli, C., Lazzari, B., DeBellis, G., Patrosso, C., Pergolizzi, R., Pellegrini, M., Maestrini, E., Rivella, S., Vezzoni, P., and Toniolo, D., 1991, The nucleotide sequence of a CpG island demonstrates the presence of the first exon of the gene encoding the human lysosomal membrane protein lamp2 and assigns the gene to Xq24, *Genomics* **9**:551–554.

Mathews, P. M., Martinie, J. B., and Fambrough, D. M., 1992, The pathway and targeting signal for delivery of the integral membrane glycoprotein LEP100 to lysosomes, *J. Cell Biol.* **118**:1027–1040.

Mattei, M.-G., Matteson, J., Chen, J. W., Williams, M. A., and Fukuda, M., 1990, Two human lysosome membrane glycoproteins, h-lamp-1 and h-lamp-2, are encoded by genes localized to chromosome 13q34 and chromosome Xq24-25, respectively, *J. Biol. Chem.* **265**:7548–7551.

Mellman, I., 1984, Membrane recycling during endocytosis in lysosomes, in *Biology and Pathology* (J. T. Dingle, R. T. Dean, and W. Sly, eds.), pp. 201–229, Elsevier, New York.

Méresse, S., and Hoflack, B., 1993, Phosphorylation of the cation-independent mannose-6-phosphate receptor is closely associated with its exit from the *trans*-Golgi network, *J. Cell Biol.* **120**:67–75.

Metzelaar, M. J., Wijngaard, L. J., Peters, P. J., Sixma, J. J., Nieuwenhuis, H. K., and Clevers, H. C., 1991, CD63 Antigen. A novel lysosomal membrane glycoprotein cloned by a screening procedure for intracellular antigens in eukaryotic cells, *J. Biol. Chem.* **266**:3239–3245.

Mulligan, M. S., Varani, J., Dame, M. K., Lane, C. W., Smith, C. W., Anderson, D. C., and Ward, P. A., 1991, Role of endothelial-leukocyte adhesion molecule 1 (ELAM-1) in neutrophil-mediated lung injury in rats, *J. Clin. Invest.* **88**:1396–1406.

Mulligan, M. S., Lowe, J. B., Larsen, R. D., Paulson, J., Walker, L., Maemura, K., Fukuda, M., and Ward, P. A., 1993, Protective effects of sialylated oligosaccharides in immune complex-induced acute lung injury, *J. Exp. Med.* **178**:623–631.

Neufeld, E. F., and Ashwell, G., 1980, Carbohydrate recognition systems for receptor-mediated pinocytosis, in *The Biochemistry of Glycoproteins and Proteoglycans* (W. J. Lennarz, ed.), pp. 241–266, Plenum Press, New York.

Nezu, J.-I., Motojima, K., Tamura, H.-O., and Ohkuma, S., 1992, Molecular cloning of a rat liver cDNA encoding the 16 kDa subunit of vacuolar H^+-ATPases: Organellar and tissue distribution of 16 kDa proteolipids, *J. Biochem.* **112**:212–219.

Nicolson, G. L., 1989, Metastatic tumor cell interactions with endothelium, basement membrane and tissue, *Curr. Opin. Cell Biol.* **1**:1009–1019.

Nieuwenhuis, H. L., Van Oosterhout, J. J. G., Rozemuller, E., Van Iwaarden, F., and Sixma, J. J., 1987, Studies with a monoclonal antibody against activated platelets: Evidence that a secreted 53,000-molecular weight lysosome-like granule protein is exposed on the surface of activated platelets in the circulation, *Blood* **70**:838–845.

Noguchi, Y., Himeno, M., Sasaki, H., Tanaka, Y., Kono, A., Sakaki, Y., and Kato, K., 1989, Isolation and sequencing of a cDNA clone encoding 96-kD sialglycoprotein in rat liver lysosomal membranes, *Biochem. Biophys. Res. Commun.* **164**:1113–1120.

Ogata, S., and Fukuda, M., 1993, Lysosomal targeting of Limp II membrane glycoprotein requires a novel Leu-Ile motif at a particular position in its cytoplasmic tail, *J. Biol. Chem.*, submitted.

Ohkuma, S., and Poole, B., 1978, Fluorescence probe measurement of the intralysosomal pH in living cells and the perturbation of pH by various agents, *Proc. Natl. Acad. Sci. USA.* **75**:3327–3331.

Oquendo, P., Hundt, E., Lawler, J., and Seed, B., 1989, CD36 directly mediates cytoadherence of *Plasmodium falciparum* infected erythrocytes, *Cell* **58**:95–101.

Pearse, B. M. F., and Robinson, M. S., 1990, Clathrin, adaptors and sorting, *Annu. Rev. Cell Biol.* **6**:151–171.

Peters, C., Braun, M., Weber, B., Wendland, M., Schmidt, B., Pohlmann, R., Waheed, A., and von Figura, K., 1990, Targeting of a lysosomal membrane protein: A tyrosine-containing endocytosis signal in the cytoplasmic tail of lysosomal acid phosphatase is necessary and sufficient for targeting to lysosomes, *EMBO J.* **9**:3497–3506.

Peters, P. J., Borst, J., Oorschot, V., Fukuda, M., Kräbenbül, O., Tschopp, J., Slot, J. W., and Geuze, H. J., 1991a, Cytotoxic T lymphocytes granules are secretory lysosomes, containing both perforin and granzymes, *J. Exp. Med.* **173**:1099–1109.

Peters, P. J., Neefjes, J. J., Oorschot, V., Ploegh, H. L., and Geuze, H. J., 1991b, Segregation of MHC class II molecules from MHC class I molecules in the Golgi complex for transport to lysosomal compartments, *Nature* **349**:669–676.

Phillips, M. L., Nudelman, E., Gaeta, F. C. A., Perez, M., Singhal, A. K., Hakomori, S.-I., and Paulson, J. C., 1990, ELAM-1 mediates cell adhesion by recognition of a carbohydrate ligand, sialyl-Le^x, *Science* **250**:1130–1132.

Pierce, M., and Arango, J., 1986, Rous sarcoma virus-transformed baby hamster kidney cells express higher levels of asparagine-linked tri- and tetraantennary glycopeptides containing [GlcNAc-β (1, 6) Man-α (1, 6) Man] and poly-*N*-acetyllactosamine sequences than baby hamster kidney cells, *J. Biol. Chem.* **261**:10772–10777.

Reggio, H., Bainton, D., Harms, E., Coudrier, E., and Louvard, D., 1984, Antibodies against lysosomal membranes reveal a 100,000-mol-wt protein that cross-reacts with purified H$^+$, K$^+$ ATPase from gastric mucosa, *J. Cell Biol.* **99**:1511–1526.

Rice, G. E., and Bevilacqua, M. P., 1989, An inducible endothelial cell surface glycoprotein mediates melanoma adhesion, *Science* **246**:1303–1306.

Robinson, M. S., and Pearse, B. M. F., 1986, Immunofluorescence localization of 100K coated vesicle proteins, *J. Cell Biol.* **102**:48–54.

Saitoh, O., Wang, W.-C., Lotan, R., and Fukuda, M., 1992, Differential glycosylation and cell surface expression of lysosomal membrane glycoproteins in sublines of a human colon cancer exhibiting distinct metastatic potentials, *J. Biol. Chem.* **267**:5700–5711.

Sawada, R., Jardine, K. A., and Fukuda, M., 1993a, The genes of major lysosomal membrane glycoproteins, lamp-1 and lamp-2: The 5′-flanking sequence of lamp-2 gene and comparison of exon organization in two genes, *J. Biol. Chem.*, **268**:9014–9022.

Sawada, R., Lowe, J. B., and Fukuda, M., 1993b, E-selectin-dependent adhesion efficiency of colonic carcinoma cells is increased by genetic manipulation of their cell surface lamp-1 expression levels, *J. Biol. Chem.*, **268**:12675–12681.

Sawada, R., Tsuboi, S., and Fukuda, M., 1993c, Differential E-selectin dependent adhesion efficiency in sublines of a human colon cancer exhibiting distinct metastatic potentials, *J. Biol. Chem.* in press.

Silverstein, R. L., and Febbraio, M., 1992, Identification of lysosome-associated membrane protein-2 as an activation-dependent platelet surface glycoprotein, *Blood* **80**:1470–1475.

Tandon, N. N., Kralisz, U., and Jamieson, G. A., 1989, Identification of GPIV (CD36) as a primary receptor for platelet-collagen adhesion, *J. Biol. Chem.* **264**:7576–7583.

Tong, P., and Kornfeld, S., 1989, Ligand interactions of the cation-dependent mannose-6-phosphate receptor: Comparison with the cation-independent mannose-6-phosphate receptor, *J. Biol. Chem.* **264**:7970–7975.

Vega, M. A., Seguí-Real, B., Garcia, J. A., Calés, C., Rodríguez, F., Vanderkerckhove, J., and Sandoval, I. V., 1991a, Cloning, sequencing, and expression of a cDNA encoding rat LIMP II, a novel 74-kDa lysosomal membrane protein related to the surface adhesion protein CD36, *J. Biol. Chem.* **266**:16818–16824.

Vega, M. A., Rodríguez, F., Seguí, B., Calés, C., Alcade, J., and Sandoval, I. V., 1991b, Targeting of lysosomal integral membrane protein LIMPII: The tyrosine-lacking carboxyl cytoplasmic tail of LIMPII is sufficient for direct targeting to lysosomes, *J. Biol. Chem.* **266**:16269–16272.

Viitala, J., Carlsson, S. R., Siebert, P. D., and Fukuda, M., 1988, Molecular cloning of cDNAs encoding lamp A, a human lysosomal membrane glycoprotein with apparent Mr ∼120,000, *Proc. Natl. Acad. Sci. USA* **85**:3743–3747.

von Figura, K., and Hasilik, A., 1986, Lysosomal enzymes and their receptors, *Annu. Rev. Biochem.* **55**:167–193.

Walz, G., Aruffo, A., Kolanus, W., Bevilacqua, M., and Seed, B., 1990, Recognition by ELAM-1 of the sialyl-Le^x determinant on myeloid and tumor cells, *Science* **250**:1132–1135.

Wickner, W. T., and Lodish, H. F., 1985, Multiple mechanisms of protein insertion into and across membranes, *Science* **230**:400–407.

Williams, A. F., and Barclay, A. N., 1988, The immunoglobulin superfamily—domains for cell surface recognition, *Annu. Rev. Immunol.* **6**:381–405.

Williams, M. A., and Fukuda, M., 1990, Accumulation of membrane glycoproteins in lysosomes requires a tyrosine residue at a particular position in the cytoplasmic tail, *J. Cell Biol.* **111**:955–966.

Yamamoto, K., Ikehara, Y., Kawamoto, S., and Kato, K., 1980, Characterization of enzymes and glycoproteins in rat liver lysosomal membranes, *J. Biochem.* **87**:237–248.

Yamashita, K., Ohkura, T., Tachibana, Y., Takasaki, S., and Kobata, A., 1984, Comparative study of the oligosaccharides released from baby hamster kidney cells and their polyoma transformant by hydrazinolysis, *J. Biol. Chem.* **259**:10634–10640.

Youakim, A., Romero, P. A., Yee, K., Carlsson, S. R., Fukuda, M., and Herscovics, A., 1989, Decrease in polylactosaminoglycans associated with lysosomal membrane glycoproteins during differentiation of CaCo-2 human colonic adencarcinoma cells, *Cancer Res.* **49**:6889–6895.

Yousefi, S., Higgins, E., Daoling, Z., Pollex-Krüger, A., Hindsgaul, O., and Dennis, J. W., 1991, Increased UDP-GlcNAc:Galβ1-3GalNAc-R (GlcNAc to GalNAc) β-1, 6-*N*-acetylglu-cosaminyltransferase activity in metastatic murine tumor cell lines: Control of polylactosamine synthesis, *J. Biol. Chem.* **266**:1772–1782.

Zot, A. S., and Fambrough, D. M., 1990, Structure of a gene for a lysosomal membrane glycoprotein (LEP100), *J. Biol. Chem.* **265**:20988–20995.

Chapter 8

Assembly of the Peroxisomal Membrane

Paul P. Van Veldhoven and Guy P. Mannaerts

1. INTRODUCTION

Peroxisomes are found in virtually all eukaryotic cells ranging from eukaryotic microorganisms to plants and animals. In addition to a variety of other enzymes, peroxisomes contain hydrogen peroxide-producing oxidases and catalase that decomposes hydrogen peroxide. The name *peroxisome* refers to the organelle's role in the production and disposition of hydrogen peroxide (de Duve and Baudhuin, 1966).

At present, more than 50 enzymes have been identified in mammalian peroxisomes. They are involved in the activation and β-oxidation of fatty acids and fatty acid derivatives, the β-oxidation of the side chain of cholesterol, which results in the formation of bile acids (hepatic peroxisomes); the synthesis of ether glycerolipids, dolichols and cholesterol; the catabolism of purines, amino acids, and polyamines; the metabolism of glyoxylate; the degradation of glucose via the hexose monophosphate pathway; and the inactivation of reactive oxygen species such as hydrogen peroxide, superoxide anions and epoxides (Tolbert, 1981; Mannaerts and Van Veldhoven, 1990, 1992a,b). Yeast peroxisomes harbor the β-oxidation enzymes but also enzymes involved in the metabolism of alcohols and alkylated amines, which can serve as carbon and/or nitrogen source (Veen-

Paul P. Van Veldhoven and Guy P. Mannaerts   Afdeling Farmacologie, Katholieke Universiteit Leuven, Campus Gasthuisberg, B-3000 Leuven, Belgium.

Subcellular Biochemistry, Volume 22: Membrane Biogenesis, edited by A. H. Maddy and J. R. Harris. Plenum Press, New York, 1994.

huis and Harder, 1987). Leaf peroxisomes participate in photorespiration (Tolbert, 1981).

Also belonging to the peroxisome family are glyoxysomes and glycosomes (Tolbert, 1981; Opperdoes, 1987; Keller et al., 1991). They contain, in addition to a number of enzymes typical of peroxisomes (e.g., the β-oxidation enzymes), some or all of the glyoxylate cycle enzymes (glyoxysomes) and part of the glycolytic sequence (glycosomes). Glyoxysomes occur in oil-rich plant seeds and in yeasts grown on fatty acids. The glyoxylate cycle bypasses part of the Krebs cycle, and in conjunction with the glyoxysomal β-oxidation enzymes converts fat into sugar. Glycosomes are found in trypanosomatids.

Most peroxisomal enzymes in mammals and perhaps also in other vertebrates have their counterparts, different enzymes that catalyze a similar reaction, in other cell compartments (Tolbert, 1981; Mannaerts and Van Veldhoven, 1992a,b). As a result of this enzyme duplication, the physiological significance of the metabolic pathways in mammalian peroxisomes is not always clear. Best known is the role of peroxisomes in β-oxidation and in ether glycerolipid synthesis. Peroxisomes—and not mitochondria—are responsible for the β-oxidation of the side chain of cholesterol during bile acid synthesis (Kase et al., 1983) and for the β-oxidation of the major portion of saturated and polyunsaturated very long chain ($> C_{20}$) fatty acids (Singh et al., 1984), dicarboxylic fatty acids (Cerdan et al., 1988; Leighton et al., 1989), isoprenoid-derived branched fatty acids (Vanhove et al., 1991), and eicosanoids such as prostaglandins and leukotrienes (Schepers et al., 1988; Jedlitschky et al., 1991). Peroxisomes—and not the endoplasmic reticulum—catalyze the major portion of the initial reactions in ether glycerolipid synthesis (Hajra and Bishop, 1982; Heymans et al., 1983). The endoplasmic reticulum, which is responsible for ester glycerolipid synthesis, catalyzes the terminal reactions of ether glycerolipid synthesis (Declercq et al., 1984).

Peroxisomes are delimited by a single membrane that encompasses the matrix. The membrane has a thickness of 6–8 nm in rodent liver and kidney peroxisomes and of 7 nm in castor bean endosperm glyoxysomes (Hruban and Rechcigl, 1969; Lazarow, 1984). The majority of the peroxisomal enzymes are soluble matrix enzymes. A few enzymes are associated with crystalloid matrical inclusions that can be found in certain cell types. The membrane-bound enzymes will be described in Section 2.4 below.

Most cells have spherical or spheroidal peroxisomes with a diameter between 0.2 and 1 μm, but in some cell types peroxisomes of various other forms (e.g., angular, tubular) can be observed. Peroxisomes that are interconnected in an irregular network have been described in germinating spores of the moss Bryum capillare (Pais and Carrapico, 1982), in sebaceous glands (Gorgas, 1984), in mouse liver (Gorgas, 1985), and in regenerating rat liver (Yamamoto and Fahimi, 1987). The possible significance of these connections for peroxisome biogenesis and for the assembly of the peroxisomal membrane is discussed in Section 3.1.

Treatment of rodents with hypolipidemic drugs of the fibrate type, with phthalate plasticizers and with certain herbicides, results in the proliferation of hepatic peroxisomes, which increase 5- to 10-fold in number (Lock *et al.*, 1989). Peroxisome proliferation is accompanied by a marked induction of several but not all peroxisomal enzymes. For instance, the activity and content of the peroxisomal β-oxidation enzymes can increase 10- to 20-fold but catalase activity is only slightly (1.5-fold) increased or not at all. Induction of the peroxisomal β-oxidation enzymes is mediated by the activation of a nuclear receptor, belonging to the thyroid/retinoid nuclear hormone receptor family (Issemann and Green, 1990; Dreyer *et al.*, 1992). It remains unknown whether the receptor is activated by the peroxisome proliferator itself, by a metabolite of the peroxisome proliferator, or by an endogenous ligand, the concentration of which is increased by treatment with the peroxisome proliferator. Smaller increases in peroxisome number and in peroxisomal β-oxidation enzyme activities are found in some extrahepatic tissues (kidney, intestinal mucosa, heart). Chronic exposure of rodents to peroxisome proliferators leads to liver cancer via a hitherto unknown mechanism. There are marked species differences in peroxisome proliferation and there is no evidence that peroxisome proliferation and its related hepatocarcinogenesis occur in humans (Lock *et al.*, 1989).

Peroxisome proliferation is also observed in certain strains of yeast when they are grown on methanol, alkylated amines, alkanes, or fatty acids (Veenhuis and Harder, 1987).

Several congenital diseases are known that are caused by a malfunctioning of the peroxisomes. These peroxisomal disorders can be subdivided into three categories: generalized peroxisomal diseases in which most if not all peroxisomal functions are lost; disorders with multiple peroxisomal enzyme deficiencies; and disorders with a single peroxisomal enzyme deficiency (Lazarow and Moser, 1989; Wanders *et al.*, 1990). The generalized peroxisomal dysfunctions (e.g., the Zellweger syndrome) will be discussed in more detail in Section 3.1. of this chapter since these dysfunctions are thought to result from a defective peroxisomal protein import. Infants with a generalized peroxisomal dysfunction have a number of anatomical (e.g., face, skull, brain, liver) and functional (e.g., severe neurological symptoms, hepatic dysfunction, metabolic disturbances) aberrations, and they usually die within their first year of life (Lazarow and Moser, 1989; Wanders *et al.*, 1990).

Several excellent reviews have been written on the peroxisomal membrane (Lazarow, 1984) and on the biogenesis of peroxisomes (Lazarow and Fujiki, 1985; Osumi and Fujiki, 1990; Subramani, 1992; Kanau and Hartig, 1992; Höhfeld *et al.*, 1992), topics that are closely related to the subject of this chapter.

2. COMPOSITION AND FUNCTION OF THE PEROXISOMAL MEMBRANE

2.1. Purification and Subfractionation of Peroxisomes

Isolation of peroxisomes is generally accomplished in three steps. After homogenization of the tissue or disruption of the cells, a subcellular fraction enriched in peroxisomes is obtained by differential centrifugation, which is then further separated by means of density gradient centrifugation (Evans, 1982; Huang *et al.*, 1983).

Owing to the apparent lability of the peroxisomal membrane, mild procedures are employed during peroxisome isolation. The homogenization medium is normally isotonic (0.25 M sucrose for mammalian tissues) and can be supplemented with various additives. Ethanol (0.1% [v/v]) is often added to prevent the formation of inactive catalase compound II (Leighton *et al.*, 1968). Low concentrations of substances buffering around a physiological pH (glycine, imidazol, tris, MOPS) or sulfhydryl protecting agents can be present as well. High salt concentrations should be avoided as they cause aggregation. The addition of chelators (EDTA or EGTA), especially during the further purification steps, tends to reduce the contamination of the purified peroxisomes with microsomal constituents (Van Veldhoven, 1986; Ghosh and Hajra, 1986a).

The homogenate is then subjected to increasing g-forces, generating subcellular fractions enriched in a particular cell constituent. For rat liver these fractions are the nuclear, the heavy and light mitochondrial fraction, the microsomal fraction, and the cytosolic fraction (de Duve *et al.*, 1955). The light mitochondrial fraction (also called the L-fraction or λ-fraction) contains mainly lysosomes and peroxisomes. Further separation of peroxisomes is achieved by means of density gradient centrifugation. Successful purification of peroxisomes via sucrose gradients requires pretreatment of the animals with Triton WR-1339 to reduce the density of the lysosomes (Leighton *et al.*, 1968). This step can be deleted with the newer gradient materials, like Metrizamide and Nycodenz (Hajra and Wu, 1985; Völkl and Fahimi, 1985; Ghosh and Hajra, 1986a; Hartl *et al.*, 1985). Because peroxisomes are permeable to sucrose (Van Veldhoven *et al.*, 1983, 1987) and apparently also to Metrizamide and Nycodenz, they band in such gradients at a relatively high density. Consequently, they are isolated under hypertonic conditions, limiting their use for protein import studies since dilution may result in membrane damage. Isolation of peroxisomes under isotonic conditions is possible, however, by means of iso-osmotic Percoll gradients (Neat *et al.*, 1980; Appelkvist *et al.*, 1981; Mannaerts *et al.*, 1982; Van Veldhoven *et al.*, 1983), but peroxisome purity is less than that obtained by means of Nycodenz gradients (Verheyden *et al.*, 1992).

Peroxisomes constitute only a minor cell compartment. According to the classical studies by Leighton *et al.* (1968) peroxisomes would account for 2.5% of the cellular protein in rat hepatocytes. Improved isolation procedures led Just and co-workers to propose a value of 1.8 to 2.0% (Hartl *et al.*, 1985).

As mentioned in the Introduction, peroxisomes are surrounded by a single membrane and may contain crystalloid inclusions. In rat liver peroxisomes, urate oxidase is associated with such crystalloid inclusions (called cores). Several procedures exist to obtain different peroxisomal subfractions. A very elegant procedure to obtain integral membrane proteins was introduced by Fujiki *et al.* (1982). By suspending the isolated peroxisomes in 0.1 M carbonate buffer at pH 11.5, peroxisomes are broken open, the membrane is stripped from peripheral membrane proteins, and the cores are solubilized. The unsealed membrane fragments are pelleted by centrifugation. The supernatant contains the matrix proteins, the peripheral membrane proteins, and the core proteins (Figure 1). In rat liver integral membrane proteins account for approximately 10% of the total peroxisomal protein (Fujiki *et al.*, 1982; Hartl *et al.*, 1985; Van Veldhoven *et al.*, 1987). Compared to other organelles this is a very low value, stressing the importance of the purity of the peroxisomal fraction when the peroxisomal membrane composition is analyzed.

For investigation of the subperoxisomal localization of an enzymatic activity, the carbonate procedure is of little value owing to the high pH. For this purpose one can rely on the sonication of the peroxisomes in hypotonic pyrophosphate buffer at pH 9. After centrifugation, more than 95% of the matrix proteins, as well as most peripheral membrane proteins, are present in the supernatant (Van Veldhoven *et al.*, 1987). Treatment of the pellet with detergent solubilizes the integral membrane proteins, but not the cores. When the sonication is performed in hypotonic buffer at a physiological pH, only the matrix proteins are released to a significant extent. Proteins bound via ionic interactions (e.g., the trifunctional protein of the peroxisomal β-oxidation) to the membrane, an interconnecting filamentous network, or the core can subsequently be released by sonication in the presence of high salt concentration (Alexson *et al.*, 1985; Van Veldhoven *et al.*, 1987).

2.2. Polypeptide and Lipid Composition of the Peroxisomal Membrane

2.2.1. Polypeptide Composition of Membranes from Mammalian Peroxisomes

Fujiki *et al.* (1982), Hashimoto *et al.* (1986), and Hartl and Just (1987) were the first to describe the polypeptide pattern of membranes obtained after carbonate treatment of highly purified peroxisomal fractions from rat liver. Carbonate treatment of peroxisomes releases the matrix and peripheral membrane proteins, leaving membranes that contain only the integral membrane proteins (see Section

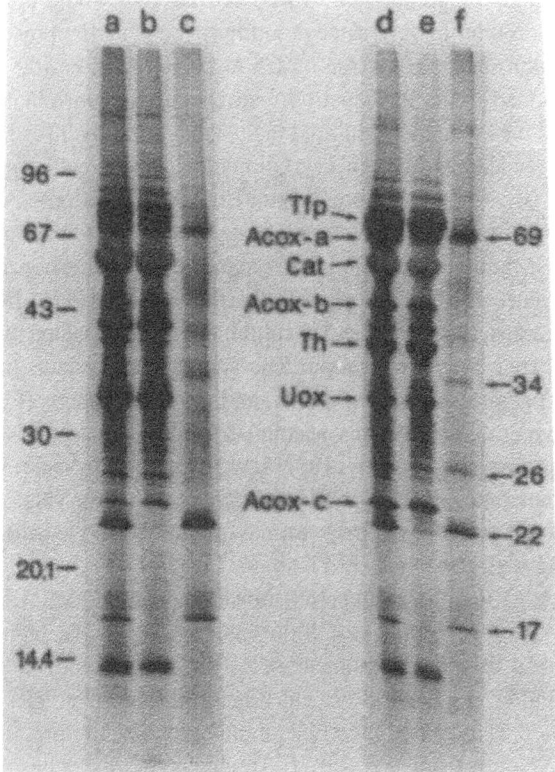

FIGURE 1. Sodium dodecyl sulfate-polyacrylamide gel electrophoresis of peroxisomal proteins. Peroxisomes were purified from livers of male control (lanes a–c) or clofibrate-treated (lanes d–f) rats by means of Percoll gradient centrifugation, followed by a Nycodenz-step gradient centrifugation as described by Verheyden *et al.* (1992). The purified peroxisomes were diluted in 0.1 M carbonate buffer pH 11.5 and analyzed by means of SDS-gel electrophoresis before (total proteins; lanes a and d) or after centrifugation to separate the carbonate-soluble proteins (lanes b and e) and integral membrane proteins (lanes c and f).

Lane a: 9.6 μg of total peroxisomal protein; lane b: carbonate-soluble protein (8.9 μg) derived from 9.6 μg of total peroxisomal protein; lane c: 1 μg of integral membrane protein; lane d: 9.8 μg of total peroxisomal protein; lane e: carbonate-soluble protein (8.7 μg) derived from 9.8 μg of total peroxisomal protein; lane f: 0.4 μg of integral membrane protein. Positions of the molecular weight markers (expressed in kDa) are indicated at the left side. In the middle, the position of well-known peroxisomal enzymes is indicated. Tfp: trifunctional protein; Acox-a/b/c: subunit A, B, or C of the inducible acyl-CoA oxidase; Cat: catalase; Th: ketoacyl-CoA thiolase; Uox: urate oxidase. At the right side, position of the major integral membrane proteins. Their molecular mass is expressed in kDa (arrows).

2.1.). The latter were separated by sodium dodecyl sulfate polyacrylamide gel electrophoresis. Fujiki *et al.* (1982) reported major peroxisomal integral membrane polypeptides with molecular masses of 70, 68, 21, and 15 kDa. Hartl and Just (1987) described polypeptides of 69, 42, 36, 28, 26, and 22 kDa. These polypeptides constituted 18% (69 kDa), 6% (42 kDa), 10% (36 kDa), 7% (28 + 26 kDa), and 14% (22 kDa) of the total peroxisomal membrane protein. As shown in Figure 1, the major integral membrane proteins of rat liver peroxisomes purified in our laboratory possess a molecular mass of 69, 34, 22, and 17 kDa. After clofibrate treatment, the presence of a 26 kDa protein becomes more apparent.

Hartl and Just (1987) mention that the 68 kDa protein described by Fujiki *et al.* (1982) is not an integral membrane protein since in their experiments it was released from the membrane by carbonate treatment. Hashimoto *et al.* (1986) consider the 68 kDa protein as a proteolytic degradation product of the 70 (69) kDa protein since incubation of liver homogenates resulted in an increase of the former protein concomitant with a decrease of the latter. The 70 kDa protein has been cloned by Kamijo *et al.* (1990), and its amino acid sequence has been reported. The protein proves to be a member of the P-glycoprotein-related ATP-binding protein superfamily. Kamijo *et al.* (1990) propose that the protein is involved in active transport of metabolites or in protein translocation across the peroxisomal membrane. Their proposition is based on the function of other members of this protein family. In rodent liver, the 70 kDa protein is induced 3- to 10-fold after treatment of the animals with peroxisome proliferators (Hashimoto *et al.*, 1986; Hartl and Just, 1987; Bodnar and Rachubinski, 1991). The protein is present not only in the rat but also in the mouse and the human (Chen *et al.*, 1988; Wilson and King, 1991; Kamijo *et al.*, 1992).

Antibodies against the 70 kDa protein react with a peroxisomal integral membrane protein of 140 kDa (Bodnar and Rachubinski, 1991), which is considered to be a dimer. They also react with the 42 (41) kDa protein (Hashimoto *et al.*, 1986; Hartl and Just, 1987) described by Hartl and Just (1987) and Hashimoto *et al.* (1986) and weakly with the 28 (27) and 26 kDa proteins (Hartl and Just, 1987) described by the same authors. Hartl and Just (1987) have provided evidence that the 42, 28, and 26 kDa polypeptides are proteolytic fragments of the 70 kDa protein that arise during peroxisome isolation.

The 70 kDa and 22 kDa proteins appear to be unique to the peroxisomal membrane. Antibodies that recognize the 36 kDa protein react with proteins of similar size present in membranes derived from endoplasmic reticulum and mitochondria (Hartl and Just, 1987; Bodnar and Rachubinski, 1991). A 15 kDa protein is present not only in peroxisomal membranes but also in those from endoplasmic reticulum, mitochondria, and lysosomes (Fujiki *et al.*, 1982). Antibodies raised against peroxisomal membranes cross-react with the 15 kDa protein of the other organelles (Bodnar and Rachubinski, 1991). The exact relationship between the peroxisomal and extraperoxisomal 36 and 15 kDa proteins remains unclear.

The 22 kDa integral membrane protein of rat liver peroxisomes has recently been cloned (Kaldi *et al.*, 1993). It probably contains four transmembrane spans and shows homology to the Mpv17 gene product. Failure to produce the Mpv17 protein in transgenic mice causes the development of a nephrotic syndrome. This is possibly caused by peroxisomal dysfunction since the Mpv17 protein has recently been localized to peroxisomes (Zwacka *et al.*, 1993).

Besides the polypeptides mentioned thus far, Bodnar and Rachubinski (1991) reported minor polypeptides of 75, 50, 30, and 24 kDa in membranes from rat liver peroxisomes. An antiserum raised against purified peroxisomal membranes cross-reacted with a 50 kDa protein and with a 36 kDa protein (already described above) present in endoplasmic reticulum. Lazarow and co-workers found that an antiserum raised against purified rat liver peroxisomal membranes reacts with a 53 kDa peroxisomal integral membrane polypeptide from human fibroblasts (Santos *et al.*, 1988). They also developed a monoclonal antibody that recognized rat liver peroxisomal membranes and more specifically a novel 57 kDa polypeptide (Imanaka *et al.*, 1991). The polypeptide is exposed to the cytosol and inducible by treatment of the animals with peroxisome proliferators. Wilson and King (1991) described a 55 kDa polypeptide in peroxisomal membranes from rats, mice, and humans but provided evidence that this polypeptide may be derived from the 70 kDa protein. Finally, Skorin *et al.* (1986) detected a minor 63 kDa integral membrane protein in rat liver peroxisomes because the protein was radioactively labeled after incubation of hepatocytes with radioactive phosphate. It could also be phosphorylated when purified peroxisomes were incubated with ATP and the catalytic subunit of cAMP-dependent protein kinase.

The peroxisomal membrane contains several enzymes involved in lipid metabolism (see Section 2.4), a pore-forming protein and possibly other proteins involved in transport (Section 2.3) and proteins involved in the translocation of matrix proteins across the membrane (Section 2.5). At present no function can be assigned to the polypeptides described in this section. An exception is the 70 kDa protein for which a putative function in membrane transport or protein translocation has been proposed (see above).

2.2.2. Polypeptide Composition of Membranes from Yeast Peroxisomes, Glyoxysomes, and Glycosomes

The laboratories of Goodman, Veenhuis, and Rachubinski analyzed the polypeptide pattern of the peroxisomal membrane of several strains of yeast grown under different conditions. Major integral membrane polypeptides of peroxisomal membranes from *Candida boidinii* that resist carbonate treatment show molecular masses of 31, 32, and 47 kDa. An abundant 20 kDa polypeptide is also present when the yeast is grown on methanol but not when it is grown on oleic acid or D-alanine (Goodman *et al.*, 1990). The major portion of the 20 kDa protein is released by carbonate treatment, suggesting that it is a peripheral membrane protein (Goodman *et al.*, 1986). Sulter *et al.* (1990) report integral membrane proteins of 31, 35, 42, 51, 57, and 68 kDa in peroxisomes from *Hansenula polymorpha* grown on methanol and ammonium sulphate. The 57 and 68 kDa polypeptides are minor components. *Saccharomyces cerevisiae* and *Candida tropicalis* grown on oleic acid contain three abundant peroxisomal integral membrane proteins of 24, 31, and 32 kDa (*S. cerevisiae;* McCammon *et al.*, 1990b) and 24, 29, and 34 kDa (*C. tropicalis;* Nuttley *et al.*, 1990). The polypeptide pattern of yeast peroxisomal membranes, which varies slightly according to the growth conditions, is very different from that of mammalian peroxisomes. All yeast strains contain abundant polypeptides with molecular masses of approximately 30 kDa. The peroxisomal 31 kDa proteins of *H. polymorpha* and *S. cerevisiae* cross-react with an antiserum against the 31 kDa mitochondrial porin from *S. cerevisiae* (W. H. Kunau and M. Veenhuis, unpublished results). The peroxisomal 31 kDa protein from *H. polymorpha* has recently been purified (Sulter *et al.*, 1993). Electrophysiological measurements on liposomes reconstituted with the protein have demonstrated that it forms a large conductance channel with a diameter of 1.5 nm in the membrane (M. Lemmens, G. J. Sulter, M. Veenhuis, G. P. Mannaerts, and E. Carmeliet, unpublished results). The function of the other peroxisomal membrane polypeptides from yeast remains unknown.

Beevers and Gonzalez (1987) report the presence of more than 20 polypeptides, in the glyoxysomal membrane from castor bean endosperm. Major components have molecular masses of 25, 36, 56, 62, 66, and 67 kDa. Glycosomes from *Trypanosoma brucei* contain only two abundant integral membrane proteins with molecular masses of 24 and 26 kDa (Aman and Wang, 1987).

2.2.3. Lipid Composition of the Peroxisomal, Glyoxysomal, and Glycosomal Membrane

Fujiki *et al.* (1982) analyzed the phospholipid content of carbonate-treated peroxisomal membranes from rat liver. They found a phospholipid/protein ratio of 204 nmol lipid phosphorus per mg of protein. This figure is extremely low as compared to the data obtained by Hardeman *et al.* (1990) and by Van Veldhoven and Mannaerts (unpublished results) for carbonate membranes from rat liver peroxisomes and by Crane and Masters (1986) for carbonate membranes from mouse liver. These authors reported values of 1054, 750, and 729 nmol lipid phosphorus per mg of protein, respectively. All authors found that phosphatidylcholine is the major peroxisomal membrane phospholipid, followed by phosphatidylethanolamine. A detailed analysis by Hardeman *et al.* (1990) provides the following percentwise phospholipid composition of the peroxisomal membrane: phosphatidylcholine: 56; phosphatidylethanolamine: 27; phosphatidylinositol: 4.7; sphingomyelin: 3.7; phosphatidylserine: 3; alkenylphosphatidylethanolamine: 2.1; and unknown: 2.6. No cardiolipin was detected. The phospholipid composition of the peroxisomal membrane resembles roughly that of the endoplasmic reticulum. Appelkvist and Dallner (1987) found that the peroxisomal membrane also contains some cholesterol and dolichol. Its cholesterol content is lower than that of other membranes, despite the fact that peroxisomes are capable of synthesizing this compound.

Nuttley *et al.* (1990) described the phospholipid composition of peroxisomal membranes from *C. tropicalis* grown on oleic acid. They found a phospholipid/protein ratio of 430 nmol/mg with phosphatidylcholine and phosphatidylethanolamine as the only detectable phospholipid species. Neutral lipids and sterols were also present in the membranes.

Beevers and Gonzalez (1987) reported that glyoxysomal membranes from castor bean endosperm contain 621 nmol phospholipid per mg of protein. The analysis was performed on KCl-washed membranes. Since KCl is not as effective as carbonate in removing peripheral membrane proteins, the phospholipid/protein ratio of carbonate-treated membranes may be somewhat higher. Luster *et al* (1988) found 703 nmol phospholipid per mg of protein for carbonate-treated membranes. As in rodent liver peroxisomes, phosphatidylcholine was the major phospholipid species, followed by phosphatidylethanolamine (Beevers and Gonzalez, 1987; Donaldson *et al.*, 1972). Small amounts of diacylglycerol, triacylglycerol, stigmasterol, and β-sitosterol were also found (between 2 and 6 nmol/mg of protein for each of these neutral lipids). Surprisingly, free fatty acids were present in an amount of approximately 100 nmol/mg of protein (Beevers and Gonzalez, 1987). Whether this high amount of fatty acids is related to the presence of a lipase in the glyoxysomal membrane (Section 2.4.) has not been investigated. Glyoxysomes in cotyledons of cotton seedlings expand strongly in volume during the first two days of postgerminative growth. Chapman and Trelease (1991) analyzed the lipid composition of carbonate-treated membranes of these enlarging glyoxysomes and found that the membranes contain an unusually high amount of nonpolar lipids, ranging from one-third to two-thirds by weight of the total membrane lipids. Approximately 50% by weight of these nonpolar lipids consisted of free fatty acids, the remainder being triacylglycerol, diacylglycerol, and monoacylglycerol. The authors also found that half of the fatty acids of the triacylglycerol and free fatty acid fractions were uncommon fatty acids that were not further identified. Phosphatidylcholine represented approximately 50% of the phospholipid fraction, the remaining percentage being about equally distributed among phosphatidylglycerol, phosphatidylinositol, and phosphatidylethanolamine.

Carbonate-treated membranes from glycosomes isolated from bloodstream trypanosomes contain 580 nmol phospholipid/mg protein. Only two major phospholipids are found, phosphatidylcholine and -ethanolamine (Opperdoes *et al.*, 1984).

2.3. Permeability of the Peroxisomal Membrane

The localization of most peroxisomal enzymes inside the peroxisome (see Introduction) necessitates that substrates, products, and possibly also cofactors traverse the membrane.

Purified rat liver peroxisomes are permeable to sucrose and various other small molecules including substrates and cofactors for peroxisomal enzymes (de Duve and Baudhuin, 1966; Van Veldhoven *et al.*, 1983, 1987; Mannaerts and Van Veldhoven, 1987). This nonselective permeability of the peroxisomal membrane explains why peroxisomes acquire a higher density in gradients made of sucrose, Nycodenz, or Metrizamide, molecules small enough to enter the peroxisomes, than in gradients composed of Percoll particles, Ficoll, or glycogen, which are too large to penetrate the organelles. Another expression of the high permeability of the membrane is the fact that isolated rat liver peroxisomes do not possess their own pools of cofactors (Mannaerts *et al.*, 1982) (except for CoA and FAD, which are bound to matrix proteins; Leighton *et al.*, 1982; Van Veldhoven and Mannaerts, 1986) and that peroxisomal enzymes do not show latency (de Duve and Baudhuin, 1966; Mannaerts and Van Veldhoven, 1987). An exception to the nonlatency rule of animal peroxisomes is catalase. However, the latency of catalase is not due to the impermeability of the peroxisomal membrane to hydrogen peroxide but to the high concentration and very high activity of the enzyme inside the peroxisome, so that diffusion at the membrane and in the matrix becomes limiting (de Duve and Baudhuin, 1966).

It has been widely assumed that the peroxisomal membrane is more labile than other biological membranes and that the permeability of isolated peroxisomes is the result of aspecific membrane damage inflicted during homogenization and isolation. A strong argument against such assumption is the observation by Van Veldhoven *et al.* (1987) that liposomes reconstituted with a peroxisomal integral membrane protein fraction from rat liver become permeable to the same molecules that penetrate purified peroxisomes. This suggests that, like the mitochondrial outer membrane, the peroxisomal membrane contains a nonselective pore-forming protein. The existence of such protein in the peroxisomal membrane from rat liver has indeed been confirmed by means of electrophysiological measurements (Labarca *et al.*, 1986; Lemmens *et al.*, 1989). The diameter of the pore is 1.7 nm, which is large enough to allow the free diffusion of substrates, products, and cofactors (Lemmens *et al.*, 1989). The rat liver protein has not been purified yet.

Douma *et al.* (1990) have shown that peroxisomes from the yeasts *H. polymorpha* and *C. boidinii* are also permeable to sucrose and other small molecules, suggesting that the occurrence of a pore-forming protein may be a more general property of the peroxisomal membrane. Electrophysiological measurements on peroxisomal membranes from *H. polymorpha* have demonstrated the presence of a pore with a diameter of 1.5 nm. The pore-forming protein from *H. polymorpha* has been purified and appears to be identical to the 31 kDa integral membrane protein (see Section 2.2.2.).

Isolated plant peroxisomes, plant glyoxysomes, and glycosomes have the same high equilibrium density in sucrose gradients as do animal peroxisomes, and various reports have mentioned that the organelles are permeable to a number of metabolites (Liang and Huang, 1983; Liang and Huang, 1984; Yu and Huang, 1986; Patthey and Deshusses, 1987). It is surprising, therefore, that several enzymes are latent in these organelles. Recently, Heupel *et al.* (1991) may have solved this discrepancy by showing that the enzyme latency in spinach leaf peroxisomes is not related to the intactness of the membrane but that it is a property of the matrix per se. They suggest that the apparent latency is the result of the dense packing of proteins in the matrix, which would restrict substrate diffusion.

As a whole, the available evidence indicates that most—if not all—peroxisomes *in vitro* are permeable to small water-soluble molecules because of the presence of proteinaceous pores in their membranes. Whether the permeability of the pores is regulated in the intact cell and whether amphiphiles, such as acyl-CoA esters, which are formed at the outer aspect of the peroxisomal membrane (Section 2.4.) and which tend to accumulate in the lipid phase of membranes, also diffuse via the pores remain unsolved questions.

Recent data from Wanders and co-workers indicate that in digitonin-permeabilized hepatocytes D-proline, urate, and glycolate freely pass the peroxisomal membrane (Verleur and Wanders, 1993). On the other hand, Wolvetang *et al.* (1990a, 1991) demonstrated that dihydroxyacetone-phosphate acyltransferase, a peroxisomal membrane enzyme, the catalytic site of which is exposed to the matrix (see Section 2.4.), shows latency in digitonin-permeabilized human fibroblasts but not in homogenized or sonicated fibroblasts. The latency in the permeabilized cells could be abolished by the addition of ATP, suggesting that in the intact cell ATP may be required for the transport of at least one of the substrates (acyl-CoA and dihydroxyacetone-phosphate) of the enzyme.

Kamijo *et al.* (1990) have speculated that the 70 kDa integral peroxisomal membrane protein from rat liver, which is a member of the P-glycoprotein-related ATP-binding protein superfamily, is involved in membrane transport, possibly that of acyl-CoAs (Section 2.2.1). Their speculation is based on the transport function of other members of the ATP-binding protein family. Veenhuis and co-workers (Nicolay *et al.*, 1987; Waterham *et al.*, 1990) provided evidence that the internal milieu of peroxisomes in *C. boidinii* and *H. polymorpha* is acidic (pH 5.8–6). They also described the presence of a proton-translocating ATPase in the peroxisomal membrane of *H. polymorpha*, which may be responsible for the acidification (Douma *et al.*, 1987). The peroxisomal ATPase activity is extremely small in comparison with that of the mitochondria, and Whitney and Bellion (1991) have argued that the ATPase activity present in peroxisomal fractions can be explained by contamination with other organelles. The acidic milieu of yeast peroxisomes is difficult to reconcile with the presence of open membrane pores. If the pores are closed in the intact yeast cell, one must postulate the presence of specific carriers. There appear to be no indications in the literature for the existence of such carriers in peroxisomes from either yeasts or other sources. ATPase activity has also been described in peroxisomal membranes from rat and bovine liver (delValle *et al.*, 1988; Makita *et al.*, 1990; Wolvetang *et al.*, 1990b; Malik *et al.*, 1991). There is no experimental reason to believe that the internal milieu of animal peroxisomes is acidic (Dunn, 1990). In animals, the function of peroxisomal ATPases, if present, remains unknown.

2.4. Enzyme Content and Metabolic Role of the Peroxisomal Membrane

The membrane of mammalian peroxisomes contains several enzymes involved in lipid metabolism. Thus far only two of these enzymes have been purified (see below). Most of them are probably minor components of the peroxisomal membrane. It is doubtful, therefore, that these enzyme activities would be associated with the known major peroxisomal membrane polypeptides.

Before a fatty acid can be degraded via β-oxidation or before it can be esterified, it needs to be activated to its CoA ester. The peroxisomal membrane contains two acyl-CoA synthetases. One enzyme prefers long chain fatty acids as the substrate (Shindo and Hashimoto, 1978; Krisans *et al.*, 1980; Mannaerts *et al.*, 1982; Bronfman *et al.*, 1984), the other one very long chain fatty acids (Singh and Poulos, 1988; Lazo *et al.*, 1990). Both enzymes are integral membrane proteins. The catalytic site of the long chain acyl-CoA synthetase is exposed to the cytosol (Mannaerts *et al.*, 1982; Lageweg *et al.*, 1991) implying that not the fatty acid but the acyl-CoA derivative crosses the membrane before it is oxidized by the matrical β-oxidation enzymes or before it is esterified at the inner side of the membrane, as will be explained later in this section. The membrane topology of the very long chain acyl-CoA synthetase is a matter of controversy; evidence has been given that the catalytic site is exposed to the cytosol (Lageweg *et al.*, 1991) but also that it is exposed to the matrix (Lazo *et al.*, 1990). Long chain acyl-CoA synthetase is present not only in the peroxisomal membrane but also in the mitochondrial outer membrane and in the endoplasmic reticulum (Shindo and Hashimoto, 1978; Krisans *et al.*, 1980; Mannaerts *et al.*, 1982). The synthetases present in these three organelles are indistinguishable with regard to their kinetic, molecular, and immunochemical properties (Miyamoto *et al.*, 1981; Miyazawa *et al.*, 1985). The molecular weight of 76 kDa, as determined by SDS-gel electrophoresis, is in close agreement with a value of 78,177 kDa, predicted from the sequence of the cloned gene (Suzuki *et al.*, 1990). Very long chain acyl-CoA synthetase is found in the peroxisomal membrane and in the endoplasmic reticulum (Singh and Poulos, 1988; Lazo *et al.*, 1990). Mitochondria lack this enzyme, and this appears to be the reason why very long chain fatty acids are not oxidized by mitochondria. In X-linked adrenoleukodystrophy, the peroxisomal very long chain acyl-CoA synthetase activity is deficient. The putative gene, responsible for this disease, has recently been cloned (Mosser *et al.*, 1993). The deduced protein sequence did not show homology to the rat sequence of acyl-CoA synthetase, but significant homology was found with the human and rat 70 kDa peroxisomal integral membrane protein. The authors speculate that this protein may be involved in the peroxisomal import of the synthetase.

Glycerolipid synthesis starts with the acylation of glycerol-3-phosphate or dihydroxyacetone-phosphate (Bell and Coleman, 1980). Endoplasmic reticulum acylates mainly glycerol-3-phosphate and is responsible for the synthesis of glycerolipids in which the fatty acids are linked to the glycerol moiety via ester bonds. These ester glycerolipids constitute by far the major portion of the cellular phospholipids and triacylglycerols. Peroxisomes acylate exclusively dihydroxyacetone-phosphate in a reaction catalyzed by dihydroxyacetone-phosphate acyltransferase (Hajra *et al.*, 1979). The catalytic site of this peroxisomal integral membrane protein is exposed to the matrix (Declercq *et al.*, 1984). The fatty acid in acyldihydroxyacetone-phosphate is then replaced by a fatty alcohol so that an ether bond is formed. The reaction is catalyzed by alkyldihydroxyacetone-phosphate synthetase, another peroxisomal integral membrane protein, the catalytic site of which is also exposed to the matrix (Bishop *et al.*, 1982). A further peroxisomal integral membrane protein is alkyldihydroxyacetone-phosphate reductase. This enzyme reduces alkyldihydroxyacetone-phosphate to 1-alkylglycerol-phosphate. Its catalytic site is exposed to the cytosol (Ghosh and Hajra, 1986b). The conversion of 1-alkylglycerol-phosphate to 1-alkyl-2-acyl phospholipids and to 1-alkyl-2,3-diacylglycerols occurs in the endoplasmic reticulum (Declercq *et al.*, 1984; Ballas *et al.*, 1984).

Dihydroxyacetone-phosphate acyltransferase (Schlossman and Bell, 1977; Declercq *et al.*, 1984), alkyldihydroxyacetone-phosphate synthetase (Hajra and Bishop, 1982), and alkyldihydroxyacetone-phosphate reductase (Ghosh and Hajra, 1986b) are also present in the endoplasmic reticulum. However, the major portion of dihydroxyacetone-phosphate acylation (Declercq *et al.*, 1984) and of alkyldihydroxyacetone-phosphate synthesis (H. Singh *et al.*, 1989) takes place in peroxisomes and, as a consequence, the latter organelles initiate the major portion of ether glycerolipid synthesis. The fatty alcohols required for ether glycerolipid synthesis are also formed by a peroxisomal integral membrane enzyme: acyl-CoA reductase. The catalytic site of this enzyme faces the cytosol (Burdett *et al.*, 1991).

Dihydroxyacetone-phosphate acyltransferase, recently purified from guinea pig liver peroxisomes, is a monomeric protein with a molecular mass of 69 kDa (SDS-gel electrophoresis) (Webber and Hajra, 1993). Based on the published enrichments, this enzyme constitutes less than 2% of the peroxisomal membrane proteins. Alkyldihydroxyacetone-phosphate synthase has only been partially purified (Horie *et al.*, 1992). Purified acyl/alkyldihydroxyacetone-phosphate reductase, isolated from guinea pig peroxisomes, migrates as a 60 kDa protein on SDS-gels. It constitutes less than 0.3% of the total peroxisomal proteins (Datta *et al.*, 1990).

Like endoplasmic reticulum, peroxisomes possess several enzymes involved in the synthesis of cholesterol and dolichols: 3-hydroxy-3-methylglutaryl-CoA reductase (Keller *et al.*, 1985, 1986), the key regulatory enzyme of sterol synthesis, and the enzymes that convert farnesyl-pyrophosphate into cholesterol on the one hand (Thompson *et al.*, 1987; Appelkvist *et al.*, 1990) and into dolichols on the other (Appelkvist, 1987; Appelkvist and Kalén, 1989). In peroxisomes, 3-hydroxy-3-methylglutaryl-CoA reductase is not membrane-bound but it is a soluble matrix enzyme (Keller *et al.*, 1986). The enzymes that catalyze the formation of cholesterol from farnesyl-pyrophosphate are associated with the membrane, most probably with the inner aspect (Appelkvist *et al.*, 1990). It has been claimed that the enzymes that convert farnesyl-pyrophosphate into dolichols are matrix enzymes (Appelkvist and Kalén, 1989) but further studies are required to substantiate this claim.

The contribution of peroxisomes to overall cholesterol and dolichol synthesis is probably small in comparison with that of the endoplasmic reticulum. The physiological function remains unknown.

There is evidence that the peroxisomal membrane may also contain a small portion of the cell's NADH cytochrome b5 reductase, cytochrome b5, and NADH cytochrome *c* reductase (for a detailed discussion, see Lazarow, 1984). The function of these enzymes in peroxisomes is not known, but it is possible that they play a role in the reduction of cytochromes P450, a group of enzymes involved in the terminal reactions of cholesterol synthesis.

Much less is known about the enzyme composition of the peroxisomal membrane in plants and in eukaryotic microorganisms. The glyoxysomal membrane from castor bean endosperm displays monoacylglycerol lipase activity (Beevers, 1982; Gonzalez *et al.*, 1987) and may also contain cytochrome b5 and NADH- and NADPH cytochrome *c* reductase (Donaldson *et al.*, 1981). Glyoxysomes from castor bean endosperm and yeast peroxisomes possess long chain acyl-CoA synthetase (Beevers, 1982; Tanaka *et al.*, 1982). It can be expected that this enzyme is membrane-bound also in these organisms.

2.5.   Membrane Polypeptides Possibly Involved in Protein Translocation and Biogenesis of Peroxisomes

Peroxisomal proteins are synthesized on free polyribosomes in the cytosol and posttranslationally imported in pre-existing peroxisomes (see Section 3.1.). Only very recently has one succeeded in identifying the first peroxisomal membrane polypeptides that seem to be part of the protein translocation machinery. Kunau and co-workers selected *S. cerevisiae* mutants deficient in peroxisome assembly (Kunau and Hartig, 1992; Höhfeld *et al.*, 1992). By functional complementation analysis they were able to clone a gene that encodes a 48 kDa peroxisomal integral membrane protein (Höhfeld *et al.*, 1991) and that restored peroxisome assembly in one of the mutants. The bulk of the protein is exposed to the cytosol. The authors propose that the protein plays a role in protein translocation, perhaps as a targeting signal receptor.

Using a similar approach of genetic complementation, Fujiki and co-workers identified a rat gene that restored peroxisome assembly in a peroxisome-deficient Chinese hamster ovary cell mutant (Tsukamoto *et al.*, 1991) and a human gene that restored peroxisome assembly in a fibroblast cell line from a Zellweger patient (Shimozawa *et al.*, 1992). Zellweger patients lack functional peroxisomes. The genes encode a 35 kDa peroxisomal integral membrane protein. The amino acid sequences of the rat and human protein are 88% homologous. Gärtner *et al.* (1992) described mutations in the 70 kDa peroxisomal integral membrane protein gene in two Zellweger patients, strengthening the proposition by Kamijo *et al.* (1990) that the 70 kDa protein may be involved in the translocation of matrix proteins.

The product of the recently cloned X-linked adrenoleukodystrophy gene (Mosser *et al.*, 1993) also belongs to the family of ABC transporters with an ATP binding site, and might be involved in peroxisomal protein translocation (see above).

Recently we described the presence of three small GTP-binding proteins (29, 27, and 25 kDa) in the peroxisomal membrane from rat liver (Verheyden *et al.*, 1992). The proteins are firmly anchored in the membrane and exposed to the cytosol. Because small GTP-binding proteins are involved in vesicle budding and fusion in the endocytotic/exocytotic compartments, one might speculate that these proteins play a role in peroxisome biogenesis via budding from or fission of pre-existing peroxisomes (see Section 3.1).

3.   ASSEMBLY OF THE PEROXISOMAL MEMBRANE

3.1.   Biogenesis of Peroxisomes

Evidently, the assembly of the peroxisomal membrane is intimately linked to peroxisome biogenesis. Therefore, we will first summarize what is currently known or thought about peroxisome biogenesis in general.

Peroxisomes do not contain DNA. Their matrix (Goldman and Blobel, 1978; Rachubinski *et al.*, 1984; Miura *et al.*, 1984; Fujiki *et al.*, 1985) and membrane (Fujiki *et al.*, 1984; Köster *et al.*, 1986; Suzuki *et al.*, 1987; Bodnar and Rachubinski, 1991) proteins are encoded by nuclear genes and synthesized on free polyribosomes in the cytosol (for a possible exception, see Section 3.2.). The newly synthesized proteins are released in the cytosol and then imported in what is believed to be pre-existing peroxisomes, which multiply by fission or budding (Lazarow and Fujiki, 1985). Practically all peroxisomal proteins are synthesized in their mature form and they are not proteolytically processed during or after translocation. An exception is mammalian acyl-CoA oxidase, the first enzyme of the peroxisomal β-oxidation of straight chain acyl-CoAs. The enzyme consists mainly of two identical 52 kDa and two identical 20 kDa subunits, which are formed from a 72 kDa subunit most probably after import (Miura *et al.*, 1984; Schram *et al.*, 1986). The proteolytical processing of the large subunit occurs independently of and is not required for import (Fujiki and Lazarow, 1985; Imanaka *et al.*, 1987; Heinemann and Just, 1992). Another exception is thiolase, which catalyzes the last reaction of peroxisomal β-oxidation. The rat enzyme consists of two identical subunits, which are synthesized as a 44 kDa precursor (Miura *et al.*, 1984; Fujiki *et al.*, 1985; Hijikata *et al.*, 1987; Bodnar and Rachubinski, 1990). Upon translocation into the peroxisome, the precursor's amino-terminal presequence is cleaved off, leaving a 41 kDa mature subunit. As is the case for acyl-CoA oxidase, processing of the precursor does not seem to be required for import (Singh *et al.*, 1991).

In a series of elegant studies, Subramani, Gould and co-workers could demonstrate that the peroxisomal targeting signal of several matrix proteins consists of three amino acids (Ser-Lys-Leu) at the carboxy terminus of the protein (Keller *et al.*, 1987; Gould *et al.*, 1988, 1989, 1990a). Conserved variants of this tripeptide (serine, alanine, cysteine at the first position; lysine, histidine, arginine at the second position; leucine at the carboxy terminus) also serve as peroxisomal targeting signals (Gould *et al.*, 1989). The carboxy-terminal signal targets proteins not only to peroxisomes of mammals, insects, plants, and yeasts but also to glyoxysomes of yeasts and plants and to glycosomes of trypanosomatids (Gould *et al.*, 1990b; Keller *et al.*, 1991; Fung and Clayton, 1991). The laboratories of Subramani and Rachubinski (Swinkels *et al.*, 1991) recently described that the cleavable amino-terminal presequence of rat peroxisomal thiolase functions as a targeting signal for this protein. An amino-terminal targeting signal has also been suggested for the glyoxysomal malate dehydrogenase of watermelon (Gietl, 1990). However, not all peroxisomal proteins (e.g., the integral membrane proteins; see Section 3.2) seem to possess either the carboxy-terminal tripeptide or an amino-terminal targeting sequence, indicating that other targeting signals remain to be discovered. This suggests that several protein import pathways exist in peroxisomes. In agreement with this contention, Kunau and co-workers isolated a *S. cerevisiae* mutant in which the import of peroxisomal thiolase is defective but not that of peroxisomal proteins carrying the carboxy-terminal tripeptide signal (Höhfeld *et al.*, 1992).

A more or less opposite situation can be found in human pathology. Cells from patients with the Zellweger syndrome lack morphologically recognizable peroxisomes. Matrix enzymes such as catalase and acyl-CoA oxidase are not found in the particulate fraction but in the cytosol, indicating a lack of import (Lazarow and Moser, 1989). Santos and a number of other workers have found that Zellweger cells contain so-called peroxisomal ghosts: vesicular structures of larger diameter and lesser density than peroxisomes and lacking most content. The ghosts contain varying but sometimes normal amounts of certain peroxisomal integral membrane proteins (22, 26, 53, 70, and 140 kDa proteins) and also the thiolase precursor (Santos *et al.*, 1988; Small *et al.*, 1988; Suzuki *et al.*, 1989; Balfe *et al.*, 1990; Gärtner *et al.*, 1991). Microinjection studies confirmed that Zellweger cell lines are not capable of importing proteins with the tripeptide carboxy-terminal targeting sequence (Walton *et al.*, 1992). It is currently not known whether the import routes differ in their signal receptors only, so that they would share the other elements of one common translocation pathway, or that several completely separate pathways exist. The latter possibility seems to be the most plausible one (see Section 3.2).

Kunau, Veenhuis, Tabak, and co-workers selected a variety of *S. cerevisiae* mutants defective in peroxisome assembly (see Kunau and Hartig, 1992). The mutants comprise 18 complementation groups, indicating that at least an equal number of gene products is involved in peroxisome biogenesis in this yeast. One of these is a 48 kDa peroxisomal integral membrane protein, which has been proposed to function as a targeting signal receptor (Section 2.5). Another one is a 117 kDa polypeptide with two putative ATP-binding sites. The subcellular localization of this protein is not known, but its hydropathy profile suggests that it is a soluble component (Erdmann *et al.*, 1991).

Human disorders of peroxisome biogenesis (Zellweger syndrome, neonatal adrenoleukodystrophy, infantile Refsum's disease, hyperpipecolic acidemia) comprise at least six complementation groups (Brul *et al.*, 1988; Roscher *et al.*, 1989), confirming the complexity of peroxisome biogenesis. Fujiki and co-workers identified a 35 kDa peroxisomal integral membrane protein involved in peroxisome biogenesis in the rat and the human (see Section 2.5). Other candidates that might play a role in peroxisome biogenesis and protein translocation in mammals are the 70 kDa peroxisomal integral membrane protein, which is mutated in some Zellweger patients (Section 2.5), the pore-forming protein, the internal diameter of which is large enough (1.7 nm) to accommodate an unfolded protein (Section 2.3) and the small GTP-binding proteins in the peroxisomal membrane, which might play a role in fission or budding (Section 2.5).

In the preceding lines we tacitly accepted that newly synthesized peroxisomal proteins are imported in pre-existing peroxisomes that grow and then multiply by fission or budding, a model advocated by Lazarow (Lazarow and Fujiki, 1985). Originally, it was thought that peroxisomes originate by budding from the endoplasmic reticulum. This model has been abandoned for several reasons: (1) Connections between peroxisomes and endoplasmic reticulum have never been unequivocally demonstrated; (2) the polypeptide content of the peroxisomal membrane differs greatly from that of the endoplasmic reticulum; and (3) peroxisomal proteins are not co-translationally but posttranslationally imported as explained above (for a discussion, see Lazarow and Fujiki, 1985). However, it is not entirely clear yet how pre-existing peroxisomes must be defined. Are they mature peroxisomes that multiply by division or are there as yet unidentified properoxisomal structures from which peroxisomes originate possibly by budding? In the methylotrophic yeasts *C. boidinii* (Veenhuis and Goodman, 1990) and *H. polymorpha* (Veenhuis *et al.*, 1979) peroxisomes appear to originate from pre-existing ones. Peroxisomes are massively induced when these yeasts are switched from a culture medium containing glucose to a medium containing methanol. During induction, single peroxisomes in *H. polymorpha* dramatically increase in volume and then multiply by segregation of small daughter organelles. In *C. boidinii* several small peroxisomes develop from a small pre-existing peroxisome before they all start growing in volume. In regenerating rat liver, Yamamoto and Fahimi (1987) found morphological evidence to suggest that peroxisomes originate by budding from a peroxisomal reticulum, an irregular network of interconnected peroxisomes. Ohno and Fujii (1990) described "peroxisome-forming sheets" in mouse hepatocytes cultured in the presence of the peroxisome proliferator clofibrate. The peroxisome-forming sheets appeared as smooth membranous structures resembling the smooth endoplasmic reticulum, from which strings of interconnected spheroid peroxisomes were seen budding. The budding peroxisomes could be stained for catalase, but the smooth membranous structures were catalase negative, suggesting that budding preceded the import of matrix proteins.

Recently, Heinemann and Just (1992) observed that rat hepatocyte homogenates contain two populations of peroxisomes: one population with the well-known density of mature peroxisomes and a less abundant population of lower density. Pulse-chase experiments in isolated hepatocytes revealed that newly synthesized peroxisomal acyl-CoA oxidase is first imported in the low density population before it reaches the population of mature peroxisomes. The observations by the groups of Ohno, Fahimi, and Just suggest that, at least in rodent liver, properoxisomal structures may exist and that mature peroxisomes may have lost their import competence. In how far the peroxisomal ghosts observed in Zellweger cell lines are related to these putative properoxisomal structures remains unknown.

In vitro translation-import assays have shown that newly synthesized peroxisomal matrix proteins first bind to peroxisomes and are then translocated across the membrane. Binding occurs at 0°C in the absence of ATP; translocation occurs only at higher temperatures (26°C) and requires ATP hydrolysis but not a membrane potential or an ion gradient across the membrane (Imanaka *et al.*, 1987). *In vitro* peroxisomal import assays lack efficiency and reliability as compared to the assays developed for mitochondria and chloroplasts. This lack of properly working *in vitro* assays has seriously slowed down the study of peroxisome biogenesis. One has argued that a leakiness of isolated peroxisomes because of membrane damage inflicted during isolation might be the reason for the malfunctioning of the *in vitro* import assays. Another reason might be that mature peroxisomes have lost most of their import competence and that *in vitro* assays have hitherto been carried out with the wrong population of peroxisomes.

3.2. Biogenesis of the Peroxisomal Membrane: Polypeptides

The laboratories of Lazarow (Fujiki *et al.*, 1984), Just (Köster *et al.*, 1986), Hashimoto (Suzuki *et al.*, 1987), and Rachubinski (Bodnar and Rachubinski, 1991) studied the *in vitro* synthesis of several integral membrane polypeptides from rat liver peroxisomes. The combined results from these laboratories have shown that the 70-, 36-, and 22 kDa proteins are synthesized on free polyribosomes in their mature form. These results indicate that the integral membrane proteins follow a route similar to that of their matrix counterparts: posttranslational insertion without further proteolytic processing. A peculiar exception to this rule was noted by Bodnar and Rachubinski (1991). They found that an antiserum raised against highly purified peroxisomal membranes from rat liver reacted not only with a 50 kDa peroxisomal membrane protein but also with a protein of equal size in the endoplasmic reticulum (see Section 2.2.1). The reacting protein(s) was (were) synthesized for 91% on membrane-bound polysomes and for only 9% on free polysomes, which would suggest a role for the endoplasmic reticulum in peroxisome biogenesis. Although intriguing, the observation does not necessarily discredit the model that all peroxisomal proteins are posttranslationally imported in pre-existing peroxisomes. The immunologically related 50 kDa proteins are not necessarily identical and the possibility remains that the peroxisomal protein is synthesized on free polyribosomes whereas its endoplasmic reticulum counterpart is synthesized on membrane-bound polyribosomes.

Thus far, the amino acid sequence is known for the following peroxisomal integral membrane proteins: the rat and human 70 kDa proteins (Kamijo *et al.*, 1990; Kamijo *et al.*, 1992); the rat and human 35 kDa proteins, which are involved in peroxisome assembly (Tsukamoto *et al.*, 1991; Shimozawa *et al.*, 1992); the 48 kDa protein involved in peroxisome assembly in *S. cerevisiae* (Höhfeld *et al.*, 1991); and the 47 kDa protein from *C. boidinii* (McCammon *et al.*, 1990a). None of these proteins carries the carboxy-terminal tripeptide targeting signal frequently found in matrix proteins.

Studies on peroxisome biogenesis in *C. boidinii*, switched from a glucose to a methanol-containing medium (Veenhuis and Goodman, 1990), and on regenerating rat liver (Lüers *et al.*, 1990), indicate that the synthesis and membrane insertion of peroxisomal integral membrane proteins precedes the synthesis and import of matrix proteins. The mechanism of protein insertion in the membrane and in exactly which structures (mature peroxisomes, properoxisomal structures; see Section 3.1) these proteins are inserted remain unknown. The same applies to the mechanism of translocation of matrix proteins across the membrane. The observations in Zellweger cell lines (Section 3.1) suggest that the insertion pathway of membrane proteins and the translocation pathway of matrix proteins may differ in more respects than merely signal receptors. Zellweger cell lines fail to import matrix proteins containing the carboxy-terminal tripeptide signal, but also other matrix proteins, the targeting sequence of which has not been identified yet (e.g., catalase). The occurrence in these cells of peroxisome ghosts, which contain certain membrane proteins (22-, 26-, 53-, 70-, 140 kDa proteins) in their membranes, and the large number of complementation groups indicate that in the Zellweger syndrome the assembly of an at least rudimentary peroxisomal membrane remains possible, and that the deficient components of the translocation pathway are not involved in the insertion of these membrane proteins.

3.3 Biogenesis of the Peroxisomal Membrane: Lipids

Our knowledge of the assembly of the lipid bilayer of the peroxisomal membrane remains extremely poor. The mammalian peroxisomal membrane contains the enzymes that catalyze the initial reactions of ether lipid synthesis: acylation of dihydroxyacetone-phosphate, conversion of acyldihydroxyacetone-phosphate to alkyldihydroxyacetone-phosphate, and reduction of the latter to 1-alkyldihydroxyacetone-phosphate. The further reactions in the synthesis of ether-linked lipids are carried out in the endoplasmic reticulum, and peroxisomes lack the enzymes involved in these reactions (see Section 2.4). The endoplasmic reticulum is also responsible for the synthesis of the ester-linked glycerolipids via the glycerol-3-phosphate acylation pathway (Bell and Coleman, 1980). Ester-linked glycerolipids are far more abundant than ether-linked glycerolipids. Peroxisomes lack not only the terminal but also the initial (e.g., glycerol-3-phosphate acyltransferase) enzymes of ester glycerolipid synthesis (Declercq *et al.*, 1984). The subcellular distribution of the glycerolipid-synthesizing enzymes, as described above, clearly indicates that peroxisomes are not capable of synthesizing their own membrane phospholipids and that phospholipids required for membrane expansion must come from the endoplasmic reticulum. However, direct proof that phospholipids are transferred from the endoplasmic reticulum to peroxisomes has not been offered yet. Evidently, how these lipids are transported from the endoplasmic reticulum to the peroxisomes—via transfer proteins or via some other mechanism—remains unknown. The endoplasmic reticulum is the site of phospholipid synthesis not only in animal cells but also in plant cells and in yeast (Huang *et al.*, 1983; Veenhuis and Harder, 1987). Of course, if proper-oxisomal structures such as the peroxisome-forming sheets described by Ohno and Fujii (1990) in mouse liver really exist, one cannot exclude the possibility that these structures might be capable of elaborating their own membrane lipids and those for budding peroxisomes. The answer to this question cannot be given before these structures have been isolated and characterized.

Chapman and Trelease (1991) have recently proposed that expanding glyox-ysomes of cotyledons of cotton seedlings acquire their membrane lipids not from the endoplasmic reticulum but from lipid storage bodies. Their proposition is based on similarities in lipid composition of the lipid bodies and the glyoxysomal membrane, on *in vivo* pulse-chase radiolabeling experiments of the phospho-lipids of glyoxysomal membranes and on *in vitro* lipid transfer experiments between isolated lipid bodies and isolated glyoxysomes.

3.4. Perspectives

In comparison with what is known about the biogenesis of other cell organ-elles, our knowledge of peroxisome assembly remains rather limited. The recent availability of mutant cell lines that are defective in peroxisome assembly—from yeast, animals, and humans—will considerably advance our knowledge of per-oxisome biogenesis in the forthcoming years. One may hope that the near future will also bring an answer to the question whether mature peroxisomes remain import competent or whether import competence is a unique feature of immature or nascent peroxisomes. This answer might allow for the isolation of the right population of peroxisomes in order to develop an efficient *in vitro* import system.

REFERENCES

Alexson, S. E. H., Fujiki, Y., Shio, H., and Lazarow, P. B., 1985, Partial disassembly of peroxi-somes, *J. Cell Biol.* **101**:294–305.

Aman, R. A., and Wang, C. C., 1987, Identification of two integral glycosomal membrane proteins in *Trypanosoma brucei*, *Mol. Biochem. Parasitol.* **25**:83–92.

Appelkvist, E.-L., 1987, Dolichol biosynthesis in rat liver peroxisomes, *Acta Chem. Scand.* **B41**:73–75.

Appelkvist, E.-L., and Dallner, G., 1987, Dolichol metabolism and peroxisomes, in *Peroxisomes in Biology and Medicine* (H. D. Fahimi and H. Sies, eds.), pp. 53–66, Springer-Verlag, Berlin, Heidelberg.

Appelkvist, E.-L., and Kalén, A., 1989, Biosynthesis of dolichol by rat liver peroxisomes, *Eur. J. Biochem.* **185**:503–509.

Appelkvist, E.-L., Brunk, U., and Dallner, G., 1981, Isolation of peroxisomes from rat liver using sucrose and Percoll gradients, *J. Biochem. Biophys. Methods* **5**:203–217.

Appelkvist, E.-L., Reinhart, M., Fischer, R., Billheimer, J., and Dallner, G., 1990, Presence of individual enzymes of cholesterol biosynthesis in rat liver peroxisomes, *Arch. Biochem. Biophys.* **282:**318–325.

Balfe, A., Hoefler, G., Chen, W. W., and Watkins, P. A., 1990, Aberrant subcellular localization of peroxisomal 3-ketoacyl-CoA thiolase in the Zellweger syndrome and rhizomelic chondrodysplasia punctata, *Pediatr. Res.* **27:**304–310.

Ballas, L. M., Lazarow, P. B., and Bell, R. M., 1984, Glycerolipid synthetic capacity of rat liver peroxisomes, *Biochim. Biophys. Acta* **795:**297–300.

Beevers, H., 1982, Glyoxysomes in higher plants, *Ann. NY Acad. Sci.* **386:**243–251.

Beevers, H., and Gonzalez, E., 1987, Proteins and phospholipids of glyoxysomal membranes from castor bean, *Methods Enzymol.* **148:**526–532.

Bell, R. M., and Coleman, R. A., 1980, Enzymes of glycerolipid synthesis in eukaryotes, *Annu. Rev. Biochem.* **49:**459–487.

Bishop, J. E., Salem, M., and Hajra, A. K., 1982, Topographical distribution of lipid biosynthetic enzymes on peroxisomes (microbodies), *Ann. NY Acad. Sci.* **386:**411–413.

Bodnar, A. G., and Rachubinski, R. A., 1990, Cloning and sequence determination of cDNA encoding a second rat liver peroxisomal 3-ketoacyl-CoA thiolase, *Gene* **91:**193–199.

Bodnar, A. G., and Rachubinski, R. A., 1991, Characterization of the integral membrane polypeptides of rat liver peroxisomes isolated from untreated and clofibrate-treated rats, *Biochem. Cell Biol.* **69:**499–508.

Bronfman, M., Inestrosa, N. C., Nervi, F. O., and Leighton, F., 1984, Acyl-CoA synthetase and the peroxisomal enzymes of β-oxidation in human liver, *Biochem. J.* **224:**709–720.

Brul, S., Westerfeld, A., Strijland, A., Wanders, R. J. A., Schram, A. W., Heymans, H. S. A., Schutgens, R. B. H., van den Bosch, H., and Tager, J. M., 1988, Genetic heterogeneity in the cerebrohepatorenal (Zellweger) syndrome and other inherited disorders with a generalized impairment of peroxisomal functions: A study using complementation analysis, *J. Clin. Invest.* **81:**1710–1715.

Burdett, K., Larkins, L. K., Das, A. K., and Hajra, A. K., 1991, Peroxisomal localization of acyl-coenzyme A reductase (long chain alcohol forming) in guinea pig intestine mucosal cells, *J. Biol. Chem.* **266:**12201–12206.

Cerdan, S., Künnecke, B., Dölle, A., and Seelig, J., 1988, In situ metabolism of 1,ω medium chain dicarboxylic acids in the liver of intact rats as detected by ^{13}C and ^{1}H NMR, *J. Biol. Chem.* **263:**11664–11674.

Chapman, K. D., and Trelease, R. N., 1991, Acquisition of membrane lipids by differentiating glyoxysomes: Role of lipid bodies, *J. Cell Biol.* **115:**995–1007.

Chen, N., Crane, D. I., and Masters, C., 1988, Analysis of the major integral membrane proteins of peroxisomes from mouse liver, *Biochim. Biophys. Acta* **945:**135–144.

Crane, D. I., and Masters, C. J., 1986, The effect of clofibrate on the phospholipid composition of the peroxisomal membrane in mouse liver, *Biochim. Biophys. Acta* **876:**256–263.

Datta, S. C., Ghosh, M. K., and Hajra, A. K., 1990, Purification and properties of acyl/alkyl dihydroxyacetone-phosphate reductase from guinea pig liver peroxisomes, *J. Biol. Chem.* **265:**8268–8274.

Declercq, P. E., Haagsman, H. P., Van Veldhoven, P., Debeer, L. J., van Golde, L. M. G., and Mannaerts, G. P., 1984, Rat liver dihydroxyacetone-phosphate acyltransferases and their contribution to glycerolipid synthesis, *J. Biol. Chem.* **259:**9064–9075.

de Duve, C., and Baudhuin, P., 1966, Peroxisomes (microbodies and related particles), *Physiol. Rev.* **46:**323–357.

de Duve, C., Pressman, B. C., Gianetto, R., Wattiaux, R., and Appelmans, F., 1955, Tissue fractionation studies. 6: Intracellular distribution patterns of enzymes in rat liver tissue, *Biochem. J.* **60:**604–617.

delValle, R., Soto, U., Necochea, C., and Leighton, F., 1988, Detection of an ATPase activity in rat liver peroxisomes, *Biochem. Biophys. Res. Commun.* **156:**1353–1359.

Donaldson, R. P., Tolbert, N. E., and Schnarrenberger, C., 1972, A comparison of microbody membranes with microsomes and mitochondria from plant and animal tissue, *Arch. Biochem. Biophys.* **152:**199–215.

Donaldson, R. P., Tully, R. E., Young, O. E., and Beevers, H., 1981, Organelle membranes from germinating castor bean endosperm. II: Enzymes, cytochromes and permeability of the glyoxysome membrane, *Plant Physiol.* **67:**21–25.

Douma, A. C., Veenhuis, M., Sulter, G. J., and Harder, W., 1987, A proton-translocating adenosine triphosphatase is associated with the peroxisomal membrane of yeasts, *Arch. Microbiol.* **147:**42–47.

Douma, A. C., Veenhuis, M., Sulter, G. J., Waterman, H. R., Verheyden, K., Mannaerts, G. P., and Harder, W., 1990, Permeability properties of peroxisomal membranes from yeasts, *Arch. Microbiol.* **153:**490–495.

Dreyer, C., Krey, G., Keller, H., Givel, F., Helftenbein, G., and Wahli, W., 1992, Control of the peroxisomal β-oxidation pathway by a novel family of nuclear hormone receptors, *Cell* **68:**879–887.

Dunn, W. A., Jr., 1990, Studies on the mechanisms of autophagy: Maturation of the autophagic vacuole, *J. Cell Biol.* **110:**1935–1945.

Erdmann, R., Wiebel, F. F., Flessau, A., Rytka, J., Beyer, A., Fröhlich, K.-U., and Kunau, W.-H., 1991, PAS1, a yeast gene required for peroxisome biogenesis, encodes a member of a novel family of putative ATPases, *Cell* **64:**499–510.

Evans, W. H., 1982, Subcellular membranes and isolated organelles: Preparative techniques and criteria for purity, *Techniques Lipid Membr. Biochem.* **B407a:**1–46.

Fujiki, Y., and Lazarow, P. B., 1985, Post-translational import of fatty acyl-CoA oxidase and catalase into peroxisomes of rat liver *in vitro*, *J. Biol. Chem.* **260:**5603–5609.

Fujiki, Y., Fowler, S., Shio, H., Hubbard, A. L., and Lazarow, P. B., 1982, Polypeptide and phospholipid composition of the membrane of rat liver peroxisomes: Comparison with endoplasmic reticulum and mitochondrial membranes, *J. Cell Biol.* **93:**103–110.

Fujiki, Y., Rachubinski, R. A., and Lazarow, P. B., 1984, Synthesis of a major integral membrane polypeptide of rat liver peroxisomes on free polysomes, *Proc. Natl. Acad. Sci. USA* **81:**7127–7131.

Fujiki, Y., Rachubinski, R. A., Mortensen, R. M., and Lazarow, P. B., 1985, Synthesis of 3-ketoacyl-CoA thiolase of rat liver peroxisomes on free polyribosomes as a larger precursor: Induction of thiolase mRNA activity by clofibrate, *Biochem. J.* **226:**697–704.

Fung, K., and Clayton, C., 1991, Recognition of a peroxisomal tripeptide entry signal by glycosomes of *Trypanosoma brucei*, *Mol. Biochem. Parasitol.* **45:**261–264.

Gärtner, J., Chen, W. W., Kelley, R. I., Mihalik, S. J., and Moser, H. W., 1991, The 22-kD peroxisomal integral membrane protein in Zellweger syndrome—presence, abundance, and association with a peroxisomal thiolase precursor protein, *Pediatr. Res.* **29:**141–146.

Gärtner, J., Moser, H., and Valle, D., 1992, Mutations in the 70 K peroxisomal membrane protein gene in Zellweger syndrome, *Nature Genet.* **1:**16–25.

Ghosh, M. K., and Hajra, A. K., 1986a, A rapid method for the isolation of peroxisomes from rat liver, *Anal. Biochem.* **159:**169–174.

Ghosh, M. K., and Hajra, A. K., 1986b, Subcellular distribution and properties of acyl/alkyl dihydroxyacetone phosphate reductase in rodent liver, *Arch. Biochem. Biophys.* **245:**523–530.

Gietl, C., 1990, Glyoxysomal malate dehydrogenase from watermelon is synthesized with an aminoterminal transit peptide, *Proc. Natl. Acad. Sci. USA* **87:**5773–5777.

Goldman, B. M., and Blobel, G., 1978, Biogenesis of peroxisomes: Intracellular site of synthesis of catalase and uricase, *Proc. Natl. Acad. Sci. USA* **75:**5066–5070.

Gonzalez, E., Brush, M. D., and Maeshima, M., 1987, The alkaline lipase of the glyoxysomal membrane is a glycoprotein, in *Peroxisomes in Biology and Medicine* (H. D. Fahimi and H. Sies, eds.), pp. 194–198, Springer-Verlag, Berlin, Heidelberg.

Goodman, J. M., Maher, J., Silver, P. A., Pacifico, A., and Sanders, D., 1986, The membrane proteins of the methanol-induced peroxisome of *Candida boidinii:* Initial characterization and generation of monoclonal antibodies, *J. Biol. Chem.* **261**:3464–3468.

Goodman, J. M., Trapp, S. B., and Huang, H., 1990, Peroxisomes induced in *Candida boidinii* by methanol, oleic acid and D-alanine vary in metabolic function but share common integral membrane proteins, *J. Cell Sci.* **97**:193–204.

Gorgas, K., 1984, Peroxisomes in sebaceous glands. V: Complex peroxisomes in the mouse preputial gland: Serial sectioning and three dimensional reconstruction studies, *Anat. Embryol.* **169**:261–270.

Gorgas, K., 1985, Serial sectioning analysis of mouse hepatic peroxisomes, *Anat. Embryol.* **172**:21–32.

Gould, S. J., Keller, G.-A., and Subramani, S., 1988, Identification of peroxisomal targeting signals located at the carboxy terminus of four peroxisomal proteins, *J. Cell Biol.* **107**:897–905.

Gould, S. J., Keller, G.-A., Hosken, N., Wilkinson, J., and Subramani, S., 1989, A conserved tripeptide sorts proteins to peroxisomes, *J. Cell Biol.* **108**:1657–1664.

Gould, S. J., Keller, G.-A., Schneider, M., Howell, S. H., Garrard, L. J., Goodman, J. M., Distel, B., Tabak, H., and Subramani, S., 1990b, Peroxisomal protein import is conserved between yeast, plants, insects and mammals, *EMBO J.* **9**:85–90.

Gould, S. J., Krisans, S., Keller, G.-A., and Subramani, S., 1990a, Antibodies directed against the peroxisomal targeting signal of firefly luciferase recognize multiple mammalian peroxisomal proteins, *J. Cell Biol.* **110**:27–34.

Hajra, A. K., and Bishop, J. E., 1982, Glycerolipid biosynthesis in peroxisomes via the acyldihydroxyacetone phosphate pathway, *Ann. NY Acad. Sci.* **386**:170–181.

Hajra, A. K., Burke, C. L., and Jones, C. L., 1979, Subcellular localization of acyl-coenzyme A: Dihydroxyacetone phosphate acyltransferase in rat liver peroxisomes (microbodies), *J. Biol. Chem.* **254**:10896–10900.

Hajra, A. K., and Wu, D., 1985, Preparative isolation of peroxisomes from liver and kidney using metrizamide density gradient centrifugation in a vertical rotor, *Anal. Biochem.* **148**:233–244.

Hardeman, D., Versantvoort, C., van den Brink, J. M., and van den Bosch, H., 1990, Studies on peroxisomal membranes, *Biochim. Biophys. Acta* **1027**:149–154.

Hartl, F.-U., and Just, W. W., 1987, Integral membrane polypeptides of rat liver peroxisomes: Topology and response to different metabolic states, *Arch. Biochem. Biophys.* **255**:109–119.

Hartl, F.-U., Just, W. W., Köster, A., and Schimassek, H., 1985, Improved isolation and purification of rat liver peroxisomes by combined rate zonal and equilibrium density centrifugation, *Arch. Biochem. Biophys.* **237**:124–134.

Hashimoto, T., Kuwabara, T., Usuda, N., and Nagata, T., 1986, Purification of membrane polypeptides of rat liver peroxisomes, *J. Biochem.* **100**:301–310.

Heinemann, P., and Just, W. W., 1992, Peroxisomal protein import: *In vivo* evidence for a novel translocation competent compartment, *FEBS Lett.* **300**:179–182.

Heupel, R., Markgraf, T., Robinson, D. G., and Heldt, H. W., 1991, Compartmentation studies on spinach leaf peroxisomes: Evidence for channeling of phosphorespiratory metabolites in peroxisomes devoid of intact boundary membrane, *Plant Physiol.* **96**:971–979.

Heymans, H. S. A., Schutgens, R. B. H., Tan, R., van den Bosch, H., and Borst, P., 1983, Severe plasmalogen deficiency in tissues of infants without peroxisomes (Zellweger syndrome), *Nature* **306**:69–70.

Hijikata, M., Ishii, N., Kagamiyama, H., Osumi, T., and Hashimoto, T., 1987, Structural analysis of cDNA for rat peroxisomal 3-ketoacyl-CoA thiolase, *J. Biol. Chem.* **262**:8151–8158.

Höhfeld, J., Veenhuis, M., and Kunau, W.-H., 1991, PAS3, a *Saccharomyces cerevisiae* gene encoding a peroxisomal integral membrane protein essential for peroxisome biogenesis, *J. Cell Biol.* **114**:1167–1178.

Höhfeld, J., Mertens, D., Wiebel, F. F., and Kunau, W. H., 1992, Defining components required for peroxisome assembly in *Saccharomyces cerevisiae*, in *Membrane Biogenesis and Protein Targeting*, New Comprehensive Biochemistry Series, Vol. 22 (W. Neupert and R. Lyll, eds.), pp. 185–207, Elsevier Science Publishing, New York.

Horie, S., Das, A. K., and Hajra, A. K., 1992, Alkyldihydroxyacetonephosphate synthase from guinea pig liver peroxisomes, *Methods Enzymol.* **209**:385–390.

Hruban, Z., and Rechcigl, M., Jr., 1969, Microbodies and related particles, in *Int. Rev. Cytol.* (Suppl. 1) Academic Press, New York.

Huang, A. H. C., Trelease, R. N., and Moore, T. S., 1983, *Plant Peroxisomes* Academic Press, New York.

Imanaka, T., Lazarow, P. B., and Takana, T., 1991, A novel 57 kDa peroxisomal membrane polypeptide detected by monoclonal antibody (PXM1a/207B), *Biochim. Biophys. Acta* **1062**:264–270.

Imanaka, T., Small, G. M., and Lazarow, P. B., 1987, Translocation of acyl-CoA oxidase into peroxisomes requires ATP hydrolysis but not a membrane potential, *J. Cell Biol.* **105**:2915–2922.

Issemann, I., and Green, S., 1990, Activation of a member of the steroid hormone receptor superfamily by peroxisome proliferators, *Nature* **347**:645–650.

Jedlitschky, G., Huber, M., Völkl, A., Müller, M., Leier, I., Müller, J., Lehmann, W.-D., Fahimi, H. D., and Keppler, D., 1991, Peroxisomal degradation of leukotrienes by β-oxidation from the ω-end, *J. Biol. Chem.* **266**: 24763–24772.

Kaldi, K., Diestelkötter, P., Stenbeck, G., Auerbach, S., Jäkle, U., Mägert, H.-J., Wieland, F. T., and Just, W. W., 1993, Membrane topology of the 22 kDa integral membrane protein, *FEBS Lett.* **315**:217–222.

Kamijo, K., Kamijo, T., Ueno, I., Osumi, T., and Hashimoto, T., 1992, Nucleotide sequence of the human 70 kDa peroxisomal membrane protein: A member of ATP-binding cassette transporters, *Biochim. Biophys. Acta* **1129**:323–327.

Kamijo, K., Taketani, S., Yokota, S., Osumi, T., and Hashimoto, T., 1990, The 70 kDa peroxisomal membrane protein is a member of the Mdr (P-glycoprotein)-related ATP-binding protein superfamily, *J. Biol. Chem.* **265**:4534–4540.

Kase, F., Björkhem, I., and Pedersen, J. I., 1983, Formation of cholic acid from 3α, 7α, 12α-trihydroxy-5β-cholestanoic acid by rat liver peroxisomes, *J. Lipid Res.* **24**:1560–1567.

Keller, G. A., Barton, M. C., Shapiro, D. J., and Singer, S. J., 1985, 3-Hydroxy-3-methylglutaryl-coenzyme A reductase is present in peroxisomes in normal rat liver cells, *Proc. Natl. Acad. Sci. USA* **82**:770–774.

Keller, G.-A., Gould, S., Deluca, M., and Subramani, S., 1987, Firefly luciferase is targeted to peroxisomes in mammalian cells, *Proc. Natl. Acad. Sci USA* **84**:3264–3268.

Keller, G.-A., Krisans, S., Gould, S. J., Sommer, J. M., Wang, C. C., Schliebs, W., Kunau, W., Brody, S., and Subramani, S., 1991, Evolutionary conservation of a microbody targeting signal that targets proteins to peroxisomes, glyoxysomes and glycosomes, *J. Cell Biol.* **114**:893–904.

Keller, G. A., Pazirandeh, M., and Krisans, S., 1986, 3-Hydroxy-3-methylglutaryl-coenzyme A reductase localization in rat liver peroxisomes and microsomes of control and cholestyramine-treated animals: Quantitative biochemical and immunoelectron microscopic analyses, *J. Cell Biol.* **103**:875–886.

Köster, A., Heisig, M., Heinrich, P. C., and Just, W. W., 1986, *In vitro* synthesis of peroxisomal membrane polypeptides, *Biochem. Biophys. Res. Commun.* **137**:626–632.

Krisans, S. K., Mortensen, R. M., Lazarow, P. B., 1980, Acyl-CoA synthetase in rat liver peroxisomes, *J. Biol. Chem.* **255**:9599–9607.

Kunau, W.-H., and Hartig, A., 1992, Peroxisome biogenesis in *Saccharomyces cerevisae, Antonie Van Leeuwenhoek* **62**:63–78.

Labarca, P., Wolff, D., Soto, U., Necochea, C., and Leighton, F., 1986, Large cation-selective pores from rat liver peroxisomal membranes incorporated to planar lipid bilayers, *J. Membr. Biol.* **94**:285–291.

Lageweg, W., Tager, J. M., and Wanders, R. J. A., 1991, Topography of very long chain fatty acid activating activity in peroxisomes from rat liver, *Biochem. J.* **275**:53–56.

Lazarow, P. B., 1984, The peroxisomal membrane, in *Membrane Structure and Function* (E. E. Bittar, ed.), Vol. 5, pp. 1–31, John Wiley, New York.

Lazarow, P. B., and Fujiki, Y., 1985, Biogenesis of peroxisomes, *Annu. Rev. Cell Biol.* **1**:489–530.

Lazarow, P. B., and Moser, H. W., 1989, Disorders of peroxisome biogenesis, in *The Metabolic Basis of Inherited Disease* (C. R. Scriver, A. L. Beaudet, W. S. Sly, and D. Valle, eds.), pp. 1479–1509, McGraw-Hill, New York.

Lazo, O., Contreras, M., and Singh, I., 1990, Topographical localization of peroxisomal acyl-CoA ligases: Differential localization of palmitoyl-CoA and lignoceroyl-CoA ligases, *J. Lipid Res.* **31**:583–595.

Leighton, F., Bergseth, S., Rørtveit, T., Christiansen, E. N., and Bremer, J., 1989, Free acetate production by rat hepatocytes during peroxisomal fatty acid and dicarboxylic acid oxidation, *J. Biol. Chem.* **264**:10347–10350.

Leighton, F., Brandan, E., Lazo, O., and Bronfman, M., 1982, Subcellular fractionation studies on the organization of fatty acid oxidation by liver peroxisomes, *Ann. NY Acad. Sci.* **386**:62–80.

Leighton, F., Poole, B., Beaufay, H., Baudhuin, P., Coffey, J. W., Fowler, S., and de Duve, C., 1968, The large-scale separation of peroxisomes, mitochondria and lysosomes from the liver of rats injected with Triton WR-1339, *J. Cell Biol.* **37**:482–513.

Lemmens, M., Verheyden, K., Van Veldhoven, P., Vereecke, J., Mannaerts, G. P., and Carmeliet, E., 1989, Single-channel analysis of a large conductance channel in peroxisomes from rat liver, *Biochim. Biophys. Acta* **984**:351–359.

Liang, Z., and Huang, A. H. C., 1983, Metabolism of glycolate and glyoxylate in intact spinach leaf peroxisomes, *Plant Physiol.* **73**:147–152.

Liang, Z., and Huang, A. H. C., 1984, Conversion of glycerate to serine in intact spinach leaf peroxisomes, *Arch. Biochem. Biophys.* **233**:393–401.

Lock, E. A., Mitchell, A. M., and Elcombe, C. R., 1989, Biochemical mechanisms of induction of hepatic peroxisome proliferation, *Annu. Rev. Pharmacol. Toxicol.* **29**:145–163.

Lüers, G., Beier, K., Hashimoto, T., Fahimi, H. D., and Völkl, A., 1990, Biogenesis of peroxisomes: Sequential biosynthesis of the membrane and matrix proteins in the course of hepatic regeneration, *Eur. J. Cell Biol.* **52**:175–184.

Luster, D. G., Bowditch, M. L., Eldridge, K. M., and Donaldson, R. P., 1988, Characterization of membrane-bound electron transport enzymes from castor bean glyoxysomes and endoplasmic reticulum. *Arch. Biochem. Biophys.* **265**:50–61.

Makita, T., Hakoi, K., and Araki, N., 1990, Cytochemical localization of Mg^{++}-ATPase and Ca^{++}-ATPase on the limiting membrane of rat liver peroxisomes, *Arch. Histochem. Cytochem.* **23**:601–611.

Malik, Z. A., Tappia, P. S., De Netto, L. A., Burdett, K., Sutton, R., and Connock, M. J., 1991, Properties of ATPase activity associated with peroxisomes of rat and bovine liver, *Comp. Biochem. Physiol.* **99B**:295–300.

Mannaerts, G. P., and Van Veldhoven, P. P., 1987, Permeability of the peroxisomal membrane, in *Peroxisomes in Biology and Medicine* (H. D. Fahimi and H. Sies, eds.), pp. 169–176, Springer-Verlag, Berlin, Heidelberg.

Mannaerts, G. P., and Van Veldhoven, P. P., 1990, The peroxisome: Functional properties in health and disease, *Biochem. Soc. Trans.* **18**:87–89.

Mannaerts, G. P., and Van Veldhoven, P. P., 1992a, Metabolic role of mammalian peroxisomes, in *Peroxisomes: Biology and Importance in Toxicology and Medicine* (G. G. Gibson and B. Lake, eds.), pp. 19–62, Taylor and Francis, London.

Mannaerts, G. P., and Van Veldhoven, P. P., 1992b, Role of peroxisomes in mammalian metabolism, *Cell Biochem. Funct.* **10**:141–151.

Mannaerts, G. P., Van Veldhoven, P., Van Broekhoven, A., Vandebroek, G., and Debeer, L. J., 1982, Evidence that peroxisomal acyl-CoA synthetase is located at the cytoplasmic side of the peroxisomal membrane, *Biochem. J.* **204**:17–23.

McCammon, M. T., Dowds, C. A., Orth, K., Moomaw, C. R., Slaughter, C. A., and Goodman, J. M., 1990a, Sorting of peroxisomal membrane protein PMP47 from *Candida boidinii* into peroxisomal membranes of *Saccharomyces cerevisiae*, *J. Biol. Chem.* **265**:20098–20105.

McCammon, M. T., Veenhuis, M., Trapp, S. B., and Goodman, J. M., 1990b, Association of glyoxylate and beta-oxidation enzymes with peroxisomes of *Saccharomyces cerevisiae*, *J. Bacteriol.* **172**:5816–5827.

Miura, S., Mori, M., Takiguchi, M., Tatibana, M., Furuta, S., Miyazawa, S., and Hashimoto, T., 1984, Biosynthesis and intracellular transport of enzymes of peroxisomal β-oxidation, *J. Biol. Chem.* **259**:6397–6402.

Miyazawa, S., Hashimoto, T., and Yokota, S., 1985, Identity of long-chain acyl-CoA synthetase of microsomes, mitochondria and peroxisomes, *J. Biochem.* **98**:723–733.

Miyamoto, A., Yamamoto, T., Kamiryo, T., and Numa, S., 1981, Purification of long-chain acyl-CoA synthetase of rat liver peroxisomes and its properties, *Proc. Jpn. Conf. Biochem. Lipids* **23**:346–349.

Mosser, J., Douar, A.-M., Sardé, C.-O., Kioschis, P., Feil, R., Moser, H., Poustka, A.-M., Mandel, J.-L., and Aubourg, P., 1993, Putative X-linked adrenoleukodystrophy gene shares unexpected homology with ABC transporters, *Nature* **361**:726–730.

Neat, C. E., Thomassen, M. S., and Osmundsen, H., 1980, Induction of peroxisomal β-oxidation in rat liver by high-fat diets, *Biochem. J.* **186**:369–371.

Nicolay, K., Veenhuis, M., Douma, A. C., and Harder, W., 1987, A ^{31}P NMR study of the internal pH of yeast peroxisomes, *Arch. Microbiol.* **147**:37–41.

Nuttley, W. M., Bodnar, A. G., Mangroo, D., and Rachubinski, R. A., 1990, Isolation and characterization of membranes from oleic acid-induced peroxisomes of *Candida tropicalis*, *J. Cell Sci.* **95**:463–470.

Ohno, S., and Fujii, Y., 1990, Three-dimensional and histochemical studies of peroxisomes in cultured hepatocytes by quick-freezing and deep-etching method, *Histochem. J.* **22**:143–154.

Opperdoes, F. R., 1987, Biogenesis of glycosomes (microbodies) in the trypanosomatidae, in *Peroxisomes in Biology and Medicine* (H. D. Fahimi and H. Sies, eds.), pp. 426–435, Springer-Verlag, Berlin, Heidelberg.

Opperdoes, F. R., Baudhuin, P., Coppens, I., De Roe, C., Edwards, S. W., Weijers, P. J., and Misset, O., 1984, Purification, morphometric analysis and characterization of the glycosomes (microbodies) of the protozoan hemoflagellate *Trypanosoma brucei*, *J. Cell Biol.* **98**:1178–1184.

Osumi, T., and Fujiki, Y., 1990, Topogenesis of peroxisomal proteins, *Bioessays* **12**:217–222.

Pais, M. S. S., and Carrapico, F., 1982, Microbodies. A membrane compartment, *Ann. NY Acad. Sci.* **306**:510–513.

Patthey, J.-P., and Deshusses, J., 1987, Accessibility of *Trypanosoma brucei* procyclic glycosomal enzymes to labeling agents of various sizes and charges, *FEBS Lett.* **210**:137–141.

Rachubinski, R. A., Fujiki, Y., Mortensen, R. M., and Lazarow, P. B., 1984, Acyl-CoA oxidase and hydratase-dehydrogenase, two enzymes of the peroxisomal β-oxidation system, are synthesized on free polysomes of clofibrate-treated rat liver, *J. Cell Biol.* **99**:2241–2246.

Roscher, A. A., Hoefler, S., Hoefler, G., Paschke, E., Paltauf, F., Moser, A., and Moser, H., 1989, Genetic and phenotypic heterogeneity in disorders of peroxisome biogenesis: A complementation study involving cell lines from 19 patients, *Pediatr. Res.* **26**:67–72.

Santos, M. J., Imanaka, T., Shio, H., and Lazarow, P. B., 1988, Peroxisomal integral membrane proteins in control and Zellweger fibroblasts, *J. Biol. Chem.* **263**:10502–10509.

Schepers, L., Casteels, M., Vamecq, J., Parmentier, G., Van Veldhoven, P. P., and Mannaerts, G. P., 1988, β-Oxidation of the carboxyl side chain of prostaglandin E_2 in rat liver peroxisomes and mitochondria, *J. Biol. Chem.* **263**:2724–2731.

Schlossman, D. M., and Bell, R. M., 1977, Microsomal sn-glycerol-3-phosphate and dihydroxyacetone phosphate acyltransferase activities from liver and other tissues, *Arch. Biochem. Biophys.* **182**:732–742.

Schram, A., Strijland, A., Hashimoto, T., Wanders, R. J. A., Schutgens, R. B. H., van den Bosch, H., and Tager, J. M., 1986, Biogenesis and maturation of peroxisomal β-oxidation enzymes in fibroblasts in relation to the Zellweger syndrome and infantile Refsum disease, *Proc. Natl. Acad. Sci. USA* **83**:6156–6158.

Shimozawa, N., Tsukamoto, T., Suzuki, Y., Orii, T., Shirayashi, Y., Mori, T., and Fujiki, Y., 1992, A human gene responsible for Zellweger syndrome that affects peroxisome assembly, *Science* **255**:1132–1134.

Shindo, Y., and Hashimoto, T., 1978, Acyl-Coenzyme A synthetase and fatty acid oxidation in rat liver peroxisomes, *J. Biochem.* **84**:1177–1181.

Singh, H., and Poulos, A., 1988, Distinct long chain and very long chain fatty acyl-CoA synthetases in rat liver peroxisomes and microsomes, *Arch. Biochem. Biophys.* **259**:382–390.

Singh, H., Usher, S., and Poulos, A., 1989, Dihydroxyacetone phosphate acyltransferase and alkyldihydroxyacetone phosphate synthase activities in rat liver subcellular fractions and human skin fibroblasts, *Arch. Biochem. Biophys.* **268**:676–686.

Singh, I., Lazo, O., Contreras, M., Stanley, W., and Hashimoto, T., 1991, Rhizomelic chondrodysplasia punctata: Biochemical studies of peroxisomes isolated from cultured skin fibroblasts, *Arch. Biochem. Biophys.* **286**:277–283.

Singh, I., Moser, A. E., Goldfischer, S., and Moser, H. W., 1984, Lignoceric acid is oxidized in the peroxisome: Implications for the Zellweger cerebro-hepato-renal syndrome and adrenoleukodystrophy, *Proc. Natl. Acad. Sci. USA* **81**:4203–4207.

Skorin, C., Soto, U., Necochea, C., and Leighton, F., 1986, Protein phosphorylation in peroxisomes, *Biochem. Biophys. Res. Commun.* **140**:188–194.

Small, G. M., Santos, M. J., Imanaka, T., Poulos, A., Danks, D. M., Moser, H. W., and Lazarow, P. B., 1988, Peroxisomal integral membrane proteins in livers of patients with Zellweger syndrome, infantile Refsum's disease and X-linked adrenoleukodystropy, *J. Inherit. Metab. Dis.* **11**:358–371.

Subramani, S., 1992, Targeting of proteins into the peroxisomal matrix, *J. Memb. Biol.* **125**:99–106.

Sulter, G. J., Looyenga, L., Veenhuis, M., and Harder, W., 1990, Occurrence of peroxisomal membrane proteins in methylotrophic yeasts grown under different conditions, *Yeast* **6**:35–43.

Sulter, G. J., Verheyden, K., Mannaerts, G. P., Harder, W., and Veenhuis, M., 1993, The in vitro permeability of yeast peroxisomal membranes is caused by a 31 kDa integral membrane protein, *Yeast*, **9**:733–742.

Suzuki, H., Kawarabayasi, Y., Kondo, J., Abe, T., Nishikawa, K., Kimura, S., Hashimoto, T., and Yamamoto, T., 1990, Structure and regulation of rat long-chain acyl-CoA synthetase, *J. Biol. Chem.* **265**:8681–8685.

Suzuki, Y., Orii, T., Takiguchi, M., Mori, M., Hijikata, M., and Hashimoto, T., 1987, Biosynthesis of membrane polypeptides of rat liver peroxisomes, *J. Biochem.* **101**:491–496.

Suzuki, Y., Shimozawa, N., Orii, T., and Hashimoto, T., 1989, Major peroxisomal membrane polypeptides are synthesized in cultured skin fibroblasts from patients with Zellweger syndrome, *Pediatr. Res.* **26**:150–153.

Swinkels, B. W., Gould, S. J., Bodnar, A. G., Rachubinski, R. A., and Subramani, S., 1991, A novel cleavable peroxisomal targeting signal at the amino-terminus of the rat 3-ketoacyl-CoA thiolase, *EMBO J.* **10**:3255–3262.

Tanaka, A., Osumi, M., and Fukui, S., 1982, Peroxisomes of alkane-grown yeast: Fundamental and practical aspects, *Ann. NY Acad. Sci.* **386**:183–198.

Thompson, S. L., Burrows, R., Laub, R. J., and Krisans, S. K., 1987, Cholesterol synthesis in rat liver peroxisomes, *J. Biol. Chem.* **262**:17420–17425.

Tolbert, N. E., 1981, Metabolic pathways in peroxisomes and glyoxysomes, *Annu. Rev. Biochem.* **50**:133–157.

Tsukamoto, T., Miura, S., and Fujiki, Y., 1991, Restoration by a 35 K membrane protein of peroxisome assembly in a peroxisome-deficient mammalian cell mutant, *Nature* **350**:77–81.

Vanhove, G., Van Veldhoven, P. P., Vanhoutte, F., Parmentier, G., Eyssen, H. J., and Mannaerts, G. P., 1991, Mitochondrial and peroxisomal β-oxidation of the branched chain fatty acid 2-methylpalmitate in rat liver, *J. Biol. Chem.* **266**:24670–24675.

Van Veldhoven, P., Debeer, L. J., and Mannaerts, G. P., 1983, Water- and solute-accessible spaces of purified peroxisomes: Evidence that peroxisomes are permeable to NAD+, *Biochem. J.* **210**:685–693.

Van Veldhoven, P. P., 1986, *The permeability of the Peroxisomal Membrane*, Ph.D. Thesis, Katholieke Universiteit Leuven, Belgium.

Van Veldhoven, P. P., Just, W. W., and Mannaerts, G. P., 1987, Permeability of the peroxisomal membrane to cofactors of β-oxidation: Evidence for the presence of a pore-forming protein, *J. Biol. Chem.* **262**:4310–4318.

Van Veldhoven, P. P., and Mannaerts, G. P., 1986, Coenzyme A in purified peroxisomes is not freely soluble in the matrix but firmly bound to a matrix protein. *Biochem. Biophys. Res. Commun.* **139**:1195–1201.

Veenhuis, M., and Goodman, J. M., 1990, Peroxisomal assembly: Membrane proliferation precedes the induction of the abundant matrix proteins in the methylotrophic yeast *Candida boidinii*, *J. Cell Sci.* **96**:583–590.

Veenhuis, M., and Harder, W., 1987, Metabolic significance and biogenesis of microbodies in yeasts, in *Peroxisomes in Biology and Medicine* (H. D. Fahimi and H. Sies, eds.), pp. 436–458, Springer-Verlag, Berlin, Heidelberg.

Veenhuis, M., Keizer, I., and Harder, W., 1979, Characterization of peroxisomes in glucose-grown *Hansenula polymorpha* and their development after the transfer of cells into methanol-containing media, *Arch. Microbiol.* **120**:167–175.

Verheyden, K., Fransen, M., Van Veldhoven, P. P., and Mannaerts, G. P., 1992, Presence of small GTP-binding proteins in the peroxisomal membrane, *Biochim. Biophys. Acta* **1109**:48–54.

Verleur, N., and Wanders, R. J. A., 1993, Permeability properties of peroxisomes in rat liver, *Biol. Cell* **77**:109.

Völkl, A., and Fahimi, H. D., 1985, Isolation and characterization of peroxisomes from the liver of normal untreated rats, *Eur. J. Biochem.* **149**:257–265.

Walton, P. A., Gould, S. J., Feramisco, J. R., and Subramani, S., 1992, Transport of microinjected proteins into peroxisomes of mammalian cells: Inability of Zellweger cell lines to import proteins with the SKL tripeptide peroxisomal targeting signal, *Mol. Cell. Biol.* **12**:531–541.

Wanders, R. J. A., van Roermund, C. W. T., Schutgens, R. B. H., Barth, P. G., Heymans, H. S. A., van den Bosch, H., and Tager, J. M., 1990, The inborn errors of peroxisomal β-oxidation: A review, *J. Inherit. Metabol. Dis.* **13**:4–36.

Waterham, H. R., Keizer-Gunnink, I., Goodman, J. M., Harder, W., and Veenhuis, M., 1990,

Immunocytochemical evidence for the acidic nature of peroxisomes in methylotrophic yeasts, *FEBS Lett.* **262**:17–19.

Webber, K. O., and Hajra, A. K., 1993, Purification of dihydroxyacetone phosphate acyltransferase from guinea pig peroxisomes, *Arch. Biochem. Biophys.* **300**:88–97.

Whitney, A. B., and Bellion, E., 1991, ATPase activities in peroxisome-proliferating yeast, *Biochim. Biophys. Acta* **1058**:345–355.

Wilson, G. N., and King, T. E., 1991, Structure and variability of mammalian peroxisomal membrane proteins, *Biochem. Med. Metab. Biol.* **46**:235–245.

Wolvetang, E. J., Tager, J. M., and Wanders, R. J. A., 1990a, Latency of the peroxisomal enzyme acyl-CoA: Dihydroxyacetone phosphate acyltransferase in digitonin-permeabilized fibroblasts: The effect of ATP and ATPase inhibitors, *Biochem. Biophys. Res. Commun.* **170**:1135–1143.

Wolvetang, E. J., Tager, J. M., and Wanders, R. J. A., 1991, Factors influencing the latency of the peroxisomal enzyme dihydroxyacetone-phosphate acyltransferase in permeabilized human skin fibroblasts, *Biochim. Biophys. Acta* **1095**:122–126.

Wolvetang, E. J., Wanders, R. J. A., Schutgens, R. B. H., Berden, J. A., and Tager, J. M., 1990b, Properties of the ATPase activity associated with peroxisome-enriched fractions from rat liver: Comparison with mitochondrial F_1F_0-ATPase, *Biochim. Biophys. Acta* **1035**:6–11.

Yamamoto, K., and Fahimi, H. D., 1987, Three-dimensional reconstruction of a peroxisomal reticulum in regenerating rat liver: Evidence of interconnections between heterogeneous segments, *J. Cell Biol.* **105**:713–722.

Yu, C., and Huang, A. H. C., 1986, Conversion of serine to glycerate in intact spinach leaf peroxisomes: Role of malate dehydrogenase, *Arch. Biochem. Biophys.* **245**:125–133.

Zwacka, R., Karasawa, M., and Weiher, H., 1993, The transgenic mouse strain Mpv17 as a model for peroxisomal disease, *Biol. Cell* **77**:116.

Chapter 9

Nuclear Envelope Assembly and Disassembly

L. S. Cox and C. J. Hutchison

INTRODUCTION

The nuclear envelope (NE) is a highly complex dynamic structure that must provide a boundary separating the cytoplasm from the nucleoplasm during interphase but not during mitosis, when chromosomes are segregated between two daughter cells. In other words, the nuclear envelope must be stable during interphase but unstable during mitosis. Understanding the processes that control nuclear envelope breakdown (NEBD) and reassembly has been the subject of intensive research during the past eight years, such that over the last year or two a consensus has started to emerge that was put succinctly by John Newport at the 1992 Lancaster symposium of the Society of Experimental Biology, when he expressed the view that NEBD and reassembly are extremes of a dynamic process involving continuous structural rearrangements of the components of the envelope. In this chapter we will review what is currently known about this process and propose a general model for the regulation of the major events involved.

L. S. Cox   CRC Cell Transformation Research Group, Department of Biochemistry, University of Dundee, Dundee DD1 4HN, Scotland.   **C. J. Hutchison**   Department of Biological Sciences, University of Dundee, Dundee DD1 4HN, Scotland.
Subcellular Biochemistry, Volume 22: Membrane Biogenesis, edited by A. H. Maddy and J. R. Harris. Plenum Press, New York, 1994.

1.1. The Structural Organization of the Nuclear Envelope

The nuclear envelope can be thought of as a tripartite structure consisting of a double bilayer of nuclear membrane interrupted periodically by nuclear pores and supported on its inner surface by a meshwork of intermediate filament proteins termed *lamins* (Figure 1; reviewed by Gerace, 1986). The envelope serves to separate nuclear components from the cytoplasm. Compartmentalization is important in separating the processes of transcription and translation, as this provides further levels of regulation of gene expression unavailable to prokaryotes and protects large chromosomes from the shear forces generated by contractile elements. In addition to providing this passive separation role, and excluding cytoplasmic components, the envelope is active in the accumulation of nuclear proteins, allowing concentration of low abundancy proteins necessary for nuclear metabolism, including transcription, RNA processing, DNA replication, and DNA repair. To permit specific uptake of nuclear proteins and exclusion of cytoplasmic components, the nucleus has pores throughout the membrane that permit diffusion of small molecules but prevent free entry of globular proteins larger than ~60 kDa (Dingwall and Laskey, 1986).

1.2. The Nuclear Membrane

The organization of nuclear membranes has been largely observed using transmission electron microscopy (TEM). Positively and negatively stained ultrathin sections of NE prepared from amphibian germinal vesicles (GV) (Scheer *et al.*, 1976) or isolated rat liver nuclei (Aaronson and Blobel, 1974, 1975) reveal the "unit membrane" pattern of opposing inner and outer nuclear membranes. Structures resembling endoplasmic reticulum (ER) are often observed to be continuous with the outer (cytoplasmic) face of the nuclear membrane, while at regular intervals the inner and outer membranes converge and are interrupted by fenestrations representing the nuclear pores (Franke and Scheer, 1970; Franke, 1974; Franke *et al.*, 1981; Roberts and Northcote, 1970; Kessel, 1983).

Surprisingly little is known of the biochemical characteristics of nuclear membranes. This results mainly from the predominant (and possibly mistaken) view that nuclear membranes are somewhat passive structures whose fate follows assembly and disassembly of the lamins (Burke and Gerace, 1986; see below). Indeed, lamin B appears to remain tightly associated with nuclear membrane vesicles during mitosis, allowing their recycling in subsequent cell cycles (Gerace and Burke, 1988; Stick *et al.*, 1988). More recently, attention has been paid to the characteristics of nuclear membrane precursors. In *Xenopus laevis* eggs, nuclear membrane precursors are stored in two pools termed NEP-A and NEP-B. Both pools are sensitive to detergents (Vigers and Lohka, 1991) and trypsin (Wilson and Newport, 1988). However, the trypsin sensitivity of these vesicles is

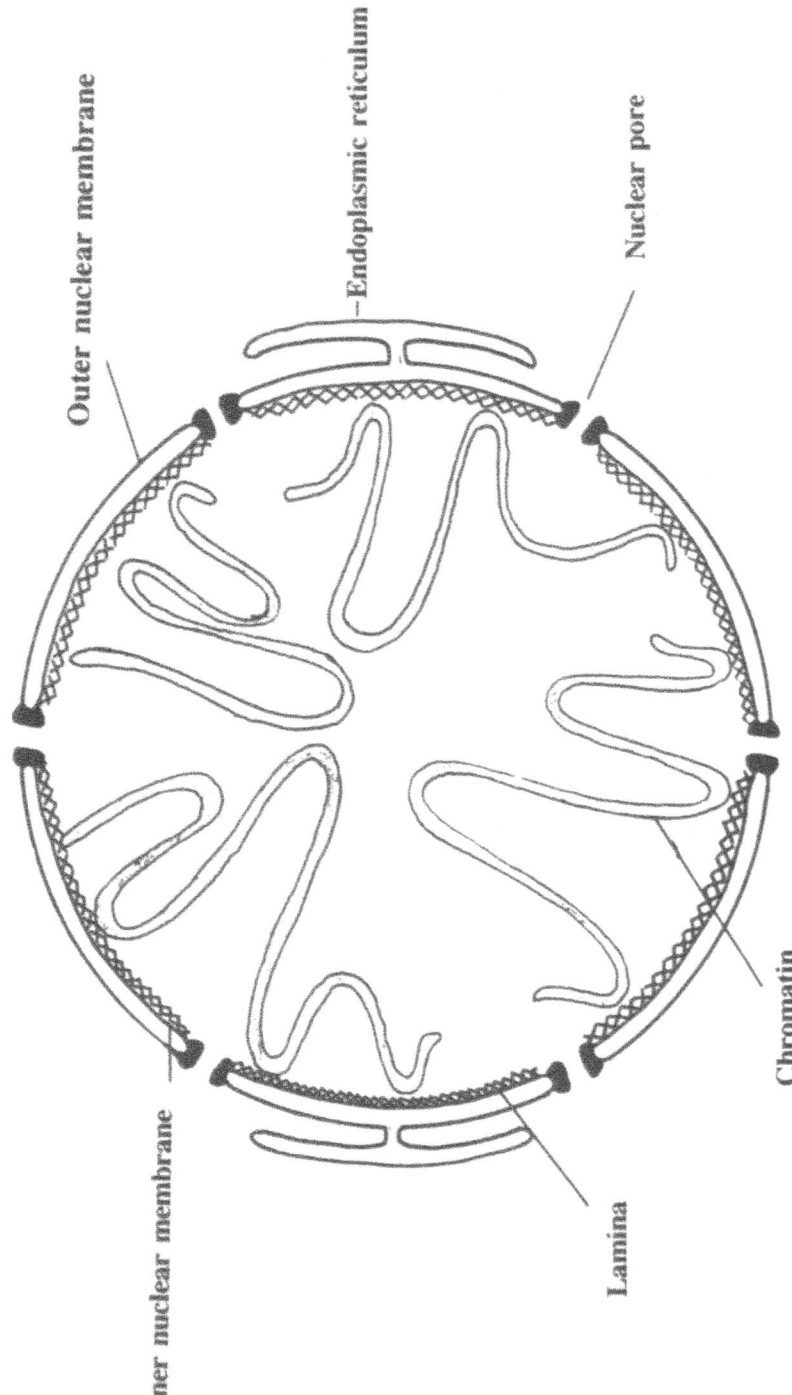

FIGURE 1. Diagrammatic representation of the structures associated with nuclear envelopes of typical eukaryotic cells.

not due to an integral lamin (Benavente *et al.*, 1985). Ultrastructural studies have been used to compare the distribution of proteins in nuclear membranes and ER. Peroxidases in differentiating thyroid, as well as antibody synthesis in differentiating plasma cells, are both initially located in the nuclear membranes, subsequently relocating to sites in the ER (Fawcett, 1966; Leduc *et al.*, 1968; Strum and Karnovsky, 1970; Zimmer *et al.*, 1992), indicating at least some similarity in composition between nuclear membranes and ER. Other studies demonstrating a role for GTP hydrolysis in nuclear membrane fusion suggest similarities with Golgi membranes (Boman *et al.*, 1992a,b). Senior and Gerace (1988) have raised a monoclonal antibody RL13 that recognizes three integral membrane proteins in rat liver nuclei located on the inner nuclear membrane but not in other organelles. Homologous proteins have also been detected in avians (Worman *et al.*, 1988; Bailer *et al.*, 1991) and *Drosophila* nuclei (Harel *et al.*, 1989). Thus, although nuclear membranes have common properties with Golgi and ER, they also have unique characteristics. The development of assays that distinguish between the binding of membrane vesicles to chromatin and their subsequent fusion will allow a more detailed characterization of the properties of these important fractions (Newport and Dunphy, 1992; Boman *et al.*, 1992a,b).

1.3. Nuclear Pores

The organization of nuclear pores is highly complex, this complexity reflecting their role in directing transport into and out of the nucleus. The most recent models suggest that the pores are constructed from three annuli, or rings, the two outer rings being mirror images of each other and both consisting of eight subunits. Every subunit of the outer ring on the cytoplasmic face of the NE appears to be topped by ribosome-like particles (Franke and Scheer, 1970; Gall 1967; Unwin and Milligan, 1982). The peripheral rings are separated by a third ring or "transporter," which consists of "spokes" surrounding a central granule (Akey, 1989, 1990; Akey and Goldfarb, 1989), which is postulated to serve as a diaphragm that opens to allow the passage of specific molecules (see Dingwall, 1990). This generalized model (Figure 2) is largely based on results from electron microscopy in which the NE is first spread over a carbon support before being disrupted to reveal residual structures attached to the grid. By processing electron microscopy (EM) images using a Fourier averaging technique, details of both the size and arrangement of nuclear pore subunits can be obtained (Unwin and Milligan, 1982; Stewart and Whytock, 1988; Akey, 1989; Reichelt *et al.*, 1990). Results from such analyses indicate that nuclear pores consist of a supramolecular assembly, estimated to be 100,000–150,000 kDa in mass (Akey, 1990; Akey and Goldfarb, 1989; Reichelt *et al.*, 1990). Each subunit of this assembly has an estimated mass, which is detailed in Table I.

The nuclear pores appear to be interconnected by a series of fibrous pro-

NUCLEAR PORE STRUCTURE

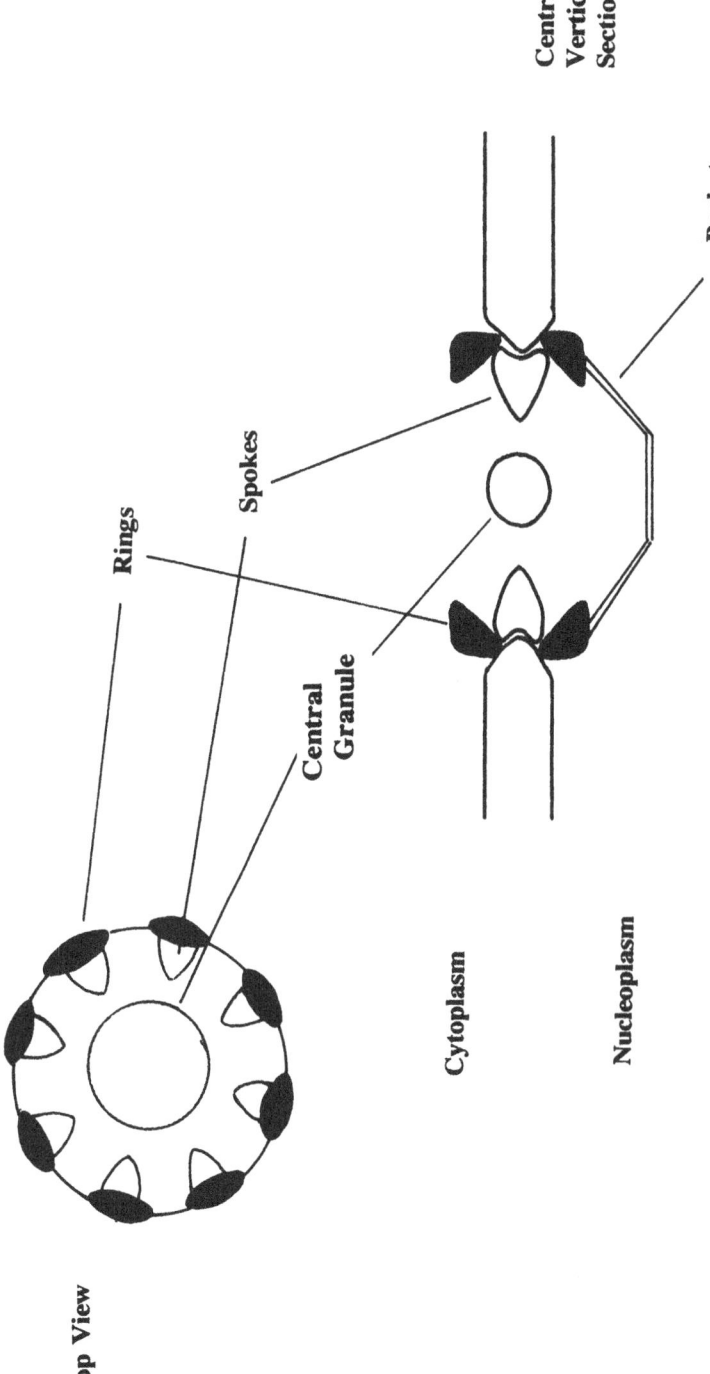

FIGURE 2. Diagrammatic representation showing the view from the top of a nuclear pore, and the view through a vertical section. The diagram illustrates the arrangement of spokes and rings with respect to each other and the nuclear membrane.

Table I
Estimated Molecular Masses of Pore
Complex Components

Pore component	Molecular mass (1000 kDa)
Nuclear pore complex	124.0 +/- 11.0
Plug	12.0 +/- 1.1
Spokes	51.7 +/- 5.3
Plug-spoke[a]	65.7 +/- 8.3
Cytoplasic rings	32.0 +/- 5.5
Nuclear ring	21.1 +/- 3.7

Moleular weights obtained following sequential disruption of NE spread
on carbon-coated grids and measured by density averaging techniques.
[a]Combined mass of plug and spokes.
See Aebi et al., (1990) and Akey (1989).

teinaceous networks (Figure 3). These networks include a filamentous lamina
that resides under the nucleoplasmic face of the envelope below the level of the
pores (Aebi et al., 1986; Stewart and Whytock, 1988) and pore-connecting
fibrils thought to be positioned in the lumen between the inner and outer membrane (Kartenbeck et al., 1971; Scheer et al., 1976; Keller and Riley, 1976;
Kirschner et al., 1977; Maul, 1977). The organization of these fibers is revealed
in TEM studies following selective removal of the inner and outer nuclear mem-

FIGURE 3. Diagrammatic representation of the interconnections between nuclear pores, suggested
from information obtained from shadowed preparations of Xenopus oocyte nuclear envelope. The
cylindrical elements represent nuclear pores, which are topped on their cytoplasmic surface by eight
granules and interconnected by pore-connecting fibrils. The network of beaded filaments, which
interact with the nucleoplasmic face of the pores, comprise the lamina. (Reproduced from Stewart
and Whytock, 1988, with kind permission from the Company of Biologists.)

branes using nonionic detergents and low-salt concentrations (Franke and Scheer, 1970; Barton *et al.*, 1971; Aaronson and Blobel, 1974; Scheer *et al.*, 1976; Maul, 1977) or more recently by presenting stereo pairs of freeze-dried or alcohol-dehydrated metal-shadowed GV envelopes (Stewart and Whytock, 1988). A more novel approach is the use of high-resolution scanning electron microscopy (SEM) (Kirschner *et al.*, 1977) to provide high-quality images of the surfaces of both the inner and outer nuclear envelope. When manually dissected and disrupted GV envelopes of the newt *Triturus cristatus* are viewed using this technique, not only is the organization of the nuclear pores on the outer surface of the NE confirmed, but the pore annulus on the inner surface of the NE is seen to be connected to the lamina via a basket-like structure consisting of eight fingers supported by a ring (Aebi *et al.*, 1990; Ris, 1991; Goldberg and Allen, 1992). The "basket" may represent an artifactual association of free filaments during the critical point freeze-drying procedure used in sample preparation, or more probably it represents a physiological structure. In favor of the physiological role is the observation that baskets are closed in the presence of divalent cations and open when such cations are removed.

Because extraction of interphase nuclei leaves nuclear pores embedded in the nuclear lamina the biochemical characterization of protein components of the pores has proved difficult. However, a recent report that annulate lamellae, consisting of pores embedded in membranes, assemble spontaneously *in vitro* (Dabauvalle *et al.*, 1991) should allow rapid progress to this end. Despite these difficulties several constituents of nuclear pores have been defined, both by genetic and biochemical techniques. To date the pore proteins that have been identified include gp210, NSP1, NUP1, some potential signal sequence receptors, and a family of novel glycoproteins.

In assigning a particular protein as part of the pore complex, ideally several criteria should be met. Stewart and Whytock (1988) have developed extraction techniques that appear to leave the structure of the pore intact, in the hope that immunogold labeling techniques will provide evidence that a particular antigen is part of the pore complex. In contrast, others have used cell-free extracts that allow nuclear transport *in vitro* or microinjection of cultured cells to test either antibodies or lectins that recognize putative pore proteins for their ability to block specifically the accumulation of karyophilic proteins (Dabauvalle *et al.*, 1988a; Finlay *et al.*, 1987; Wolff *et al.*, 1988; Yoneda *et al.*, 1987).

The pore proteins NSP1 and NUP1 of *Saccharomyces cerevisiae* have both been assigned on the basis of function. Both are localized at the nuclear envelope and display temperature-sensitive conditional lethality in some mutants. In addition, null mutants of NSP1 are incapable of transporting karyophilic proteins (Aris and Blobel, 1989; Davis and Fink, 1990; Hurt, 1988, 1990; Nehrbass *et al.*, 1990). In addition to these proteins, a family of 8–10 proteins that all possess a single O-linked N-acetylglucosamine residue have been identified in the nuclear envelopes of a range of species (Davis and Blobel, 1986; Finlay *et al.*, 1987; Hanover *et al.*, 1987; Holt and Hart, 1986; Holt *et al.*, 1987; Park *et al.*, 1987; Schindler *et al.*, 1987; Snow *et al.*, 1987; Hart *et al.*, 1989), some of which have been shown to be required for transport.

The O-linked glycoproteins were originally identified by their ability to bind ^{125}I-wheat germ agglutinin (WGA) in protein blots (Finlay *et al.*, 1987) and were found to correspond to other proteins detected using monoclonal antibodies (MAb) raised against rat nuclear lamina-pore complex fractions (Davis and Blobel, 1986, 1987; Hanover *et al.*, 1987; Holt *et al.*, 1987; Snow *et al.*, 1987). This protein family (the nucleoporins) includes polypeptides at Mr 45, 54, 58, 62, 100, 145, 180, 210, and 270 kDa, of which p62 is the most abundant (Snow *et al.*, 1987). Some of the members of this family of proteins, originally discovered in rat, are related to a family of three glycoproteins in *Xenopus* eggs [Mr 60/68, 97, and 200 kDa (Featherstone *et al.*, 1988; Finlay and Forbes, 1990; Scheer *et al.*, 1988)], as the rat proteins are able to rescue nuclear transport in *Xenopus* egg extracts depleted of endogenous nucleoporins (Finlay and Forbes, 1990).

Characterization of these glycoproteins as components of the pore complex derived from the observation that the lectin WGA would inhibit nuclear transport when microinjected into cells or when added to cell-free nuclear transport extracts (Dabauvalle *et al.*, 1988a; Finlay *et al.*, 1987; Wolff *et al.*, 1988; Yoneda *et al.*, 1987). Similar results have also been obtained with a MAb that reacts with *N*-acetylglucosamine-linked peptides (Featherstone *et al.*, 1988). However, as these reagents recognize a number of proteins (up to 10), they cannot be used to determine requirements for specific proteins. More recently, antibody reagents that specifically recognize a p62 (p60/68 in *Xenopus*) have been shown to block nuclear transport (Dabauvalle *et al.*, 1988b), demonstrating directly its involvement in the pore complex. As of now, p62 has been cloned and sequenced (Shaw *et al.*, 1990; D'Onofrio *et al.*, 1991), but while its primary structure is dominated by an alpha-helical coiled-coil domain at the carboxy-terminus, indicating that it is capable of forming duplex structures, few clues have been obtained to account for its function (D'Onofrio *et al.*, 1988; Starr *et al.*, 1990).

The gp210 is an abundant transmembrane protein found in rats and *Drosophila* (Gerace *et al.*, 1982; Filson *et al.*, 1985; Wozniak *et al.*, 1989) that spans the outer membrane, having a large domain located in the lumen, and a short cytoplasmic tail (Greber *et al.*, 1990). This protein is thought to anchor the pore complex to the nuclear membrane (Greber *et al.*, 1990). A number of other putative pore proteins have been identified in cross-linking studies in which radiolabeled nuclear transport signal sequences have been used to probe protein blots (Adam *et al.*, 1989; Lee and Melese, 1989; Meier and Blobel, 1990; Yoneda *et al.*, 1988). However, it remains to be seen whether these proteins are components of the nuclear pores or whether they are chaperonins (Laskey *et al.*. 1978: Newmeyer and Forbes. 1990).

1.4. The Lamina

Extraction of envelopes of rat nuclei with nonionic detergents and salt reveals the major residual structural component to be the lamina. which remains associated with both nuclear pores and interphase chromatin (Aaronson and Blobel. 1975: Dwyer and Blobel. 1976: Fawcett. 1966). Electron microscopy analyses of the inner surface of *Xenopus* GV envelopes indicate that the lamina is composed of beaded filaments of 10 nm diameter. which form a regular basket-weave pattern over the entire surface of interphase nuclei (Aebi *et al.*. 1986). However. other studies involving deconvoluted fluorescence images of *Drosophila* nuclei suggest that the lamina may be interrupted at regular intervals (Paddy *et al.*. 1990). Although the lamina was previously thought to be located solely at the nuclear periphery. new immunofluorescence confocal microscopy data of synchronized human dermal fibroblasts show that G1 nuclei contain lamin clusters and filaments throughout the nucleus (Bridger *et al.*. 1993). possibly representing stages in the reformation of the nuclear lamina following mitosis.

The lamina of most mammalian cells is composed primarily of three polypeptides of molecular weights 70. 68. and 60 kDa. which have been termed. respectively. lamins A. B. and C. Peptide mapping and immunological cross-reactivity indicate that lamins A and C are related whereas lamin B is distinct (Gerace and Blobel. 1980: Shelton *et al.*. 1980). Based on the size of the lamina filaments revealed by electron microscope analyses of isolated *Xenopus laevis* oocyte GV. it was proposed that lamins were intermediate filament (IF)-like proteins (Aebi *et al.*. 1986). This was confirmed by cross-reactivity of lamins with the anti-IF antibody IFA (Osborn and Weber. 1987). the spontaneous polymerization of purified or expressed lamins into 10-nm filaments. revealing a head-to-tail assembly. their assembly into striated paracrystals with a 23-nm axial repeat (Aebi *et al.*. 1986: Moir *et al.*. 1991) and sequence analysis of cDNAs encoding a variety of lamins (Fisher *et al.*. 1986: McKeon *et al.*. 1986: Franke. 1987).

Although a lamina appears to be present in most but not all eukaryotic cells (Stick and Schwarz. 1982. 1983). its major constituents. the lamins. vary considerably between organisms. Typically. invertebrates contain only one lamin polypeptide (Dessev and Goldman. 1990: Gruenbaum *et al.*. 1988). whereas vertebrates express a variety of lamin polypeptides at different developmental stages (reviewed by Krohne and Benavente. 1986). Indirect evidence also indicates the presence of lamins in yeast (Georgatos *et al.*. 1989). As sequence analysis indicates that lamin C is derived from the same primary transcript as lamin A, by alternative splicing (Fisher *et al.*, 1986), the vertebrate lamins can be grouped into A-type and B-type (Hoger *et al.*, 1988; Krohne *et al.*, 1987; Lehner *et al.*, 1986; Peter *et al.*, 1989; Vorburger *et al.*, 1989b). Expression of one or two B-type lamins appears to be constitutive in most somatic cells (Lehner *et al.*, 1987; Vorburger *et al.*, 1989a) and embryos (Lehner *et al.*, 1987; Rober *et al.*, 1989; Stewart and Burke, 1987; Wolin *et al.*, 1987), while expression of A-type lamins is highly regulated during development, suggesting that the A-type may have a role in cell differentiation (Benavente *et al.*, 1985; Stick and Hausen, 1985; Lehner *et al.*, 1987; Rober *et al.*, 1989). The linear sequence of the lamin protein, in common with all IF proteins, consists of a tripartite domain structure (Figure 4), with a central rod region comprising three α-helical coiled coils characterized by a heptad repeat of hydrophobic amino acids (Doring and Stick, 1990; Steinert and Roop, 1988; Weber *et al.*, 1988, 1989).

However, other features of the proteins, especially in the globular head and tail domains, distinguish them from cytoplasmic IF proteins. Lamins possess a nuclear localization signal that directs them to the nuclear compartments (Loewinger and McKeon, 1988). They also have a C-terminal motif CaaX (C, cystein; a, aliphatic amino acid; X, any amino acid), which is also found in yeast-mating factors and *ras* proteins (Hancock *et al.*, 1989). The CaaX motif serves as a recognition sequence for posttranslational isoprenylation as well as for proteolytic processing (Beck *et al.*, 1988; Vorburger *et al.*, 1989b; Wolda and Glomset, 1988). In B-type lamins the isoprenylation site persists and allows direct association of these species with membranes (Gerace and Blobel, 1980; Holtz *et al.*, 1989; Krohne *et al.*, 1989; Stick *et al.*, 1988). However, in A-type lamins the isoprenylation sequence is removed by proteolysis prior to or during nuclear import (Vorburger *et al.*, 1989b; Krohne *et al.*, 1989). Lamins are also modified by carboxyl methylation at the CaaX motif (Chelsky *et al.*, 1987, 1989) and phosphorylation (Gerace and Blobel, 1980; Ottaviano and Gerace, 1985; Miake-Lye and Kirschner, 1985). These modifications have been implicated in mediating mitotic changes in the polymeric state of lamins (Burke and Gerace, 1986; Chelsky *et al.*, 1989; Nakagawa *et al.*, 1989; Smith and Fisher, 1989; Suprynowicz and Gerace, 1986) and will be considered in detail later in this chapter.

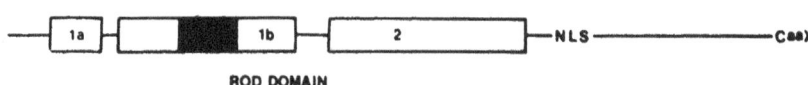

ROD DOMAIN

FIGURE 4. Schematic representation of a lamin monomer showing a central rod domain comprising coil 1a, 1b, and 2, the position of the nuclear localisation sequence (NLS) and the carboxy-terminal CaaX motif. The shaded area in coil 1b represents the additional heptad repeats, which distinguish lamins from other intermediate filament proteins.

1.5. Cytoplasmic and Nucleoplasmic Filaments That Interact with the Nuclear Envelope

Two distinct networks of IFs appear to converge at the inner and outer faces of the NE. Specific details of the arrangement of these filaments have been revealed by thin resinless section EM of extracted nuclei (Fey *et al.*, 1984; He *et al.*, 1990; Jackson and Cook, 1988; Nickerson *et al.*, 1989). At the cytoplasmic face of the NE, IF filaments interact with the lamina via the nuclear pores, while at its nucleoplasmic face intranuclear filaments, which are defined as IF by virtue of their size (Jackson and Cook, 1988), directly abut the inner surface of the lamina (reviewed by Cook, 1988). These studies correlate well with immunofluorescence analyses that clearly demonstrate a concentration of cytoplasmic intermediate filaments in the region close to the nucleus (reviewed by Goldman *et al.*, 1986) and biochemical studies indicating attachment of vimentin to the nuclear membrane (Georgatos and Blobel, 1987a,b). The significance of these associations is not yet clear.

1.6. Control of Entry into and Exit from Mitosis

Various studies have shown that entry into mitosis and exit from mitosis is controlled by the activation and inactivation of a major regulatory protein kinase complex. It has long been known that entry into mitosis is marked by an increase in the phosphorylation of several intracellular proteins. Although many of these proteins have yet to be characterized, lamins (Gerace and Blobel, 1980; Ottaviano and Gerace, 1985), vimentin (Evans and Fink, 1982), histone H1 (Bradbury *et al.*, 1973, 1974), and histone H3 (Gurley *et al.*, 1978) have all been shown to become highly phosphorylated. Indeed, histone H1 phosphorylation has long been used to characterize the proliferative status of cells, the so-called growth-associated H1 kinase (GAK), displaying a periodicity of activity that peaks at mitosis (Bradbury *et al.*, 1974). These observations have led to the idea that entry into mitosis is controlled by protein kinases (Bradbury *et al.*, 1973).

Identification of the gene products responsible for entry into mitosis has involved both genetic and biological assays (for reviews see Fantes, 1988; Hunt, 1989; Lohka, 1989). The biological activity of proteins that induce entry into mitosis can be assayed either by cell-fusion studies (Rao and Johnson, 1970) or by microinjection of cytoplasmic extracts into mature oocytes of frogs and starfish (Masui and Markert, 1971; Smith and Ecker, 1971; Meijer and Guerrier, 1984). In particular, microinjection into oocytes has been used to characterize M-phase-promoting factor (MPF), an activity that has been detected in unfertilized eggs, in mitotically dividing sea urchin, starfish, and amphibian embryos, in mammalian tissue culture cells and budding yeast (reviewed by Maller, 1985) that is able to induce NEBD, chromosome condensation, and spindle reorganization (Miake-Lye *et al.*, 1983; Newport and Kirschner, 1984; Lohka and Maller, 1985). Genetic analyses of cell division in the fission yeast (*S. pombe*) and the budding yeast (*S. cerevisiae*) have also identified a gene product required both for the transition from G1 to S-phase and from G2 to mitosis (reviewed by Nurse 1985). In *S. pombe*, this is the product of the cdc2 gene, which encodes a 34 kDa protein kinase (reviewed by Lee and Nurse, 1988). The equivalent gene in *S. cerevisiae* is CDC28, which also encodes a 34 kDa protein kinase (Reed *et al.*, 1985) that is functionally identical to cdc2 (Beach *et al.*, 1982). A cdc2 homologue, cdc2(Hs), has also been identified in human cells (Lee and Nurse, 1987), which shares 60% amino acid identity with both cdc2 and CDC28.

1.7. M-Phase Promoting Factor (MPF)

M-phase promoting factor from unfertilized *Xenopus* eggs, and histone H1 kinase activity from starfish eggs, have both been purified (Arion *et al.*, 1988; Labbe *et al.*, 1988; Lohka *et al.*, 1988). Both activities display extensive serine-threonine kinase activity and contain major proteins of 32–34 kDa (Labbe *et al.*, 1988; Lohka *et al.*, 1988). Purified *Xenopus* MPF also contains a 45 kDa peptide (Lohka *et al.*, 1988). Antibodies that recognize the conserved regions of cdc2 (Draetta *et al.*, 1987; Lee and Nurse, 1987; Mendenhall *et al.*, 1987) have been used to demonstrate that the 34 kDa peptides identified in purified fractions of MPF and H1 kinase are identical to cdc2 (Gautier *et al.*, 1988; Labbe *et al.*, 1988; Langan *et al.*, 1989) and that cdc2(Hs) is associated with a major histone H1 kinase activity in mitotic HeLa cells (Draetta and Beach, 1988). Furthermore, depletion of cdc2 from cytoplasmic extracts containing MPF leads to their inactivation (Dunphy and Newport, 1989).

Despite evidence that the cdc2 kinase is an essential component of MPF, it alone cannot account for the properties of MPF. Both the levels of MPF and its associated protein kinases activities oscillate throughout the cell cycle (Bradbury *et al.*, 1973; Gerhart *et al.*, 1984). In contrast, cdc2 protein does not vary in its abundance (Draetta and Beach, 1988; Gautier *et al.*, 1988; Simanis and Nurse, 1986). However, the levels of two members of the cyclin family, cyclins A and B, do vary considerably in their abundance during the progress of the cell cycle. In starfish and sea-urchin embryos, metabolic labeling experiments have identified a group of proteins that accumulated throughout interphase and early mitosis before being destroyed precipitously at anaphase (Evans *et al.*, 1983; Standart *et al.*, 1985). Because of their behavior they were termed *cyclins* (Evans *et al.*, 1983). Following the cloning and sequencing of these proteins, other cyclins were identified. Cyclins are now thought to form a large family of proteins that are required for regulation of each stage of the cell cycle. However, only cyclins A and B appear to be required for mitosis (reviewed by Lew and Reed, 1992). Cyclin A message is able to induce completion of meiosis when injected into

Xenopus oocytes (Swenson *et al.*, 1986), whereas cyclin B is required for MPF activity in amphibian embryos (Minshull *et al.*, 1989; Murray and Kirschner, 1989). Of the two members of the cyclin family associated with entry into mitosis, proteins homologous to cyclin B have now been found to be complexed with p34^{cdc2} in MPF preparations from amphibian embryos, HeLa cells, fission yeast, and budding yeast (reviewed by Hunt, 1991).

These studies provide a simple model in which the association of cyclin B with p34^{cdc2} regulates the rate of entry into mitosis. The p34^{cdc2}/cyclin B complexes constitute active MPF, which provides a potent protein kinase (p34^{cdc2}-kinase) that modifies, either directly or indirectly, structures within the cell leading to their breakdown or reorganization during mitosis. The rate at which this complex forms is regulated both by the rate of cyclin synthesis and by direct modification of p34^{cdc2} by other kinases and phosphatases (Krek and Nigg, 1991; Morla *et al.*, 1989; Norbury *et al.*, 1991; Solomon *et al.*, 1990). Similarly, exit from mitosis can be achieved by destruction of cyclin, which leads to the inactivation of p34^{cdc2}-kinase (A. W. Murray *et al.*, 1989).

2. NUCLEAR ENVELOPE ASSEMBLY

Nuclear reassembly following mitosis is thought to be a stepwise process in which one event must be completed before the next can occur (Newport, 1987). It should be noted that there appear to be differences between the sequence of events in amphibian egg extracts (Lohka and Masui, 1983, 1984; Blow and Laskey, 1986; Newport, 1987) and mitotic mammalian cell extracts (Burke and Gerace, 1986). These variations may arise as a consequence of the differences in cell cycle time and availability of nuclear envelope components between rapidly dividing embryonic cells and more slowly growing somatic cells. In addition, the points to be discussed below represent those events associated with the most extreme form of mitosis. In most tissues in higher eukaryotes, the NE breaks down entirely at mitosis and must reform completely in telophase and early G1 phase. This process of envelope assembly therefore occurs once per cell cycle and is referred to as *open mitosis*. In some invertebrates such as *Drosophila*, the NE is not completely broken down at mitosis; instead, nuclear pores are lost and larger fenestrations in the envelope are observed (Stafstrom and Staehlin, 1984). Thus, in this instance, reassembly of the NE postmitosis requires reformation of nuclear pore complexes (NPCs) but not entire rebuilding of the nuclear membrane. Lower eukaryotes such as yeast do not undergo NE breakdown at mitosis (for review see Heath, 1980), but changes in membrane permeability are detected (Nasmyth, 1990), indicating loss of integrity of the NE during cell division. This is referred to as *closed mitosis* but may still involve subtle changes in the distribution of components of the NE.

2.1. Systems for Analysis of Nuclear Envelope
Assembly and Disassembly

Morphological aspects of mammalian somatic NE assembly were initially analyzed by TEM of, for example, cultured pig embryo kidney (PK) cells (Zatsepina *et al.*, 1982), showing that partially decondensed chromosomes were the preferred substrate for NE formation. A combination of TEM and SEM of HeLa S3 cells suggested that the NE is reestablished early in telophase on a shell-like configuration of chromatids arranged in a double concentric layer (Welter *et al.*, 1985). In addition to such purely observational studies, cell fusion experiments have increased our understanding of the stages of NE formation in mammalian HeLa cells (e.g., Ghosh and Pawaletz, 1987). However, further analysis only became possible on the development of a cell-free system derived from mitotic CHO cells that supported morphologically normal NE reformation (Burke and Gerace, 1986). Such extracts are amenable to a range of experimental manipulations and have proven to be a great source of information on the components and regulation of NE assembly and NEBD in somatic cells.

Amphibian eggs have provided a valuable source of material for detailed study of NE dynamics. Frog oocytes have very large nuclei with high pore density and are relatively easy to dissect and prepare for EM analysis. In addition, *Xenopus* eggs are arrested in meiotic metaphase II, with the female DNA complement present on a spindle close to the animal pole of the egg. Sperm entry into the egg results in a calcium transient that releases meiotic arrest, and the cell enters its first interphase. The egg contains stockpiles of components necessary for the assembly of thousands of nuclei (Laskey, 1985) and is therefore an ideal starting material for study of the process of NE formation. Purified DNA injected into *Xenopus* is found to become assembled into nucleus-like structures (Forbes *et al.*, 1983), and cell-free extracts prepared from *Rana* (Lohka and Masui, 1983) or *Xenopus* (Blow and Laskey, 1986) eggs still show nuclear formation capacity on exogenous DNA templates. Such extracts have the advantage over whole cells of relative ease of manipulation, and have been used for the analysis of nuclear formation (Blow and Laskey, 1986, 1988; Hutchison *et al.*, 1987, 1988; Newport, 1987; Cox and Laskey, 1991; Leno and Laskey, 1991) and NEBD and mitosis (Hutchinson *et al.*, 1988, 1989; Murray and Kirschner, 1989; Minshull *et al.*, 1989).

The fruitfly *Drosophila melanogaster* has a similar advantage to *Xenopus* in that components for assembly of many nuclei are stockpiled in the egg and early embryo (Steller and Pirotta, 1985), permitting rapid and synchronous early cell cycles (Zalokar and Erk, 1976). Electron microscopy studies of embryos have provided information concerning NE assembly and, surprisingly, showed the presence of a second membrane around reforming nuclei, the so-called spindle envelope (e.g., Stafstrom and Staehlin, 1984). Cell-free extracts have been prepared from early *Drosophila* embryos that support NE formation around exogenous *Xenopus* or chicken sperm or λ phage DNA templates (Ulitzer and Gruenbaum, 1989; Berrios and Avilion, 1990; Ulitzer *et al.*, 1992). Although these reformed nuclei can support physiological processes such as DNA replication (Crevel and Cotterill, 1991), the low levels of DNA synthesis suggest that new "nuclei" are not completely normal.

Other organisms have also been used as experimental systems for analysis of NE dynamics. Many of the data concerning lamin structure and polymerization have derived from work in chickens. *Spisula* and starfish embryos have yielded much information on the regulation of NEBD, although rather less on NE formation (reviewed by Dessev, 1992). We predict that the majority of advances in the study of NE formation will be made using highly purified cell-free systems (e.g., Pfaller *et al.*, 1991) or from manipulation of NE components by molecular biological techniques.

2.2. Preparation of Chromatin Substrate for Nuclear Assembly

Gametic pronuclei are the physiological substrate for NE assembly in eggs. An early event in pronucleus formation is the decondensation of the female meiotic chromosomes and highly condensed sperm nucleus, a process brought about by the abundant acidic protein nucleoplasmin (Philpott *et al.*, 1991). The necessity for decondensation is suggestive that the preferred template for nuclear envelope formation *in vivo* is decondensed chromatin. Subsequent cell cycles in the early amphibian embryo are rapid and synchronous, and nuclear envelope assembly proceeds around mitotic chromosomes, as in somatic cells.

In somatic cells and during the cleavage stage in embryos, the substrates for NE reformation *in vivo* are mitotic chromosomes. The first traces of surrounding envelope can be detected in telophase. However, discrepancies exist in the literature with respect to the preferred chromatin state for NE reformation in somatic cells. Electron microscopy observations of pig embryo kidney (PK) cells suggest that partial decondensation of chromosomes (postanaphse) promotes nuclear envelope formation: Any artificially induced delay in chromosome decondensation results also in a delay in NE assembly, while enhancement of decondensation permits more rapid envelope formation (Zatsepina *et al.*, 1982).

In contrast, studies using multinucleate HeLa or PtK$_1$ cells from different stages of the cell cycle, fused to give so-called heterophasic heterokaryons, suggest that NE assembly depends more on the cell cycle status of the cytoplasm than on the stage of chromosome decondensation. Using this approach, nuclear membranes can be observed to reform around metaphase chromosomes (Ghosh and Pawaletz, 1987) instead of the usual substrate of telophase chromosomes. Because chromosomes are complex mixes of protein and nucleic acid in a highly ordered structure, molecular analysis of NE formation around these substrates is difficult. Therefore, model systems have been developed using purified DNA to determine the minimum substrate requirement for NE assembly. Extracts of amphibian eggs have provided one of the best-characterized *in vitro* systems for analysis of the steps of NE formation (Lohka and Masui, 1983, 1984; Blow and Laskey, 1986; Newport, 1987; Sheehan *et al.*, 1988; Hutchinson *et al.*, 1988). Addition of bacteriophase λ DNA to an extract of activated *Xenopus* eggs results in the formation of intact nuclei after 2 hr (Newport, 1987). Such "pseudonuclei" are fully functional for semiconservative DNA replication (Blow and Sleeman, 1990) and are indistinguishable from normal eukaryotic nuclei with respect to the organization of replication sites within the nuclei (Cox and Laskey, 1991). However, NE formation was not observed to occur around the purified DNA immediately on addition of bacteriophage DNA. Instead, a lag period of 80 min was noted during which time the DNA was first assembled into chromatin by the extract and then was further condensed before membrane association could be detected (Newport, 1987). This result suggests that nuclear membranes require more than DNA, or first-order chromatin, to assemble an envelope structure, with the implication that some component(s) from the extract binds to the chromatin and acts as a receptor for membrane vesicle binding.

2.3. Vesicle Association with Chromatin

In amphibian embryos before the mid-blastula transition (MBT), the NE is reformed after each mitosis from pre-existing components; *de novo* synthesis is not necessary, as large maternal stockpiles of nuclear assembly components are present (reviewed by Laskey, 1985). In somatic cells, however, synthesis of new envelope components occurs during G1 and S-phase. In G1, the NE surface area remains relatively constant, but through S-phase, the nuclear volume doubles with the replication of the DNA, and a concomitant increase in the NE surface area can be measured (Maul *et al.*, 1972; Maul, 1977). Thus, following NEBD at mitosis, envelope reassembly does not require new membrane synthesis: The DNA content of the nucleus has been halved during division and sufficient membrane components are present in each daughter cell to permit formation of a new intact envelope.

Various types of vesicles exist in meiotic and mitotic cells, derived from NE, Golgi membranes, and ER. During subsequent NE reassembly, do all membrane vesicles contribute equally to the reformation of membrane-bound organelles, or are these different vesicle populations distinguished so that only ex-nuclear membrane forms new NE? Analysis of membrane composition from different organelles has shown similarities and differences between nuclear membrane and Golgi membrane and ER (Pathak *et al.*, 1986; Puddington *et al.*, 1985). Thus, it is impossible to determine purely by protein and lipid analysis which membranes will contribute to NE assembly.

Requirements for specific vesicle populations in nuclear membrane formation have been studied more productively using an extract of *Xenopus* eggs fractionated by high-speed centrifugation into soluble and vesicular components (Sheehan *et al.*, 1988). Addition of the vesicular to the soluble fraction at the appropriate *in vivo* ratio (1:10) results in the assembly of intact NE around decondensed sperm chromatin. The vesicular pool has been fractionated using both discontinuous sucrose gradient (Wilson and Newport, 1988) and high-speed ultracentrifugation (Vigers and Lohka, 1991). Separation on sucrose gradients leads to the recovery of two fractions termed *heavy* and *light*. Only the light fraction can support nuclear membrane assembly in reconstitution experiments, the heavy fraction being enriched in mitochondria and large granular vesicles. Electron microscopy analysis of the light fraction reveals that it is a heterogeneous collection of vesicles ranging from 100–400 nm in diameter, this being comparable to the range of sizes of vesicle that bind to chromatin during the early stages of NE assembly in extracts of *Rana* eggs (Lohka and Masui, 1984). The vesicles are enriched with BiP and α-glucosidase, indicating that they are related to the ER. However, vesicles derived solely from ER are unable to substitute for the light vesicle fraction in reconstitution experiments (Wilson and Newport, 1988; K. Labib, personal communication). Vigers and Lohka (1991) were able to separate the nuclear membrane precursor pool into two distinct fractions, nuclear envelope precursor (NEP-A and NEP-B), by high-speed ultracentrifugation. Based on EM analysis, NEP-A was indistinguishable from NEP-B. However, the specific activity of α-glucosidase was ten times greater in NEP-A compared with NEP-B. Furthermore, NEP-A was sensitive to treatment with *N*-ethylmaleimide, whereas NEP-B was not; also, NEP-B was inactivated with high salt but NEP-A was not.

2.4. Vesicle Targeting to Chromatin

Targeting of vesicles to chromatin may differ between embryonic and somatic cells, in that vesicles have been shown to interact directly with chromatin in *Xenopus* embryonic systems in defined buffer systems without detectable lamins, whereas membrane association with chromatin in somatic cells is probably mediated via nuclear lamins. This possible difference is discussed in more detail later (see Section 2.8).

In reconstitution experiments in *Xenopus* egg extracts, Vigers and Lohka (1991) have shown that NEP-B binds to chromatin during the initial stages of nuclear membrane assembly, while NEP-A is required for membrane fusion and expansion of the NE. As both fractions are only active when vesicular, it has been proposed that NEP-B first binds to chromatin, then NEP-A binds to NEP-B, initiating membrane fusion. This would result in NEP-B being distributed primarily at the inner nuclear membrane and NEP-A at the outer membrane and would account for the close relationship of NEP-A with the ER (Figure 5).

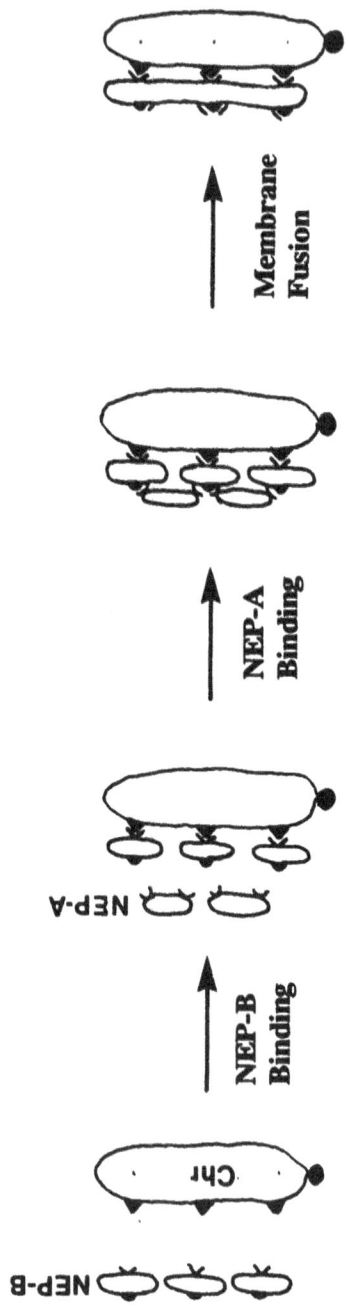

FIGURE 5. Model to explain the sequential involvement of NEP-B and NEP-A in nuclear membrane assembly. The inclusion of elements representing putative receptors is to emphasize how sequential binding may lead to the preferential orientation of NEP-B at the inner nuclear membrane and NEP-A at the outer nuclear membrane (see Vigers and Lohka, 1991).

Implicit in the model proposed by Vigers and Lohka (1991) is the existence of two distinct types of membrane receptor, one found only in NEP-B, which mediates vesicle-chromatin interactions, and others found in both NEP-A and NEP-B, which mediate vesicle-vesicle interactions. These receptors appear to be proteinaceous since pretreatment of vesicles with trypsin can inactivate them as nuclear membrane precursors in reconstitution experiments (Wilson and Newport, 1988). Furthermore, NEP-B binding to chromatin can be prevented by phosphorylation with a protein kinase (Vigers and Lohka, 1992) that is regulated by p34cdc2 (Pfaller et al., 1991; see below). At present, these receptors remain uncharacterized although several candidate proteins have been identified (e.g., see Table II). However, future progress seems likely following refinement of Xenopus egg extracts to allow membrane vesicles to bind to decondensed chromatin and then to fuse in a defined buffer (Newport and Dunphy, 1992). In this system, isolated vesicles of diameter ~74 nm readily bind to decondensed sperm heads at an approximate density of 1 vesicle per 100,000 base pairs of DNA (Newport and Dunphy, 1992). In the presence of ATP and GTP, membrane fusion then occurs (Newport and Dunphy, 1992; Boman et al., 1992a,b). That GTP

Table II
Lamin Receptors

Name	Organism	Comments	Reference
Chromatin-associated receptors			
Peripherin	Drosophila		Chaley et al., 1984
Perichromin	Human	33 kDa	McKeon et al., 1984
p62		Nuclear scaffold	Fields and Shaper, 1988
Membrane-associated receptors			
p58	Yeast	Lamin B receptor	Georgatos et al., 1989
Otefin	Drosophila	53 kDa, one transmembrane domain; expression parallels lamins	Padan et al., 1990
p58	Avian	Putative 73 kDa, 8 trans-membrane domains; lamin B receptor	Worman et al., 1988
p54	Chick	Phosphorylated by p34cdc2	Bailer et al., 1991
p75	Rat	INM inner face	Senior and Gerace, 1988
p68	Rat		Senior and Gerace, 1988
p55	Rat		Senior and Gerace, 1988
p55	Rat	Required as lamin A receptor in interspecific hybrids	Powell and Burke, 1990
p52	Rat	Found by cross-reaction with antibodies to avian p58	Worman et al., 1988

hydrolysis is required suggests that nuclear membrane precursors might be coated vesicles that must be uncoated before fusion can occur (Mayorga *et al.*, 1989; Beckers and Balch, 1989; Melancon *et al.*, 1987; Boman *et al.*, 1992a,b).

By contrast, somatic NE assembly appears to begin with interaction of chromatin and lamins. Two major classes of lamins are known in mammalian cells. A-type lamins are developmentally regulated, and are solubilized at mitosis. B-type lamins, by contrast, remain associated with nuclear membrane vesicles during mitosis (Gerace and Blobel, 1980; Stick *et al.*, 1988). Analysis of NE re-formation in a cell-free extract of mitotic and postmitotic Chinese hamster ovary (CHO) cells (Burke and Gerace, 1986) suggests that soluble lamins first bind to a protein receptor on chromatin (Glass and Gerace, 1990; Burke, 1990) and as they start to polymerize, they subsequently incorporate membrane-associated lamin B (Figure 6). Hence, vesicles are targeted to chromatin indirectly, by virtue of lamin-lamin interactions. The identity of the high-affinity chromatin receptor for lamin A is again unclear, although candidates include perichromin (McKeon *et al.*, 1984), chicken p54 (Bailer *et al.*, 1991), and *Drosophila* p40 (Wataya-Kaneda *et al.*, 1987) (summarized in Table II).

2.5. Assembly of Nuclear Pore Complexes

Nuclear pore complex (NPC) assembly in the nuclear envelope is a continuous process, occurring at telophase using pre-existing NPC subunits, and during subsequent nuclear envelope growth in G1 and S-phase of the cell cycle (Maul *et al.*, 1972; Gerace and Burke, 1988; Lohka, 1988). In somatic cells, this growth requires new synthesis of NPC components during interphase, in addition to the use of an existing small cytoplasmic pool (Davis and Blobel, 1986). Nuclear pores are present in the NE at relatively low frequency during telophase in somatic cells, and they increase in density in early G1. The number of pores then doubles during S-phase (Maul *et al.*, 1972), probably in preparation for mitosis, so that each daughter cell will inherit sufficient pore complexes to reassemble a functionally transporting nuclear envelope immediately after anaphase.

By contrast, in early amphibian and *Drosophila* embryos, maternal stockpiles of these components supply the requirements for NPCs in nuclear envelope assembly at telophase and for subsequent growth (Laskey, 1985). To date, the process of NPC assembly has not been fully clarified in any experimental system. However, progress has been made recently in determining intermediate stages in NPC assembly, using physical and biochemical fractionation of nuclear reconstitution extracts derived from cultured mammalian cells (Burke and Gerace, 1986) and *Xenopus* eggs (Sheehan *et al.*, 1988; Finlay and Forbes, 1990; Finlay *et al.*, 1991). A combination of genetic and biochemical approaches in yeast has also yielded information on the structure of NPC components and

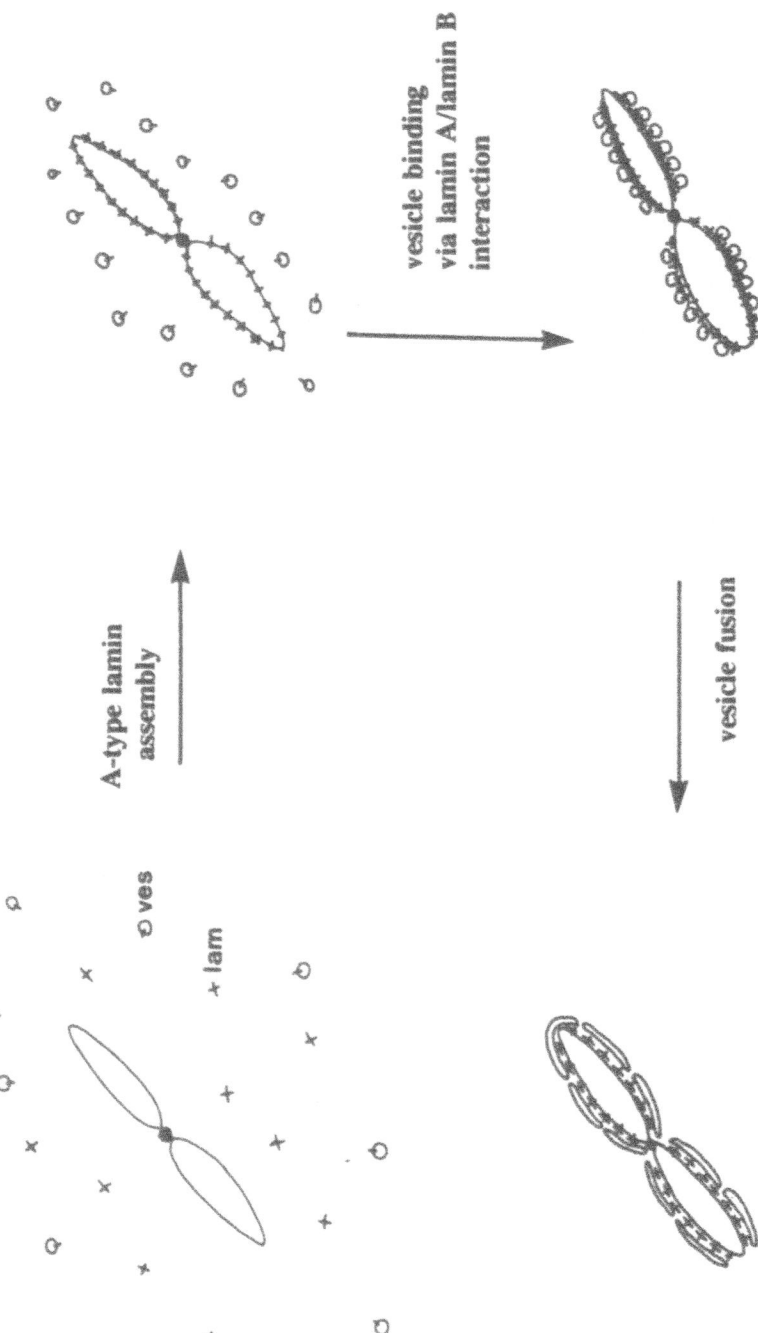

FIGURE 6. Stages in the assembly of a nuclear envelope, in somatic cells. lam = lamin monomers; ves = membrane vesicles with integral B-type lamins. In the model, A-type lamins first bind to the surface of chromosomes and polymerize. B-type lamins then bind to the A-type lamins and in so doing target membrane vesicles to chromosome surfaces. Finally, vesicles fuse to give rise to a nuclear membrane (see Burke and Gerace, 1986; Glass and Gerace, 199).

shows potential for analysis of the pathway of NPC assembly (Nehrbass et al., 1990; Davis and Fink, 1990).

2.5.1. Nuclear Pore Complex Precursor Assembly

In *Xenopus*, NPCs may be assembled from large soluble precursors representing a complex of various pore proteins. Immunoprecipitation of the major nucleoporin, p68, from *Xenopus* egg extracts showed co-precipitation of two higher molecular weight components (Dabauvalle et al., 1990). Further analysis found p68 to be stably associated with other proteins in a large globular complex of Mr 254 kDa, with a Stoke's radius of 5.2 nm that sediments at 11.3S (Dabauvalle et al., 1990). Immunoprecipitation of p62 from ^{35}S-met-labeled BRL cells suggested, by contrast, that rat p62 was synthesized as a soluble 61 kDa precursor that remained monomeric until incorporation into a post-mitochondrial pellet (presumably the NE) with a $t_{1/2}$ of 6 hr (Davis and Blobel, 1986). Incorporation of p62 into a large complex similar to that described in *Xenopus* (Dabauvalle et al., 1990) may therefore occur at the time of NPC insertion into membranes, possibly at S-phase (Maul et al., 1972). Finlay et al. (1991) have shown that p62 exists as a large complex in rat liver nuclei (as opposed to cytoplasm). Using wheat germ agglutinin (WGA) extraction followed by gel filtration, these workers have isolated from rat liver nuclei a complex of 550–600 kDa that contains three nuclear pore proteins, p62, p58, and p54 in a 4:1:4 ratio, respectively.

2.5.2. Models for Nuclear Pore Complex Formation

Three models have been proposed for the formation of NPCs in the assembly of nuclear envelopes. The vesicle fusion model, proposed by Lohka (1988), suggests that NPCs form following association of NE-specific vesicles with chromatin, and subsequent flattening of these vesicles. Pores are presumed to form on the fusion of the inner and outer membranes at positions within a single flattened vesicle. Hence, NPCs may be derived from components within individual vesicles by "top-to-bottom" fusion, rather than arising on "side-to-side" fusion of adjacent vesicles. In support of this model is the observation that pores can occur closer together than the span of a single vesicle—i.e., that there are insufficient "gaps" between adjacent vesicles to account for observed pore densities.

Additionally, Lohka (1988) reports that pores can form in membranes derived from a single vesicle. Mechanistically, this model is possible if some component of the NPC remains vesicle-associated throughout mitosis and then serves as a site for intravesicle membrane fusion once the vesicles have become bound to chromatin in interphase. A large glycoprotein, gp210, probably located toward the periphery of the NPC, is indeed membrane-anchored during mitosis via one or possibly two short transmembrane domains, HS1 and HS2 (Wozniak et al., 1989; Greber et al., 1990; Wozniak and Blobel, 1992). The peripheral situation of gp210 and its lack of involvement with the central transporter are suggestive of a role for gp210 in organizing NPC structure (Wozniak et al., 1989; Greber et al., 1990). In addition, we propose that the very long lumenal N-terminal domain (1783 amino acids, compared with only 58 for the cytoplasmic tail: Greber et al., 1990) may be involved in the type of intravesicle membrane fusion required by the vesicle fusion model of NPC assembly.

An alternative is the "pre-pore" model proposed by Sheehan et al. (1988) on the basis of EM examination of NE formation in Xenopus extract reconstituted in vitro from soluble and vesicular fractions. Sheehan et al. (1988) suggested that immature "pre-pores," representing the halves of pores usually associated with the inner nuclear membrane, bind directly to decondensing chromatin. These pores may then serve as receptors for membrane vesicles bearing other pore components. Once such vesicles are bound to the chromatin via the pre-pores, it is proposed that they flatten and fuse, to form the double membrane structure of the nuclear envelope, and the two halves of the pore likewise fuse to give a continuous channel across both membranes (Sheehan et al., 1988).

Both the vesicle fusion model and the pre-pore model require the presence of chromatin, either as the site of vesicle accumulation and fusion, or as the substrate for the assembly of pre-pores. However, analysis of the formation of annulate lamellae (reviewed by Merisko, 1989) has clearly shown that NPC assembly into double membrane structures requires neither chromatin nor lamins (Dabauvalle et al., 1991), suggesting a third mechanism of NPC formation. Vesicles containing electron-dense material have been observed in Xenopus egg extracts, which in the absence of chromatin fuse to form long cisternae called annulate lamellae, with an electron-dense coat (Dabauvalle et al., 1991). Structures indistinguishable from NPCs are present at high density in these membrane stacks, and it is possible that NPCs form from this electron-dense material at intervals along the cisternae. Interestingly, mature annulate lamellae (AL) that contain many NPCs lack the electron-dense coat. In this system, chromatin successfully competes for vesicles and NPCs such that at high chromatin concentrations, AL are not observed. However, Xenopus lamin Liii may be necessary in providing the high affinity of chromatin for vesicles, since anti-lamin Liii antibodies abolish the competitive effect of chromatin and permit AL assembly (Dabauvalle et al., 1991). The AL probably do not serve as pools of components that can be mobilized for NE formation, but rather form when such components are in excess. Instead, at mitosis, AL break down, like other cellular membranes, and their interphase re-formation is competitive with nuclear envelope assembly depending on the concentration of chromatin present. Although neither Lohka's vesicle fusion model nor the pre-pore model (Sheehan et al., 1988) can account for AL formation, it is interesting to note that gp210 is present in both NE and AL, leading to the suggestion of Wozniak *et al.* (1989) that it may play a similar role in NPC formation in both membrane structures.

2.6. Role of Nuclear Pore Complexes in Nuclear Envelope Assembly

The existence of soluble NPC precursors in *Xenopus* eggs suggests that they may play a role in NE formation. However, although the complex isolated by Dabauvalle *et al.* (1990) is necessary for NPC formation *in vitro,* double nuclear membranes lacking pores can form in extracts depleted of the complex. Hence, the soluble pore complex is used in NPC formation but is not required for nuclear membrane association with chromatin. Similarly, Finlay and Forbes (1990) have shown that nuclear envelopes can form in *Xenopus* egg extracts depleted of nucleoporins by WGA treatment. In this case, however, NPCs were observed that could be made functional for nuclear transport when the WGA-binding fraction was re-added. NE formation is reported by Finlay *et al.* (1991) on addition of untreated *Xenopus* membranes to *Xenopus* egg extracts depleted by WGA adsorbtion but supplemented with rat pore proteins lacking the 550 kDa complex. By contrast, Cox (1992) reports an inability to form intact nuclear envelopes around sperm chromatin in egg low-speed supernatant in the presence of WGA. It is difficult to reconcile these disparate results into a general model of the role of nuclear pores in NE assembly. However, it should be noted that Finlay and Forbes (1990) and Finlay *et al.* (1991) used a *Xenopus* egg extract fractionated into soluble and vesicular components. They depleted the cytosol of nucleoporins by WGA treatment but did not expose the vesicular fraction to the lectin. Although they show no significant binding of WGA to their membrane fraction, in our hands the washed vesicular fraction from *Xenopus* eggs contains WGA-binding proteins of 67, 95, and 105 kDa (K. Labib and L. Cox, unpublished data). Differences may also exist in the way membranes are formed around various chromatin templates, and pseudonuclei might have different receptors for vesicle binding than do mitotic chromosomes or gametic pronuclei.

Because different experimental systems suggest different pathways for NPC assembly, it is impossible at present to propose a single all-encompassing model for NPC assembly, or for the role of NPCs in nuclear envelope formation.

2.7. Lamina Assembly

The nuclear lamina during interphase is a highly polymerized orthogonal network of 10–20 nm filaments subadjacent to the inner nuclear membrane, as visualized by EM (Aebi *et al.*, 1986). The lamina is probably important in maintaining higher-order structure of the chromatin during interphase, as it provides structural continuity with the NE (Paddy *et al.,* 1990). In mitosis, the lamina becomes solubilized prior to nuclear envelope breakdown (NEBD; see below). Therefore, after each mitosis, the nuclear lamina must be reassembled from the depolymerized components inherited by each daughter cell, and possibly from newly synthesized lamin proteins. The lamina associates with the nuclear membrane either directly by hydrophobically modified regions of the protein, or via integral membrane proteins that act as lamin receptors (see below and Table II). Lamins have been cloned and sequenced from various species and are found to be members of the intermediate filament protein family (McKeon *et al.*, 1986; Fisher *et al.*, 1986; Franke, 1987; Stick, 1988, 1992). Like other intermediate filament proteins, lamins at high concentrations are capable of self-assembly *in vitro*, into head-to-tail filaments, and into laterally associated paracrystals (Aebi *et al.*, 1986; Moir *et al.*, 1990). However, *in vivo*, nuclear lamina assembly proceeds from soluble dimeric or tetrameric molecules (lamins A and C) or vesicle-associated proteins (lamin B), rather than from pre-polymerized networks (Burke and Gerace, 1986; Glass and Gerace, 1990). Once correctly localized on decondensing chromatin or within intact nuclei, the lamins then polymerize. It is important to remember that lamina assembly, unlike NEBD, is a continuous process, starting with NE re-formation in telophase and continuing with NE growth up to the end of S-phase (Maul *et al.*, 1972; Fry, 1976; Gerace *et al.*, 1984; Loewinger and McKeon, 1988). Thus, lamin molecules must be capable of assembling around chromatin in soluble systems, and must also be able to enter intact nuclei, to contribute to NE growth.

2.7.1. Lamin Polymerization

Modifications of lamin proteins leading to lamin polymerization/depolymerization have been most widely studied in mammalian and avian systems, where two major lamin types, A and B, exist (reviewed by Goldman *et al.*, 1991). Sites in both the amino- and carboxy-terminal domains of the protein are important for polymerization of lamin dimers into head-to-tail filaments, and this polymerization is interrupted by phosphorylation of specific N-terminal sites by the mitotic kinase p34^{cdc2} (see below) and also by phosphorylation of C-terminal sites (Peter *et al.*, 1990; Ward and Kirschner, 1990; Heald and McKeon, 1990). Thus, dephosphorylation is necessary in telophase to permit lamina re-formation (Gerace and Blobel, 1980; Burke and Gerace, 1986). Upon dephosphorylation, lamin polymerization may be triggered by correct association with chromatin, or by reaching a critical concentration (Glass and Gerace, 1990). The central rod domain, characteristic of all intermediate filament proteins, is not modified during the cell cycle. Network formation, by lateral association of lamin filaments into paracrystals *in vitro* or orthogonal structures *in vivo*, is probably controlled by the carboxy-terminal domain. Phosphorylation of the carboxy-terminal domain in interphase, by an unknown kinase, leads to paracrystal formation *in vitro* and inhibits head-to-tail filament formation. Deletion of this region actually promotes head-to-tail association of lamin molecules to assemble long filaments, but paracrystal formation is inhibited (Moir *et al.*, 1990). In mammalian, avian, and amphibian cells, interphase levels of lamin phosphorylation are three to four times lower than in mitosis (Ottaviano and Gerace, 1985; Miake-Lye and Kirschner, 1985; Peter *et al.*, 1990), whereas in *D. melanogaster*, the overall levels of phosphorylation do not change but instead phosphorylation sites alter from serine only in mitosis, to serine and threonine in interphase (Smith and Fisher, 1989). Subsequent growth of the lamina after intact nuclear envelope formation may require localized relaxation of lamin-lamin interactions, possibly by phosphorylation, to allow insertion of additional dimers (Ottaviano and Gerace, 1985; Burke and Gerace, 1986). Thus, regulation of phosphorylation at both the amino- and carboxy-domains of lamin proteins is important throughout interphase for filament assembly and network formation.

In addition to interphase phosphorylation of the carboxy-terminal domain, lamina assembly requires specific modifications of the carboxy-terminal motif CaaX (C, cysteine; a, aliphatic amino acid; X, any amino acid) by isoprenylation (Beck *et al.*, 1988; Wolda and Glomset, 1988; Pollard *et al.*, 1990). Chicken lamin B_2 is modified by a mevalonic acid derivative (Farnsworth *et al.*, 1989). Three carboxy-terminus amino acids are then proteolyzed from lamin B_2. Lamin A is also proteolytically processed after isoprenylation by a nuclear-envelope-associated protease (Lehner *et al.*, 1986) to remove the modified cysteine (Weber *et al.*, 1989; Beck *et al.*, 1990). The retention of a hydrophobic carboxy-terminus capable of direct membrane association presumably ensures the association of lamin B with membrane vesicles at NEBD (Gerace and Blobel, 1980). The cysteine of lamin B_2 CaaX motif is also methylated (Chelsky *et al.*, 1987), increasing its hydrophobicity. However, the role of methylation in interphase is still unclear as it is correlated with NEBD (Chelsky *et al.*, 1989). Interestingly, the proto-oncoprotein, *ras*, is likewise isoprenylated and methylated at a C-terminal CaaX motif, but these modifications direct *ras* to the plasma membrane (Hancock *et al.*, 1989). Therefore, signals in addition to CaaX modification must act in the lamin proteins to target them to the NE. Putative nuclear localization sequences (NLSs) of lamins have been defined by homology to other characteristics NLSs, e.g., SV40 large T antigen (Kalderon *et al.*, 1984; Lanford and Butel, 1984), and by mutations within the C-terminal domain that result in aberrant cytoplasmic assembly of the lamins (Loewinger and McKeon, 1988).

By examination of certain mutant forms, CaaX modifications have been shown to be important in correct assembly of the nuclear lamina. Mutation of the cysteine residue to alanine in the CaaX motif of chicken lamin B_2 prevents isoprenylation and methylation, and although the transfected lamin is still expressed and imported into the nuclei of host mouse L cells, it is not correctly associated with the nuclear membrane (Kitten and Nigg, 1991). Site-specific mutagenesis of human lamin A CaaX motif (sequence CSIM) suggests that this sequence acts in conjunction with the nuclear localization sequence to direct the lamin to the nucleus (Holtz *et al.*, 1989). This raises the question of how such proteolyzed lamins can be relocated to the nucleus during subsequent telophases. By injecting modified recombinant mammalian lamins into *Xenopus* oocyte cytoplasm and analyzing subsequent nuclear distribution, it has been shown that fusion of lamin A CaaX (CAIM) sequence to lamin C is sufficient to confer nuclear association, whereas lamin C alone is cytoplasmic in this system (Krohne *et al.*, 1989). Lamin C lacks a CaaX motif and may be correctly localized within the nucleus by association with other lamin types (McKeon *et al.*, 1986; Fisher *et al.*, 1986).

Stable association of lamins with the NE requires more than a nuclear localization sequence and CaaX motif. Interactions via the N-terminus and central rod domains with previously polymerized lamins, or with lamin receptors in the nuclear membrane and chromatin, may also be involved. These interactions will be considered below.

2.7.2. Lamin Receptors

Various candidate proteins have been suggested as chromatin-associated lamin receptors, including peripherin (Chaley *et al.*, 1984), perichromin (McKeon *et al.*, 1984), and a nuclear scaffold protein of 62 kDa (Fields and Shaper, 1988). In addition, various integral membrane proteins have been characterized that possibly act as lamin receptors in the nuclear membrane. Rat liver nuclear envelopes that have been treated to remove membrane and pore complexes retain three proteins of Mr 75, 68, and 55 kDa that may function as lamin attachment sites (Senior and Gerace, 1988). Antibodies against these proteins show that putative receptors occur at $\sim$1.5–4 copies per 50 nm length of lamin filament and are arranged discontinuously along the nucleoplasmic face of the inner nuclear membrane (Senior and Gerace, 1988). Subsequently, a 54 kDa integral membrane protein from chicken nuclear envelopes has been detected that interacts with the nuclear lamina in a cell-cycle dependent manner, possibly controlled by a p34[cdc2]-dependent phosphorylation (Bailer *et al.*, 1991). *In vitro* binding and competition assays using ^{125}I-labeled avian lamin B show associations with an integral membrane protein of apparent Mr 58 kDa in lamin-depleted avian red blood cells (Worman *et al.*, 1988). The p58 cDNA encodes a putative 73 kDa protein that has eight potential transmembrane domains, and the protein is proposed to act as a membrane receptor for lamin B (Worman *et al.*, 1990). The equivalent lamin B receptor in yeast is also 58 kDa (Georgatos *et al.*, 1989). Antibodies against p58 also react with a 52 kDa protein from rat liver nuclear envelopes (Worman *et al.*, 1988). Similarly, a 55 kDa protein from rat nuclei has been suggested to associate with lamin A, since the protein equilibrates between nuclei of interspecific heterokaryons if and only if the individual nuclei possess lamin A (Powell and Burke, 1990). In *D. melanogaster*, a distinct inner nuclear membrane protein of 53 kDa, called *otefin* (Harel *et al.*, 1989), has been cloned, and its expression is found to parallel that of *Drosophila* lamin protein (Padan *et al.*, 1990). Although antibodies against otefin also recognize a 58 kDa mammalian protein, otefin is distinct from putative lamin B receptor, p58, on the basis of primary sequence (Padan *et al.*, 1990; Worman *et al.*, 1990). Thus, various integral membrane proteins are potential candidates for lamin receptors and may serve to stabilize the lamina-membrane contacts once vesicles have been targeted to chromosomes by interaction of membrane-associated lamin B with chromatin-bound lamins A and C (see Table II for summary).

2.8. The Role of Lamins in Nuclear Envelope Formation

In somatic cells, a large body of evidence points toward a critical role for lamins in the early stages of NE reassembly (e.g., Burke and Gerace, 1986; Benavente and Krohne, 1986). In contrast, conflicting evidence exists for the role of lamins in the formation of embryonic cell nuclear envelopes (Meier *et al.*, 1991; Newport *et al.*, 1990; Dabauvalle *et al.*, 1991; Jenkins *et al.*, 1993) that at present can only be partially reconciled with the accepted pathways in somatic cells. Here, we postulate a unified model with similar requirements for lamins in the formation of embryonic and somatic cell nuclear envelopes.

Although lamina assembly and nuclear membrane formation occur in telophase and early G1, it must not be forgotten that the NE continues to grow, especially during S-phase when the twofold doubling in nuclear volume requires a concomitant increase in the surface area of the nucleus. It is highly probable that these two stages of lamina assembly differ, the primary stage of telophase formation possibly occurring by association of lamins with chromatin, which may then lead to vesicle binding to chromatin via interactions between lamins and vesicle proteins. Subsequent stages of lamina growth, once the nuclear membrane is intact and functional for transport, probably occur by transport of soluble lamin monomers or dimers through NPCs, followed by polymerization into a locally destabilized lamin network. These two distinct stages of lamina assembly and growth are not necessarily distinguished in some experimental reports, and the use of different systems for analysis may have led to slightly misleading interpretations with respect to the putative role of lamins in early stages of NE assembly. We shall discuss the experimental data and attempt to reconcile different views on the importance of lamins at various stages of NE reconstitution.

Somatic lamins are incorporated very rapidly into insoluble structures, presumably the NPC-lamina, suggesting a requirement at the earliest possible stages of postmitotic nuclear reorganization. In living mammalian cells, lamin A inserts into a postmitochondrial pellet with a $t_{1/2}$ of only 5 min, with a $t_{1/2}$ of 60 min for lamins B and C (Gerace *et al.*, 1984). This can be dissected *in vitro* using cell-free extracts of Chinese hamster ovary (CHO) cells, where NE formation appears to follow physiological pathways, since EM analysis of intermediates in nuclear envelope assembly show morphologically normal structures (Burke and Gerace, 1986). These authors reported that lamin formation in the cell-free extract was inhibited by high concentrations of ATP or other conditions that prevent de-phosphorylation, demonstrating that dephosphorylation is necessary for lamina polymerization, in agreement with earlier studies (Gerace and Blobel, 1980).

To determine whether lamins are required for vesicles to associate with chromatin, and for nuclear envelope assembly, affinity-purified anti-lamin anti-bodies were injected into dividing PtK2 cells (rat kangaroo epithelium). The presence of anti-lamin antibodies resulted in cells arresting in telophase, with condensed chromosomes and only partially formed nuclear envelopes (Bena-vente and Krohne, 1986), indicating a requirement for lamin function in tel-ophase nuclear envelope reassembly. As an alternative to antibody addition, CHO homegenates were immunodepleted by preincubation with affinity-purified anti-lamin antibodies coupled to killed *Staphylococcus aureus* cells (Burke and Gerace, 1986). Selective removal of lamin B (and associated membrane vesicles) still permitted association of lamins A and C with chromatin, as determined by immunofluorescence microscopy, but EM analysis showed that vesicle associa-tion with chromatin was substantially inhibited. When lamins A and C were immunoadsorbed, NE formation and association of lamin B with chromatin were both prevented. These data suggest a model whereby lamins A and C first bind to chromatin, and then association with membrane-bound lamin B brings vesicles to the chromatin. These vesicles then fuse. By the stage of vesicle fusion, nuclear pores can be observed (see Figure 6).

To test this model, the interactions of lamins A and C with chromatin have been examined in cell-free extracts (Burke, 1990) and defined systems (Glass and Gerace, 1990). Purified rat lamins A and C were shown to interact with mitotic chromosomes, in a defined buffer, to promote lamina assembly at lamin concentrations that are .8-fold lower than those necessary for self-assembly *in vitro* (Glass and Gerace, 1990). These data therefore suggest that high affinity of lamins for chromatin may be important in targeting lamins to chromosomes in telophase NE re-formation. Direct interaction of lamins with DNA or histones is not thought to occur, since lamins do not associate with histone or DNA-histone-coupled microspheres (Glass and Gerace, 1990). However, brief treatment of chromatin with either trypsin (a protease) or micrococcal nuclease (which cleaves DNA in the linker region between nucleosomes) abolishes lamina assem-bly around chromatin (Glass and Gerace, 1990), thereby suggesting that a higher-order chromatin structure is necessary for lamina assembly.

Direct analysis of *in vivo* NE formation by immunofluorescence confocal microscopy of synchronized human diploid fibroblasts (Bridger *et al.*, 1993) has yielded further information regarding the stages of lamina re-formation after mitosis, in a system not prone to the possible artifacts experienced using purely *in vitro* assays (e.g., Glass and Gerace, 1990). Bridger *et al.* (1993) report that early G1 nuclei contain A-type lamins dispersed as clusters and fibers throughout the nucleus, without significant peripheral staining, while at later cell-cycle stages, lamins become redistributed to the nuclear periphery.

Further addition of lamins to an extant polymerized lamina is probably required for NE growth during G1 and S-phases of the cell cycle. Goldman *et al.* (1992) analyzed possible pathways of lamina growth by microinjecting biotinylated recombinant lamin A into 3T3 fibroblasts. They observed by fluorescence confocal microscopy that lamin A first accumulated as foci deep within nuclei, which only subsequently become localized to the nuclear periphery (Goldman *et al.*, 1992). Two caveats exist with interpretation of these data. First, the high concentration of endogenous unlabeled lamin A may effectively compete out the biotinylated lamin A, leading to aberrant nuclear accumulation without peripheral polymerization. Second, the very high concentrations of injected lamins, which far exceeds the critical concentration for self-assembly, may result in polymerization of the injected lamins to give visible clusters at nonphysiological sites. However, this experiment is the first attempt to analyze processes involved in NE growth as distinct from initial NE formation.

In *Drosophila* cell-free nuclear assembly extracts, lamina formation can only be detected by immunofluorescence microscopy at late stages of nuclear assembly, when phase-dense nuclear envelopes are present (Ulitzer and Gruenbaum, 1989). However, in these *Drosophila* extracts, the demembranated *Xenopus* sperm nuclear template may contribute its own lamin (Liv) to the process of lamina formation (Berrios and Avilion, 1990). A recent study using cell-free extracts of early *Drosophila* embryos has suggested that lamin function is necessary for early NE reassembly, including a requirement for lamins prior to vesicle association with chromatin (Ulitzer *et al.*, 1992). *Drosophila* is reported to possess only one lamin protein (Gruenbaum *et al.*, 1988). However, this exists in two different isotypes of 74 kDa and 76 kDa, modified by phosphorylation, and in a further modified mitotic form (75 kDa) that can be distinguished by differential sucrose centrifugation (Smith and Fisher, 1989; Ulitzer *et al.*, 1992). Polyclonal antibodies raised against recombinant *Drosophila* lamin were reported to prevent lamin association with chromatin, if both membrane and cytosolic fractions were treated with the anti-lamin antibodies. By using fluorescent membrane dyes or electron microscopy, Ulitzer *et al.* (1992) observed that their anti-lamin antibody added to crude embryo extracts blocked the association of membrane vesicles with chromatin. Despite problems with the clarity of the data, and caveats in interpretation, because *Drosophila* nuclear membranes and laminae do not completely disassemble at mitosis *in vivo* (e.g., Harel *et al.*, 1989), this suggests that NE reassembly in *Drosophila* embryos is similar to that in somatic cells, requiring the association of lamin with chromatin to permit membrane vesicle binding to the chromatin.

2.9. Differences in *Xenopus* Embryos

Until very recently, *Xenopus* has appeared to differ in the pathway for NE formation, at least in early embryos and cell-free extracts prepared from activated eggs. It is unclear why the early stages of nuclear envelope formation might differ so considerably between *Xenopus* embryos and other cells, but the large maternal stockpile of nuclear envelope components and rapid rate of cell division in embryos may permit and require a faster process of NE formation than is necessary in somatic cells. By contrast to somatic cells, early amphibian embryos have been reported to possess only one type of nuclear lamin protein, Liii (Stick and Hausen, 1985; Benavente *et al.*, 1985), that is homologous to, but distinct from, B-type lamins (Stick, 1988). Two-dimensional immunoblot analysis with monoclonal antibody L6-8A7 showed lamin Liii to be the only lamin present in germinal vesicles of oocytes, and in 10th cleavage stage embryos, with Liv present in sperm nuclei (Stick, 1988). Other lamin types become expressed later; for example, Li becomes expressed at the mid-blastula transition (stage 8), and Lii after gastrulation (Stick and Hausen, 1985). Do amphibian embryos really differ from somatic cells in the role played by lamins in nuclear envelope assembly? It appears that this may be so, since immunofluorescence microscopy indicates that lamina assembly occurs *after* nuclear membrane formation (Hutchison *et al.*, 1988) in *Xenopus* embryos, whereas somatic laminae form before the NE is complete (see above). Further evidence that lamina assembly requires transport of soluble lamin molecules into intact nuclear membranes comes from the observation that *Xenopus* egg extracts depleted of nuclear pore proteins can assemble double membrane structures around purified bacteriophage λ DNA that lack an associated lamina (Dabauvalle *et al.*, 1990). Furthermore, an intact membrane with functional nuclear pores has been reported to form around demembranated sperm nuclei or purified DNA (assembled into chromatin *in vitro*), when lamin Liii is depleted from *Xenopus* egg extracts (Meier *et al.*, 1991; Newport *et al.*, 1990). However, other workers have reported that Liii is required for nuclear membrane formation in *Xenopus* egg extracts (Dabauvalle *et al.*, 1991).

The discrepancies between these results can be accounted for by the differing experimental procedures used. Meier *et al.* (1991) functionally depleted lamin Liii from *Xenopus* egg extracts using the MAb L6-5D5. Such extracts still supported nuclear membrane assembly around demembranated sperm chromatin but not lamina assembly. The resulting nuclei appeared fragile and did not grow after initial stages of NE formation. However, the nuclear envelopes of depleted nuclei did contain nuclear pore complexes and could transport karyophilic proteins including the proliferating cell nuclear antigen (PCNA) and lamin Liii/antibody complexes, although correct subnuclear localization was disrupted (Jenkins et al., 1993). In contrast, using the same approach but with a different MAb (S49H2), Dabauvalle et al. (1991) have shown by EM that nuclear membrane assembly around chromatin is incomplete in lamin-depleted extracts.

In a similar series of experiments, Newport et al. (1990) removed lamin Liii from Xenopus egg extracts using anti-Liii antibody bound to S. aureus cells via rabbit-anti-mouse IgG. After readdition of untreated membranes (see Sheehan et al., 1988) to the depleted extract, nuclear membrane formation and pore-complex assembly was observed. However, nuclear growth was again restricted to one quarter to one third that of controls, while pore density was two to three times lower than controls. In contrast to somatic cells, lamins were not found to associate with chromatin in the absence of membrane vesicles (Newport et al., 1990). Lamina formation could also be reversibly prevented by blocking nuclear transport with the lectin WGA (Finlay et al., 1987); on addition of the competing sugar triacetylchitotriose, lamin-depleted nuclei started to accumulate a normal lamina (Newport et al., 1990). Similarly, removal of nucleoporins from egg extract leads to the formation of double membrane structures around λ DNA, which lacks nuclear pores and a lamina (Dabauvalle et al., 1990), suggesting that translocation across the nuclear membrane may be a prerequisite for lamina assembly.

Can these contrasting data be reconciled? Perhaps in embryos, lamins might be required initially as receptors to mediate associations between membranes and chromatin, and polymerization and network formation are late events requiring the accumulation of a phosphatase within the nuclear compartment (Figure 7). It is possible that the antibody used by Dabauvalle et al. (1991) interferes with the receptor function of Liii, whereas the antibody used by Meier et al. (1991) interferes with polymerization. The results obtained by Newport and co-workers (1990) might then be explained if an undetected pool of Liii or another lamin in the undepleted membrane fraction was sufficient to provide the receptor function. It has recently been reported that two isoforms of Liii, differing by only 12 amino acids in the C-terminus and generated by alternative splicing, co-exist in early frog embryos (Doring and Stick, 1990). Both forms contain a CaaX motif, and it has been suggested that the minor form may be membrane associated. However, in recent experiments, Jenkins et al. (1993) used anti-lamin antibodies linked to immunobeads to deplete unfractionated egg extracts, as opposed to the cytosol depleted by Newport et al. (1990), with the result that NE assembly was permitted even in the absence of detectable lamins.

Additionally, lamin Liii may exist in different phosphorylated forms in egg cytoplasmic and nuclear fractions, since two-dimensional gel electrophoresis and immunoblotting show that while cytoplasmic Liii migrates as a single spot,

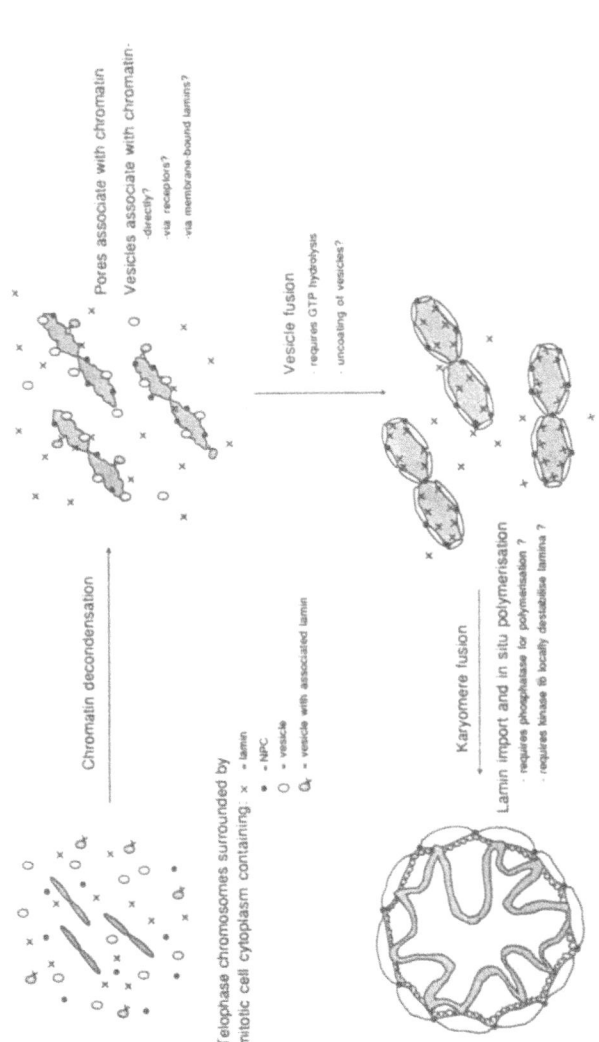

FIGURE 7. Stages in the assembly of a nuclear envelope in embryonic cells. lam = lamin Liii monomers; ves = membrane vesicles; npc = nuclear pore complex precursors. In the model, chromatin is remodeled, giving limited decondensation. Membranes and pore complex precursors then bind to the surface of individual chromosomes, giving rise to karyomeres consisting of nuclear envelopes surrounding an individual chromosome. Lamins are imported through the nuclear pores and polymerize on the inner surface of the nuclear membrane. The karyomeres fuse to form the embryonic nucleus (see Philpott *et al.*, 1991; Newport and Dunphy, 1992; Sheehan *et al.*, 1988; Newport *et al.*, 1990; Meier *et al.*, 1991; Hutchison *et al.*, 1988).

nuclear Liii migrates as three spots, each more basic than the cytoplasmic species (Crompton, Codd, and Hutchison, unpublished results). It was therefore predicted that lamina assembly should be prevented in egg extracts treated with phosphatase inhibitors, and it was observed that the protein phosphatase inhibitor microcystin-LR (MacKintosh et al., 1990) blocks lamina assembly in egg extracts at concentrations that permit nuclear membrane assembly (Crompton et al., unpublished results). These data support a model in which polymerization of Liii in the nucleus may be mediated by a nuclear protein phosphatase (analogous to the dis2 gene product of S. pombe) that dephosphorylates the lamin proteins once they are positioned at the chromatin-inner nuclear membrane interface. However, this would not exclude a role for Liii as a ligand that mediates chromatin/vesicle associations.

In conclusion, lamins are known to be essential for NE assembly in somatic cells, and might play a similar role in the early stages of NE formation in embryonic cells. Nuclear envelope assembly in the somatic cell systems described above appears to start with an association between decondensing chromatin and lamins A and C. This is followed by interaction of vesicle-associated lamin B with bound lamins A and C, then further vesicle addition and fusion to form an intact envelope (Figure 6). However, as many somatic cells lack lamins A and C, their expression being developmentally regulated, the universality of this model remains somewhat contentious. In embryos, faster nuclear reassembly in very rapid cell cycles may require only minimal association of lamins with chromatin in order to mediate vesicle binding. Subsequent lamina growth is then dependent on transport of soluble precursors into the intact nucleus, followed by in situ polymerization of these precursors, requiring the activity of nuclear protein phosphatases. Insertion of soluble lamins into the pre-existing polymerized lamina may require localized destabilization of the network, which could be mediated by the action of specific lamin kinases (see Figure 7).

3. NUCLEAR ENVELOPE DISASSEMBLY

Nuclear envelope disassembly has been studied extensively in a number of organisms, and the details vary widely between higher and lower eukaryotes (McIntosh and Koonce, 1989). Lower eukaryotes such as fission yeast undergo a so-called closed mitosis in which no discernible breakdown of the nuclear membrane is observed (McIntosh and Koonce, 1989; Tanaka and Kanbe, 1986). Nonetheless, structural proteins associated with the nuclear envelope are dispersed (Enoch et al., 1991; Hurt, 1988) possibly to accommodate the rapid distortion of nuclear shape that takes place as the nucleus is stretched along the length of the cell before dividing (Toda et al., 1981).

In contrast to fission yeast, partial NE breakdown (NEBD) is observed in some fungi (Zickler, 1981), dipteran species (Stafstrom and Staehlin, 1984), and differentiated cell types of higher vertebrates (R. G. Murray *et al.*, 1965). This involves apparent disruption of the envelope at the spindle poles (Stafstrom and Staehlin, 1984) accompanied by extensive fenestration of the membrane at other points (R. G. Murray *et al.*, 1965; Stafstrom and Staehlin, 1984; Zickler, 1981). Fenestration is apparently due to dissociation of nuclear pore complexes (Stafstrom and Staehlin, 1984) accompanied by breakdown of the nuclear lamina (Fuchs *et al.*, 1983), although other envelope proteins remain associated with the membrane (Harel *et al.*, 1989).

In many higher eukaryotes, complete NEBD is the norm. The NEBD occurs at prophase and follows a well-defined pathway. Initially the envelope appears stressed and blebs can be observed at its surface (Wilson, 1925; Dessev and Goldman, 1988; Stafstrom and Staehlin, 1984). This apparent weakening of the membrane accompanies spindle assembly and has been attributed to the effects of the growing aster (Wilson, 1925). Concurrent with these events the lamina disassembles (Dessev and Goldman, 1988; Gerace and Blobel, 1980; Miake-Lye and Kirschner, 1985; Ottaviano and Gerace, 1985) and the nuclear membrane dissociates from the chromatin before fragmenting into vesicles, which, based on their size, are indistinguishable from mitotic Golgi and ER (Coleman *et al.*, 1985; Kessel and Subtelny, 1981; R. G. Murray *et al.*, 1965; Zeligs and Wollman, 1979). The nuclear pores are also dispersed throughout the cytoplasm (Maul, 1977) as complexes, individual components, or membrane-associated components (Dabauvalle *et al.*, 1990; Finlay *et al.*, 1991).

The observations described above pose several important questions, namely, How is nuclear envelope breakdown controlled? What is the fate of the elements that make up the nuclear envelope? And what structural relationships might account for the differences between open and closed mitosis? Over the past eight years much has been learned about what controls nuclear lamina disassembly, a little has been discovered about the dynamics of nuclear membrane disassembly, but almost nothing has thus far been revealed about the disassembly or fate of the components of the nuclear pore. Because of the availability of extensive panels of antibodies to the lamins and because of their relative abundance as nuclear envelope proteins, it is not surprising that nuclear lamina disassembly has been so well studied, nor perhaps that its role in nuclear envelope breakdown has been seen as central (Gerace and Blobel, 1980; Burke and Gerace, 1986). In the following section we will review what is currently known about the mechanisms controlling lamina disassembly and question whether the loss of this structure can account for the subsequent dissolution of the nuclear pores and nuclear membranes.

3.1. Nuclear Lamina Breakdown

The nuclear lamina is resistant to extraction when isolated interphase nuclei are treated with detergents, salt, and nucleases (Aaronson and Blobel, 1975; Dwyer and Blobel, 1976; Ely *et al.*, 1978; Gerace *et al.*, 1978; Krohne *et al.*,

1978a,b; Scheer *et al.*, 1976). However, in mitotic cells (Ely *et al.*, 1978; Gerace and Blobel, 1980; Gerace *et al.*, 1978; Stick and Hausen, 1980) or in extracts of mitotic cells that induce NEBD of exogenous nuclei (Dessev *et al.*, 1988; Miake-Lye and Kirschner, 1985; Nakagawa *et al.*, 1989; Newport and Spann, 1987; Suprynowicz and Gerace, 1986), the lamins become soluble. Following NEBD, lamin A is solubilized to monomers, whereas lamin B can be recovered in a microsomal pellet (Gerace and Blobel, 1980) and can be observed by immu-nogold labeling to be associated with Golgi-like vesicles (Stick *et al.*, 1988). These observations have led to the suggestion that, although A-type lamins are dispersed at mitosis, B-type lamins remain associated with nuclear membranes throughout the cell cycle (Gerace and Blobel, 1980; Stick *et al.*, 1988).

In many organisms lamina disassembly is temporally correlated with a 4- to 7-fold increase in the steady-state level of lamin phosphorylation, as judged by quantitative incorporation of radiolabeled phosphate into lamin proteins and increases in the acidity of lamins on two-dimensional gels (Gerace and Blobel, 1980; Ottaviano and Gerace, 1985). In addition, conditions preventing lamin dephosphorylation also prevent telophase reassembly of the lamina (Burke and Gerace, 1986). Following these results a simple hypothesis was presented in which lamin phosphorylation or hyperphosphorylation by mitotically active ki-nases leads directly to lamina disassembly. Furthermore, the loss of the lamina results in structural weakening of the rest of the NE and, as a consequence, its dissolution (Burke and Gerace, 1986; Dessev *et al.*, 1988; Gerace and Blobel, 1980; Ottaviano and Gerace, 1985).

The hypothesis presented above can be subdivided into three distinct ques-tions. First, does increased phosphorylation of lamin proteins lead directly to disassembly of lamin filaments? Second, is increased lamin phosphorylation at mitosis a result of phosphorylation at specific mitotic sites or hyperphosphoryla-tion at sites that are phosphorylated throughout the cell cycle? Third, does lamina breakdown lead directly to NE dissolution? In conjunction with these central questions is the equally important question of which cellular kinases are respon-sible for lamin phosphorylation during mitosis. Each of these questions has recently been investigated using a range of biochemical and genetic approaches and will be dealt with in turn.

3.2. Does Lamin Phosphorylation Lead Directly to Lamina Disassembly?

Lamin phosphorylation and lamina disassembly have been studied exten-sively in cell-free extracts of mitotic cells (Nakagawa *et al.*, 1989; Newport and Spann, 1987; Suprynowicz and Gerace, 1986) and in extracts of interphase cells that can be induced to enter mitosis upon the addition of MPF (Dessev *et al.*, 1989; Lohka and Maller, 1985; Miake-Lye and Kirschner, 1985). Extracts are prepared in buffers containing calcium chelators (EGTA), phosphatase inhibitors (β-glycerophosphate), and Mg^{2+} ions, from cells lysed by centrifugation (Lohka and Maller, 1985; Miake-Lye and Kirschner, 1985; Newport and Spann, 1987) or homogenization (Dessev et al., 1989; Nakagawa et al., 1989; Suprynowicz and Gerace, 1986). To date these extracts have been prepared from parthenogenically activated and unfertilised Xenopus eggs (Lohka and Maller, 1985; Miake-Lye and Kirschner, 1985; Newport and Spann, 1987), Spisula (clam) oocytes (Dessev et al., 1989), CHO cells (Suprynowicz and Gerace, 1986), and chicken hepatoma cells (Nakagawa et al., 1989).

In each type of extract the dynamics of NEBD are broadly similar although some differences are observed, especially between somatic and embryonic extracts. Following the addition of exogenous nuclei, NEBD and chromosome condensation are induced in extracts prepared from Xenopus eggs at times varying between 30 and 60 minutes (Lohka and Maller, 1985; Miake-Lye and Kirschner, 1985; Newport and Spann, 1987). When NEBD is induced by the addition of MPF, both soluble and insoluble fractions of the extract are required to support NEBD (Lohka and Maller, 1985; Miake-Lye and Kirschner, 1985) whereas when endogenous MPF activity is present in cell extracts, postmicrosomal fractions are sufficient (Nakagawa et al., 1989; Newport and Spann, 1987; Suprynowicz and Gerace, 1986).

In all of the extracts described above, nuclear membrane dissolution is preceded by nuclear lamina disassembly, as judged by either immunoblotting or loss of anti-lamin immunofluorescence. Concomitant with the loss of anti-lamin immunofluorescence, increased lamin phosphorylation is also detectable by changes in the mobility of lamins on two-dimensional gels (Dessev et al., 1989; Miake-Lye and Kirschner, 1985; Suprynowicz and Gerace, 1986) and by radioactive phosphate labeling (Nakagawa et al., 1989; Newport and Spann, 1987; Suprynowicz and Gerace, 1986). Indeed, the changes in mobility that have been reported in both Xenopus and CHO cells are entirely due to phosphorylation and can be reversed by treatment with alkaline phosphatases (Miake-Lye and Kirschner, 1985; Ottaviano and Gerace, 1985). Furthermore, hyperphosphorylated lamins are always associated with the soluble rather than the insoluble fraction following detergent extraction of nuclei (Dessev et al., 1989; Ottaviano and Gerace, 1985; Suprynowicz and Gerace, 1986). Conditions preventing lamin phosphorylation (e.g., addition of nonhydrolysable analogues of ATP) also prevent lamina disassembly (Nakagawa et al., 1989; Suprynowicz and Gerace, 1986) while addition of γ-S-ATP to the extracts inhibits lamina reassembly at telophase (Burke and Gerace, 1986).

Although these results are highly suggestive of a primary role for phosphorylation in lamina disassembly, they do not in themselves constitute formal proof. Such proof was provided by Heald and McKeon (1990), who demonstrated that cDNA clones of lamins A and C, which were mutated to prevent phosphorylation at serine positions 22 and 392, produced mitotically stable lamins when transfected in CHO cells. Despite this finding, induction of mitotic phosphorylation patterns in interphase lamins following treatment of chicken lymphoma cells with the phosphatase inhibitor okadaic acid is not sufficient to induce lamina disassembly (Luscher et al., 1991). Thus, while phosphorylation of lamins A and C is required for lamina disassembly, it may not be sufficient. It has been noted that carboxyl methylation of lamins also occurs at mitosis (Chelsky et al., 1987, 1989), indicating that other modifications may be required for the completion of lamina disassembly.

3.3. Sites of Lamin Phosphorylation

In CHO cells and in *Xenopus* embryos, the level of lamin phosphorylation associated with interphase nuclei is 0.3–0.5 mol phosphate/mol lamin. In contrast, at mitosis, the level of phosphorylation increases to between 1.4 and 2.2 mol phosphate/mol lamins (Burke and Gerace, 1986; Miake-Lye and Kirschner, 1985; Ottaviano and Gerace, 1985). Tryptic mapping of the sites of phosphorylation in CHO cells also revealed that most were common to both interphase and mitotic lamins (Ottaviano and Gerace, 1985). These results have led to the suggestion that at mitosis, lamins become hyperphosphorylated at sites that may also be phosphorylated during interphase (Ottaviano and Gerace, 1985).

In contrast to these findings, the single *Drosophila* lamin species is phosphorylated at either one of two sites (Ser and Thr) during interphase but at a unique site (Ser) during mitosis, implying that different kinases are regulating phosphorylation at distinct sites in different phases of the cell cycle (Smith and Fisher, 1989). Newport and Spann (1987) have reported that lamin phosphorylation in mitotic extracts of *Xenopus* eggs is due to stable phosphorylation of new sites rather than to an increase in the steady-state level of previously phosphorylated sites. However, these results contrast with other studies using *Xenopus* cell-free extracts (Ward and Kirschner, 1990), where tryptic maps of $^{32}PO_4^{3}$-labeled exogenous lamin C indicated that there was one site of phosphorylation unique to mitosis, with four other sites phosphorylated during both interphase and mitosis. Furthermore, the amount of phosphate associated with the mitosis-specific site could not account for the overall increase in the steady-state level of phosphorylation (Ward and Kirschner, 1990). Directly contradictory results have been obtained from tryptic maps of chicken B-type lamins, which indicate that, although interphase sites of phosphorylation are suppressed at prophase, there is a rapid increase in phosphorylation at several new sites during the onset of mitosis (Luscher et al., 1991; Peter et al., 1990). The apparent discrepancies between these reports may reflect differences in the regulation of A-type and B-type lamins or perhaps between somatic and embryonic cells.

The sites of lamin phosphorylation in mitosis are predominantly serine residues (Dessev and Goldman, 1988; Ottaviano and Gerace, 1985; Smith and Fisher, 1989). In the case of human lamin C, some of these sites have been sequenced and are located in the N-terminal head domain (Ser 22) and the C-terminal tail domain (serines 392, 404, 406) on either side of the major α-helical coiled-coil domain (Ward and Kirschner, 1990). As lamin polymerization is known to occur by head-to-tail interactions (Georgatos *et al.*, 1988; Moir *et al.*, 1991), it is tempting to postulate that phosphorylation at these sites may contribute to lamina disassembly through electrostatic repulsion. By this mechanism, limited lamin phosphorylation in interphase would allow dynamic exchange of lamins between soluble and insoluble fractions. During mitosis, however, hyperphosphorylation at these sites would lead to total lamina disassembly. A precedent for such a mechanism has been observed in the mitotic disassembly of cytoplasmic intermediate filaments (Inagaki *et al.*, 1987; Evans, 1988; Geisler and Weber, 1988). In further support of this hypothesis, peptides corresponding to N- and C-terminal lamin phosphorylation sites have been shown to compete with lamins as substrates for mitotic kinases, thereby preventing lamin disassembly in cell-free extracts (Peter *et al.*, 1990; Ward and Kirschner, 1990).

Additionally, mutations of lamin A at Ser-22 and Ser-392 act cooperatively to prevent lamina disassembly in mitotic cells (Heald and McKeon, 1990). We have also observed that prolonged incubation of exogenous human nuclei in interphase *Xenopus* egg extracts leads to the disassembly of the existing lamina and its replacement with a new lamina consisting of the endogenous lamin Liii (Crompton and Hutchison, unpublished data). These data are compatible with the above hypothesis. When exogenous nuclei are added to egg extracts this type of exchange would eventually lead to replacement of the existing lamina, since lamin Liii is in vast excess. In addition, an overlap of phosphorylation patterns between interphase and mitotic lamins would also be expected (Ward and Kirschner, 1990). Thus, although there is evidence to suggest that B-type lamins at least are disassembled by specific phosphorylation mechanisms (Peter *et al.*, 1990; Luscher *et al.*, 1991), the hyperphosphorylation mechanism suggested by Ottaviano and Gerace (1985) still provides an attractive conceptual framework to account for the cell-cycle dynamics of A-type and embryonic lamins.

3.4. Does Lamina Disassembly Initiate Nuclear Envelope Breakdown?

In their original model, Gerace and Blobel (1980) suggested that lamina dissolution at prophase led directly to NEBD. A strong temporal correlation exists between lamina disassembly and nuclear membrane disassembly in some systems. For example, in extracts of *Spisula* oocytes induced to complete meiosis, lamin solubility increases linearly over a 40-min period while blebbing of the nuclear membrane, followed by its shrinkage and dissolution, occurs over the same period of time (Dessev *et al.*, 1989). In contrast to these results, Miake-

Lye and Kirschner (1985) have reported that in extracts of *Xenopus* eggs induced to enter mitosis, the membranes of exogenous CHO nuclei remain largely intact for 25 min after lamins have become fully phosphorylated. This apparent discrepancy can be accommodated if it is assumed that the dynamics of nuclear membrane disassembly are dependent not only upon lamin phosphorylation, which presumably leads to lamina disassembly, but also upon other titratable factors. Indeed, this has been demonstrated by Newport and Spann (1987), who report that nuclear membrane dissolution is inhibited in mitotic extracts of *Xenopus* eggs at nuclear concentrations that still permit lamina disassembly (see below).

It has recently been demonstrated that when chicken B-type lamins are expressed exogenously in *S. pombe*, they assemble into a nuclear lamina. This lamina breaks down at mitosis and reassembles during interphase under apparently physiological control mechanisms (Enoch *et al.*, 1991), but without leading to nuclear membrane breakdown during mitosis. These data imply that mitotic progression in fission yeast involves disassembly of some nuclear envelope structures, but not dissolution of membranes. This would be explicable if *S. pombe* lacked stoichiometric binding factors required for membrane dissolution. Additional support for a titratable vesicularization factor was recently obtained when it was shown that chicken fibroblast nuclei underwent lamina disassembly but not membrane dissolution on addition to a defined system consisting of buffer supplemented with purified cyclin B-cdc2 kinase (Peter *et al.*, 1990). This defined system might be unable to disassemble membranes if it lacks the putative vesicularization factor, which may be a clathrin-like molecule, but an alternative explanation is the absence of the putative membrane kinase. In contrast to these results, *Drosophila*, which has fenestrated nuclear membranes at mitosis, retains some lamin structures (T. Gonzalez and D. M. Glover, personal communication). In this instance the residual lamin structure may stabilize membranes that would otherwise break down.

The data described above imply that, although lamina disassembly may be necessary, it is not sufficient for nuclear membrane breakdown. To test whether lamina disassembly is a prerequisite for nuclear membrane dissolution, a mitotically stable lamina would have to be generated. Unfortunately, studies in which lamina disassembly at mitosis has been prevented, either by mutation (Heald and McKeon, 1990) or following the addition to cell-free extracts of peptides that compete with lamins as substrates for mitotic kinases (Peter *et al.*, 1990; Ward and Kirschner, 1990), have failed to address whether nuclear membrane dissolution still occurs. In studies where the lamina is fragmented by digestion of nuclear envelopes with trypsin (Dwyer and Blobel, 1976; Feldherr *et al.*, 1984; Hoeijmakers *et al.*, 1974), NPCs and pore-connecting fibrils become unstable on treatment with non-ionic detergents (Whytock *et al.*, 1990), suggesting, indeed, that lamina disassembly does destabilize other NE structures.

3.5. Serine-Threonine Kinase Activities Responsible for Lamin Phosphorylation at Mitosis

It has been noted that the majority of lamin phosphorylation occurs at serine residues (Dessev and Goldman, 1988; Ottaviano and Gerace, 1985), thus implying that serine-threonine kinases are responsible for regulating lamina assembly and disassembly. Because some sites of lamin phosphorylation have been identified (Luscher et al., 1991; Ottaviano and Gerace, 1985; Peter et al., 1990; Ward and Kirschner, 1990), it has been possible to test which mitotic kinases may account for lamin phosphorylation. One approach has been to incubate either bacterially expressed lamins or immunoprecipitated lamins with purified protein kinases to determine whether these kinases phosphorylate lamins at mitosis-specific sites (Luscher et al., 1991; Peter et al., 1990; Ward and Kirschner, 1990). An alternative approach is to test whether highly purified kinases are able to induce lamina disassembly (Dessev et al., 1991). Previous reports have implicated protein kinase C and the cAMP-dependent kinase in cell-cycle dependent modulation of cytoplasmic intermediate filament organization (Evans, 1988; Geisler and Weber, 1988; Inagaki et al., 1987; Lamb et al., 1989). However, although protein kinase C was able to phosphorylate both human lamin C and chicken lamins B_1 and B_2 in vitro, and the catalytic subunit of cAMP-dependent kinase was able to phosphorylate chicken B-type lamins, neither enzyme phosphorylated physiological sites (Ward and Kirschner, 1990; Peter et al., 1990). In contrast, Xenopus S6 kinase II specifically phosphorylated human lamin C at Ser 404, one of the sites implicated in disassembly (Ward and Kirschner, 1990). Moreover, sequences surrounding this residue conform to the consensus sequence for phosphoacceptor sites of S6 kinase (Erikson and Maller, 1988). However, purified rat S6 kinase was not able to phosphorylate either of the chicken B-type lamins (Peter et al., 1990).

Several authors have reported that p34[cdc2]-kinase specifically phosphorylates B-type lamins. Peter et al. (1990) observed that purified starfish cdc2-kinase was able to phosphorylate all of the mitosis-specific sites on chicken lamins B_1 and B_2, whereas p34[cdc2]-kinase from human fibroblasts was only able to phosphorylate some of the sites required for depolymerization of chicken lamin B_2 (Luscher et al., 1991). Spisula oocyte GVs contain only a single lamin species L67, which can be efficiently disassembled in vitro (Dessev et al., 1989) and upon incubation with purified p34[cdc2]-kinase (Dessev et al., 1991). Support for the notion that p34[cdc2]-kinase is a physiological lamin kinase was also derived from studies in which chicken B-type lamins were expressed exogenously in S. pombe. Extracts prepared from wild-type cells, but not cdc2[ts] mutant cells incubated at the restrictive temperature, readily incorporated phosphate into lamin-B molecules (Enoch et al., 1991). These data suggest that p34[cdc2]-kinase may be sufficient to account for lamin phosphorylation at mitosis. Despite this evidence, only the N-terminal phosphorylation site (Ser 22 in human lamin C; Ser 16 in chicken lamin B) conforms to the consensus sequence for phosphoacceptor sites of p34[cdc2]-kinase (Shenoy et al., 1989; Peter et al., 1990). As many of the studies described above use isolated nuclear material as substrates for kinase assays, it is possible that this contains endogenous protein kinases activated by p34[cdc2]-kinase. Indeed, evidence shows that some protein kinases are specifically associated with NE preparations (Lam and Kasper, 1979). Thus, while it is clear that cdc2-kinase is responsible for direct phosphorylation of at least some of the sites associated with lamina disassembly, another kinase, possibly S6-kinase, may also be involved.

3.6. Nuclear Membrane Disassembly

Studies in which nuclear membrane dynamics have been observed in cell-free extracts of Xenopus eggs have pointed to the existence of a mitotic kinase activity other than p34[cdc2], which catalyzes nuclear membrane breakdown by modifying integral membrane proteins. When the nuclear envelope precursor pool NEP-B (see above) is incubated in mitotic extracts in the presence of ATPγS, vesicles in the pool become irreversibly phosphorylated, and are subsequently incapable of binding to chromatin in reconstitution experiments. However, highly purified fractions of p34[cdc2]-kinase, S6-kinase, and mitogen-activated protein kinase (MAP-kinase) were all incapable of phosphorylating NEP-B in vitro (Vigers and Lohka, 1992), so the NEP-B kinase has still to be identified. More recently, Newport and colleagues have developed an in vitro system for the identification of kinases and phosphatases necessary for modifying those integral membrane proteins required for interaction with chromatin (Pfaller et al., 1991; Newport and Dunphy, 1992). Inhibition of protein phosphatase activity in this system leads to the dissolution of the nuclear membrane, again indicating that increased phosphorylation of the nuclear membrane is required for its dissolution (Pfaller et al., 1991). These authors also demonstrate that p34[cdc2] is not directly involved in membrane dissolution but do show that it is required for the activation of the "membrane kinase" at mitosis. Although a membrane kinase has not yet been identified in Xenopus egg extracts, a Ca^{2+}-camodulin-dependent kinase has been shown to be required for NEBD in sea urchin eggs (Batinger et al., 1990). As calcium transients coincide with NEBD in amphibian embryos, calcium-dependent kinases may prove to be universally required for NEBD.

As yet there are no data to indicate the target protein of the "membrane kinase." Several potential nuclear membrane proteins have been reported, which include a lamin B receptor (Worman et al., 1988), an isoform of Xenopus lamin Liii (Doring and Stick, 1990) and a family of integral membrane proteins whose function is unclear (Senior and Gerace, 1988; Padan et al., 1990). Membrane-associated putative lamin receptors (see Table II) or NPC proteins that bind DNA directly (nup153; Sukegawa and Blobel, 1993) are also possible targets for phosphorylation by the membrane kinase. Comparisons between phosphorylation of these proteins during interphase and mitosis have thus far proved ambiguous (Vigers and Lohka, 1992).

Although the putative membrane kinase may initiate membrane dissolution, other factors are also involved in the mechanism of vesicularization. Newport and Spann (1987) have reported that nuclear membrane dissolution becomes inhibited in *Xenopus* egg extracts at nuclear concentrations equivalent to those expected at the mid-blastula transition. If membrane dissolution was catalyzed by a high-turnover enzyme, upon increasing nuclear concentrations, a steady-state level of dissolution should be reached. Inhibition of membrane dissolution implies the depletion of a titratable stoichiometric binding factor (Newport, 1987) such as a clathrin-like molecule, in a reaction analogous to Golgi vesicularization (Griffiths *et al.*, 1985; Orci *et al.*, 1986), further suggesting that nuclear membrane vesicles may become "coated" during mitosis. Lack of a stoichiometric binding factor may explain why organisms such as *S. pombe* undergo closed mitoses. In some cells that have been arrested in mitosis, structures known as confronting cisternae have been reported (reviewed by Ghadially, 1988). These structures, normally associated with cellular pathology, consist of tubes of membrane cisternae that form a sandwich partially surrounding an electron-dense core (Wheatley, 1990, 1992). The origin of these structures is unclear, but it has been speculated that they derive from prophase nuclear membranes and that the electron-dense material between confronting cisternae is proteinaceous (Barton *et al.*, 1971; R. G. Murray *et al.*, 1965). Perhaps this proteinaceous material represents the remnants of a membrane coat in cells that are trying unsuccessfully to reform nuclear membranes.

3.7. Nuclear Envelope Proteins That Segregate to the Perichromosomal Region at Mitosis

At least two components of human fibroblast NEs have been reported to segregate to the perichromosomal region during mitosis. These include perichromin, a 33 kDa peptide identified using a systemic lupus erythematosis (SLE) autoimmune serum, which associates with the lamina during interphase but coats the outer surface of chromosomes at mitosis (McKeon *et al.*, 1984), and a 40 kDa protein identified using MAbs (Wataya-Kaneda *et al.*, 1987). Cell-fusion studies, in which premature chromosome condensation is induced in S-phase, G1, and G2 cells fused to mitotic cells, indicated that the association of NE antigens with chromosomes can occur only when DNA replication is complete (Wataya-Kaneda *et al.*, 1987). This supports earlier ultrastructural data suggesting that the initial steps of chromosome condensation in late G2 or early prophase occur at a position close to the nuclear envelope (Comings and Okada, 1971; Gurley et al., 1978; Foe and Alberts, 1983). Interphase chromosomes are organized into supercoiled domains of 30–80 kb (Ide et al., 1975; Benyajati and Worcel, 1976; Paulson and Laemmli, 1977; Vogelstein et al., 1980), which are immobilized at the NE (Fawcett, 1966; Gurley et al., 1978; Olins and Olins, 1979; Hancock and Hughes, 1982). It has been suggested that proteins such as perichromin bind to these domains during mitosis to ensure their reassociation with the NE at telophase (McKeon et al., 1984). More recent data have suggested that although chromosome domains are organized at the nuclear periphery, these are not directly associated with the lamina (Paddy et al., 1990). These authors proposed that proteins such as perichromin mediate an association between the lamina and DNA. Studies using cell-free extracts of Rana eggs have emphasized the importance of interactions between the surface of chromosomes and membrane vesicles during NE reassembly (Lohka and Masui, 1984). Perhaps the segregation of perichromin to the chromosomal fraction and lamin B with membrane vesicles ensures the proper reorganization of interphase chromosomes.

4. CONCLUSIONS

4.1. Stages of Nuclear Envelope Formation

The order of steps in NE assembly and the final structures formed are morphologically almost indistinguishable between embryonic and somatic cells. Figure 6 shows schematic representation of the events of NE assembly in somatic cells, and Figure 7 collates data on NE formation from embryonic systems. The fundamental stages in each are very similar: Mitotic chromosomes or gametic pronuclei first decondense (Philpott et al., 1991), possibly becoming coated with lamins (Krohne and Benavente, 1986; Ulitzer et al., 1992), pre-pores (Sheehan et al., 1988), or other receptor proteins for various membrane proteins. Membrane vesicles then bind to the decondensing chromatin (Pfaller et al., 1991; Boman et al., 1992a,b; Newport and Dunphy, 1992) directly or via interactions either between membrane proteins and chromatin, or between chromatin-bound and membrane-associated lamins. Vesicles then fuse to form an intact membrane tightly surrounding the chromatin (Pfaller et al., 1991; Boman et al., 1992a,b; Newport and Dunphy, 1992). Further NE growth involves import of soluble lamin dimers or tetramers via nuclear pore complexes (NPCs) (Dabauvalle et al., 1990; Newport et al., 1990; Meier et al., 1991), and polymerization of these lamins within the nucleus to form the submembranous network of the nuclear lamina. The timing of NPC assembly during this sequence is unclear, but it is known that the number of NPCs doubles during S-phase, when NPCs are added to the intact NE (Maul *et al.*, 1972) by an unknown mechanism. The pathway of NE reassembly differs only slightly on telophase chromosomes. Here, individual chromosomes are first enclosed by membrane to form karyomeres. Each karyomere then fuses with its neighbors to form a highly invaginated nucleus (Montag *et al.*, 1988).

4.2. Stages in Nuclear Envelope Breakdown

The stages of NEBD are summarized in Figure 8. The p34^{cdc2}-kinase is activated by association with cyclin B and phosphorylation of Thr161 accompanied by dephosphorylation of Thr14 and Tyr15 (Hunt, 1991). Active p34^{cdc2}-kinase may directly phosphorylate the lamins, initiating their disassembly (Nigg, 1992) and it also activates other kinases that phosphorylate nuclear membranes (Pfaller *et al.*, 1991; Vigers and Lohka, 1992). Nuclear membrane dissolution is completed on vesicle formation, possibly by association of NE membranes with clathrin-like molecules (Newport and Spann, 1987). The regulation of nuclear pore disassembly is as yet unclear.

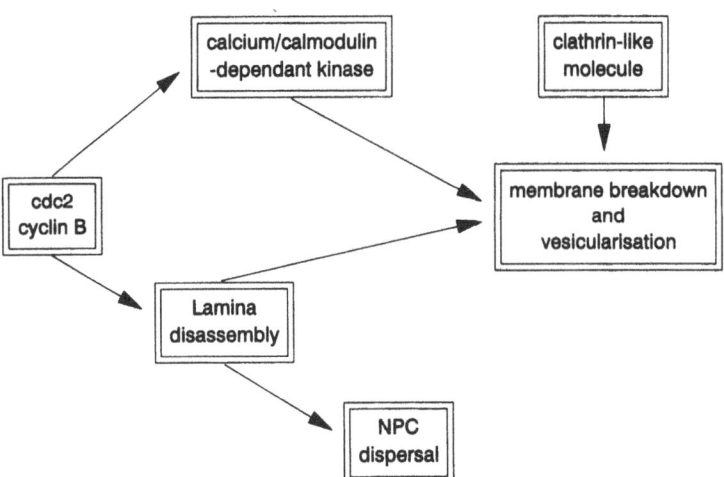

FIGURE 8. Stages in nuclear envelope breakdown. The p34^{cdc2}/cyclin B initiates lamin phosphorylation and activates the membrane kinase. Lamina disassembly follows phosphorylation at sites in the N- and C-terminal domains, leading to the destabilization of nuclear membranes and nuclear pores. The membrane kinase inhibits the interaction of membranes with chromatin, and following association with a clathrin-like molecule vesicles are formed. NPC dispersal occurs through a poorly characterized mechanism (see Peter *et al.*, 1990; Vigers and Lohka, 1992; Pfaller *et al.*, 1991; Boman *et al.*, 1992a,b).

4.3. Importance of Dynamic Changes in Nuclear Envelope Integrity

The interphase nuclear envelope provides a passive barrier against mixing of components of the nucleus and cytoplasm, and protects fragile, extended DNA molecules from cytoplasmic shearing forces, especially in motile or contractile cells. In addition, the NE permits selective accumulation of components necessary for nuclear metabolism. The importance of an intact NE is shown in the inability of *Xenopus* sperm pronuclei to support DNA replication when the NE is incomplete, lacking a lamina, or artificially permeabilized (Cox and Leno, 1990; Newport *et al.*, 1990; Meier *et al.*, 1991; Cox, 1992). In mitosis of higher eukaryotes, the NE breaks down to permit segregation of chromosomes to daughter cells, then re-formation of two daughter nuclei takes place.

The acts of NEBD and NE re-formation in the past have been viewed as purely physical processes. However, more recent hypotheses of cell-cycle control have ascribed a regulatory role to the dynamic changes in the state of the nuclear envelope. The licensing factor (LF) model first proposed by Blow and Laskey (1988) suggests that the onset of DNA replication in the S-phase of interphase is regulated by a positively acting factor that can gain access to the chromatin only during mitosis when the NE is broken down (Figure 9). Such a factor would be usually cytoplasmic and unable to enter an intact interphase nucleus, presumably because it lacks a nuclear localization sequence (or, alternatively, by being cytoplasmically anchored). This factor would be stable within the nucleus if and only if it were bound to chromatin. Sites where the factor is bound would then serve as "origins" for DNA replication, once the NE had reformed and was intact. Initiation of replication or passage of a replication fork would cancel the factor, so that at the end of S-phase the nucleus lacks LF. Hence, no further DNA replication can occur until the NE has once again broken down, permitting access of LF to the chromatin.

Support for the licensing factor model has derived from observational and experimental approaches. In artificial giant nuclei composed of clusters of de-membranated chick erythrocyte nuclei incubated in *Xenopus* egg extracts, a single, all-encompassing NE, derived from *Xenopus* egg vesicles, is found to define the unit of DNA replication (Leno and Laskey, 1991). Early cell-fusion studies by Rao and Johnson (1970) suggested that entry into S-phase was controlled by a positively acting factor, and it was observed that G2 nuclei were refractory to the effects of such a factor. When nuclei with intact nuclear envelopes are prepared from synchronous mammalian cell cultures, only those in G1 (containing LF?) are found to be capable of DNA synthesis in *Xenopus* egg extracts, whereas those in G2 (having used up all LF?) cannot undergo DNA synthesis unless their nuclear envelopes are artificially permeabilized (Leno *et al.*, 1992). Candidates for LF are being sought, and the product of a cell-cycle gene CDC46 from *S. cerevisiae* has been noted to show the cell-cycle-dependent

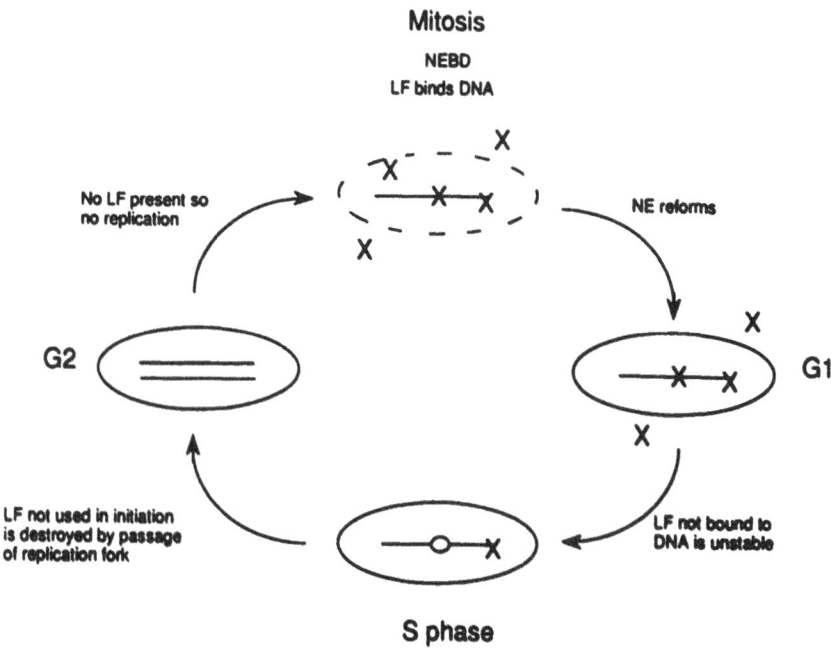

FIGURE 9. Licensing factor (LF) model (adapted from Blow and Laskey, 1988). Licensing factor (X) binds to DNA during mitosis. The NE re-forms in telophase, and further LF cannot enter the nucleus. Only LF bound to DNA is stable; this LF can support initiation of DNA replication. LF is destroyed on initiation or on passage of a replication fork, leaving G2 nuclei devoid of any LF. Therefore, DNA replication cannot occur again until nuclei have undergone NEBD at mitosis, when they acquire fresh LF.

changes in distribution predicted for LF (Hennessey *et al.*, 1990) and is involved in the initiation of DNA replication. Triggering of LF bound to specific chromatin sites must occur at the G1/S-phase border, and may result in a change in the phosphorylation or oxidation status of LF, possibly brought about by a cell-cycle-regulated, cyclin-dependent protein kinase such as cyclin E-cdk2. Very recent work has suggested that passage of replication forks may negatively mark replicated DNA to prevent illegitimate re-replication, and that a positively acting factor, possibly LF, is required to cancel this negative label (Cox, Madine, and Laskey, in preparation). Candidates for this negative marker include RCC1 (first identified as the product of the BN2ts gene by Nishimoto *et al.*, 1978; for review of RCC1, see Dasso, 1993), whose yeast homologue, pim1, is present at approximately one copy per nucleosome (Matsumoto and Beach, 1991).

These recent observations have emphasized the importance of correct NE dynamics in regulation of the cell cycle: Rather than being a passive accompaniment to cell-cycle changes, NEBD and NE re-formation may play a crucial regulatory role.

4.4. Future Prospects

Refinement of the techniques of electron microscopy, especially in the field of scanning EM, should allow the internal and external structures of the nuclear envelope to be clarified at even greater resolution. Already, high-resolution SEM has shown the presence of inner basket structures not previously suspected (Goldberg and Allen, 1992). The interaction between the lamina and nuclear pore complexes, and between individual lamin filaments should soon become visible, and consequently elucidated.

Studies are already underway to identify the receptors on membrane vesicles and chromatin that permit NE assembly, and regulation of NE assembly and disassembly by modification—for example, phosphorylation—of such receptors has already been proposed (Pfaller et al., 1991; Vigers and Lohka, 1992; Newport and Dunphy, 1992). Once such receptors have been identified at the molecular level, the processes of NE assembly and NEBD should be amenable to careful dissection, possibly yielding information on precise cell-cycle control mechanisms governing NE dynamics.

The assembly of nuclear pore complexes is only partially understood, and even less is currently known concerning NPC disassembly at mitosis. The cloning of several presumptive pore components including p62, gp210, NSP1, and NUP1 and recent identification of further NPC components (Wente et al., 1992; Wimmer et al., 1992) should allow fine analysis of the stages of NPC assembly and disassembly. The finding of a presumptive pore protein that interacts with DNA directly in a zinc-dependent manner (Sukegawa and Blobel, 1993) suggests exciting possibilities for new mechanisms in mitotic dissociation of the NE from chromatin. The availability of methodologies that permit construction of nuclei lacking a lamina (Newport et al., 1990; Meier et al., 1991; Jenkins et al., 1993) now provides possibilities for reconstruction of different type of intermediate filament networks at the inner surface of the nuclear envelope. This approach may present opportunities to modify the organization of chromatin domains, normally anchored at the NE, and therefore to analyze their function.

As to the role of the NE in cell-cycle regulation, especially in the onset of S-phase, further more stringent tests of the licensing factor model are required. These would include the use of recent liposome technology to allow the resealing of artificially permeabilized nuclei. It is expected that most progress will be made using cell-free systems in which nuclei can be manipulated with relative ease. Xenopus egg extracts, which have already yielded so much information concerning the processes of nuclear envelope assembly and disassembly, may prove to be the source of many more interesting observations. It is highly likely

that the next two years will also see the purification of the kinase and phosphatase regulating vesicle association with chromatin, and a component of the vesicle fusion pathway has recently been identified (Boman *et al.*, 1992b). Similar
cell-free extracts of *Drosophila* embryos should also prove to be powerful analytical tools when prepared from mutant strains defective in execution of various
cell-cycle events. Such extracts have already indicated that reported differences
between embryonic and somatic cells in the stages of NE reassembly following
mitosis are not as distinct as previously thought (compare Ulitzer *et al.*, 1992
with Newport *et al.*, 1990; Meier *et al.*, 1991).

Overall, it is to be expected that major advances will be made in the next
few years in the field of nuclear envelope dynamics, and that an integrated view
of an evolutionarily conserved mechanism for nuclear envelope assembly, as well
as disassembly, will emerge.

5. REFERENCES

Aaronson, R. P., and Blobel, G., 1974, On the attachment of the nuclear pore complex, *J. Cell Biol.*
62:746–754.

Aaronson, R. P., and Blobel, G., 1975, Isolation of pore complexes in association with a lamin,
Proc. Natl. Acad. Sci. USA 72:1007–1011.

Adam, S. A., Lobi, T. J., Mitchell, M. A., and Gerace, L. 1989, Identification of specific binding
proteins for a nuclear location sequence, *Nature* 337:276–279.

Aebi, U., Cohn, J., Buhle, L., and Gerace, L., 1986, The nuclear lamina is a meshwork of
intermediate-type filaments, *Nature* 323:560–564.

Aebi, U., Jarnik, M., Reichelt, R., and Engel, A., 1990, Structural analysis of the nuclear pore
complex by conventional and scanning transmission electron microscopy, *EMSA Bull.* 20:69–
76.

Akey, C. W., 1989, Interactions and structure of the nuclear pore complex revealed by cryo-electron
microscopy, *J. Cell Biol.* 109:955–970.

Akey, C. W., 1990, Visualization of transport-related configurations of the nuclear pore transporter,
Biophys. J. 58:341–355.

Akey, C. W., and Goldfarb, D. S., 1989, Protein import through the nuclear pore complex is a
multistep process, *J. Cell Biol.* 109:971–982.

Arion, D., Meier, L., Brizuela, L., and Beach, D., 1988, cdc2 is a component of the M phase-
specific histone H1-kinase: Evidence for identity with MPF, *Cell* 55:371–378.

Aris, J. P., and Blobel, G., 1989, Yeast nuclear envelope proteins cross-react with an antibody
against mammalian pore complex proteins, *J. Cell Biol.* 108:2059–2067.

Bailer, S. M., Eppenberger, H. M., Griffiths, G., and Nigg, E. A., 1991, Characterisation of a 54
kD protein of the inner nuclear membrane: Evidence for cell cycle-dependent interaction with
the nuclear lamina, *J. Cell Biol.* 114:389–400.

Barton, A. D., Kisieleski, W. E., Wasserman, F., and Mackevicius, F., 1971, Experimental modification of structures at the periphery of the liver cell nucleus, *Z. Zellforsch. Mikrosk. Anat.*
115:299–306.

Batinger, C., Alderton, J., Poerric, M., Schulman, H., and Steinhardt, R. A., 1990, Involvement of
Ca^{++}/calmodulin-dependent kinase in NEBD in sea urchin eggs, *J. Cell Biol.* 111:1763–1773.

Beach, D. H., Durkacz, B., and Nurse, P. A., 1982, Functionally homologous cell cycle control genes in budding and fission yeast, *Nature* **300**:706–709.

Beck, L. A., Hosick, T. J., and Sinensky, M., 1988, Incorporation of a product of mevalonic acid metabolism into proteins of Chinese hamster ovary cell nuclei, *J. Cell Biol.* **107**:1307–1316.

Beck, L. A., Hosick, T. J., and Sinensky, M., 1990, Isoprenylation is required for the processing of the lamin A precursor, *J. Cell Biol.* **110**:1489–1499.

Beckers, C.J.M., and Balch, E. W., 1989, Calcium and GTP: Essential components in vesicular trafficking between the endoplasmic reticulum and the Golgi apparatus, *J. Cell Biol.* **108**:1245–1256.

Benavente, R., and Krohne, G., 1986, Involvement of nuclear lamina in post-mitotic reorganization of chromatin as demonstrated by microinjection of lamin antibodies, *J. Cell Biol.* **103**:1847–1854.

Benavente, R., Krohne, G., and Franke, W. W., 1985, Cell type-specific expression of nuclear lamina proteins during development of *Xenopus* laevis, *Cell* **41**:177–190.

Benyajati, C., and Worcel, A., 1976, Isolation, characterization and structure of the folded interphase genome of *Drosophila melanogaster*, *Cell* **9**:393–407.

Berrios, M., and Avilion, A. A., 1990, Nuclear formation in a *Drosophila* cell-free system, *Exp. Cell Res.* **191**:64–70.

Blow, J. J., and Laskey, R. A., 1986, Initiation of DNA replication in nuclei and purified DNA by a cell-free extract of *Xenopus* eggs, *Cell* **47**:577–587.

Blow, J. J., and Laskey, R. A., 1988, A role for the nuclear envelope in controlling DNA replication within the cell cycle, *Nature* **332**:546–548.

Blow, J. J., and Sleeman, A. M., 1990, Replication of purified DNA in *Xenopus* egg extract is dependent on nuclear assembly, *J. Cell Sci.* **95**:383–391.

Boman, A. L., Delannoy, M. R., and Wilson, K. L., 1992a, GTP hydrolysis is required for vesicle fusion during nuclear envelope assembly *in vitro*, *J. Cell Biol.* **116**:281–294.

Boman, A. L., Taylor, T. C., Melancon, P., and Wilson, K. L., 1992b, A role for ADP-ribosylation factor in nuclear vesicle dynamics, *Nature* **358**:512–514.

Bradbury, E. M., Inglis, R. J., Matthews, H. R., and Sarner, N., 1973, Phosphorylation of very-lysine-rich histone in *Physarum polycephalum*, *Eur. J. Biochem.* **33**:131–139.

Bradbury, E. M., Inglis, R. J., and Matthews, H. R., 1974, Control of cell division by very-lysine-rich histone (f1) phosphorylation, *Nature* **247**:257–261.

Bridger, J. M., Kill, I. R., O'Farrell, M., and Hutchison, C. J., 1993, Internal lamin structures within G1 nuclei of human dermal fibroblasts, *J. Cell Sci.* **104**:297–306.

Burke, B., 1990, On the cell-free association of lamins A and C with metaphase chromosomes, *Exp. Cell Res.* **186**:169–176.

Burke, B., and Gerace, L., 1986, A cell-free system to study reassembly of the nuclear envelope at the end of mitosis, *Cell* **44**:639–652.

Chaly, N., Bladon, G., Setterfield, J. E., Little, J. G., Kaplan, G., and Brown, D. L., 1984, Changes in distribution of nuclear matrix antigens during the mitotic cell cycle, *J. Cell Biol.* **99**:661–671.

Chelsky, D., Olson, J. F., and Koshland, D. E., 1987, Cell-cycle-dependent methyl esterification of lamin B, *J. Biol. Chem.* **262**:4304–4309.

Chelsky, D., Sobotka, C., and O'Neill, C. L., 1989, Lamin B methylation and assembly into the nuclear envelope, *J. Biol. Chem.* **264**:7637–7643.

Coleman, A., Jones, E., and Heasman, J., 1985, Meiotic maturation in *Xenopus* oocytes: A link between the cessation of protein secretion and the polarized disappearance of Golgi apparati, *J. Cell Biol.* **101**:313–318.

Comings, D. E., and Okada, T. A., 1971, Condensation of chromosomes onto the nuclear membrane during prophase, *Exp. Cell Res.* **63**:471–473.

Cook, P. R., 1988, The nucleoskeleton: Artifact, passive framework or active site? *J. Cell Sci.* **90**: 1–6.

Cox, L. S., 1992, DNA replication in cell-free extracts from *Xenopus* eggs is prevented by disrupting nuclear envelope function, *J. Cell Sci.* **101**:43–53.

Cox, L. S., and Laskey, R. A., 1991, DNA replication occurs at discrete sites in pseudonuclei assembled from purified DNA *in vitro, Cell* **66**:271–275.

Cox, L. S., and Leno, G., 1990, Extracts of eggs and oocytes of *Xenopus laevis* differ in capacities for nuclear assembly and DNA replication, *J. Cell Sci.* **97**:177–184.

Crevell, G., and Cotterill, S., 1991, DNA replication in cell-free extracts from *Drosophila melanogaster, EMBO J.* **10**:4361–4369.

Dabauvalle, M-C., Shultz, B., Scheer, U., and Peters, R., 1988a, Inhibition of nuclear accumulation of karyophilic proteins in living cells by microinjection of lectin wheat germ agglutinin, *Exp. Cell Res.* **174**:291–296.

Dabauvalle, M-C., Benavente, R., and Chaly, N., 1988b, Monoclonal antibodies to a Mr. 68,000 pore complex glycoprotein interfere with nuclear uptake in *Xenopus* oocytes, *Chromsoma* **97**:193–197.

Dabauvalle, M-C., Loos, K., and Scheer, U., 1990, Identification of a soluble precursor complex essential for nuclear pore assembly *in vitro, Chromosoma* **100**:56–66.

Dabauvalle, M-C., Loos, K., Merkert, H., and Scheer, U., 1991, Spontaneous assembly of pore complex-containing membranes (annulate lamellae) in *Xenopus* egg extract in the absence of chromatin, *J. Cell Biol.* **112**:1073–1082.

Dasso, M., 1993, RCC1 in the cell cycle: The regulator of chromosome condensation takes on new roles, *Trends Biochem. Sci.* **18**:96–101.

Davis, L. I., and Blobel, G., 1986, Identification and characterization of a nuclear pore complex protein, *Cell* **45**:699–709.

Davis, L. I., and Blobel, G., 1987, Nuclear pore complex contains a family of glycoproteins that includes p62: Glycosylation through a previously unidentified cellular pathway, *Proc. Natl. Acad. Sci. USA* **84**:7552–7556.

Davis, L. I., and Fink, G. R., 1990, The NUP1 gene encodes an essential component of the yeast nuclear pore complex, *Cell* **61**:965–978.

Dessev, G., 1992, Nuclear envelope structure, *Curr. Opin. Cell Biol.* **4**:430–453.

Dessev, G., and Goldman, R. D., 1988, Meiotic breakdown of nuclear envelope in oocytes of *Spisula solidissima* involves phosphorylation and release of nuclear lamins, *Dev. Biol.* **130**:543–550.

Dessev, G., and Goldman, R. D., 1990, The oocyte lamin persists as a single major component of the nuclear lamina during embryonic development of the surf clam, *Int. J. Dev. Biol.* **34**:267–274.

Dessev, G., Iorchera, C., Tasheva, B., and Goldman, R., 1988, Protein kinase activity associated with the nuclear laminae, *Proc. Natl. Acad. Sci. USA* **85**:2994–2998.

Dessev, G., Palazzo, R., Rebhun, L., and Goldman, R., 1989, Disassembly of the nuclear envelope of *Spisula* oocytes in a cell-free system, *Dev. Biol.* **131**:496–504.

Dessev, G., Iorchera-Dessev, C., Bischoff, J. R., Beach, D., and Goldman, R., 1991, A complex containing p34cdc2 and cyclin C phosphorylates the nuclear lamina and disassembles nuclei of clam oocytes *in vitro, J. Cell Biol.* **112**:523–533.

Dingwall, C., 1990, Plugging the nuclear pore, *Nature* **346**:512–514.

Dingwall, C., and Laskey, R. A., 1986, Protein import into the cell nucleus, *Annu. Rev. Cell Biol.,* **2**:367–390.

D'Onofrio, M., Lee, M. D., Starr, C. M., Miller, M., and Hanover, J. A., 1991, The gene encoding rat nuclear pore glycoprotein p62 is intronless, *J. Biol. Chem.* **266**:11980–11985.

D'Onofrio, M., Starr, C. M., Park, M. K., Holt, G. D., Haltiwanger, R. S., Hart, G. W., and Hanover, J. A., 1988, Partial cDNA sequence encoding a nuclear pore protein modified by O-linked *N*-acetylglucosamine, *Proc. Natl. Acad. Sci. USA* **85**:9595–9599.

Doring, V., and Stick, R., 1990, Gene structure for nuclear lamin Liii of *Xenopus laevis:* A model for the evolution of IF proteins from a lamin-like ancestor, *EMBO J.* **9**:4073–4081.

Draetta, G., and Beach, D., 1988, Activation of cdc2 protein kinase during mitosis in human cells: Cell cycle-dependent phosphorylation and subunit rearrangement, Cell **54**:17–26.

Draetta, G., Brizuela, L., Potashkin, J., and Beach, D., 1987, Identification of p34 and p13, human homologs of the cell cycle regulators of fission yeast encoded by cdc2+ and suc1+, *Cell* **50**:17–26.

Dunphy, W. G., and Newport, J. W., 1989, Fission yeast p13 blocks mitotic activation and tyrosine dephosphorylation of the *Xenopus* cdc2 protein kinase, *Cell* **58**:181–191.

Dwyer, N., and Blobel, G., 1976, A modified procedure for the isolation of a pore complex-lamina fraction from rat liver nuclei, *J. Cell Biol.* **70**:581–591.

Ely, S., D'Arcy, A., and Jost, E., 1978, Interaction of antibodies against envelope associated proteins from rat liver nuclei with rodent and human cells, *Exp. Cell Res.* **116**:325–331.

Enoch, T., Matthias, P., Nurse, P., and Nigg, E. A., 1991, p34[cdc2] acts as a lamin kinase in fission yeast, *J. Cell Biol.* **112**:797–807.

Erikson, E., and Maller, J. L., 1988, Substrate specificity of ribosomal protein kinase II from *Xenopus* eggs, *Second Messengers Phosphoproteins* **12**:135–143.

Evans, R. M., and Fink, L. M., 1982, An alteration in the phosphorylation of vimentin-type intermediate filaments is associated with mitosis in cultured mammalian cells, *Cell* **29**:43–52.

Evans, R. M., 1988, Cyclic AMP-dependent protein kinase-induced vimentin disassembly involves modification of the N-terminal domain of intermediate filament sub-units, *FEBS Lett.* **234**:73–78.

Evans, T., Rosenthal, E., Youngbloom, J., Distel, D., and Hunt, T., 1983, Cyclin: A protein specified by maternal mRNA in sea urchin eggs that is destroyed at each cleavage division, *Cell* **33**:389–396.

Fantes, P., 1988, Intersecting cell-cycles, *Trends Genet.* **4**:275–276.

Farnsworth, C. C., Wolda, S. L., Gelb, M. H., and Glomset, J. A., 1989, Human lamin B contains a farneslyated cysteine residue, *J. Biol. Chem.* **264**:20422–20429.

Fawcett, D. W., 1966, On the occurrence of fibrous lamina on the inner aspect of the nuclear envelope in certain cells of vertebrates, *Am. J. Anat.* **119**:129–146.

Featherstone, C., Darby, M. K., and Gerace, L., 1988, A monoclonal antibody against the nuclear pore complex inhibits nucleocytoplasmic transport of protein and RNA *in vivo*, *J. Cell Biol.* **107**:1289–1297.

Feldherr, C. M., Kallenback, E., and Schultz, N., 1984, Movement of a karyophilic protein through the nuclear pores of oocytes, *J. Cell Biol.* **99**:2216–2222.

Fey, E. G., Wan, K. M., and Penman, S., 1984, Epithelial cytoskeletal framework and nuclear matrix-intermediate filament scaffold: Three-dimensional organization and protein composition, *J. Cell Biol.* **98**:1973–1984.

Fields, A., and Shaper, J., 1988, A major 62-kD intranuclear matrix polypeptide is a component of metaphase chromosomes, *J. Cell Biol.* **107**:833–840.

Filson, A. J., Lewis, A., Blobel, G., and Fisher, P. A., 1985, Monoclonal antibodies against nuclear envelope-associated proteins from rat liver nuclei with rodent and human cells, *Exp. Cell Res.* **116**:325–332.

Finlay, D. R., and Forbes, D. J., 1990, Reconstitution of biochemically altered nuclear pores: Transport can be eliminated and restored, *Cell* **60**:17–29.

Finlay, D. R., Newmeyer, D. D., Price, T. M., and Forbes, D. J., 1987, Inhibition of *in vitro* nuclear transport by a lectin that binds to nuclear pores, *J. Cell Biol.* **104:**189–200.

Finlay, D. R., Meier, E., Bradley, P., Horecka, J., and Forbes, D. J., 1991, A complex of nuclear pore proteins required for pore function, *J. Cell Biol.* **114:**169–183.

Fisher, D. Z., Chaudhary, N., and Blobel, G., 1986, cDNA sequencing of nuclear lamins A and C reveals primary and secondary structural homology to intermediate filaments, *Proc. Natl. Acad. Sci. USA* **83:**6450–6454.

Foe, V. E., and Alberts, B. M., 1983, Studies of nuclear and cytoplasmic behavior during the five mitotic cycles that precede gastrulation in *Dropophila* embryogenesis, *J. Cell Sci.* **6:**31–70.

Forbes, D. J., Kirschner, M. W., and Newport, J. W., 1983, Spontaneous formation of nucleus-like structures around bacteriophage DNA microinjected into *Xenopus* eggs, *Cell* **34:**13–23.

Franke, W. W., 1974, Structure, biochemistry and functions of the nuclear envelope, *Int. Rev. Cytol.* **4 (Suppl):**71–236.

Franke, W. W., 1987, Nuclear lamins and cytoplasmic intermediate filament proteins: A growing multigene family, *Cell* **48:**3–4.

Franke, W. W., and Scheer, U., 1970, The ultrastructure of the nuclear envelope of amphibian oocytes: A reinvestigation, *J. Ultrastruc. Res.* **30:**288–316.

Franke, W. W., and Scheer, U., 1974, Structures and functions of the nuclear envelope, *Cell Nucleus* **1:**219–347.

Franke, W. W., Scheer, U., Krohne, G., and Jarasch, E., 1981, The nuclear envelope and the architecture of the nuclear periphery, *J. Cell Biol.* **91:**39s–50s.

Fry, D. J., 1976, The nuclear envelope in mammalian cells, in *Mammalian Cell Membranes* (G. A. Jameson and D. M. Robinson, eds.), pp. 197–264, Butterworth, Woburn, Mass.

Fuchs, J. P., Giloh, H., Kuo, C. H., Saumweber, H., and Sedat, J. W., 1983, Nuclear structure: Determination of the fate of the nuclear envelope in *Drosophila* during mitosis using monoclonal antibodies, *J. Cell Sci.* **64:**331–349.

Gall, J. G., 1967, Octagonal nuclear pores, *J. Cell Biol.* **32:**391–399.

Gautier, J., Norbury, C., Lohka, M., Nurse, P., and Maller, J., 1988, Purified maturation promoting factor contains the product of a *Xenopus* homolog of the fission yeast cell cycle control gene cdc2+, *Cell* **54:**433–439.

Geisler, N., and Weber, K., 1988, Phosphorylation of desmin *in vitro* inhibits the formation of intermediate filaments: Identification of three kinase A sites in the amino terminal head domain, *EMBO J.* **7:**15–20.

Georgatos, S. D., and Blobel, G., 1987a, Two distinct attachment sites for vimentin along the plasma membrane and the nuclear envelope in avian erythrocytes: A basis for vectorial assembly of intermediate filaments, *J. Cell Biol.* **105:**105–115.

Georgatos, S. D., and Blobel, G., 1987b, Lamin B constitutes an intermediate filament attachment site at the nuclear envelope, *J. Cell Biol.* **105:**117–125.

Georgatos, S. D., Stournavos, C., and Blobel, G., 1988, Heteotypic and homotypic associations between the nuclear lamins: Site specificity and control by phosphorylation, *Proc. Natl. Acad. Sci. USA* **85:**4325–4329.

Georgatos, S. D., Marouakon, I., and Blobel, G., 1989, Lamin A, lamin B, and lamin B receptor analogues in yeast, *J. Cell Biol.* **108:**2069–2082.

Gerace, L., 1986, Nuclear lamina and organization of nuclear architecture, *Trends Biochem. Sci.* **11:**443–446.

Gerace, L., and Blobel, G., 1980, The nuclear envelope lamina is reversibly depolymerized during mitosis, *Cell* **19:**277–287.

Gerace, L., and Burke B., 1988, Functional organization of the nuclear envelope, *Annu. Rev. Cell Biol.* **4:**335–374.

Gerace, L., Blum, A., and Blobel, G., 1978, Immunocytochemical localization of the major polypeptides of the pore complex-lamina fraction: Interphase and mitotic distribution, *J. Cell Biol.* **79**:546–566.

Gerace, L., Ottaviano, Y., and Kondor-Koch, C., 1982, Identification of a major polypeptide of the nuclear pore complex, *J. Cell Biol.* **95**:826–837.

Gerace, L., Comeau, C., and Benson, M., 1984, Organization and modulation of nuclear lamina structure, *J. Cell Sci.* **1(suppl)**:137–160.

Gerhart, J., Wu, M., and Kirschner, M., 1984, Cell cycle dynamics of an M-phase specific cytoplasm factor in *Xenopus laevis* oocytes and eggs, *J. Cell Biol.* **98**:1247–1255.

Ghadially, F. N., 1988, *Ultrastructural Pathology of the Cell Matrix*, 3rd ed. pp. 462–475, Butterworth, London.

Ghosh, S., and Pawaletz, N., 1987, Is nuclear envelope formation an inducible event? *Exp. Cell Res.* **171**:243–249.

Glass, J. R., and Gerace, L., 1990, Lamins A and C bind and assemble at the surface of mitotic chromosomes, *J. Cell Biol.* **111**:1047–1057.

Goldberg, M. W., and Allen, T. D., 1992, High-resolution scanning electron microscopy of the nuclear envelope: Demonstration of a new, regular, fibrous lattice attached to the baskets of the nucleoplasmic face of the nuclear pores, *J. Cell Sci.* **119**:1429–1440.

Goldman, A. E., Maul, G., Steinert, P. M., Yang, H. Y., and Goldman, R. D., 1986, Keratin-like proteins that coisolate with intermediate filaments of BHK-21 cells are nuclear lamins, *Proc. Natl. Acad. Sci. USA* **83**:3839–3843.

Goldman, A. E., Moir, R. D., Montag-Lowy, M., Stewart, M., and Goldman, R. D., 1992, Pathway of incorporation of microinjected lamin A into the nuclear envelope, *J. Cell Biol.* **119**:725–735.

Goldman, R. D., Chou, Y-H., Dessev, C., Dessev, G., Eriksson, J., Goldman, A., Khoun, S., Kohnken, R., Lowy, M., Miller, R., Murphy, K., Opal, P., Skalli, O., and Straube, K., 1991, Dynamic aspects of cytoskeletal and karyoskeletal intermediate filament systems during the cell-cycle, *Cold Spring Harb. Symp. Quant. Biol.* **56**:629–642.

Greber, V. F., Senior, A., and Gerace, L., 1990, A major glycoprotein of the nuclear pore complex is a membrane-spanning polypeptide with a large lumenal domain and a small cytoplasmic tail, *EMBO J.* **9**:1495–1502.

Griffiths, G., Pfeiffer, S., Simons, K., and Matlin, K., 1986, Exit of newly synthesized membrane proteins from the *trans* cisterna of the Golgi complex to the plasma membrane, *J. Cell Biol.* **101**:949–964.

Gruenbaum, Y., Landesman, Y., Dree, B., Barre, J., Saumweber, H., Paddy, M., Sedat, J., Smith, D., Benton, B., and Fisher, P., 1988, *Drosophila* nuclear lamin precursor Dmo is translated from either of two developmentally regulated mRNA species apparently encoded by a single gene, *J. Cell Biol.* **106**:585–596.

Gurley, L. R., D'Anna, J. A., Bartham, S. S., Deaven, L. L., and Tobey, R., 1978, Histone phosphorylation and chromatin structure during mitosis in Chinese hamster cells, *Eur. J. Biochem.* **84**:1–15.

Hancock, R., and Hughes, M., 1982, Organization of DNA in the interphase nucleus, *Biol. Cell* **44**:201–212.

Hancock, J. F., Magee, A. I., Childs, J. E., and Marshall, C. J., 1989, All ras proteins are polyadenylated but only some are palmitoylated, *Cell* **57**:1167–1177.

Hanover, J. A., Cohen, C. K., Willingham, M. C., and Park, M. K., 1987, O-linked N-acetylglucosamine is attached to proteins of the nuclear pore: Evidence for cytoplasmic glycosylation, *J. Biol. Chem.* **262**:9887–9894.

Harel, A., Zlotkin, E., Nainudel-Epszteyn, S., Feinstein, N., Fisher, P. A., and Gruenbaum, Y., 1989, Persistence of major nuclear envelope antigens in an envelope-like structure during mitosis in *Drosophila melanogaster* embryos, *J. Cell Sci.* **94**:463–470.

Hart, G. W., Haltiwanger, G. D., Holt, D., and Kelly, W. G., 1989, Glycosylation in the nucleus and cytoplasm, *Annu. Rev. Biochem.* **58**:841–874.

He, D., Nickerson, J. A., and Penman, S., 1990, Core filaments of the nuclear matrix, *J. Cell Biol.* **110**:569–580.

Heald, R., and McKeon, F., 1990, Mutations of phosphorylation sites in lamin A that prevent nuclear lamina disassembly in mitosis, *Cell* **61**:579–589.

Heath, I. B., 1980, Variant mitoses in lower eukaryotes: Indicators of the evolution of mitosis, *Int. Rev. Cytol.* **64**:1–80.

Hennessey, K. H., Clark, C. D., and Bolstein, D., 1990, Subcellular localization of yeast CDC46 varies with the cell cycle, *Genes Dev.* **4**:2252–2263.

Hoeijimakers, J. H. J., Schell, J. H. N., and Wanka, F., 1974, Structure of the nuclear pore complex in mammalian cells, *Exp. Cell Res.* **87**:195–206.

Hoger, J. H., Krohne, G., and Franke, W. W., 1988, Amino acid sequence and molecular characterization of murine lamin B as deduced from cDNA clones, *Eur. J. Cell Biol.* **47**:283–290.

Holt, G. D., and Hart, G. W., 1986, The subcellular distribution of terminal *N*-acetylglucosamine moieties: Localization of a novel protein-saccharide linkage, O-linked GlcNAc, *J. Biol Chem.* **261**:8049–8057.

Holt, G. D., Snow, C. M., Senior, A., Haltiwanger, R. S., Gerace, L., and Hart, G. W., 1987, Nuclear pore glycoproteins contain cytoplasmically disposed O-linked *N*-acetylglucosamine, *J. Cell Biol.* **104**:1157–1164.

Holtz, D., Tanaka, R. A., Hartwig, J., and McKeon, F., 1989, The CaaX motif of lamin A functions in conjunction with the nuclear localization signal to target assembly to the nuclear envelope, *Cell* **59**:969–977.

Hunt, T., 1989, Maturation promoting factor, cyclin and the control of M-phase, *Curr. Opin. Cell Biol.* **1**:268–274.

Hunt, T., 1991, Cyclins and their partners: From a simple idea to complicated reality, *Semin. Cell Biol.* **2**:213–222.

Hurt, E. C., 1988, A novel nucleoskeletal-like protein located at the nuclear periphery is required for the life cycle of *Saccharomyces cerevisiae*, *EMBO J.* **7**:4323–4334.

Hurt, E. C., 1990, Targeting of a cytosolic protein to the nuclear periphery, *J. Cell Biol.* **111**:2829–2837.

Hutchison, C. J., Cox, R., Drepaul, S., Gomperts, M., and Ford, C. C., 1987, Periodic DNA synthesis in cell-free extracts of *Xenopus* eggs, EMBO J. **6**:2003–2010.

Hutchison, C. J., Cox, R., and Ford, C. C., 1988, The control of DNA synthesis in a cell-free extract that recapitulates a basic cell cycle *in vitro*, *Development* **103**:553–566.

Hutchison, C. J., Brill, D., Cox, R., Gilbert, J., Kill, I. R., and Ford, C. C., 1989, DNA replication and cell-cycle control in *Xenopus* egg extracts, *J. Cell Sci.* **12(suppl)**:197–212.

Ide, T., Nakane, M., Anzai, K., and Andohl, T., 1975, Supercoiled DNA folded by non-histone proteins in cultured mammalian cells, *Nature* **194**:495–496.

Inagaki, M., Nishi, Y., Nishizawa, K., Matsushi, M., and Sato, C., 1987, Site-specific phosphorylation induces disassembly of vimentin filaments *in vitro*, *Nature* **328**:649–652.

Jackson, D. A., and Cook, P. R., 1988, Visualization of a filamentous nucleoskeleton with a 23nm axial repeat, *EMBO J.* **7**:3667–3677.

Jenkins, H., Lyons, C., Lane, B., Stick, R., and Hutchison, C., 1993, Nuclei which lack a lamina accumulate karyophilic proteins and assemble a nuclear matrix, *J. Cell Sci.*, in press.

Kalderon, D., Richardson, W. D., Markham, A. F., and Smith, A. E., 1984, Sequence requirements for nuclear location of simian virus 40 large T antigen, *Nature* **311**:499–509.

Kartenbeck, J., Jarasch, E-D., and Franke W. W., 1971, The nuclear envelope in freeze-etching, *Adv. Anat. Embryol. Cell Biol.* **45**:1–55.

Keller, J. M., and Riley, D. E., 1976, Nuclear ghosts: A nonmembranous structural component of mammalian cell nuclei, *Science* **193**:399–401.

Kessel, R. G., 1983, The structure and function of annulate lamellae: Porous cytoplasmic and intranuclear membranes, *Int. Rev. Cytol.* **82**:181–303.

Kessel, R., and Subtelny, S., 1981, Alteration of annulate lamellae in the *in vitro* progesterone-treated, full grow *Rana pipiens* oocyte, *J. Exp. Zool.* **217**:119–135.

Kirschner, R. H., Rusli, M., and Martin, T. E., 1977, Characterization of the nuclear envelope, pore complexes, and dense lamina of mouse liver nuclei by high-resolution scanning electron microscopy, *J. Cell Biol.* **72**:118–132.

Kitten, G. T., and Nigg, E. A., 1991, The CaaX motif is required for isoprenylation, carboxyl methylation, and nuclear membrane association of lamin B2, *J. Cell Biol.* **113**:13–23.

Krek, W., and Nigg, E. A., 1991, Mutations of p34cdc2 phosphorylation sites induce premature mitotic events in HeLa cells: Evidence for a double block to p34cdc2 kinase activation in vertebrates, *EMBO J.* **10**:3331–3341.

Krohne, G., and Benavente, R., 1986, The nuclear lamins: A multigene family of proteins in evolution and differentiation, *Exp. Cell Res.* **162**:1–10.

Krohne, G., Franke, W. W., and Scheer, U., 1978a, The major polypeptides of the nuclear pore complex, *Exp. Cell Res.* **116**:85–102.

Krohne, G., Franke, W. W., Ely, S., D'Arcy, A., and Jost, E., 1978b, Localization of a nuclear envelope-associated protein by indirect immunofluorescence microscopy using antibodies against a major polypeptide from rat liver fractions enriched in nuclear envelope-associated material, *Cytobiologie* **18**:22–38.

Krohne, G., Wolin, S. L., McKeon, F. D., Franke, W. W., and Kirschner, M. W., 1987, Nuclear lamin LI of *Xenopus laevis:* cDNA cloning, amino acid sequence and binding specificity of a member of the lamin B subfamily, *EMBO J.* **6**:3801–3808.

Krohne, G., Waizenegger, I., and Hoger, T. H., 1989, The conserved carboxy-terminal cysteine of nuclear lamins is essential for lamin association with the nuclear envelope, *J. Cell Biol.* **109**:2003–2011.

Labbe, J. C., Lee, M. G., Nurse, P., Picard, A., and Doree, M., 1988, Activation at M-phase of a protein kinase encoded by a starfish homologue of the cell cycle control gene cdc2+, *Nature* **335**:251–254.

Lam, K., and Kasper, C., 1979, Selective phosphorylation of a nuclear envelope polypeptide by an endogenous protein, *Biochemistry* **18**:307–311.

Lamb, N. J. C., Fernandez, A., Feramisco, J. R., and Welch, W. J., 1989, Modulation of vimentin containing intermediate filament distribution and phosphorylation in living fibroblasts by the cAMP-dependent protein kinase, *J. Cell Biol.* **108**:2409–2422.

Lanford, R. E., and Butel, J. S., 1984, Construction of an SV40 mutant defective in nuclear transport of T antigen, *Cell* **37**:801–813.

Langan, T. A., Gautier, J., Lohka, M., Hollingsworth, R., Moreno, S., Nurse, P., Maller, J., and Sclafani, R. A., 1989, Mammalian growth associated H1 kinase: A homolog of cdc2+/CDC28 protein kinases controlling mitotic entry in yeast and frog cells, *Mol. Cell Biol.* **9**:3860–3868.

Laskey, R. A., 1985, Chromosome replication in early development of *Xenopus laevis*, *J. Embryol. Exp. Morphol.* **89**:285–295.

Laskey, R. A., Honda, B. M., Mills, A. D., and Finch, J. T., 1978, Nucleosomes are assembled by an acidic protein which binds histones and transfers them to DNA, *Nature* **275**:416–420.

Leduc, E. H. S., Avrameas, M., and Bouteille, M., 1968, Ultrastructural localization of antibody in differentiating plasma cells, *J. Exp. Med.* **127**:109–131.

Lee, M. G., and Nurse, P., 1988, Complementation used to clone a human homologue of the fission yeast cell cycle control gene cdc2, *Nature* **327**:31–35.

Lee, M., and Nurse, P., 1988, Cell cycle control genes in fission yeast and mammalian cells, *Trends in Genetics* **4**:287–290.

Lee, W., and Melese, T., 1989, Identification and characterization of a nuclear localization sequence-binding protein in yeast, *Proc. Natl. Acad. Sci. USA* **86**:8808–8812.

Lehner, C. F., Kurer, V., Eppenberger, H. M., and Nigg, E. A., 1986, The nuclear lamin protein family in higher vertebrates; identification of quantitatively minor lamin proteins by monoclonal antibodies, *J. Biol. Chem.* **261**:13293–13301.

Lehner, C. F., Stick, R., Eppenberger, H. M., and Nigg, E. A., 1987, Differential expression of nuclear lamin proteins during chicken development, *J. Cell Biol.* **105**:577–587.

Leno, G. H., Downes, C. S., and Laskey, R. A., 1992, The nuclear membrane prevents replication of human G2 nuclei but not G1 nuclei in *Xenopus* egg extract, *Cell* **69**:151–158.

Leno, G. H., and Laskey, R. A., 1991, The nuclear membrane determines the timing of DNA replication in *Xenopus* egg extracts, *J. Cell Biol.* **112**:557–566.

Lew, D. J., and Reed, S. I., 1992, A proliferation of cyclins, *Trends Cell Biol.* **2**:77–81.

Loewinger, L., and McKeon, F., 1988, Mutations in the nuclear lamin proteins resulting in their aberrant assembly in the cytoplasm, *EMBO J.* **7**:2301–2309.

Lohka, M. J., 1988, The reconstitution of nuclear envelopes in cell-free extracts, *Cell Biol. Int. Rep.* **12**:833–848.

Lohka, M. J., 1989, Mitotic control by metaphase-promoting factor and cdc proteins, *J. Cell Sci.* **92**:131–135.

Lohka, M. J., and Maller, J. L., 1985, Induction of nuclear envelope breakdown, chromosome condensation and spindle formation in cell-free extracts, *J. Cell Biol.* **101**:518–523.

Lohka, M. J., and Masui, Y., 1983, Formation *in vitro* of sperm pronuclei and mitotic chromosomes induced by amphibian ooplasmic components, *Science* **220**:719–721.

Lohka, M. J., and Masui, Y., 1984, Roles of cytosol and cytoplasmic particles in nuclear envelope assembly and sperm pronuclear formation in cell-free preparations from amphibian eggs, *J. Cell Biol.* **98**:1222–1230.

Lohka, M. J., Hayes, M. K., and Maller, J. M., 1988, Purification of maturation promoting factor, an intracellular regulator of early mitotic events, *Proc. Natl. Acad. Sci. USA* **85**:3009–3013.

Luscher, B., Brizuela, L., Beach, D., and Eiseman, R. N., 1991, A role for the p34cdc2 kinase and phosphatases in the regulation of phosphorylation and disassembly of lamin B$_2$ during the cell cycle, *EMBO J.* **10**:865–875.

MacKintosh, C., Beattie, K. A., Klumpp, S., Cohen, P., and Codd, G. A., 1990, Cyanobacterial microcystin-LR is a potent and specific inhibitor of protein phosphatases 1 and 2A from both mammals and plants, *FEBS Lett.* **264**:187–192.

Maller, J. L., 1985, Regulation of amphibian oocyte maturation, *Cell Different.* **16**:211–220.

Masui, Y., and Markert, C. L., 1971, Cytoplasmic control of nuclear behaviour during meiotic maturation of frog oocytes, *J. Exp. Zool.* **117**:129–146.

Matsumoto, T., and Beach, D., 1991, Premature initiation of mitosis in yeast lacking RCC1 or an interacting GTPase, *Cell* **66**:347–360.

Maul, G. G., 1977, The nuclear and cytoplasmic pore complex: Structure, dynamics, distribution and evolution, *Int. Rev. Cytol.* **6(suppl)**:75–186.

Maul, G. G., Maul, H. M., Scogna, J. E., Lieberman, M. W., Stein, G. S., Hsu, Y-L., and Borun, T. W., 1972, Time sequence of nuclear pore formation in phytohemagglutinin-stimulated lymphocytes and in HeLa cells during the cell cycle, *J. Cell Biol.* **55**:433–447.

Mayorga, L. S., Diaz, R., and Stahl, P. D., 1989, Regulatory role for GTP-binding proteins in endocytosis, *Science* **244**:1475–1477.

McIntosh, J. R., and Koonce, M. P., 1989, Mitosis, *Science* **246**:622–628.

McKeon, F. D., Tuffanelli, D. L., Kobayashi, S., and Kirschner, M. W., 1984, The redistribution of a conserved nuclear envelope protein during the cell cycle suggests a pathway for chromosome condensation, *Cell* **36**:83–92.

McKeon, F. D., Kirschner, M. W., and Caput, D., 1986, Homologies in both primary and secondary structure between nuclear envelope and intermediate filament proteins, *Nature* **319**:463–468.

Meier, J., Campbell, K.H.S., Ford, C. C., Stick, R., and Hutchison, C. J., 1991, The role of lamin Liii in nuclear assembly and DNA replication, in cell-free extracts of *Xenopus* eggs,. *J. Cell Sci.* **98**:271–279.

Meier, U.T., and Blobel, G., 1990, A nuclear localization signal binding protein in the nucleolus, *J. Cell Biol.* **111**:1775–1783.

Meijer, L., and Guerrier, P., 1984, Maturation and fertilization in starfish oocytes, *Int. Rev. Cytol.* **86**:129–149.

Melancon, P., Glick, J. G., Malhotra, V., Weidman, P. J., Serafini, T., Gleason, M. L., Orci, L., and Rothman, J. E., 1987, Involvement of GTP-binding "G" proteins in transport through the Golgi stack, *Cell* **51**:1053–1062.

Mendenhall, M. D., Jones, C. A., and Reed, S. I., 1987, Dual regulation of the yeast CDC28-p40 protein kinase complex: Cell cycle, pheromone and nutrient limitation effects, *Cell* **50**:927–935.

Merisko, E. M., 1989, Annulate lamellae: An organelle in search of a function, *Tissue Cell* **21**:343–354.

Miake-Lye, R., and Kirschner, M. W., 1985, Induction of early mitotic events in a cell-free system, *Cell* **41**:165–175.

Miake-Lye, R., Newport, J., and Kirschner, M., 1983, Maturation-promoting factor induces nuclear envelope breakdown in cycloheximide-arrested embryos of *Xenopus laevis*, *Cell* **37**:731–742.

Minshull, J., Blow, J. J., and Hunt, T., 1989, Translation of cyclin mRNA is necessary for extracts of activated *Xenopus* eggs to enter mitosis, *Cell* **56**:947–956.

Moir, R. D., Quinlan, R. A., and Stewart, M., 1990, Expression and characterization of human lamin C, FEBS Lett. **268**:463–468.

Moir, R. D., Donaldson, A. E., and Stewart, M., 1991, Expression in *Escherichia coli* of human lamins A and C: Influence of head and tail domains on assembly properties and paracrystal formation, *J. Cell Sci.* **99**:363–372.

Montag, M., Spring, H., and Trendelenberg, M. F., 1988, Structural analysis of the mitotic nucleus in pre-gastrula *Xenopus* embryos, *Chromosoma* **96**:187–196.

Morla, A. O., Draetta, G., Beach, D., and Wang, J. Y., 1989, Reversible tyrosine phosphorylation of cdc2: Dephosphorylation accompanies activation during entry into mitosis, *Cell* **58**:193–203.

Murray, A. W., and Kirschner, M. W., 1989, Cyclin synthesis drives the early embryonic cell cycle, *Nature* **339**:287–292.

Murray, A. W., Solomon, M. J., and Kirschner, M. W., 1989, The role of cyclin synthesis and degradation in the control of MPF activity, *Nature* **339**:280–286.

Murray, R. G., Murray, A. S., and Pizzo, A., 1965, The fine structure of mitosis in rat thymic lymphocytes, *J. Cell Biol.* **26**:601–619.

Nakagawa, J., Kitten, G. T., and Nigg, E. A., 1989, A somatic cell-derived system for studying both early and late mitotic events *in vitro*, *J. Cell Sci.* **94**:449–462.

Nasmyth, K., Adolf, G., Lydall, D., and Seddon, A., 1990, The identification of a second cell cycle control on the HO promoter in yeast: Cell cycle regulation of SW15 nuclear entry, *Cell* **62**:631–647.

Nehrbass, U., Kern, H., Mutvei, A., Horstmann, H., Marshallsay, B., and Hurt, E. C., 1990, NSP1: A yeast nuclear envelope protein localized at the nuclear pores exerts its essential function by its carboxy-terminal domain, *Cell* **61**:979–989.

Newmeyer, D. D., and Forbes, D. J., 1990, An *N*-ethylmaleimide-sensitive cytosolic factor necessary for nuclear protein import: Requirement in signal-mediated binding to the nuclear pore, *J. Cell Biol.* **110**:547–557.

Newport, J. W., 1987, Nuclear reconstitution *in vitro:* Stages of assembly around protein-free DNA, *Cell* **48**:205–217.

Newport, J., and Kirschner, M., 1984, Regulation of the cell cycle during early *Xenopus* development, *Cell* **37**:731–742.

Newport, J. W., and Spann, T., 1987, Disassembly of the nucleus in mitotic extracts: Membrane vesicularization, lamin disassembly and chromosome condensation are independent processes, *Cell* **48**:219–230.

Newport, J. W., Wilson, K. L., and Dunphy, W. G., 1990, A lamin-independent pathway for nuclear envelope assembly, *J. Cell Biol.* **111**:2247–2259.

Newport, J. W., and Dunphy, W. G., 1992, Characterization of the membrane binding and fusion events during nuclear envelope assembly using purified components, *J. Cell Biol.* **116**:295–306.

Nickerson, J. A., Krockmalnic, G., Wan, K. M., and Penman, S., 1989, Chromatin architecture and nuclear RNA, *Proc. Natl. Acad. Sci. USA* **86**:177–181.

Nigg, E. A., 1992, Assembly and cell cycle dynamics of the nuclear lamina, *Semin. Cell Biol.* **3**:245–253.

Nishimoto, T., Eilen, E., and Basilico, C., 1978, Premature chromosome condensation in a ts DNA mutant of BHK cells, *Cell* **15**:475–483.

Norbury, C., Blow, J., and Nurse, P., 1991, Regulatory phosphorylation of the p34^{cdc2} protein kinase in vertebrates, *EMBO J.* **10**:3321–3329.

Nurse, P., 1985, Cell cycle control genes in fission yeast, *Trends Genet.* **1**:51–55.

Olins, A. L., and Olins, D. E., 1979, Stereo electron microscopy of the 25-nM chromatin fibers of isolated nuclei, *J. Cell Biol.* **81**:260–265.

Orci, L., Glick, B. S., and Rothman, J. E., 1986, A new type of coated vesicular carrier that appears not to contain clathrin: its possible role in protein transport within the Golgi Stack, *Cell* **46**:171–184.

Osborn, M., and Weber, K., 1987, Cytoplasmic intermediate filament proteins and the nuclear lamins A, B and C share the IFA epitope, *Exp. Cell Res.* **170**:195–203.

Ottaviano, Y., and Gerace, L., 1985, Phosphorylation of the nuclear lamins during interphase and mitosis, *J. Cell Biol.* **101**:518–523.

Padan, R., Nainudel-Epszteyn, S., Goitein R., Fainsod, A., and Gruenbaum, Y., 1990, Isolation and characterization of the *Drosophila* nuclear envelope otefin cDNA, *J. Biol. Chem.* **265**:7808–7813.

Paddy, M. R., Belmont, A. S., Saunweber, H., Agard, D. A., and Sedat, J. W., 1990, Interphase nuclear lamins form a discontinuous network that interacts with only a fraction of the chromatin in the nuclear periphery, *Cell* **62**:89–106.

Park, M. K., D'Onofrio, M., Willingham, M. C., and Hanover, J. A., 1987, A monoclonal antibody against a family of nuclear pore proteins (nucleoporins) recognizes a shared determinant: O-linked *N*-acetylglucosamine, *Proc. Natl. Acad. Sci. USA* **84**:6462–6466.

Pathak, R. K., Luskey, K. L., and Anderson, R.G.W., 1986, Biogenesis of the crystalloid endoplasmic reticulum in UT-1 cells: Evidence that newly formed endoplasmic reticulum emerges from the nuclear envelope, *J. Cell Biol.* **102**:2158–2168.

Paulson, J. R., and Laemmli, U. K., 1977, The structure of histone-depleted metaphase chromosomes, *Cell* **12**:817–828.

Peter, M., Kitten, G. T., Lehner, C. F., Vorburger, K., Bailer, S. M., Maridor, G., and Nigg, E. A., 1989, Cloning and sequencing of cDNA clones encoding chicken lamins A and B1 and comparison of the primary structures of vertebrate A- and B-type lamins, *J. Mol. Biol.* **208**:393–404.

Peter, M., Nakagawa, J., Doree, M., Labbe, J. C., and Nigg, E. A., 1990, *In vitro* disassembly of the nuclear lamina and M-phase-specific phosphorylation of lamins by cdc2 kinase, *Cell* **61**:591–602.

Pfaller, R., Smythe, C., and Newport, J. W., 1991, Assembly/disassembly of the nuclear envelope membrane: Cell cycle-dependent binding of nuclear membrane vesicles to chromatin *in vitro*, *Cell* **65**:209–217.

Philpott, A., Leno, G. H., and Laskey, R. A., 1991, Sperm decondensation in *Xenopus* egg cytoplasm is mediated by nucleoplasmin, *Cell* **65**:569–578.

Pollard, K. M., Chan, E. K. L., Grant, B. J., Sullivan, K. F., Tan, E. M., and Glass, C. A., 1990, *In vitro* post-translational modification of lamin B cloned from a human T-cell line, *Mol. Cell. Biol.* **10**:2164–2175.

Powell, L., and Burke, B., 1990, Internuclear exchange of an inner nuclear membrane protein (p55) in heterokaryons: *In vivo* evidence for the interaction of p55 with the nuclear lamina, *J. Cell Biol.* **111**:2225–2234.

Puddington, L., Lively, O. M., and Lyles, D. S., 1985, Role of the nuclear envelope in synthesis, processing and transport of membrane glycoproteins, *J. Biol. Chem.* **26**:5641–5647.

Rao, P. N., and Johnson, R. T., 1970, Mammalian cell fusion: Studies on the regulation of DNA synthesis and mitosis, *Nature* **225**:159–164.

Reed, S. I., Hadwiger, J. A., and Locrincz, A. T., 1985, Protein kinase activity with the product of the yeast cell division cycle gene CDC28, *Proc. Natl. Acad. Sci. USA* **82**:4055–4059.

Reichelt, R., Holzenburg, A., Buhle, E. L., Jarnik, M., Engel, A., and Aebi, U., 1990, Correlation between structure and mass of the nuclear pore complex and of distinct pore complex components, *J. Cell Biol.* **110**:883–894.

Ris, H., 1991, The three-dimensional structure of the nuclear pore complex as seen by high-voltage electron microscopy and high-resolution low-voltage scanning electron microscopy, *EMSA Bull.* **21**:54–56.

Rober, R-A., Weber, K., and Osborn, M., 1989, Differential timing of nuclear lamin A/C expression in the various organs of the mouse embryo and the young animal: A developmental study, *Development* **105**:365–378.

Roberts, K., and Northcote, D. H., 1970, Structure of the nuclear pore in higher plants, *Nature* **228**:385–386.

Scheer, U., Kartenbeck, J., Trendelenberg, M. F., Stadler, J., and Franke, W. W., 1976, Experimental disintegration of the nuclear envelope: Evidence for pore-connecting fibrils, *J. Cell Biol.* **69**:1–18.

Scheer, U., Dabauvalle, M-C., Merkert, H., and Benavente, R., 1988, The nuclear envelope and the organization of nuclear pore complexes, *Cell Biol. Int. Rep.* **12**:669–689.

Schindler, M., Hogan, M., Miller, R., and DeGaetano, D., 1987, A nuclear specific glycoprotein representative of a unique pattern of glycosylation, *J. Biol. Chem.* **262**:1254–1260.

Senior, A., and Gerace, L., 1988, Integral membrane proteins specific to the inner nuclear membrane and associated with the nuclear lamina, *J. Cell Biol.* **107**:2029–2036.

Shaw, C. M., D'Onofrio, M., Park, M. K., and Hanover, J. A., 1990, Primary sequence and heterologous expression of nuclear pore glycoprotein p62, *J. Cell Biol.* **110**:1861–1871.

Sheehan, M. A., Mills, A. D., Sleeman, A. M., Laskey, R. A., and Blow, J. J., 1988, Steps in the assembly of replication-competent nuclei in a cell-free system from *Xenopus* eggs, *J. Cell Biol.* **106**:1–12.

Shelton, K. R., Cobbs, C. S., Povloshock, J. T., and Brkat, R. K., 1980, Nuclear envelope fraction proteins: Isolation and comparison with the nuclear protein of the avian erythrocyte, *Arch. Biochem. Biophys.* **174**:177–186.

Shenoy, S., Choi, J-K., Bagrodia, S., Copeland, T. D., Maller, J. L., and Shalloway, D., 1989, Purified maturation promoting factor phosphorylates pp60c-src at the sites phosphorylated during fibroblast mitosis, *Cell* **57**:763–774.

Simanis, V., and Nurse, P., 1986, The cell cycle control gene cdc2+ of fission yeast encodes a protein kinase potentially regulated by phosphorylation, *Cell* **45**:261–268.

Smith, D. E., and Fisher, P. A., 1989, Interconversion of *Drosophila* nuclear lamin isoforms during the five mitotic cycles that precede gastrulation in *Drosophila* embryogenesis, *J. Cell Biol.* **108**:255–266.

Smith, L. D., and Ecker, R. E., 1971, The interaction of steroids with *Rana pipiens* oocytes in the induction of mautration, *Dev. Biol.* **25**:232–247.

Snow, C. M., Senior, A., and Gerace, L., 1987, Monoclonal antibodies identify a group of nuclear pore complex glycoproteins, *J. Cell Biol.* **104**:1143–1156.

Solomon, M. J., Glotzer, M., Lee, T. H., Philippe, M., and Kirschner, M. W., 1990, Cyclin activation of p34-cdc2, *Cell* **63**:1013–1024.

Stafstrom, J. P., and Staehlin, L. A., 1984, Dynamics of the nuclear envelope and of nuclear pore complexes during mitosis in the *Drosophila* embryo, *Eur. J. Cell Biol.* **34**:179–189.

Standart, N. M., Bray, S. J., George, E. L., Hunt, T., and Ruderman, J. V., 1985, The small subunit of ribonucleotide reductase is encoded by one of the most abundant translationally regulated maternal RNA's in clam and sea urchin eggs, *J. Cell Biol.* **100**:1968–1976.

Starr, C. M., D'Onofrio, M., Park, M. K., and Hanover, J. A., 1990, Primary sequence and heterologous expression of nuclear pore glycoprotein, p62, *J. Cell Biol.* **110**:1861–1871.

Steinert, G., Baltus, J., Hancoq-Quertier, J., and Brachet, J., 1974, Ultrastructure of *Xenopus laevis* oocytes after injection of an extract from progesterone-treated oocytes, *J. Ultrastruct. Res.* **49**:188–210.

Steinert, P. M., and Roop, R., 1988, The molecular and cellular biology of intermediate filaments, *Annu. Rev. Biochem.* **57**:593–625.

Steller, H., and Pirotta, V., 1985, Fate of DNA injected into early *Drosophila* embryos, *Dev. Biol.* **109**:54–62.

Stewart, C., and Burke, B., 1987, Teratocarcinoma stem cells and early mouse embryos contain only a single major lamin polypeptide closely resembling lamin B, *Cell* **51**:383–392.

Stewart, M., and Whytock, S., 1988, The structure and interactions of components of nuclear envelopes from *Xenopus* oocyte germinal vesicles observed by heavy metal shadowing, *J. Cell Sci.* **90**:409–423.

Stick, R., 1988, cDNA cloning of the developmentally regulated lamin Liii of *Xenopus laevis*, *EMBO J.* **7**:3189–3197.

Stick, R., 1992, The gene structure of *Xenopus* nuclear lamin A: A model for the evolution of A-type from B-type lamins by exon shuffling, *Chromosoma* **101**:566–574.

Stick, R., and Hausen, P., 1980, Immunological analysis of nuclear lamin proteins, *Chromosoma* **80**:219–236.

Stick, R., and Schwarz, H., 1982, The disappearance of the nuclear lamina during spermatogenesis: An electron microscopic and immunofluorescence study, *Cell Differ.* **11**:235–243.

Stick, R., and Schwarz, H., 1983, Disappearance and reformation of the nuclear structure during specific stages of meiosis in oocytes, *Cell* **33**:949–958.

Stick, R., and Hausen, P., 1985, Changes in the nuclear lamina composition during early development of *Xenopus laevis*, *Cell* **41**:191–200.

Stick, R., Angres, B., Lehner, C. F., and Nigg, E. A., 1988, The fates of chicken nuclear lamin proteins during mitosis: Evidence for a reversible redistribution of lamin B2 between inner nuclear membrane and elements of the endoplasmic reticulum, *J. Cell Biol.* **107**:397–406.

Strum, J. M., and Karnovsky, M. J., 1970, Cytochemical localization of endogenous peroxidase in thyroid follicular cells, *J. Cell Biol.* **44**:655–666.

Sukegawa, J., and Blobel, G., 1993, A nuclear pore complex protein that contains zinc finger motifs, binds DNA, and faces the nucleoplasm, *Cell* **72**:29–38.

Suprynowicz, F. A., and Gerace, L., 1986, A fractionated cell-free system for analysis of prophase nuclear disassembly, *J. Cell Biol.* **103**:2073–2081.

Swenson, K. I., O'Farrell, K. M., and Ruderman, J. V., 1986, The clam embryo protein cyclin A induces entry into M-phase and resumption of meiosis in *Xenopus* oocytes, *Cell* **47**:861–870.

Tanaka, K., and Kanbe, T., 1986, Mitosis in fission yeast *S. pombe* as revealed by freeze-substitution electron microscopy, *J. Cell Sci.* **80**:253–268.

Toda, T., Yamamot, M., and Yanagida, M., 1981, Sequential alterations in the nuclear chromatin region during mitosis of the fission yeast *S. pombe:* Video fluorescence microscopy of synchronously growing wild-type and cold-sensitive cdc mutants by using DNA-binding probe, *J. Cell Sci.* **62**:271–287.

Ulitzer, N., and Gruenbaum, Y., 1989, Nuclear envelope assembly around sperm chromatin in cell-free preparations from *Drosophila* embryos, *FEBS Lett.* **259**:113–116.

Ulitzer, N., Harel, A., Feinstein, N., and Gruenbaum, Y., 1992, Lamin activity is essential for nuclear envelope assembly in a *Drosophila* embryo cell-free system. *J. Cell Biol.* **119**:17–25.

Unwin, P. N. T., and Milligan, R. A., 1982, A large particle associated with the perimeter of the nuclear pore complex, *J. Cell Biol.* **93**:63–75.

Vigers, G. P. A., and Lohka, M. J., 1991, A distinct vesicle population targets membranes and pore complexes to the nuclear envelope in *Xenopus* eggs, *J. Cell Biol.* **112**:545–554.

Vigers, G. P. A., and Lohka, M. J., 1992, Regulation of nuclear envelope precursor functions during cell division, *J. Cell Sci.* **102**:273–284.

Vogelstein, B., Pardoll, D. M., and Coffey, D. S., 1980, Supercoiled loops and eukaryotic DNA replication. *Cell* **22**:79–85.

Vorburger, K., Lehner, C. F., Kitten, G., Eppenberger, H. M., and Nigg, E. A., 1989a, A second higher vertebrate B-type lamin: cDNA sequence determination and in vitro processing of chicken lamin B2, *J. Mol. Biol.* **208**:405–415.

Vorburger, K., Kitten, G. T., and Nigg, E. A., 1989b, Modification of nuclear lamin proteins by a mevalonic acid derivative occurs in reticulocyte lysates and requires the cysteine residue of the C-terminal CXXM motif, *EMBO J.* **8**:4007–4013.

Ward, G. E., and Kirschner, M. W., 1990, Identification of cell cycle-regulated phosphorylation sites on nuclear lamin C, *Cell* **61**:561–577.

Wataya-Kaneda, M., Kaneda, Y., Sakurai, T., Sugawa, H., and Uchida, T., 1987, A monoclonal antibody against the nucleus reveals the presence of a common protein in the nuclear envelope, the perichromosomal region, and cytoplasmic vesicles, *J. Cell Biol.* **104**:1–7.

Weber, K., Plessman, U., Dodernont, H., and Kossmagk-Stephan, K., 1988, Amino acid sequences and homoploymer-forming ability of the intermediate filament proteins from an invertebrate epithelium, *EMBO J.* **7**:2995–3001.

Weber, K., Plessman, U., and Traub, P., 1989, Maturation of nuclear lamin A involves a specific carboxy-terminal trimming, which removes the polyisoprenylation site from the precursor; implications for the structure of the nuclear lamina, *FEBS Lett.* **257**:411–414.

Welter, D. A., Black, D. A., and Hodge, L. D., 1985, Nuclear reformation following metaphase in HeLa S3 cells: Three-dimensional visualization of chromatid rearrangements, *Chromosoma* **93**:57–68.

Wente, S. R., Rout, M. P., and Blobel, G., 1992, A new family of yeast nuclear pore complex proteins, *J. Cell Biol.* **119**:705–723.

Wheatley, D. N., 1990, Mitosis and protein synthesis: 3. Organelle redistribution during normal and colcemid arrested M-phase in HeLa S-3 cell, *Cytobios.* **63**:109–130.

Whytock, S., and Stewart, M., 1988, Preparation of shadowed nuclear envelopes from *Xenopus* oocyte germinal vesicles for electron microscopy, *J. Microsc.* **151**:115–126.

Whytock, S., Moir, R. D., and Stewart, M., 1990, Selective digestion of nuclear envelopes from

Xenopus oocyte germinal vesicles: Possible structural role for the nuclear lamina, *J. Cell Sci.* **97:**571–580.

Wilson, E. B., 1925, *The Cell in Development and Heredity*. Macmillan, New York.

Wilson, K. L., and Newport, J., 1988, A trypsin-sensitive receptor on membrane vesicles is required for nuclear envelope formation *in vitro*, *J. Cell Biol.* **107:**57–68.

Wimmer, C., Doye, V., Grandi, P., Nehrbass, U., and Hurt, E. C., 1992, A new subclass of nucleoporins that functionally interact with nuclear pore protein NSP1, *EMBO J.* **11:**5051–5061.

Wolda, S. L., and Glomset, J. A., 1988, Evidence for modification of lamin B by a product of mevalonic acid, *J. Biol. Chem.* **263:**5997–6000.

Wolin, S. L., Krohne, G., and Kirschner, M. W., 1987, A new lamin in *Xenopus* somatic tissues displays strong homology to human lamin A, *EMBO J.* **6:**3809–3818.

Worman, H. J., Yuan, J., Blobel, G., and Georgatos, S. D., 1988, A lamin B receptor in the nuclear envelope, *Proc. Natl. Acad. Sci. USA* **85:**8531–8534.

Worman, H. J., Evans, C. D., and Blobel, G., 1990, The lamin B receptor of the nuclear envelope inner membrane: A polytopic protein with eight potential transmembrane domains, *J. Cell Biol.* **111:**1535–1542.

Wozniak, R. W., Bartnik, E., and Blobel, G., 1989, Primary structure analysis of an integral membrane glycoprotein of the nuclear pore, *J. Cell Biol.* **108:**2083–2092.

Wozniak, R. W., and Blobel, G., 1992, The single transmembrane segment of gp210 is sufficient for sorting to the pore membrane domain of the nuclear envelope, *J. Cell Biol.* **119:**1441–1449.

Wolff, B., Willingham, M. C., and Hanover, J. A., 1988, Nuclear protein import: Specificity for transport across the nuclear pore, *Exp. Cell Res.* **178:**318–334.

Yoneda, Y., Imamoto-Sonobe, N., Yamaizumi, M., and Uchida, T., 1987, Reversible inhibition of protein import into the nucleus by wheat germ agglutinin injected into cultured cells, *Exp. Cell Res.* **173:**586–595.

Yoneda, Y., Imamoto-Sonobe, N., Matsuoka, Y., Imatoto, R., Kiho, Y., and Uchida, T., 1988, Antibodies to Asp-Asp-Glu-Asp can inhibit transport of nuclear proteins into the nucleus, *Science* **242:**276–278.

Zalokar, M., and Erk, I., 1976, Division and migration of nuclei during early embryogenesis of *Drosophila melanogaster*, *J. Microsc. Biol. Cell* **25:**97–106.

Zatsepina, O. V., Polyakov, V. Y., and Chentsov, Y. S., 1982, Nuclear envelope formation around metaphase chromosomes: Chromosome decondensation and nuclear envelope reconstitution during mitosis, *Eur. J. Cell Biol.* **26:**277–283.

Zeligs, J. D., and Wollman, S. H., 1979, Mitosis in rat thyroid epithelial cells *in vivo*, *Int. J. Ultrastruc. Res.* **66:**53–77.

Zickler, D., 1981, Mechanisms of nuclear division in fungi, in *The Fungal Nucleus* (K. Gull and S. G. Oliver, eds.), pp. 85–113, Cambridge University Press, Cambridge, England.

Zimmer, K-P., Hengst, K., Carayon, P., Bramswig, J., and Harms, E., 1992, Different concentrations of thyroid peroxidase and thyroglobulin in the nuclear envelope and the endoplasmic reticulum throughout the cytoplasm, *Eur. J. Cell Biol.* **57:**12–20.

Chapter 10

Membrane Assembly in Bacteria

Matthias Müller and Juan MacFarlane

1. INTRODUCTION

Membrane biogenesis requires *de novo* synthesis of protein as well as phospholipid. This chapter focuses on the molecular events by which newly synthesized proteins are incorporated into bacterial membranes. By far the most detailed knowledge of these processes stem from studies performed with gram-negative bacteria, in particular with *E. coli*. The following description will also include some recent results obtained with a different gram-negative bacterium, *Rhodobacter capsulatus*, which from an evolutionary point of view is rather distantly related to *E. coli* (Woese, 1987).

The cell envelope of gram-negative bacteria is composed of two membrane layers, the plasma or inner membrane, hereafter called *cytoplasmic membrane*, and the outer membrane. The two membranes are separated by the intermediate periplasmic space. Whether the two membranes intimately associate at some places (so-called adhesion sites, or Bayer bridges, as revealed by electron microscopy), at least during the process of cell division and septation (periseptal anuli), is still a matter of debate (Kellenberger, 1990). In any case, several lines of evidence suggest that an outer membrane protein has first to overcome the obstacle of the plasma membrane before it integrates into the lipid bilayer of the

Matthias Müller and Juan MacFarlane   Institut für Physikalische Biochemie der Universität München, D-80366 München, Germany.

Subcellular Biochemistry, Volume 22: Membrane Biogenesis, edited by A. H. Maddy and J. R. Harris. Plenum Press, New York, 1994.

outer membrane. Consequently, integration into the two membranes of gram-negative bacteria represent distinct processes.

Mechanistically speaking, the integration of a particular membrane protein results from an interrupted translocation process. In this concept (Blobel, 1980) translocation of a (part of a) protein is initiated by distinct signals, termed *signal sequences,* which are present on the polypeptide chain itself. Signal sequences lead to a complete translocation of the protein unless information to stop the transfer (stop-transfer signal) is provided by sections of the protein downstream of the signal sequence. Stop-transfer signals cause a halt in translocation with a concomitant or subsequent incorporation of the protein segment encoding this information into the membrane lipids.

In eukaryotic organisms, protein translocation into the endoplasmic reticulum as well as integration into the endoplasmic reticular membrane of many membrane proteins depend on the function of the signal recognition particle, SRP (reviewed in Saier *et al.,* 1989). Thus, both processes involve the same SRP-catalyzed membrane-targeting steps. In gram-negative bacteria the situation is less clear. The targeting to the cytoplasmic membrane of those proteins that leave the cytoplasm and are exported to one of the compartments of the cell envelope is so far the best-understood event during bacterial protein export. The question as to what extent this targeting mechanism also applies to integral membrane proteins is now being addressed. This chapter will start (Section 2.1) with a description of how secretory proteins are targeted to the cytoplasmic membrane of *E. coli,* followed by an account of the known determinants of protein integration into the cytoplasmic membrane of *E. coli.* This latter section will also list the methods used to study membrane assembly of bacterial integral membrane proteins. Section 2.2 summarizes what is known about the integration of proteins into the outer membrane of *E. coli.* Finally, unique features of protein export in *R. capsulatus* will be discussed (Section 3).

2. PROTEIN INTEGRATION INTO THE MEMBRANES OF *ESCHERICHIA COLI*

Proteins whose final destination is one of the three layers of the cell envelope of gram-negative bacteria are termed *exported proteins.* The process of protein export encompasses (1) recognition of these proteins during or after their synthesis within the cytoplasm; (2) their targeting to the cytoplasmic membrane; (3) their complete or partial translocation across the cytoplasmic membrane; and (4) where appropriate, their targeting to the outer membrane.

2.1. Cytoplasmic Membrane of *E. coli*

In view of the similarities established for the initial steps of protein translocation across, and integration into, the membrane of the endoplasmic reticulum (ER) (see Introduction), this chapter begins with a description of the molecular mechanisms involved in targeting of secretory proteins to, and their complete translocation across, the cytoplasmic membrane of *E. coli*. The reader is referred to recent reviews of this topic (Saier *et al.*, 1989; Schatz and Beckwith, 1990; Wickner *et al.*, 1991). The second part of the chapter will deal with the question of the extent that protein integration into the *E. coli* cytoplasmic membrane resembles the complete translocation of proteins across this membrane and what kind of molecular determinants are known to lead to anchorage of proteins within the cytoplasmic membrane of *E. coli*.

2.1.1. Factors Involved in Protein Translocation across the Cytoplasmic Membrane

Detailed genetic analyses have revealed that the export of periplasmic and outer-membrane proteins across the cytoplasmic membrane of *E. coli* is dependent on a distinct set of Sec-proteins (Sec stands for secretion; Bieker *et al.*, 1990; Schatz and Beckwith, 1990). The basic role Sec-proteins play in protein export in *E. coli* has been illustrated biochemically by the reconstitution of transmembrane transport of one precursor protein (preOmpA) from SecA, ATP, SecY/SecE-containing liposomes, and purified precursor (Brundage *et al.*, 1990; Akimura *et al.*, 1991). While SecA and SecB are required for targeting of precursor proteins to the cytoplasmic membrane, SecY most likely functions as a translocator in the membrane in concert with at least SecE. SecD and SecF are presumably involved in late translocation steps. The transmembrane movements of polypeptide chains require two sources of energy, ATP and the H^+-motive force.

2.1.1a. SecB: A Targeting Chaperone or a Chaperoning Targeting Factor?   SecB (reviewed in Kumamoto, 1990) is a 17 kDa soluble protein that elutes from gelfiltration columns as multimer (Weiss *et al.*, 1988; Kumamoto *et al.*, 1989; Watanabe and Blobel, 1989a). A *secB*[null]-mutation interferes with the export of maltose-binding protein (MBP), OmpF, lambda receptor (LamB), OmpA, and PhoE, but does not affect that of alkaline phosphatase, lipoprotein, ribose-binding protein, and β-lactamase (Kumamoto and Beckwith, 1985; Watanabe *et al.*, 1988; Kusukawa *et al.*, 1989; Kusters *et al.*, 1989; Collier *et al.*, 1990; Altman *et al.*, 1990b; Laminet *et al.*, 1991). Mutants of the ribose-binding protein exhibiting a diminished export efficiency, however, were found to be dependent on SecB (Kim *et al.*, 1992). Sec B is essential only for *E. coli* cells growing on rich medium (Kumamoto and Beckwith, 1985). A direct molecular interaction of SecB with preproteins has frequently been described (Collier *et al.*, 1988; Kumamoto, 1989; Lecker *et al.*, 1989; Watanabe and Blobel, 1989b; Lecker *et al.*, 1990). Both the signal sequence itself (Kumamoto, 1989; Watanabe and Blobel, 1989b; Altman *et al.*, 1990b) and regions outside of it (Collier *et al.*, 1988; Trun *et al.*, 1988; Watanabe *et al.*, 1988; Randall *et al.*, 1990; Altman *et al.*, 1990a; de Cock *et al.*, 1992) were reported to be recognized by SecB. Even nonsecretory proteins interact with SecB if they are in a nonnative conformation (Randall, 1992), the specificity of SecB for exported proteins being explained by reduced folding kinetics of signal sequence-bearing proteins (Hardy and Randall, 1991; MacIntyre *et al.*, 1991). The selective binding of SecB to nascent precursors of MBP, LamB, OmpA, and OmpF was recently demonstrated by Kumamoto and Francetić (1993).

As to the function of SecB, an antifolding, (i.e., chaperoning) activity is suggested by the coincidence of a SecB-dependence of a particular precursor protein and its tendency to fold rapidly into a transport-incompetent structure (Collier *et al.*, 1988; Weiss *et al.*, 1988; Collier and Bassford, 1989; Liu *et al.*, 1989). Direct biophysical analysis, however, revealed little effect of SecB on the actual folding of the precursor of PhoE, but rather showed that the formation of a stabile SecB-prePhoE complex prevents the precursor from aggregating (Breuking *et al.*, 1992). In contrast to SecB, the general folding modulator, GroEL, is not specific for exported proteins (Bochkareva *et al.*, 1988; Phillips and Silhavy, 1990; Kumamoto, 1991). Only the export of β-lactamase was found to be affected by a GroEL-deficiency (Kusukawa *et al.*, 1989). On the other hand, the demonstration of a competition between SecB and the eukaryotic signal recognition particle (SRP) prompted Watanabe and Blobel (1989b) to suggest a signal recognition function for SecB. Recent evidence indicates that the main function of SecB might be the targeting of precursor proteins to sites at the cytoplasmic membrane of *E. coli* (comprising at least Sec A, SecY, SecE; Hartl *et al.*, 1990), from which a productive transmembrane translocation takes place (Swidersky *et al.*, 1990; Hartl *et al.*, 1990; de Cock and Tommassen, 1992). In mediating contact between precursor and membrane, SecB would certainly help to prevent a premature folding of the precursor.

2.1.1b. SecA: An Autoregulated Targeting Factor. SecA (reviewed in Oliver *et al.*, 1990b) is a 102 kDa protein that is found attached to the plasma membrane, in a soluble form (Cabelli *et al.*, 1991), and ribosome-associated (Liebke, 1987; Swidersky *et al.*, 1990). A SecA-defect is lethal, and conditional *secA* mutants revealed a requirement of SecA for the export of all periplasmic and outer-membrane proteins tested thus far (Oliver *et al.*, 1990b). SecA exhibits a low ATPase-activity (Lill *et al.*, 1989) stimulated by precursor and cytoplasmic membrane vesicles. Blocking the ATP-binding sites of SecA with 8-azido-ATP (Lill *et al.*, 1989) and sodium azide (Oliver *et al.*, 1990a; Fortin *et al.*, 1990) abolishes the translocation-promoting function of SecA. A direct demonstration, however, of a coupling between ATP-hydrolysis and the translocation-promoting function of SecA has so far remained elusive (Schiebel *et al.*, 1991; Driessen, 1992a). Interaction of SecA with precursor proteins via the signal sequence is suggested by the isolation of *secA (prlD)* mutants, which suppress a number of signal sequence mutations of maltose binding protein (Fikes and Bassford, 1989) and by a signal sequence-dependent cross-linking of SecA to the model export protein OmpF-Lpp (Akita *et al.*, 1990). SecA further associates with the membrane-embedded SecY (Fandl *et al.*, 1988; Hartl *et al.*, 1990; Bieker-Brady and Silhavy, 1992).

An involvement of acidic phospholipids in the fixation of SecA at the plasma membrane has also been demonstrated (Lill *et al.*, 1990; Cabelli *et al.*, 1991; Hendrick and Wickner, 1991; Shinkai *et al.*, 1991; Ulbrandt *et al.*, 1992). The level of SecA is regulated by the protein export capability of the cell with the SecA protein serving as autogenous repressor that directly interacts with its own mRNA (Schmidt *et al.*, 1991), a process that also requires ATP (Dolan and Oliver, 1991). Blockage of protein export by mutations in *secA, secD, secE*, and *secY* by nontranslocatable hybrid proteins, and by a changed lipid composition of the cytoplasmic membrane, all lead to derepression of SecA (Oliver *et al.*, 1990b; Kusters *et al.*, 1992).

From these properties it appears that SecA fulfills three major functions in the cell: (1) regulation of its own synthesis, which might involve an ATP-dependent RNA-helicase activity (Koonin and Gorbalenya, 1992); (2) targeting of precursors to the SecY-containing export sites of the plasma membrane (Cunningham *et al.*, 1989; Hartl *et al.*, 1990; Swidersky *et al.*, 1990); as with SecB, the binding of SecA to precursor proteins might simultaneously serve a certain chaperoning function, as indicated by the fact that a SecA defect can be suppressed by GroE (Ueguchi and Ito, 1992) as well as by Skp (see Section 2.2.3); (3) participation in the transmembrane movement of a polypeptide chain potentially by preventing reverse translocation (Schiebel *et al.*, 1991). A functional mapping of the SecA protein revealing the binding sites for ATP, precursor molecules, membrane, and mRNA has been initiated by several groups (Matsuyama *et al.*, 1990; Cabelli *et al.*, 1991; Jarosik and Oliver, 1991; Kimura *et al.*, 1991).

2.1.1c. SecY: Presumably a Major Constituent of the Translocation Pore.

SecY is a 49 kDa protein embedded in the lipid bilayer of the cytoplasmic membrane of *Escherichia coli*. It is essential for the export of periplasmic and outer membrane proteins of *E. coli* since mutations in *secY* result in temperature-sensitive export defects. This was shown both *in vivo* (Ito *et al.*, 1983, 1989; Shiba *et al.*, 1984) and *in vitro* (Fandl and Tai, 1987; Baba *et al.*, 1990), although under certain *in vitro* conditions SecY has been found to be dispensable (Watanabe *et al.*, 1990). A model of its membrane topography predicts ten hydrophobic, transmembrane segments with the short NH_2- and COOH-terminal peptides facing the cytoplasm (Akiyama and Ito, 1987, 1989). Such a transmembrane orientation of SecY is consistent with the formation of, or at least participation in the formation of, a hydrophilic pore across the membrane to allow translocation of exported proteins. The translocation pore-forming function of SecY is strongly supported by the selective cross-linking of a translocating polypeptide chain to SecY (Joly and Wickner, 1993). Moreover, SecY was recently reported to exhibit significant structural homology to the SEC61 protein of the endoplasmic reticular membrane, which intimately interacts with nascent translocating proteins (Görlich et al., 1992).

The cytoplasmic loops of the molecule, including the NH_2- and COOH-terminal peptides, are involved in precursor binding. This is demonstrated by a decreased targeting of preproteins to cytoplasmic membrane vesicles that had been coated with anti-SecY-antibodies specific for cytoplasmic epitopes (Watanabe and Blobel, 1989c; Tokuda et al., 1990; Swidersky et al., 1992) or prepared form the secY24 mutant (Swidersky et al., 1992) carrying the mutation in one of the cytoplasmic loops (Shiba et al., 1984). The targeting of precursor molecules to SecY appears to be mediated by SecA, as several indications exist for a molecular interaction between SecY and SecA (see 2.1.1b).

One of the main issues about the function of SecY is whether it directly recognizes the signal sequences of precursor proteins. Such an interaction is suggested by the suppression of the export defects of various signal sequence mutations by the prlA4-allele of secY as demonstrated in vivo (Emr et al., 1981; Emr and Bassford, 1982; Stader et al., 1986; Sako and Iino, 1988; Puziss et al., 1992) and recently also in vitro (Swidersky et al., 1992). The interpretation of these data in terms of a binding of signal sequences to SecY (PrlA) has, however, been questioned (Randall et al., 1987) because of a lack of allele-specificity— i.e., one suppressor has been found to be active on more than one signal sequence mutation (Emr et al., 1981; Emr and Bassford, 1982; Stader et al., 1986). However, such a clear-cut allele-specificity might not be expected in view of the fact that the function of signal sequences is not mediated by a consensus sequence but rather by some common tertiary-structure motif (for a discussion of this matter see also Bieker et al., 1990; Ito, 1990).

On the contrary, prlA mutants were shown to allow for export of proteins deprived of their entire signal sequences (Derman et al., 1993), thus arguing against an interaction between SecY and signal sequences as an essential feature of protein export in E. coli. Simon and Blobel (1992) demonstrated that a signal peptide is able to open large transmembrane aqueous channels in the plasma membrane of E. coli probably by directly binding to the mouth of the channel (pore) protein(s). This finding reinforces the idea of a direct interaction of precursors with membrane-embedded pore-proteins such as SecY. A functional interaction of SecY with SecE (Section 2.1.1d) is suggested by the co-purification of both proteins (Brundage *et al.*, 1990) and the titration of the SecE-activity with inactive SecY (Bieker and Silhavy, 1990).

2.1.1d. SecE. The *secE*-locus was first identified by a mutation causing a pleiotropic defect in protein export and a cold-sensitive growth defect (Riggs *et al.*, 1988). It encodes a 13.6 kDa protein integrated into the plasma membrane via three transmembrane segments (Schatz *et al.*, 1989). A distinct part of the molecule encompassing only one of the transmembrane segments appears to be sufficient for the activity of SecE (Schatz *et al.*, 1991). Two of the *prlG*-alleles of *secE*, whose phenotype is characterized by the suppression of signal sequence mutations of LamB and MalE (Stader *et al.*, 1989), map to the essential trans-membrane segment of SecE (Schatz *et al.*, 1991). Identification of *prlG*-suppressors suggests an interaction of SecE with signal sequences similarly as detailed for SecY. Because signal sequence-bearing hybrid proteins that enter the export site without being completely translocated are not proteolytically processed in *prlG*-mutants as opposed to *prlA*-mutants, it was proposed that SecE acts at a step before SecY during protein export (Bieker and Silhavy, 1990; Bieker-Brady and Silhavy, 1992). Conversely, inactivation of SecY causes a decrease in precursor binding, indicating that SecE cannot function as the primary membrane receptor for exported proteins when SecY is inhibited (Swidersky *et al.*, 1992).

2.1.1e. SecD/SecF. Mutations in the *secD*-locus result in severe defects in protein export and in a cold-sensitivity for growth (Gardel *et al.*, 1987). They fall into two complementation groups, termed *secD* and *secF*. The two genes are co-transcribed. Both encode integral membrane proteins with large periplasmic domains (Gardel *et al.*, 1990). This feature, in combination with the failure to isolate alleles of *secD/F* that suppress signal sequence mutations, suggests an involvement of SecD/F in a later step of translocation across the cytoplasmic membrane. In fact, blockage of SecD at the periplasmic site of the plasma membrane by anti-SecD antibodies interferes with the release of MBP and OmpA, leading to an accumulation of both proteins within the cell (Matsuyama *et al.*, 1993). Recent genetic evidence suggests an interaction of SecD/F with SecY (Bieker-Brady and Silhavy, 1992).

2.1.1f. Energy Requirement. Two sources of energy are utilized to drive translocation of exported proteins across the cytoplasmic membrane of *E. coli* (recently summarized in Geller, 1991; Driessen, 1992b; see individual references therein): ATP-hydrolysis and the H^+-motive force, $\Delta\mu H^+$. The export process can be separated into single steps, each one having an individual energy requirement. Membrane binding of precursor proteins occurs in the absence of ATP (Swidersky *et al.*, 1990). Translocation of the first 40–50 amino acids leading to the formation of a hairpin-loop structure (Inouye and Halegoua, 1980) (such that the NH_2-terminus of the signal peptide remains on the cytoplasmic side and a turn around the cleavage site is located on the periplasmic face of the cytoplasmic membrane) requires ATP and is unaffected by $\Delta\mu H^+$. Nonhydrolyzable ATP binding to SecA was shown to be sufficient to generate such a translocation intermediate. A complete translocation involves further steps that are dependent on ATP-hydrolysis and/or $\Delta\mu H^+$. SecA-mediated ATP-hydrolysis (see Section 2.1.1b.) has not been shown to be strictly coupled to net protein translocation. Circumstantial evidence suggests that ATP-hydrolysis by SecA might cause dissociation of the precursor from SecA to allow further translocation (Schiebel *et al.*, 1991; Bassilana *et al.*, 1992). Both components of $\Delta\mu H^+$, the electrical gradient $\Delta\psi$, and the chemical H^+-gradient ΔpH, can meet the requirements for $\Delta\mu H^+$ (Bakker and Randall, 1984; Driessen, 1992a). While translocation of certain precursors proceed in the complete absence of $\Delta\mu H^+$, albeit at a reduced rate, SecA can substitute *in vitro* for the need of $\Delta\mu H^+$. The $\Delta\psi$ might exert an electrophoretic effect on the transmembrane movement of polypeptide chains and/or influence the rate of the reaction by changing the kinetic parameters of membrane components. Dependence on ΔpH suggests an involvement of H^+-transfer reactions. This view is corroborated by the finding that translocation is significantly retarded in D_2O relative to H_2O. Incidentally, overproduction of some truncated forms of SecY dissipates $\Delta\mu H^+$ (Ito, 1990), pointing toward the possibility of a H^+-transfer activity of SecY. It should be kept in mind that many of the results described here were obtained by *in vitro* studies conducted with one particular precursor protein only. The energy requirement of protein export, however, has been shown to vary dramatically with the precursor studied (Yamada *et al.*, 1989; Ernst *et al.*, in preparation).

    2.1.1g. Open Questions. Although transmembrane translocation of one precursor protein (preOmpA) has been reconstituted from a few defined components (Section 2.1.1) future work will have to address several unresolved problems: (1) the translocation efficiency of these minimal *in vitro* systems is low (Brundage *et al.*, 1990; Akimura *et al.*, 1991), suggesting the need for auxiliary components; (2) SecY and SecE appear to act in concert with other membrane proteins, such as band 1 (Brundage *et al.*, 1992), and Ydr (Ito, 1992), whose precise functions, however, are unknown; (3) certain precursor proteins appear to translocate into membrane vesicles independently of SecY (Watanabe *et al.*, 1990); (4) the signal recognition process is poorly understood; whereas SecA seems to recognize positive charges of the signal sequence (Akita *et al.*, 1990), the signal sequence binding subunit of the eukaryotic SRP binds via hydrophobic interactions (Zopf *et al.*, 1990; High and Dobberstein, 1991); (5) *E. coli* contains a ribonucleoprotein analogous to SRP with an as yet ill-defined physiological function (Poritz *et al.*, 1990; Ribes *et al.*, 1990); specific cross-linking of the *E. coli* ribonucleoprotein to a eukaryotic signal sequence suggests that it interacts with signal sequences also in *E. coli* (Luirink *et al.*, 1992); depletion of the *E. coli* homolog of the 54 kDa subunit of SRP causes a defect in the export of several precursor proteins (Phillips and Silhavy, 1992); and (6) most precursor proteins can translocate posttranslationally in *E. coli,* but it is poorly understood whether components of the export apparatus associate with a nascent precursor molecule.

2.1.2.   Determinants of Protein Integration into the Cytoplasmic Membrane

Integral membrane proteins (IMPs) are characterized by one or several stretches of mostly hydrophobic amino acids that are long enough (about 20 residues) to span the plane of cellular membranes. These transmembrane segments anchor the protein within the lipid bilayer of the membrane. The topography of an individual IMP is given by the number of transmembrane segments (single and multiply spanning), the localization and size of their extramembrane, hydrophilic domains, and the orientation of the NH_2-terminus, being either on the *trans*-side of the membrane (so-called class I IMP) or alternatively on the *cis*-side (class II IMP).

Unraveling the molecular prerequisites for the integration of IMPs is generally hampered by the difficulty in experimentally assessing proper integration (resulting in an authentic topography and, consequently, functional activity of an IMP) as opposed to its mere association with the structures of the membrane. In contrast to completely translocated proteins, no proteolytic processing—i.e., removal of a signal sequence—accompanies the integration of an IMP into the membrane. In bacteria, glycosylation of the translocated domains of an IMP does not take place. Therefore, no natively occurring covalent modification serves as an indication for the correct integration of IMPs into the bacterial cytoplasmic membrane (an exception is the processing of the NH_2-terminal Met discussed in 2.1.2a).

2.1.2a.   **Methodological Strategies to Assess the Topography of Bacterial IMPs.**   The direct embedment of a protein within the phospholipids of cellular membranes (not involving other proteins) renders this protein resistant toward extraction with protein denaturants such as Na_2CO_3 at alkaline pH (Fujiki *et al.,* 1982) or 0.1 *N* NaOH (Russel and Model, 1982). Applied to bacterial membranes, a considerable amount of nonlipid anchored proteins turns out to be also nonextractable by Na_2CO_3 (Ahrem *et al.,* 1989) unless the method is properly modified (Ito and Akiyama, 1991). Similarly, 5–6 *M* urea releases peripherally associated membrane proteins, whereas lipid-embedded proteins are not affected (Gilmore and Blobel, 1985). This method applied to *in vitro* assembled bacterial IMPs gives results comparable to other procedures that indirectly assay membrane integration of IMPs (Troschel *et al.,* 1992; Werner *et al.,* 1992). At 9 *M* urea, however, even lipid-anchored proteins are solubilized (Tadros *et al.,* 1989), indicating that the effective concentration of urea needs to be determined individually.

Limited proteolysis of an IMP should result in a distinct set of proteolytic fragments protected against further degradation by the membrane lipids. The pattern of fragments thus created should be different from that of a nonintegrated or an incorrectly integrated form of the protein, which has in fact been demonstrated for bacterial IMPs (Ahrem *et al.*, 1989; Werner *et al.*, 1992). Both the kind and the number of fragments generated depend on the particular protease used and the conditions of incubation chosen (Stochaj *et al.*, 1986). Furthermore, the stability of a newly inserted IMP monitored by a pulse/chase approach allows conclusion of whether or not correct, i.e., stable, integration has occurred (Bibi *et al.*, 1991; Richter and Drews, 1991).

In contrast to the aforementioned operational criteria, a direct probing of the membrane topography of an IMP can be achieved by assessing the degree of proteolytic degradation when the protease is added *in situ* from either side of the cytoplasmic membrane. Thus, periplasmic localization of parts of an IMP is concluded from partial proteolysis following addition of the protease to pulse-labeled *E. coli* cells whose outer membrane was permeabilized and therefore allows access to the periplasmic space (for example, Eckert and Beck, 1989; Kuhn *et al.*, 1990; Nilsson and von Heijne, 1990; Yamane *et al.*, 1990). Conversely, cytoplasmic moieties of an IMP are rendered accessible to proteolysis by disrupting the spheroplasts or using inside-out plasma membrane vesicles (Seckler and Wright, 1984; Eckert and Beck, 1989). The transmembrane topography of an IMP is also probed using epitope-specific antibodies (Seckler *et al.*, 1986). Blockage of the NH_2-terminus, reflected by the inability to perform Edman-degradation, is an indication for a periplasmic localization of the NH_2-terminus because the formylated NH_2-terminal Met is removed when retained in the cytoplasm (von Heijne, 1989). The sidedness of the NH_2-terminus can be determined also by chemical modification of the free α-NH_2-group as shown for the tetracycline resistance protein (Eckert and Beck, 1989).

Another way to analyze the transmembrane topography of an IMP is the molecular insertion of foreign protein domains whose membrane sidedness is easily tested. This has been verified by introducing a small antigenic peptide, the first 18 amino acids of the Pf3 bacteriophage coat protein, at the beginning of *E. coli* signal peptidase I to immuno-localize the NH_2-terminus (Lee *et al.*, 1992). Furthermore, a widely used technique is the fusion of (parts of) the gene encoding the IMP with that of signal sequence-free alkaline phosphatase (*phoA*) (Manoil *et al.*, 1990). PhoA exhibits enzymatic activity only when translocated to the periplasmic space where it can assemble into the dimeric structure required for activity (Manoil *et al.*, 1990). In addition, periplasmic PhoA, as opposed to a cytoplasmically located enzyme, is protease-resistant. Thus, PhoA-activity of an IMP-PhoA-fusion protein is always an indication that the IMP-derived, NH_2-terminal fraction of the fusion protein must contain topogenic information for the translocation of the downstream PhoA-moiety. Instead of the *phoA*-gene the *bla-*

gene encoding periplasmic β-lactamase is also used to create these kinds of diagnostic fusion proteins (Zhang and Broome-Smith, 1990).

Finally, an unequivocal proof of an IMP acquiring its authentic tertiary structure within the lipid environment of the membrane is its enzymatic activity, such as the catalysis of transmembrane transport processes (Bibi *et al.*, 1991; Ehrmann and Beckwith, 1991; Werner *et al.*, 1992). In the case of IMPs synthesized *in vitro* and assembled into membrane vesicles, this approach is more difficult to realize because of the minute amounts of IMP made by the usual *in vitro* systems. Nevertheless, enzymatic activity of two *in vitro* synthesized IMPs was demonstrated by a functional reconstitution of membrane vesicles that, prior to the experiment, lacked this activity (Ahrem *et al.*, 1989; Swidersky *et al.*, 1992). Correct membrane assembly of *in vitro* synthesized photosynthetic proteins of *R. capsulatus* (Troschel and Müller, 1990) was indicated by the spectroscopically verified formation of supramolecular complexes between protein and pigment molecules.

2.1.2b. Topogenic Sequence Determinants of IMPs. The signals responsible for the insertion of IMPs are called *topogenic sequences* (Blobel, 1980) and reside in and around the transmembrane segments of the membrane proteins (reviewed in Hartmann *et al.*, 1989; Saier *et al.*, 1989; Boyd and Beckwith, 1990; Dalbey, 1990). They are defined as discrete regions that allow initiation (signal sequences) and termination (stop-transfer sequences) of polypeptide chain translocation. A stop-transfer sequence always leads to anchorage of the polypeptide chain within the lipid bilayer, whereas a signal sequence does so only if not cleaved during assembly of the protein (signal anchor or internal signal sequence). The final topography of an IMP spanning the membrane several times was proposed to result from the concerted action of pairs of signal anchor and stop-transfer sequences, the first one specifying translocation of downstream sequences, the second one halting this translocation process and thereby leading to the generation of cytoplasmic domains (Blobel, 1980). The functionality of multiple topogenic sequences within an IMP becomes manifest when these are individually fused to reporter proteins. This is the most direct way to disclose the topogenic information inherent to a given polypeptide chain segment.

The topogenic information of prokaryotic IMPs is contained in their hydrophobic transmembrane stretches and adjacent charged residues. The orientation of a topogenic sequence within the membrane bilayer is usually such that positive charges remain on the *cis*-side (i.e., cytoplasmic side) of the membrane. This was originally revealed by a statistical analysis of the distribution of charged amino acids within the *cis*- and *trans*-loops of IMPs ("positive-inside rule"; von Heijne, 1986). The fact that positively charged amino acids are usually not translocated across the membrane bilayer applies also to eukaryotic IMPs (von Heijne and Gavel, 1988) and is also verified for cleavable signal sequences whose positively charged NH_2-terminus remains cytoplasmic (reviewed in Müller, 1992) and which become inactive upon introduction of positive charges at their COOH-terminal end (Yamane and Mizushima, 1988; Puziss et al., 1989; Summers et al., 1989; Summers and Knowles, 1989).

In general, the distribution of positive charges around a transmembrane segment determines its orientation and consequently its topogenic information: Preceding positive charges render the following transmembrane segment a signal sequence (cleavable or internal); subsequent clusters of basic residues confer stop-transfer information on the preceding transmembrane segment. This has been experimentally verified by introducing positively charged amino acids before and after the transmembrane segments of prokaryotic (Nilsson and von Heijne, 1990; Yamane et al., 1990) and eukaryotic (Beltzer et al., 1991; Parks and Lamb, 1991) IMPs and determining the transmembrane orientation of the modified proteins. It was thereby shown that it is possible to reverse the orientation of a transmembrane segment and that, if both boundaries of a transmembrane segment contain positive charges, the dominating part of the topogenic sequence determining the orientation of the transmembrane segment appears to be the one that has a higher net positive charge and contains more Arg than Lys (charge-difference rule by Hartmann et al., 1989; Nilsson and von Heijne, 1990; McGovern et al., 1991 and references therein). The distance between the hydrophobic string and the basic amino acids influences the strength of the topogenic signal with a weakening concomitant with an increase in distance (discussed in Boyd and Beckwith, 1990). It should be mentioned, however, that exceptions to these rules have been described (Hartmann et al., 1989; Kuhn et al., 1990; Andrews et al., 1992; Krijnse Locker et al., 1992).

Central to the question of how a polypeptide chain assembles into a polytopic IMP is whether the topogenic information of the most NH_2-terminal topogenic sequence dominates in such a way that incorporation of downstream transmembrane segments follows passively in a sequential order. This problem was addressed experimentally by removing selected transmembrane segments of polytopic IMPs of E. coli and probing the topography of the truncated IMP using PhoA-fusions, protease-accessibility, and activity measurements (Bibi et al., 1991; Ehrmann and Beckwith, 1991; McGovern et al., 1991). The results are in line with previous reports on eukaryotic IMPs (Audigier et al., 1987), demonstrating that COOH-terminal transmembrane segments contain topogenic information that is decoded separately from preceding topogenic sequences. Thus, whereas in certain instances removal of a transmembrane segment early in the protein causes inversion of the orientation of subsequent portions, compatible with a sequential mode of assembly (McGovern et al., 1991), strong COOH-terminal topogenic sequences counteract inversion in other constructs (Bibi et al., 1991; McGovern et al., 1991). In the latter cases the experimental strategy employed to demonstrate topogenic information of COOH-terminal trans-membrane segments involved IMP-PhoA-fusions. This allows an accurate assessment of the sidedness only of those parts of the IMP that follow the late topogenic signal. Therefore, it remains essentially unclear as to how an IMP that contains an even number of transmembrane segments, so exposing NH_2- and COOH-termini to the same side of the membrane, partially reorientates when one of the transmembrane segments is removed. Such a construct may now either adopt a new orientation with both termini pointing to opposite sides of the membrane or retain its original orientation, thereby forcing one hydrophobic segment to appose to, rather than traverse, the membrane.

Although such truncated IMPs are artificial constructs (which might not even yield a homogeneous population of IMP molecules with the same topography, McGovern *et al.*, 1991), the results obtained have an impact on the idea how IMPs might assemble into the bacterial cytoplasmic membrane. If late topogenic information exists within a polypeptide chain, the assembly process must account for this by allowing late (i.e., stepwise) decoding of individual topogenic sequences. This might, at least potentially in the case of truncated IMPs, require correcting (reversing) of the transmembrane orientation of upstream stretches. This is difficult to reconcile with a co-translational insertion mode that usually might be the preferred one because of the hydrophobicity of IMPs (Ahrem *et al.*, 1989), unless a flip-flop mechanism is postulated, enabling postinsertional reorientation of transmembrane segments. Either such a situation does not occur in nature because the topogenic sequences have been selected in such a way that signal anchor sequences are always followed by stop-transfer sequences, a combination that does not require late corrections or, alternatively, the final insertion into the lipid is achieved from an insertion intermediate that might conceivably form on the surface of a soluble or membrane-bound chaperone. Accordingly, recent examination of the membrane topography of the *E. coli* tetracycline resistance protein employing PhoA-fusions suggested long-range interactions at least between the two adjacent transmembrane segments to guarantee proper insertion (Allard and Bertrand, 1992). These authors reported that fusions of PhoA to three predicted periplasmic segments of the tetracycline resistance protein gave low, rather than high, PhoA activity because aspartate residues near the middle of the preceding transmembrane segments prevented the latter from functioning as efficient insertion signal. In the full-length molecule, these weak topogenic signals must then be somehow compensated for by flanking sequences provided that the proposed topography of this protein is correct.

2.1.2c. Does the Integration of Proteins into the Cytoplasmic Membrane of *E. coli* Depend on Sec-Proteins?   As mentioned in the Introduction, eukaryotic IMPs are integrated into the ER membrane in an SRP-dependent fashion, suggesting that the ER-targeting mechanism is the same for both translocated and integrated proteins. Extending this notion to prokaryotic cells, one would expect that the integration of IMPs into the cytoplasmic membrane of *E. coli* requires the Sec-proteins.

In a global analysis Baker *et al.* (1987) found that in two *secA*[ts]-mutants the rate of integration of newly synthesized total protein into outer- and inner-membrane fractions decreased by about 70% at the restrictive temperature. The Sec-dependence of several distinct cytoplasmic membrane proteins of *E. coli* has been investigated in more detail. In some cases fusion proteins consisting of (parts of) the IMP under investigation and a reporter protein such as PhoA (Gebert *et al.*, 1988; Akiyama and Ito, 1989; Kuhn *et al.*, 1990; McGovern and Beckwith, 1991) and others (Cobet *et al.*, 1989; Kuhn *et al.*, 1990; McGovern and Beckwith, 1991) were used for these studies. Only the fusion proteins made from MalF, a cytoplasmic membrane protein involved in maltose uptake, were reported to integrate properly in the absence of Sec-functions (McGovern and Beckwith, 1991), whereas all other fusions studied required Sec-proteins for membrane assembly. In these experimental conditions it is difficult to discern between the Sec-dependence of the integration of the IMP-moiety and that of the export of the reporter domain. The finding that the *E. coli* signal peptidase I (SPase I, leader peptidase) requires SecA and SecY for membrane assembly only if its large periplasmic domain has to be translocated across the plasma membrane (see below) suggests that the Sec-dependence of the fusion proteins listed above might actually reflect that of the fused hydrophilic portions.

Sec-dependent integration has also been investigated for authentic IMPs of *E. coli*, including SPase I, lactose permease (LacY), MalF, mannitol permease (MtlA), and SecY. Membrane-embedded SPase I has the NH_2-terminus in the periplasm followed by a hydrophobic transmembrane segment, a 25-amino-acid-long cytoplasmic loop, a second transmembrane segment, and a large periplasmic domain comprising about two-thirds of the protein including the COOH-terminus (San Millan *et al.*, 1989). Its membrane insertion requires SecA and SecY (Wolfe *et al.*, 1985). However, the insertion of SPase I becomes Sec-independent if the orientation of the protein is inversed (von Heijne, 1989), which can be achieved by reducing the positive net charge of the cytoplasmic loop of the protein (Nilsson and von Heijne, 1990). In this case the large periplasmic domain remains in the cytoplasm. Such constructs with inversed topography can be again rendered Sec-dependent if the now periplasmically located loop between the two transmembrane segments is increased in size. The degree to which translocation of the lengthened constructs requires SecA and SecY is linearly correlated with the length of the loop within a range of 25–55 amino acid residues (Andersson and von Heijne, 1993). Consequently, IMPs with less extended extramembrane domains might not use the Sec-dependent export pathway at all. Recent reports seem to confirm this notion. LacY, which spans the membrane 12 times with both NH_2- and COOH-termini in the cytoplasm (Foster *et al.*, 1983; Seckler *et al.*, 1983; Calamia and Manoil, 1990), was found to be enzymatically active in *secA*- and *secY*-mutant cells (Yamato, 1992). Using extractibility by alkali as a criterion for a lack of membrane integration, however, SecY was reported to be required for the membrane integration of LacY (Ito and Akiyama, 1991).

The pattern of proteolytic fragments derived from MalF, an eightfold transmembrane protein of the cytoplasmic membrane with an approximate 180-amino-acid-long periplasmic loop, was the same whether or not cells had been treated with sodium azide, which is known to inhibit SecA-functions (Section 2.1.1b). This finding suggests a SecA-independent integration of MalF (McGovern and Beckwith, 1991). Similarly, a combined *in vitro* and *in vivo* approach employing partial proteolysis and measurement of enzymatic activity to demonstrate proper membrane integration revealed that the assembly of mannitol permease in the cytoplasmic membrane of *E. coli* proceeds in the absence of SecA, whereas SecY, either directly or indirectly, appears to be required for this process (Werner *et al.*, 1992). MtlA is an interesting IMP in that it possesses an NH_2-terminal amphipathic targeting sequence preceding the first hydrophobic transmembrane segment (Yamada *et al.*, 1991). Finally, SecY (for a model of its secondary structure see 2.1.1c) was recently found to integrate into cytoplasmic membrane vesicles lacking endogenous SecY. This was shown by a functional reconstitution of *secY*-mutant membrane vesicles with *in vitro* synthesized wild-type SecY (Swidersky *et al.*, 1992). Collectively, these studies point to a SecA-independent integration of predominantly hydrophobic IMPs of the *E. coli* cytoplasmic membrane, while the dependence on SecY appears to vary with the protein studied.

The demonstration of a Sec-independent membrane assembly of some *E. coli* cytoplasmic membrane proteins raises the interesting, yet ultimately unresolved, question as to which feature of an exported protein determines (i.e., mediates) Sec-dependence. As discussed earlier in this chapter, the translocation of large, hydrophilic domains appears to require Sec-functions. In accordance, the small bacteriophage M13 major coat protein inserts independently of SecA/SecY into the cytoplasmic membrane of *E. coli* during phage assembly. The 73-amino-acid-long precursor inserts into the membrane to give a topography with the NH_2-terminus remaining cytoplasmic, the hydrophobic core of the signal sequence traversing the membrane, a 20-amino-acid-long periplasmic loop, a stop-transfer sequence, and a short, positively charged COOH-terminal peptide also facing the cytoplasm. However, the insertion of the precursor of the coat protein becomes Sec-dependent when its periplasmic loop is extended by a 173-residue-long fragment of OmpA (Kuhn, 1988). These and other results have been interpreted such that the precursor of the M13 coat protein consisting of the two transmembrane segments and the short intervening sequence contains all the information necessary to spontaneously insert into the cytoplasmic membrane of *E. coli* without the need for other proteins to mediate integration (Wickner, 1988). The self-integrating tripartite structure of the M13 precursor protein was

termed *insertion domain* (Dalbey, 1990), the term *insertion sequence* being originally proposed to mean a signal that initiates integration of proteins into the lipid
bilayer without protein effectors (Blobel, 1980).

Similarly, the first transmembrane segment of SPase I and the subsequent
positively charged cluster was shown to comprise a Sec-independent insertion
signal (Lee *et al.*, 1992), whereas the remainder of the molecule requires SecA
and SecY for assembly. According to the proposed models (Wickner, 1988;
Kuhn *et al.*, 1990; Nilsson and von Heijne, 1990) the topographies of M13 coat
protein and SPase I are generated by an initial attachment of the two hydrophobic
transmembrane segments of each protein to the surface of the cytoplasmic membrane from where spontaneous insertion occurs (M13 coat protein) or, alternatively, where interaction with SecA/SecY takes place to catalyze inser-
tion/translocation (SPase I, COOH-terminal part). The latter situation would
demand a similar integration intermediate, presumably on the surface of a molecular chaperone as proposed above (Section 2.1.2b) to avoid reorientation of
the transmembrane topography of IMP-constructs in which a late topogenic
signal might supersede earlier ones.

The findings discussed in the previous paragraph invoke the following question: Do IMPs of the *E. coli* cytoplasmic membrane, whose membrane integration was shown to proceed in the absence of SecA and SecY, consist of a string of
insertion domains similar to those described for the M13 coat protein? Alternatively, does independence of SecA/SecY necessarily imply a nonprotein catalyzed integration mechanism? The involvement of translocators (pore- or
channel-forming membrane structures) (Singer and Yaffe, 1990; Lingappa,
1991) in the temporary accommodation of translocating transmembrane segments of ER membrane proteins prior to their release laterally into the lipid
bilayer has been suggested to account for both the existence of protein-
conducting channels (Simon and Blobel, 1991) and the cross-linking of nascent
IMPs to ER membrane proteins (High *et al.*, 1991; Thrift *et al.*, 1991). Although
similar pores have recently been detected in the *E. coli* cytoplasmic membrane
(Simon and Blobel, 1992), the demonstration that proteins of this membrane can
be cross-linked to a traversing IMP is still lacking. In view of the SecY-
independent integration of some IMPs, this particular experimental approach
should ultimately yield a clue as to the molecular environment of an IMP *in statu
integrandi*.

2.2. Outer Membrane of *E. coli*

The outer membrane of gram-negative bacteria contains a relatively small
set of IMPs (porins) whose function are mostly related to the uptake of nutrients
(for a comprehensive review see Benz, 1988; Nikaido, 1992). Except for the
murein lipoprotein (see below), which is anchored within the outer membrane by

a lipid moiety, outer-membrane proteins are polytopic transmembrane proteins. The native conformation of most outer-membrane proteins consists of trimers. Each monomer spans the membrane in multiple, antiparallel β-strands, giving rise to the formation of an amphiphilic β-sheet barrel as the structural basis for the pore function of these proteins (Walian and Jap, 1990; Weiss *et al.*, 1991).

2.2.1. Topogenic Sequences of Outer-Membrane Proteins

Intensive studies have been performed to pinpoint stretches within the primary sequence of outer-membrane proteins that are required for localization to, and incorporation into, the outer membrane. Whereas deletions throughout the molecule of PhoE all interfere with proper localization to the outer membrane (Bosch *et al.*, 1986, 1988), the most COOH-terminal β-strand of OmpA was elucidated as the important determinant for assembly of this protein into the outer membrane (Klose *et al.*, 1989; Francisco *et al.*, 1992). Although stretches of conserved amino acid sequences had been originally detected in several outer-membrane proteins and consequently regarded as candidates for topogenic signals (Nikaido and Wu, 1984), it is unlikely that such localization signals are defined by the primary protein sequence. Rather, sequence determinants are required for the folding into the tertiary structure within the lipid bilayer, and this event might be severely altered if single amino acids are exchanged: Both introduction of a strand-breaking Pro into the most COOH-terminal β-strand of OmpA (Klose *et al.*, 1988) and the substitution of a turn-forming Gly in PhoE (de Cock *et al.*, 1991) were found to affect membrane incorporation of the respective proteins. This Gly is the only strictly conserved residue in the segment of high homology between various outer-membrane proteins (Nikaido and Wu, 1984). Many outer-membrane proteins of *E. coli* harbor a Phe-residue at their COOH-terminus, which, when removed, leads to a decrease in membrane integration (Struyvé *et al.*, 1991).

2.2.2. Assembly Pathway of Outer-Membrane Proteins

A body of evidence has been presented for the existence of soluble periplasmic intermediates in the assembly process of outer-membrane proteins. Subfractionation of pulse/chase-labeled cells revealed soluble, periplasmic intermediates of TonA (FhuA, involved in uptake of iron chelates, Jackson *et al.*, 1986) and the morphogenetic, gene IV-product of the filamentous phage f1, located to the outer membrane of the host cell and required for virus assembly (Brisette and Russel, 1990). Truncated forms of OmpA (Freudl *et al.*, 1985) and PhoE (Bosch *et al.*, 1986), constructed via deletion mutagenesis, were shown by immuno-electronmicroscopy to accumulate within the periplasm of *E. coli*. OmpF was secreted as water-soluble protein into the medium by spheroplasts of

E. coli (Sen and Nikaido, 1990), and an *in vitro* synthesized, signal-sequence-free form of PhoE was not recognized by monoclonal anti-PhoE antibodies specific for cell-surface-exposed conformational epitopes (de Cock *et al.* 1990). These results suggest that outer-membrane proteins pass through a soluble, periplasmic intermediate before they acquire their authentic conformation within the outer membrane. Transition from the soluble intermediate to the outer-membrane-embedded form of outer-membrane proteins is accompanied by the formation of trimers from monomeric subunits, possibly via dimers (Vos-Scheperkeuter and Witholt, 1984; Reid *et al.*, 1988; de Cock *et al.*, 1990) as shown by use of epitope-specific antibodies and via the different temperature-dependent susceptibility of monomers and trimers toward denaturation by SDS. A processed but still membrane-associated form of OmpA, designated imp-OmpA (immature processed), was also described (Freudl *et al.*, 1986).

It should be mentioned that a different view has also been advanced. According to this, outer-membrane proteins such as LamB are first integrated into the cytoplasmic membrane of *E. coli* from where they might reach the outer membrane by lateral diffusion via contact sites. This was inferred from the finding that LamB, after translocation *in vitro* into inside-out plasma membrane vesicles of *E. coli,* remained Na_2CO_3-resistant as expected for a lipid-embedded protein. In contrast, LamB segregated within microsomal vesicles prepared from dog pancreas could be extracted by Na_2CO_3 (Watanabe *et al.*, 1986).

2.2.3. Molecular Determinants of Outer-Membrane Assembly

Few details are known about the molecular events following translocation of outer-membrane proteins across the cytoplasmic membrane prior to their sorting to the outer membrane. It is conceivable that oligomerization of outer-membrane proteins proceeds on the surface of molecular chaperones similar to what has been elucidated for the oligomerization of mitochondrial and chloroplast proteins by GroE-type chaperonins or of ER-proteins by BiP-hsp70 class chaperones (reviewed in Ellis and Hemmingsen, 1989). The existence of a periplasmic chaperone binding to outer-membrane proteins prior to their membrane assembly was inferred from a pleiotropic inhibition of outer-membrane protein synthesis by a truncated form of OmpA, which is thought to titrate the presumed chaperone by irreversibly binding to its free form (Ried *et al.*, 1990b). Current knowledge on distinct periplasmic chaperones, however, is scarce. A periplasmic peptidyl-prolyl-*cis-trans*-isomerase, also termed *rotamase*, has been identified and structurally characterized (Liu and Walsh, 1990; Hayano *et al.*, 1991). A periplasmic protein involved in disulfide bond formation (protein disulfide isomerase) has also been discovered (Bardwell *et al.*, 1991; Kamitani *et al.*, 1992). Whereas these enzymes presumably catalyze protein folding like their eukaryotic counterparts, a direct involvement of these chaperones in the oligomerization process of outer-membrane proteins has not been demonstrated.

Recently a protein was identified within the periplasm of *E. coli* whose functional properties might well be compatible with that of an outer-membrane protein-specific chaperone. The protein Skp (*seventeen kilodalton protein*) was purified to homogeneity from a ribosomal salt-extract of *E. coli* by its ability to partially compensate the defect in translocation of preLamB into cytoplasmic membrane vesicles caused by a lack of SecA. A partial NH_2-terminal sequence analysis revealed its identity with Skp of *E. coli* and OmpH of *Salmonella typhimurium* (Thome *et al.*, 1990). The nucleotide sequence of *skp* disclosed the existence of a classical NH_2-terminal signal sequence (Holck and Kleppe, 1988), suggesting a localization of Skp to the cell envelope. This was subsequently confirmed by subfractionation studies and the demonstration of a SecA/SecY-dependent translocation of Skp across the cytoplasmic membrane of *E. coli* (Thome and Müller, 1991). Despite the periplasmic destination of Skp within a living cell, its SecA-like activity *in vitro* suggests a possible interaction with the precursor of LamB and probably other outer-membrane proteins. Such a functional interaction of Skp with a precursor protein is also indicated by the finding that translocation of preLamB into cytoplasmic membrane vesicles proceeded with decreased efficiency upon immunodepletion of the cytosolic extract of *E. coli* used to synthesize preLamB (Thome *et al.*, 1990). Interaction of Skp with precursor proteins *in vitro*, however, might simply reflect the potential association with loosely folded forms of an outer-membrane protein in the periplasm *in vivo*. The tendency of Skp to form oligomers and the mapping of its gene on the *E. coli* chromosome to a region specifying outer-membrane components and functions might all point toward a physiological function of Skp as a periplasmic chaperone involved in the posttranslocational biogenesis of outer-membrane proteins. A definite proof of this hypothesis, however, has not yet been arrived at.

An involvement of outer-membrane lipopolysaccharide (LPS) in the biogenesis of the oligomeric forms of outer-membrane proteins has also been demonstrated, yet appears to be protein-specific. The monomeric form of OmpF secreted by spheroplasts is taken up by outer-membrane preparations and correctly converted into a trimeric conformation when Triton X-100 or, alternatively, intact LPS is added (Sen and Nikaido, 1990, 1991), indicating that LPS functions as a soluble integration factor of OmpF. In accordance with this, the antibiotic cerulenin, which blocks fatty acid synthesis and thereby that of LPS, inhibits synthesis of OmpF (and OmpC) (Bocquet-Pages *et al.*, 1981). Conversely, synthesis and membrane incorporation of OmpA was not affected by the antibiotic (Ried *et al.*, 1990a), and denatured OmpA refolded correctly into liposomes in the absence of LPS (Dornmair *et al.*, 1990).

2.2.4. Lipoproteins

The major outer-membrane (Braun's) lipoprotein of *E. coli* is a prototype of lipid-modified proteins of bacterial cell envelopes. It is first synthesized as a precursor protein that is translocated in a SecA-, SecD-, SecE-, SecF-, and SecY-dependent manner across the cytoplasmic membrane. Prior to signal sequence cleavage the precursor is modified by a glyceryl-moiety attached to the prospective NH_2-terminal Cys of the mature protein. The glyceride is subsequently *O*-acylated before a lipoprotein-specific signal peptidase (SPase II; for a recent review see Müller, 1992), which is unique in its inhibition by the antibiotic globomycin, removes the signal sequence. The emerging α-NH_2-group is finally *N*-acylated to form mature lipoprotein, which is anchored within the outer membrane by its lipid moiety and which is also covalently coupled via its COOH-terminal Lys to the murein layer (for a detailed discussion and references, see Hayashi and Wu, 1990). Interestingly, the lipoprotein precursor that accumulates in various *sec* mutants is not modified with glyceride, indicating that the Sec protein-dependent steps of lipoprotein's export precede its modification reactions (Sugai and Wu, 1992).

3. PROTEIN INTEGRATION INTO THE MEMBRANES OF α-PURPLE BACTERIA

Members of the group of facultatively photoheterotrophic bacteria, such as *Rhodobacter sphaeroides* and *R. capsulatus*, are capable of growth by aerobic and anaerobic respiration, fermentation, and anoxygenic photosynthesis. They have a typical gram-negative cell envelope when growing aerobically. Removal of oxygen from such a culture triggers a differentiation event of the cytoplasmic membrane, resulting in multiple invaginations termed *intracytoplasmic membranes* (ICM). The ICM, although contiguous to the cytoplasmic membrane, represent specialized domains containing specifically the photosynthetic pigment-protein complexes (reaction center, RC; and two light harvesting complexes, LH I and LH II), whereas other proteins such as substrate carriers appear to be restricted to the cytoplasmic membrane. Thus, photosynthetic membranes are induced in the absence of light, upon lowering the O_2-partial pressure (gratuitous event). Decreasing incident light intensities also regulate ICM synthesis by increasing the intracellular amount of ICM and changing their molecular composition—i.e., inserting pigment-protein complexes into pre-existing ICM (nongratuitous event). The different ways of induction of ICM render this group of α-purple bacteria obviously an excellent system in which to study membrane assembly and differentiation. Although the physiology of ICM, the genetic organization of structural genes for ICM components, their regulation, and the three-dimensional structure of the reaction center complexes have been extensively reviewed (Drews, 1985, 1986; Kiley and Kaplan, 1988; Deisenhofer and Michel, 1989), few details are known about the assembly of photosynthetic and other proteins in these bacteria.

3.1. Protein Targeting in *Rhodobacter capsulatus*

No export-defective mutants have been isolated from *Rhodospirillaceae*, and only very recently was the translocation of a precursor protein reproduced *in vitro* (Wieseler and Müller, 1992). Using a cell-free synthesis/export system, which allows translocation of *R. capsulatus* precytochrome c_2 into ICM, and crosswise exchanging components of this system with a comparable one prepared from *E. coli*, no soluble *R. capsulatus* proteins were found that can functionally substitute for *E. coli* SecA and SecB. Nor was active SecA, being partially membrane-associated in *E. coli*, detected on *R. capsulatus* membrane vesicles. In accordance with the lack of a SecB-function, precytochrome c_2 of *R. capsulatus* does not require SecB for translocation across the *E. coli* plasma membrane. Conversely, SecA has to be supplemented to achieve efficient translocation of the *E. coli* preOmpA across the *R. capsulatus* plasma membrane. Translocation of the *R. capsulatus* precursor, however, is independent of SecA under these experimental conditions, as expected for an organism devoid of a SecA-analogue. These results seem to suggest that SecA- and SecB-like proteins are not involved in protein export of the α-purple bacterium *R. capsulatus* (Wieseler *et al.*, submitted).

3.2. Assembly of Reaction Center Complexes

Reaction centers (RC) consist of $1:1:1$ complexes of two pigment-binding proteins, L and M, each spanning the membrane five times, and the nonpigment-binding protein H, which is anchored within the membrane by an NH_2-terminal transmembrane segment. Because the H subunit is found in aerobically grown cells that do not have detectable amounts of the other RC subunits and because its total cellular content is in excess of L and M, it was suggested that H might fulfill a critical role in the membrane assembly of the two other subunits (summarized in Kiley and Kaplan, 1988). It is not known whether L and M form a complex prior to insertion or whether they are individually incorporated into the lipid bilayer of ICM.

3.3. Topogenic Sites of Light-Harvesting Complex I Proteins
of *R. capsulatus*

The two proteins α and β of LH I (B870) are single-spanning IMPs with the two NH_2-terminal peptides remaining in the cytoplasm. The opposite net charge of both NH_2-termini suggests a direct molecular interaction of both subunits. By deleting one of the two subunits and using Na_2CO_3-extraction of pulse/chase-labeled membrane proteins to assay for membrane integration, it was demonstrated that β inserts in the absence of α but does not remain as stably integrated as in the presence of α, whereas α alone could not be appreciably detected in the membrane. Pulse/chase-labeling experiments also revealed that when both proteins are co-expressed, the incorporation of β appears to precede that of α (Richter and Drews, 1991).

Replacing the positively charged amino acids of the NH_2-terminal α-peptide by negative residues using site-directed mutagenesis abolished assembly of LH I as judged by spectroscopic measurements and pulse/chase-labeling of the two proteins. Conversely, reversing the negative charge of the β-subunit to a positive one resulted in a measurable complex formation of somewhat reduced stability (Stiehle et al., 1990). Similarly, replacement of the Trp-residue at position 8 of the LH I α-subunit, which is a highly conserved amino acid among α-peptides of LH I and LH II complexes of Rhodospirillaceae, prevented membrane integration of the complex (Richter et al., 1991). This kind of study was extended to a more general mutagenesis of the NH_2-terminal portion of LH II α, which altogether interfered with stable complex formation, indicating that the NH_2-termini of LH-peptides contains important topogenic information (Babst et al., 1991; Richter et al., 1992).

4. REFERENCES

Ahrem, B., Hoffschulte, H. K., and Müller, M., 1989, In vitro membrane assembly of a polytopic, transmembrane protein results in an enzymatically active conformation, J. Cell Biol. 108:1637–1646.

Akimura, J., Matsuyama, S., Tokuda, H., and Mizushima, S., 1991, Reconstitution of a protein translocation system containing purified SecY, SecE, and SecA from Escherichia coli, Proc. Natl. Acad. Sci. USA 88:6545–6549.

Akita, M., Sasaki, S., Matsuyama, S., and Mizushima, S., 1990, SecA interacts with secretory proteins by recognizing the positive charge at the amino terminus of the signal peptide in Escherichia coli, J. Biol. Chem. 265:8164–8169.

Akiyama, Y., and Ito, K., 1987, Topology analysis of the SecY protein, an integral membrane protein involved in protein export in Escherichia coli, EMBO J. 6:3465–3470.

Akiyama, Y., and Ito, K., 1989, Export of Escherichia coli alkaline phosphatase attached to an integral membrane protein, SecY, J. Biol. Chem. 264:437–442.

Allard, J. D., and Bertrand, K. P., 1992, Membrane topology of the pBR322 tetracycline resistance protein, J. Biol. Chem. 267:17809–17819.

Altman, E., Bankaitis, V. A., and Emr, S. D., 1990a, Characterization of a region in mature LamB protein that interacts with a component of the export machinery of Escherichia coli, J. Biol. Chem. 265:18148–18153.

Altman, E., Emr, S. D., and Kumamoto, C. A., 1990b, The presence of both the signal sequences and a region of mature LamB protein is required for the interaction of LamB with the export factor SecB, *J. Biol. Chem.* **265**:18154–18160.

Andersson, H., and von Heijne, G., 1993, Sec-dependent and *sec*-independent assembly of *E. coli* inner membrane proteins: The topological rules depend on chain length, *EMBO J.* **12**:683–691.

Andrews, D. W., Young, J. C., Mirels, L. F., and Czarnota, G. J., 1992, The role of the N region in signal sequence and signal-anchor function, *J. Biol. Chem.* **267**:7761–7769.

Audigier, Y., Friedlander, M., and Blobel, G., 1987, Multiple topogenic sequences in bovine opsin, *Proc. Natl. Acad. Sci. USA* **84**:5783–5787.

Baba, T., Jacq, A., Brickman, E., Beckwith, J., Taura, T., Ueguchi, C., Akiyama, Y., and Ito, K., 1990, Characterization of cold sensitive *secY* mutants of *Escherichia coli*, *J. Bacteriol.* **172**:7005–7010.

Babst, M., Albrecht, H., Wegmann, I., Brunisholz, R., and Zuber, H., 1991, Single amino acid substitutions in the B870 α and β light-harvesting polypeptides of *Rhodobacter capsulatus*, *Eur. J. Biochem.* **202**:277–284.

Baker, K., Mackman, N., Jackson, M., and Holland, I. B., 1987, Role of SecA and SecY in protein export as revealed by studies of TonA assembly into the outer membrane of *Escherichia coli*, *J. Mol. Biol.* **198**:693–703.

Bakker, E. P., and Randall, L. L., 1984, The requirement for energy during export of β-lactamase in *Escherichia coli* is fulfilled by the total protonmotive force, *EMBO J.* **3**:895–900.

Bardwell, J. C. A., McGovern, K., and Beckwith, J., 1991, Identification of a protein required for disulfide bond formation *in vivo*, *Cell* **67**:581–589.

Bassilana, M., Arkowitz, R. A., and Wickner, W., 1992, The role of the mature domain of proOmpA in the translocation ATPase reaction, *J. Biol. Chem.* **267**:25246–25250.

Beltzer, J. P., Fiedler, K., Fuhrer, C., Geffen, I., Handschin, C., Wessels, H. P., and Spiess, M., 1991, Charged residues are major determinants of the transmembrane orientation of a signal-anchor sequence, *J. Biol. Chem.* **266**:973–978.

Benz, R., 1988, Structure and function of porins from Gram-negative bacteria, *Annu. Rev. Microbiol.* **42**:359–393.

Bibi, E., Verner, G., Chang, C. Y., and Kaback, H. R., 1991, Organization and stability of a polytopic membrane protein: Deletion analysis of the lactose permease of *Escherichia coli*, *Proc. Natl. Acad. Sci. USA* **88**:7271–7275.

Bieker, K. L., and Silhavy, T. J., 1990, PrlA (SecY) and PrlG (SecE) interact directly and function sequentially during protein translocation in *E. coli*, *Cell* **61**:833–842.

Bieker, K. L., Phillips, G. J., and Silhavy, T. J., 1990, The *sec* and *prl* genes of *Escherichia coli*, *J. Bioenerg. Biomembr.* **22**:291–310.

Bieker-Brady, K., and Silhavy, T. J., 1992, Suppressor analysis suggests a multistep, cyclic mechanism for protein secretion in *Escherichia coli*, *EMBO J.* **11**:3165–3174.

Blobel, G., 1980, Intracellular protein topogenesis, *Proc. Natl. Acad. Sci. USA* **77**:1496–1500.

Bochkareva, E. S., Lissin, N. M., and Girshovich, A. S., 1988, Transient association of newly synthesized unfolded proteins with the heat-shock GroEL protein, *Nature* **336**:254–257.

Bocquet-Pages, C., Lazdunski, C., and Lazdunski, A. 1981, Lipid-synthesis-dependent biosynthesis (or assembly) of major outer-membrane proteins of *Escherichia coli*, *Eur. J. Biochem.* **118**:105–111.

Bosch, D., Leunissen, J., Verbakel, J., de Jong, M., van Erp, H., and Tommassen, J., 1986, Periplasmic accumulation of truncated forms of outer-membrane PhoE protein of *Escherichia coli* K-12, *J. Mol. Biol.* **189**:449–455.

Bosch, D., Voorhout, W., and Tommassen, J., 1988, Export and localization of N-terminally truncated derivatives of *Escherichia coli* K-12 outer-membrane protein PhoE, *J. Biol. Chem.* **263**:9952–9957.

Boyd, D., and Beckwith, J., 1990, The role of charged amino acids in the localization of secreted and membrane proteins, *Cell* **62**:1031–1033.

Breukink, E., Kusters, R., and de Kruijff, B., 1992, *In vitro* studies on the folding characteristics of the *Escherichia coli* precursor protein prePhoE, *Eur. J. Biochem.* **208**:419–425.

Brissette, J. L., and Russel, M., 1990, Secretion and membrane integration of a filamentous phage-encoded morphogenetic protein, *J. Mol. Biol.* **211**:565–580.

Brundage, L., Hendrick, J. P., Schiebel, E., Driessen, A.J.M., and Wickner, W., 1990, The purified *E. coli* integral membrane protein SecY/E is sufficient for reconstitution of SecA-dependent precursor protein translocation, *Cell* **62**:649–657.

Brundage, L., Fimmel, C. J., Mizushima, S., and Wickner, W., 1992, SecY, SecE, and Band 1 form the membrane-embedded domain of *Escherichia coli* preprotein translocase, *J. Biol. Chem.* **267**:4166–4170.

Cabelli, R. J., Dolan, K. M., Qian, L., and Oliver, D. B., 1991, Characterization of membrane associated and soluble states of SecA protein from wild-type and *secA51(ts)* mutant strains of *Escherichia coli*, *J. Biol. Chem.* **266**:24420–24427.

Calamia, J., and Manoil, C., 1990, Lac permease of *Escherichia coli:* Topology and sequence elements promoting membrane insertion, *Proc. Natl. Acad. Sci. USA* **87**:4937–4941.

Cobet, W. W. E., Mollay, C., Müller, G., and Zimmermann, R., 1989, Export of honeybee pre-promellitin in *Escherichia coli* depends on the membrane potential but does not depend on proteins SecA and SecY, *J. Biol. Chem.* **264**:10169–10176.

Collier, D. N., and Bassford Jr., P. J., 1989, Mutations that improve export of maltose-binding protein in SecB⁻ cells of *Escherichia coli*, *J. Bacteriol.* **171**:4640–4647.

Collier, D. N., Bankaitis, V. A., Weiss, J. B., and Bassford Jr., P. J., 1988, The antifolding activity of SecB promotes the export of the *E. coli* maltose-binding protein, *Cell* **53**:273–283.

Collier, D. N., Strobel, S. M., Bassford Jr., P. J., 1990, SecB-independent export of *Escherichia coli* ribose-binding protein (RBP): Some comparisons with export of maltose-binding protein (MBP) and studies with RBP-MBP hybrid proteins, *J. Bacteriol.* **172**:6875–6884.

Cunningham, K., Lill, R., Crooke, E., Rice, M., Moore, K., Wickner, W., and Oliver, D., 1989, SecA protein, a peripheral protein of the *Escherichia coli* plasma membrane, is essential for the functional binding and translocation of proOmpA, *EMBO J.* **8**:955–959.

Dalbey, R. E., 1990, Positively charged residues are important determinants of membrane protein topology, *Trends Biochem. Sci.* **15**:253–257.

de Cock, H., and Tommassen, J., 1992, SecB-binding does not maintain the translocation-competent state of prePhoE, *Mol. Microbiol.* **6**:599–604.

de Cock, H., Hendriks, R., de Vrije, T., and Tommassen, J., 1990, Assembly of an in vitro synthesized *Escherichia coli* outer membrane porin into its stable trimeric conformation, *J. Biol. Chem.* **265**:4646–4651.

de Cock, H., Quaedvlieg, N., Bosch, D., Scholten, M., and Tommassen, J., 1991, Glycine-144 is required for efficient folding of outer membrane protein PhoE of *Escherichia coli* K 12, *FEBS Lett.* **279**:285–288.

de Cock, H., Overeem, W., and Tommassen, J., 1992, Biogenesis of outer-membrane protein PhoE of *Escherichia coli:* Evidence for multiple SecB-binding sites in the mature portion of the PhoE protein, *J. Mol. Biol.* **224**:369–379.

Deisenhofer, J., and Michel, H., 1989, Das photosynthetische Reaktionszentrum des Purpurbakteriums *Rhodopseudomonas viridis* (Nobel-Vortrag), *Angew. Chem.* **101**:872–892.

Derman, A. I., Puziss, J. W., Bassford, P. J., and Beckwith, J., 1993, A signal sequence is not required for protein export in *prlA* mutants of *Escherichia coli*, *EMBO J.* **12**:879–888.

Dolan, K. M., and Oliver, D. B., 1991, Characterization of *Escherichia coli* SecA protein binding to a site on its mRNA involved in autoregulation, *J. Biol. Chem.* **266**:23329–23333.

Dornmair, K., Kiefer, H., and Jähnig, F., 1990, Refolding of an integral membrane protein, *J. Biol. Chem.* **265**:18907–18911.

Drews, G., 1985, Structural and functional organization of light-harvesting complexes and photo-chemical reaction centers in membranes of phototrophic bacteria, *Microbiol. Rev.* **49**:59–70.

Drews, G., 1986, Adaptation of the bacterial photosynthetic apparatus to different light intensities, *Trends Biochem. Sci.* **11**:255–257.

Driessen, A. J. M. 1992a, Precursor protein translocation by the *Escherichia coli* translocase is directed by the protonmotive force, *Embo J.* **11**:847–853.

Driessen, A. J. M., 1992b, Bacterial protein translocation: Kinetic and thermodynamic role of ATP and the protonmotive force, *Trends Biochem. Sci.* **17**:219–223.

Eckert, B., and Beck, C. F., 1989, Topology of the transposon Tn10-encoded tetracycline resistance protein within the inner membrane of *Escherichia coli*, *J. Biol. Chem.* **264**:11663–11670.

Ehrmann, M., and Beckwith, J., 1991, Proper insertion of a complex membrane protein in the absence of its amino-terminal export signal, *J. Biol. Chem.* **266**:16530–16533.

Ellis, R. J., and Hemmingsen, S. M., 1989, Molecular chaperones: Proteins essential for the biogenesis of some macromolecular structures, *Trends Biochem. Sci.* **14**:339–342.

Emr, S. D., and Bassford Jr., P. J., 1982, Localization and processing of outer-membrane and periplasmic proteins in *Escherichia coli* strains harboring export-specific suppressor mutations, *J. Biol. Chem.* **257**:5852–5860.

Emr, S. D., Hanley-Way, S., and Silhavy, T. J., 1981, Suppressor mutations that restore export of a protein with a defective signal sequence, *Cell* **23**:79–88.

Fandl, J. P., and Tai, P. C., 1987, Biochemical evidence for the *secY24* defect in *Escherichia coli* protein translocation and its suppression by soluble cytoplasmic factors, *Proc. Natl. Acad. Sci. USA* **84**:7448–7452.

Fandl, J. P., Cabelli, R., Oliver, D., and Tai, P. C., 1988, SecA suppresses the temperature-sensitive SecY24 defect in protein translocation in *Escherichia coli* membrane vesicles, *Proc. Natl. Acad. Sci. USA* **85**:8953–8957.

Fikes, J. D., and Bassford Jr., P. J., 1989, Novel secA alleles improve export of maltose-binding protein synthesized with a defective signal peptide, *J. Bacteriol.* **171**:402–409.

Fortin, Y., Phoenix, P., and Drapeau, G. R., 1990, Mutations conferring resistance to azide in *Escherichia coli* occur primarily in the *secA* gene, *J. Bacteriol.* **172**:6607–6610.

Foster, D. L., Boublik, M., and Kaback, H. R., 1983, Structure of the *lac* carrier protein of *Escherichia coli*, *J. Biol. Chem.* **258**:31–34.

Francisco, J. A., Earhart, C. F., and Georgiou, G., 1992, Transport and anchoring of β-lactamase to the external surface of *Escherichia coli*, *Proc. Natl. Acad. Sci. USA* **89**:2713–2717.

Freudl, R., Schwarz, H., Klose, M., Rao Movva, N., and Henning, U., 1985, The nature of information, required for export and sorting, present within the outer-membrane protein OmpA of *Escherichia coli* K-12, *EMBO J.* **4**:3593–3598.

Freudl, R., Schwarz, H., Stierhof, Y. D., Gamon, K., Hindennach, I., and Henning, U., 1986, An outer-membrane protein (OmpA) of *Escherichia coli* K-12 undergoes a conformational change during export, *J. Biol. Chem.* **261**:11355–11361.

Fujiki, Y., Hubbard, A. L., Fowler, S., and Lazarow, P. B., 1982, Isolation of intracellular mem-branes by means of sodium carbonate treatment: Application to endoplasmic reticulum, *J. Cell Biol.* **93**:97–102.

Gardel, C., Benson, S., Hunt, J., Michaelis, S., and Beckwith, J., 1987, *secD*, a new gene involved in protein export in *Escherichia coli*, *J. Bacteriol.* **169**:1286–1290.

Gardel, C., Johnson, K., Jacq, A., and Beckwith, J., 1990, The *secD* locus of *E. coli* codes for two membrane proteins required for protein export, *EMBO J.* **9**:3209–3216.

Gebert, J. F., Overhoff, B., Manson, M. D., and Boos, W., 1988, The Tsr chemosensory transducer of *Escherichia coli* assembles into the cytoplasmic membrane via a *secA*-dependent process, *J. Biol. Chem.* **263**:16652–16660.

Geller, B. L., 1991, Energy requirements for protein translocation across the *Escherichia coli* inner membrane, *Mol. Microbiol.* **5**:2093–2098.

Gilmore, R., and Blobel, G., 1985, Translocation of secretory proteins across the microsomal membrane occurs through an environment accessible to aqueous perturbants, *Cell* **42**:497–505.

Görlich, D., Prehn, S., Hartmann, E., Kalies, K. U., and Rapoport, T. A., 1992, A mammalian homolog of SEC61p and SECYp is associated with ribosomes and nascent polypeptides during translocation, *Cell* **71**:489–503.

Hardy, S.J.S., and Randall, L. L., 1991, A kinetic partitioning model of selective binding of nonnative proteins by the bacterial chaperone SecB, *Science* **251**:439–443.

Hartl, F.-U., Lecker, S., Schiebel, E., Hendrick, J. P., and Wickner, W., 1990, The binding cascade of SecB to SecA to SecY/E mediates preprotein targeting to the *E. coli* plasma membrane, *Cell* **63**:269–279.

Hartmann, E., Rapoport, T. A., and Lodish, H. F., 1989, Predicting the orientation of eukaryotic membrane-spanning proteins, *Proc. Natl. Acad. Sci. USA* **86**:5786–5790.

Hayano, T., Takahashi, N., Kato, S., Maki, N., and Suzuki, M., 1991, Two distinct forms of peptidylprolyl-*cis-trans*-isomerase are expressed separately in periplasmic and cytoplasmic compartments of *Escherichia coli* cells, *Biochemistry* **30**:3041–3048.

Hayashi, S., and Wu, H. C., 1990, Lipoproteins in bacteria, *J. Bioenerg. Biomembr.* **22**:451–471.

Hendrick, J. P., and Wickner, W., 1991, SecA protein needs both acidic phospholipids and SecY/E protein for functional high-affinity binding to the *Escherichia coli* plasma membrane, *J. Biol. Chem.* **266**:24596–24600.

High, S., and Dobberstein, B., 1991, The signal sequence interacts with the methionine-rich domain of the 54-kD protein of signal recognition particle, *J. Cell Biol.* **113**:229–233.

High, S., Görlich, D., Wiedmann, M., Rapoport, T. A., and Dobberstein, B., 1991, The identification of proteins in the proximity of signal-anchor sequences during their targeting to and insertion into the membrane of the ER, *J. Cell Biol.* **113**:35–44.

Holck, A., and Kleppe, K., 1988, Cloning and sequencing of the gene for the DNA-binding 17 K protein of *Escherichia coli*, *Gene* **67**:117–124.

Inouye, M., and Halegoua, S., 1980, Secretion and membrane localization of proteins in *Escherichia coli*, *Crit. Rev. Biochem.* **7**:339–371.

Ito, K., 1990, Structure, function, and biogenesis of SecY, an integral membrane protein involved in protein export, *J. Bioenerg. Biomembr.* **22**:353–367.

Ito, K., 1992, SecY and integral membrane components of the *Escherichia coli* protein translocation system, *Mol. Microbiol.* **6**:2423–2428.

Ito, K., and Akiyama, Y., 1991, *In vivo* analysis of integration of membrane proteins in *Escherichia coli*, *Mol. Microbiol.* **5**:2243–2253.

Ito, K., Hirota, Y., and Akiyama, Y., 1989, Temperature-sensitive *sec* mutants of *Escherichia coli*: Inhibition of protein export at the permissive temperature, *J. Bacteriol.* **171**:1742–1743.

Ito, K., Wittekind, M., Nomura, M., Shiba, K., Yura, T., Miura, A., and Nashimoto, H., 1983, A temperature-sensitive mutant of E. coli exhibiting slow processing of exported proteins, *Cell* **32**:789–797.

Jackson, M. E., Pratt, J. M., and Holland, I. B., 1986, Intermediates in the assembly of the TonA polypeptide into the outer membrane of *Escherichia coli* K 12, *J. Mol. Biol.* **189**:477–486.

Jarosik, G. P., and Oliver, D. B., 1991, Isolation and analysis of dominant *secA* mutations in *Escherichia coli*, *J. Bacteriol.* **173**:860–868.

Joly, J. C., and Wickner, W., 1993, The SecA and SecY subunits of translocase are the nearest neighbors of a translocating preprotein, shielding it from phospholipids, *EMBO J.* **12**:255–263.

Kamitani, S., Akiyama, Y., and Ito, K., 1992, Identification and characterization of an *Escherichia coli* gene required for the formation of correctly folded alkaline phosphatase, a periplasmic enzyme, *EMBO J.* **11**:57–62.

Kellenberger, E., 1990, The "Bayer bridges" confronted with results from improved electron microscopy methods, *Mol. Microbiol.* **4**:697–705.

Kiley, P. J., and Kaplan, S., 1988, Molecular genetics of photosynthetic membrane biosynthesis in *Rhodobacter sphaeroides*, *Microbiol. Rev.* **52**:50–69.

Kim, J., Lee, Y., Kim, C., and Park, C., 1992, Involvement of SecB, a chaperone, in the export of ribose-binding protein, *J. Bacteriol.* **174**:5219–5227.

Kimura, E., Akita, M., Matsuyama, S., and Mizushima, S., 1991, Determination of a region in SecA that interacts with presecretory proteins in *Escherichia coli*, *J. Biol. Chem.* **266**:6600–6606.

Klose, M., MacIntyre, S., Schwarz, H., and Henning, U., 1988, The influence of amino substitutions within the mature part of an *Escherichia coli* outer-membrane protein (OmpA) on assembly of the polypeptide into its membrane, *J. Biol. Chem.* **263**:13297–13302.

Klose, M., Jähnig, F., Hindennach, I., and Henning, U., 1989, Restoration of membrane incorporation of an *Escherichia coli* outer membrane protein (OmpA) defective in membrane insertion, *J. Biol. Chem.* **264**:21842–21847.

Koonin, E. V., and Gorbalenya, A. E., 1992, Autogenous translation regulation by *Escherichia coli* ATPase SecA may be mediated by an intrinsic RNA helicase activity of this protein, *FEBS Lett.* **298**:6–8.

Krijnse Locker, J., Rose, J. K., Horzinek, M. C., and Rottier, P.J.M., 1992, Membrane assembly of the triple-spanning coronavirus M protein, *J. Biol. Chem.* **267**:21911–21918.

Kuhn, A., 1988, Alterations in the extracellular domain of M13 procoat protein make its membrane insertion dependent on *secA* and *secY*, *Eur. J. Biochem.* **177**:267–271.

Kuhn, A., Zhu, H. Y., and Dalbey, R., 1990, Efficient translocation of positively charged residues of M13 procoat protein across the membrane excludes electrophoresis as the primary force for membrane insertion, *EMBO J.* **9**:2385–2389.

Kumamoto, C. A., 1989, *Escherichia coli* SecB protein associates with exported protein precursors *in vivo*, *Proc. Natl. Acad. Sci. USA* **86**:5320–5324.

Kumamoto, C. A., 1990, SecB protein: A cytosolic export factor that associates with nascent exported proteins, *J. Bioenerg. Biomembr.* **22**:337–351.

Kumamoto, C. A., 1991, Molecular chaperones and protein translocation across the *Escherichia coli* inner membrane, *Mol. Microbiol.* **5**:19–22.

Kumamoto, C. A., and Beckwith, J., 1985, Evidence for specificity at an early step in protein export in *Escherichia coli*, *J. Bacteriol.* **163**:267–274.

Kumamoto, C. A., Chen, L., Fandl, J., and Tai, P. C., 1989. Purification of the *Escherichia coli* secB gene product and demonstration of its activity in an *in vitro* protein translocation system, *J. Biol. Chem.* **264**:2242–2249.

Kumamoto, C. A., and Francetić, O., 1993, Highly selective binding of nascent polypeptides by an *Escherichia coli* chaperone protein *in vivo*, *J. Bacteriol.* **175**:2184–2188.

Kusters, R., de Vrije, T., Breukink, E., and de Kruijff, B., 1989, SecB protein stabilizes a translocation-competent state of purified prePhoE protein, *J. Biol. Chem.* **264**:20827–20830.

Kusters, R., Huijbregts, R., and de Kruijff, B., 1992, Elevated cytosolic concentrations of SecA compensate for a protein translocation defect in *Escherichia coli* cells with reduced levels of negatively charged phospholipids, *FEBS Lett.* **308**:97–100.

Kusukawa, N., Yura, T., Ueguchi, C., Akiyama, Y., and Ito, K., 1989. Effects of mutations in heat-shock genes groES and groEL on protein export in *Escherichia coli*, *EMBO J.* **8**:3517–3521.

Laminet, A. A., Kumamoto, C. A., and Plückthun, A., 1991, Folding *in vitro* and transport *in vivo* of pre-β-lactamase are SecB independent, *Mol. Microbiol.* **5**:117–122.

Lecker, S., Lill, R., Ziegelhoffer, T., Georgopoulos, C., Bassford Jr., P. J., Kumamoto, C. A., and

Wickner, W., 1989, Three pure chaperone proteins of *Escherichia coli*—SecB, trigger factor and GroEL—form soluble complexes with precursor proteins *in vitro*, *EMBO J.* **8**:2703–2709.

Lecker, S., Driessen, A.J.M., and Wickner, W., 1990, ProOmpA contains secondary and tertiary structure prior to translocation and is shielded from aggregation by association with SecB protein, *EMBO J.* **9**:2309–2314.

Lee, J. I., Kuhn, A., and Dalbey, R. E., 1992, Distinct domains of an oligotopic membrane protein are sec-dependent and sec-independent for membrane insertion, *J. Biol. Chem.* **267**:938–943.

Liebke, H. H., 1987, Multiple SecA protein isoforms in *Escherichia coli*, *J. Bacteriol.* **169**:1174–1181.

Lill, R., Cunningham, K., Brundage, L. A., Ito, K., Oliver, D., and Wickner, W., 1989, SecA protein hydrolyzes ATP and is an essential component of the protein translocation ATPase of *Escherichia coli*, *EMBO J.* **8**:961–966.

Lill, R., Dowhan, W., and Wickner, W., 1990, The ATPase activity of SecA is regulated by acidic phospholipids, SecY, and the leader and mature domains of precursor proteins, *Cell* **60**:271–280.

Lingappa, V. R., 1991, More than just a channel: Provocative new features of protein traffic across the ER membrane, *Cell* **65**:527–530.

Liu, J., and Walsh, C. T., 1990, Peptidyl-prolyl *cis-trans*-isomerase from *Escherichia coli*: A periplasmic homolog of cyclophilin that is not inhibited by cyclosporin A, *Proc. Natl. Acad. Sci. USA* **87**:4028–4032.

Liu, G., Topping, T. B., and Randall, L. L., 1989, Physiological role during export for the retardation of folding by the leader peptide of maltose-binding protein, *Proc. Natl. Acad. Sci. USA* **86**:9213–9217.

Luirink, J., High, S., Wood, H., Giner, A., Tollervey, D., and Dobberstein, B., 1992, Signal-sequence recognition by an *Escherichia coli* ribonucleoprotein complex, *Nature*, **359**:741–743.

McGovern, K., and Beckwith, J., 1991, Membrane insertion of the *Escherichia coli* MalF protein in cells with impaired secretion machinery, *J. Biol. Chem.* **266**:20870–20876.

McGovern, K., Ehrmann, M., and Beckwith, J., 1991, Decoding signals for membrane protein assembly using alkaline phosphatase, *EMBO J.* **10**:2773–2782.

MacIntyre, S., Mutschler, B., and Henning, U., 1991, Requirement of the SecB chaperone for export of a non-secretory polypeptide in *Escherichia coli*, *Mol. Gen. Genet.* **227**:224–228.

Manoil, C., Mekalanos, J. J., and Beckwith, J., 1990, Alkaline phosphatase fusions: sensors of subcellular location, *J. Bacteriol.* **172**:515–518.

Matsuyama, S., Fujita, Y., and Mizushima, S., 1993, SecD is involved in the release of translocated secretory proteins from the cytoplasmic membrane of *Escherichia coli*, *EMBO J.* **12**:265–270.

Matsuyama, S., Kimura, E., and Mizushima, S., 1990, Complementation of two overlapping fragments of SecA, a protein translocation ATPase of *Escherichia coli*, allows ATP binding to its amino-terminal region, *J. Biol. Chem.* **265**:8760–8765.

Müller, M., 1992, Proteolysis in protein import and export: Signal peptide processing in eu- and prokaryotes, *Experientia* **48**:118–129.

Nikaido, H., 1992, Porins and specific channels of bacterial outer membranes, *Mol. Microbiol.* **6**:435–442.

Nikaido, H., and Wu, H.C.P., 1984, Amino acid sequence homology among the major outer-membrane proteins of *Escherichia coli*, *Proc. Natl. Acad. Sci. USA* **81**:1048–1052.

Nilsson, I., and von Heijne, G., 1990, Fine-tuning the topology of a polytopic membrane protein: Role of positively and negatively charged amino acids, *Cell* **62**:1135–1141.

Oliver, D. B., Cabelli, R. J., Dolan, K. M., and Jarosik, G. P., 1990a, Azide-resistant mutants of *Escherichia coli* alter the SecA protein, an azide-sensitive component of the protein export machinery, *Proc. Natl. Acad. Sci. USA* **87**:8227–8231.

Oliver, D. B., Cabelli, R. J., and Jarosik, G. P., 1990b, SecA protein: Autoregulated initiator of secretory precursor protein translocation across the *E. coli* plasma membrane, *J. Bioenerg. Biomembr.* **22**:311–336.

Parks, G. D., and Lamb, R. A., 1991, Topology of eukaryotic type II membrane proteins: Importance of N-terminal positively charged residues flanking the hydrophobic domain, *Cell* **64**:777–787.

Phillips, G. J., and Silhavy, T. J., 1990, Heat-shock proteins DnaK and GroEL facilitate export of LacZ hybrid proteins in *E. coli, Nature* **344**:882–884.

Phillips, G. J., and Silhavy, T. J., 1992, The *E. coli ffh* gene is necessary for viability and efficient protein export, *Nature,* **359**:744–746.

Poritz, M. A., Bernstein, H. D., Strub, K., Zopf, D., Wilhelm, H., and Walter, P., 1990, An *E. coli* ribonucleoprotein containing 4.5S RNA resembles mammalian signal recognition particle, *Science* **250**:1111–1117.

Puziss, J. W., Fikes, J. D., and Bassford Jr., P. J., 1989, Analysis of mutational alterations in the hydrophilic segment of the maltose-binding protein signal peptide, *J. Bacteriol.* **171**:2303–2311.

Puziss, J. W., Strobel, S. M., and Bassford Jr., P. J., 1992, Export of maltose-binding protein species with altered charge distribution surrounding the signal peptide hydrophobic core in *Escherichia coli* cells harboring *prl* suppressor mutations, *J. Bacteriol.* **174**:92–101.

Randall, L. L., 1992, Peptide binding by chaperone SecB: Implications for recognition of nonnative structure, *Science* **257**:241–245.

Randall, L. L., Hardy, S. J. S., and Thom, J. R., 1987, Export of protein: A biochemical view, *Annu. Rev. Microbiol.* **41**:507–541.

Randall, L. L., Topping, T. B., and Hardy, S. J. S., 1990, No specific recognition of leader peptide by SecB, a chaperone involved in protein export, *Science* **248**:860–863.

Reid, J., Fung, H., Gehring, K., Klebba, P. E., and Nikaido, H., 1988, Targeting of porin to the outer membrane of *Escherichia coli, J. Biol. Chem.* **263**:7753–7759.

Ribes, V., Römisch, K., Giner, A., Dobberstein, B., and Tollervey, D., 1990, *E. coli* 4.5S RNA is part of a ribonucleoprotein particle that has properties related to signal recognition particle, *Cell* **63**:591–600.

Richter, P., and Drews, G., 1991, Incorporation of light-harvesting complex I α and β polypeptides into the intracytoplasmic membrane of *Rhodobacter capsulatus, J. Bacteriol.* **173**:5336–5345.

Richter, P., Cortez, N., and Drews, G., 1991, Possible role of the highly conserved amino acids Trp-8 and Pro-13 in the N-terminal segment of the pigment-binding polypeptide LHI α of *Rhodobacter capsulatus, FEBS-Lett.* **285**:80–84.

Richter, P., Brand, M., and Drews, G., 1992, Characterization of LHI⁻ and LHI⁺ *Rhodobacter capsulatus pufA* mutants, *J. Bacteriol.* **174**:3030–3041.

Ried, G., Hindennach, I., and Henning, U., 1990a, Role of lipopolysaccharide in assembly of *Escherichia coli* outer-membrane proteins OmpA, OmpC, and OmpF, *J. Bacteriol.* **172**:6048–6053.

Ried, G., MacIntyre, S., Mutschler, M., and Henning, U., 1990b, Export of altered forms of an *Escherichia coli* K-12 outer membrane protein (OmpA) can inhibit synthesis of unrelated outer membrane proteins, *J. Mol. Biol.* **216**:39–47.

Riggs, P. D., Derman, A. I., and Beckwith, J., 1988, A mutation affecting the regulation of a *secA-lacZ* fusion defines a new *sec* gene, *Genetics* **118**:571–579.

Russel, M., and Model, P. 1982, Filamentous phage pre-coat is an integral membrane protein: Analysis by a new method of membrane preparation, *Cell* **28**:177–184.

Saier Jr., M. H., Werner, P. K., and Müller, M., 1989, Insertion of proteins into bacterial membranes: Mechanism, characteristics, and comparisons with the eukaryotic process, *Microbiol. Rev.* **53**:333–366.

Sako, T., and Iino, T., 1988, Distinct mutation sites in *prlA* suppressor mutant strains of *Escherichia coli* respond either to suppression of signal peptide mutations or to blockage of staphylokinase processing, *J. Bacteriol.* **170:**5389–5391.

San Millan, J. L., Boyd, D., Dalbey, R., Wickner, W., and Beckwith, J., 1989, Use of PhoA fusions to study the topology of the *Escherichia coli* inner membrane protein leader peptidase, *J. Bacteriol.* **171:**5536–5541.

Schatz, P. J., and Beckwith, J., 1990, Genetic analysis of protein export in *Escherichia coli, Annu. Rev. Genet.* **24:**215–248.

Schatz, P. J., Riggs, P. D., Jacq, A., Fath, M. J., and Beckwith, J., 1989, The *secE* gene encodes an integral membrane protein required for protein export in *Escherichia coli, Genes Dev.* **3:**1035–1044.

Schatz, P. J., Bieker, K. L., Ottemann, K. M., Silhavy, T. J., and Beckwith, J., 1991, One of three transmembrane stretches is sufficient for the functioning of the SecE protein, a membrane component of the *E. coli* secretion machinery, *EMBO J.* **10:**1749–1757.

Schiebel, E., Driessen, A. J. M., Hartl, F. U., and Wickner, W., 1991, $\Delta\mu_H^+$ and ATP function at different steps of the catalytic cycle of preprotein translocase, *Cell* **64:**927–939.

Schmidt, M. G., Dolan, K. M., and Oliver, D. B., 1991, Regulation of *Escherichia coli secA* mRNA translation by a secretion-responsive element, *J. Bacteriol.* **173:**6605–6611.

Seckler, R., and Wright, J. K., 1984, Sidedness of native membrane vesicles of *Escherichia coli* and orientation of the reconstituted lactose: H^+ carrier, *Eur. J. Biochem.* **142:**269–279.

Seckler, R., Wright, J. K., and Overath, P., 1983, Peptide-specific antibody locates the COOH-terminus of the lactose carrier of *Escherichia coli* on the cytoplasmic side of the plasma membrane, *J. Biol. Chem.* **258:**10817–10820.

Seckler, R., Möröy, T., Wright, K. J., and Overath, P., 1986, Anti-peptide antibodies and proteases as structural probes for the lactose/H^+ transporter of *Escherichia coli:* A loop around amino acid residue 130 faces the cytoplasmic side of the membrane, *Biochemistry* **25:**2403–2409.

Sen, K., and Nikaido, H., 1990, *In vitro* trimerization of OmpF porin secreted by spheroplasts of *Escherichia coli, Proc. Natl. Acad. Sci. USA* **87:**743–747.

Sen, K., and Nikaido, H., 1991, Lipopolysaccharide structure required for *in vitro* trimerization of *Escherichia coli* OmpF porin, *J. Bacteriol.* **173:**926–928.

Shiba, K., Ito, K., Yura, T., and Ceretti, D. P., 1984, A defined mutation in the protein export gene within the *spc* ribosomal protein operon of *Escherichia coli:* Isolation and characterization of a new temperature-sensitive *secY* mutant, *EMBO J.* **3:**631–635.

Shinkai, A., Mei, L. H., Tokuda, H., and Mizushima, S., 1991, The conformation of SecA, as revealed by its protease sensitivity, is altered upon interaction with ATP, presecretory proteins, everted membrane vesicles, and phospholipids, *J. Biol. Chem.* **266:**5827–5833.

Simon, S. M., and Blobel, G., 1991, A protein-conducting channel in the endoplasmic reticulum, *Cell* **65:**371–380.

Simon, S. M., and Blobel, G., 1992, Signal peptides open protein-conducting channels in *E. coli, Cell* **69:**677–684.

Singer, S. J., and Yaffe, M. P., 1990, Embedded or not? Hydrophobic sequences and membranes, *Trends Biochem. Sci.* **15:**369–373.

Stader, J., Benson, S. A., and Silhavy, T. J., 1986, Kinetic analysis of *lamB* mutants suggests the signal sequence plays multiple roles in protein export, *J. Biol. Chem.* **261:**15075–15080.

Stader, J., Gansheroff, L. J., Silhavy, T. J., 1989, New suppressors of signal-sequence mutations, *prlG*, are linked tightly to the *secE* gene of *Escherichia coli, Genes Dev.* **3:**1045–1052.

Stiehle, H., Cortez, N., Klug, G., and Drews, G., 1990, A negatively charged N terminus in the α polypeptide inhibits formation of light-harvesting complex I in *Rhodobacter capsulatus, J. Bacteriol.* **172:**7131–7137.

Stochaj, U., Bieseler, B., and Ehring, R., 1986, Limited proteolysis of lactose permease from *Escherichia coli, Eur. J. Biochem.* **158:**423–428.

Struyvé, M., Moons, M., and Tommassen, J., 1991, Carboxy-terminal phenylalanine is essential for the correct assembly of a bacterial outer membrane protein, *J. Mol. Biol.* **218**:141–148.

Sugai, M., and Wu, H. C., 1992, Export of the outer membrane lipoprotein is defective in *secD*, *SecE*, and *secF* mutants of *Escherichia coli*, *J. Bacteriol.* **174**:2511–2516.

Summers, R. G., and Knowles, J. R., 1989, Illicit secretion of a cytoplasmic protein into the periplasm of *Escherichia coli* requires a signal peptide plus a portion of the cognate secreted protein, *J. Biol. Chem.* **264**:20074–20081.

Summers, R. G., Harris, C. R., and Knowles, J. R., 1989, A conservative amino acid substitution, arginine for lysine, abolishes export of a hybrid protein in *Escherichia coli*, *J. Biol. Chem.* **264**:20082–20088.

Swidersky, U. E., Hoffschulte, H. K., and Müller, M., 1990, Determinants of membrane-targeting and transmembrane translocation during bacterial protein export, *EMBO J.* **9**:1777–1785.

Swidersky, U. E., Rienhöfer-Schweer, A., Werner, P. K., Ernst, F., Benson, S. A., Hoffschulte, H. K., and Müller, M., 1992, Biochemical analysis of the biogenesis and function of the *Escherichia coli* export factor SecY, *Eur. J. Biochem.* **207**:803–811.

Tadros, M. H., Garcia, A. F., Gad'on, N., and Drews, G., 1989, Characterization of a pseudo-B870 light-harvesting complex isolated from the mutant strain Ala+Pho− of *Rhodobacter capsulatus* which contains B800-850-type polypeptides, *Biochim. Biophys. Acta* **976**:161–167.

Thome, B. M., and Müller, M., 1991, Skp is a periplasmic *Escherichia coli* protein requiring SecA and SecY for export, *Mol. Microbiol.* **5**:2815–2821.

Thome, B. M., Hoffschulte, H. K., Schiltz, E., and Müller, M., 1990, A protein with sequence identity to Skp (FirA) supports protein translocation into plasma membrane vesicles of *Escherichia coli*, *FEBS Lett.* **269**:113–116.

Thrift, R. N., Andrews, D. W., Walter, P., and Johnson, A. E., 1991, A nascent membrane protein is located adjacent to ER membrane proteins throughout its integration and translation, *J. Cell Biol.* **112**:809–821.

Tokuda, H., Shiozuka, K., and Mizushima, S., 1990, Reconstitution of translocation activity for secretory proteins from solubilized components of *Escherichia coli*, *Eur. J. Biochem.* **192**:583–589.

Troschel, D., and Müller, M., 1990, Development of a cell-free system to study the membrane assembly of photosynthetic proteins of *Rhodobacter capsulatus*, *J. Cell Biol.* **111**:87–94.

Troschel, D., Eckhardt, S., Hoffschulte, H. K., and Müller, M., 1992, Cell-free synthesis and membrane-integration of the reaction center subunit H from *Rhodobacter capsulatus*, *FEMS Microbiol. Lett.* **91**:129–134.

Trun, N. J., Stader, J., Lupas, A., Kumamoto, C., and Silhavy, T. J., 1988, Two cellular components, PrlA and SecB, that recognize different sequence determinants are required for efficient protein export, *J. Bacteriol.* **170**:5928–5930.

Ueguchi, C., and Ito, K., 1992, Multicopy suppression: An approach to understanding intracellular functioning of the protein export system, *J. Bacteriol.* **174**:1454–1461.

Ulbrandt, N. D., London, E., and Oliver, D. B., 1992, Deep penetration of a portion of *Escherichia coli* SecA protein into model membranes is promoted by anionic phospholipids and by partial unfolding, *J. Biol. Chem.* **267**:15184–15192.

von Heijne, G., 1986, The distribution of positively charged residues in bacterial inner membrane proteins correlates with the trans-membrane topology, *EMBO J.* **5**:3021–3027.

von Heijne, G., 1989, Control of topology and mode of assembly of a polytopic membrane protein by positively charged residues, *Nature* **341**:456–458.

von Heijne, G., and Gavel, Y., 1988, Topogenic signals in integral membrane proteins, *Eur. J. Biochem.* **174**:671–678.

Vos-Scheperkeuter, G. H., and Witholt, B., 1984, Assembly pathway of newly synthesized LamB protein, an outer membrane protein of *Escherichia coli* K-12, *J. Mol. Biol.* **175**:511–528.

Walian, P . J., and Jap, B. K., 1990, Three-dimensional electron diffraction of PhoE Porin to 2.8 Å resolution, *J. Mol. Biol.* **215:**429–438.

Watanabe, M., and Blobel, G., 1989a, Cytosolic factor purified from *Escherichia coli* is necessary and sufficient for the export of a preprotein and is a homotetramer of SecB, *Proc. Natl. Acad. Sci. USA* **86:**2728–2732.

Watanabe, M., and Blobel, G., 1989b, SecB functions as a cytosolic signal recognition factor for protein export in *E. coli, Cell* **58:**695–705.

Watanabe, M., and Blobel, G., 1989c, Site-specific antibodies against the PrlA (SecY) protein of *Escherichia coli* inhibit protein export by interfering with plasma membrane binding of prepro- teins, *Proc. Natl. Acad. Sci. USA* **86:**1895–1899.

Watanabe, M., Hunt, J. F., and Blobel, G., 1986, *In vitro* synthesized bacterial outer membrane protein is integrated into bacterial membranes but translocated across microsomal membranes, *Nature* **323:**71–73.

Watanabe, T., Hayashi, S., and Wu, H. C., 1988, Synthesis and export of the outer membrane lipoprotein in *Escherichia coli* mutants defective in generalized protein export, *J. Bacteriol.* **170:**4001–4007.

Watanabe, M., Nicchitta, C. V., and Blobel, G., 1990, Reconstitution of protein translocation from detergent-solubilized *Escherichia coli* inverted vesicles: PrlA protein-deficient vesicles effi- ciently translocate precursor proteins, *Proc. Natl. Acad. Sci. USA* **87:**1960–1964.

Weiss, J. B., Ray, P. H., and Bassford Jr., P. J., 1988, Purified SecB protein of *Escherichia coli* retards folding and promotes membrane translocation of the maltose-binding protein *in vitro,* *Proc. Natl. Acad. Sci. USA* **85:**8978–8982.

Weiss, M. S., Abele, U., Weckesser, J., Welte, W., Schiltz, E., and Schulz, G. E., 1991, Molecular architecture and electrostatic properties of a bacterial porin, *Science* **254:**1627–1630.

Werner, P. K., Saier, M. H., Jr., and Müller, M., 1992, Membrane insertion of the mannitol permease of *Escherichia coli* occurs under conditions of impaired SecA function, *J. Biol. Chem.* **267:**24523–24532.

Wickner, W., 1988, Mechanisms of membrane assembly: General lessons from the study of M13 coat protein and *Escherichia coli* leader peptidase, *Biochemistry* **27:**1081–1086.

Wickner, W., Driessen, A. J. M., and Hartl, F. U., 1991, The enzymology of protein translocation across the *Escherichia coli* plasma membrane, *Annu. Rev. Biochem.* **60:**101-124.

Wieseler, B., and Müller, M., 1992, Translocation of precytochrome c_2 into intracytoplasmic mem- brane vesicles of *R. capsulatus* requires a peripheral membrane protein, *Mol Microbiol.* **7:**167– 176.

Wieseler, B., Troschel, D., Potgeter, M., Hoffschulte, H. K., and Müller, M., 1993, Protein export in the α-purple bacterium *R. capsulatus:* Lack of SecA- and SecB-like activities as revealed by *in vitro* analysis, submitted.

Woese, C. R., 1987, Bacterial evolution, *Microbiol. Rev.* **51:**221–271.

Wolfe, P. B., Rice, M., and Wickner, W., 1985, Effects of two *sec* genes on protein assembly into the plasma membrane of *Escherichia coli, J. Biol. Chem.* **260:**1836–1841.

Yamada, H., Tokuda, H., and Mizushima, S., 1989. Proton motive force-dependent and -indepen- dent protein translocation revealed by an efficient in vitro assay system of *Escherichia coli, J. Biol. Chem.* **264:**1723–1728.

Yamada, Y., Chang, Y. Y., Daniels, G. A., Wu, L. F., Tomich, J. M., Yamada, M., and Saier Jr., M. H., 1991, Insertion of the mannitol permease into the membrane of *Escherichia coli, J. Biol. Chem.* **266:**17863–17871.

Yamane, K., and Mizushima, S., 1988, Introduction of basic amino acid residues after the signal peptide inhibits protein translocation across the cytoplasmic membrane of *Escherichia coli, J. Biol. Chem.* **263:**19690–19696.

Yamane, K., Akiyama, Y., Ito, K., and Mizushima, S., 1990, A positively charged region is a determinant of the orientation of cytoplasmic membrane proteins in *Escherichia coli*, *J. Biol. Chem.* **265**:21166–21171.

Yamato, I., 1992, Membrane assembly of lactose permease of *Escherichia coli*, *J. Biochem.* **111**:444–450.

Zhang, Y., and Broome-Smith, J. K., 1990, Correct insertion of a simple eukaryotic plasma membrane protein into the cytoplasmic membrane of *Escherichia coli*, *Gene* **96**:51–57.

Zopf, D., Bernstein, H. D., Johnson, A. E., and Walter, P., 1990, The methionine-rich domain of the 54 kd protein subunit of the signal recognition particle contains an RNA binding site and can be crosslinked to a signal sequence, *EMBO J.* **9**:4511–4517.

Chapter 11

Ischemic Effects on the Structure and Function of the Plasma Membrane

Simone Wattiaux-De Coninck and Robert Wattiaux

1. PLASMA MEMBRANE: STRUCTURE AND FUNCTION

The main function of the plasma membrane is to provide a barrier between the intracellular milieu and the extracellular environment. Selective ion transport and cellular ion homeostasis depend on plasma membrane ion channels, transporters, and pumps. The integrity of the plasma membrane rests on the underlying cytoskeleton connected via integrins with the extracellular matrix or the basement membrane. Ischemia-induced irreversible injury is associated with plasma membrane defects, resulting in the release of intracellular components and disruption of ion, and especially of calcium, homeostasis.

1.1. The Bilayer

The role of the plasma membrane as a selective barrier between the extracellular and intracellular compartments depends on its molecular organization (Fortunati and Bianchi, 1989). The cell membrane works best when the lipid bilayer is in the liquid crystalline or fluid phase. This state depends on its

Simone Wattiaux-De Coninck and Robert Wattiaux   Laboratoire de Chimie Physiologique, Facultés Universitaires Notre-Dame de la Paix, B-5000 Namur, Belgium.
Subcellular Biochemistry, Volume 22: Membrane Biogenesis, edited by A. H. Maddy and J. R. Harris. Plenum Press, New York, 1994.

composition. Cholesterol increases the order of fatty acyl chains in the liquid crystalline phase, decreases membrane permeability, and alters the rate of transport of a solute across the lipid bilayer. Proteins may be integrated into or associated with the membrane. The membrane surface occupied by the proteins may be as extensive as the surface occupied by the lipids in the bilayer (Yeagle, 1989). Some proteins of the outer surface of the cell are bound by covalent linkage to a glycosylphosphatidylinositol, to myristic or palmitic acids on the cytoplasmic surface (Beaudoin and Grondin, 1991; Low, 1989). Lipid composition, cholesterol level, and unsaturation of fatty acids not only determine the fluidity of the bilayer but also the cell membrane activity. Lipid-protein interactions might involve a specific binding of the phospholipid on the protein or a nonspecific interaction in the form of a lipid annulus or a surface-surface interaction between the surface of the bilayer and the extramembranous surface of the protein (Yeagle, 1989; Cheng et al., 1986). Cholesterol has been shown to regulate the ion pump activity of $Na^+ K^+$ ATPase, which is responsible for pumping Na^+ out of the cell and K^+ into the cell against their respective concentration gradients.

Another characteristic of the bilayer is the maintenance of membrane asymmetry. In the erythrocytes, membrane phosphatidylserine and phosphatidylethanolamine are located on the cytoplasmic face, and glycolipids and sphingomyelin on the exterior face, whereas phosphatidylcholine is disproportionately distributed toward the exterior (Beaudoin and Grondin, 1991; Jain, 1985). Cholesterol is also distributed asymmetrically. Establishment and maintenance of this asymmetry requires energy. The interaction between plasma membrane components and the submembranous network modifies the lateral topographic distribution of both proteins and lipids, resulting in the formation of microdomains (Beaudoin and Grondin, 1991).

The lipid bilayer controls cell permeability. Passive permeability is slow for most solutes (except for water) so that transporters control the entry and exit of ions and nutrients (Yeagle, 1989). Intracellular ion concentration is controlled by ion channels, antiporters and symporters, and ATP-driven ion pumps. Loss of the integrity of the bilayer makes the cell freely permeable and leads to cell death (Yeagle, 1989).

1.2. Ion Transport

Ion channels open briefly and then close again in response to specific perturbations of the membrane. The main types of perturbations known to cause channels to open are a change in the voltage across the membrane, the binding of a ligand, or mechanical stimulation. Ion channels are distinguished from ion carriers (Brown, 1991) by the high rates at which they transport ions once they are opened. They are generally highly selective for a given ion species, Na^+, K^+, Ca^{2+}, or Cl^-. Calcium flux is unique because in addition to producing an electrical signal, it also acts as a cytoplasmic messenger.

Voltage-gated Na^+ channels of muscle cells and neurons are responsible for membrane depolarization (Gordon, 1990; Alberts et al., 1989). There are at least three distinct classes of voltage-gated Ca^{2+} channels: the low-threshold, rapidly inactivated T-type, which displays little or no sensitivity to dihydropyridine; the high-threshold L-type, which is dihydropyridine, verapamil, and diltiazem sensitive, present in most cell types, though more abundant in heart and smooth muscle cells (Wier, 1990); and the N and P types, which are only present in neurons and both of which are resistant to dihydropyridine. The L-type calcium channel plays a crucial part in excitation-contraction coupling in cardiac and smooth muscle cells (Nargeot, 1991; Sher e al., 1991; Tsien et al., 1991; Rampe and Triggle, 1990; Zernig, 1990; Bean, 1989). One important characteristic of calcium channels is their sensitivity to adrenergic transmitters (Schultz et al., 1990; Bean, 1989).

The three ligand-activated ion channels—nicotinic acetylcholine receptor, GABA, and glycine receptors—all of which belong to one superfamily, are located at the synapses between nerve cells and their target cells (Gordon, 1990; Alberts et al., 1989). Chloride channels are characterized by a wide variability in their functional characteristics. Some are gated by agonists and others by voltage (Franciolini and Petris, 1990).

Antiporters are found in many different cell types, including excitable cells. They can be divided in cation antiporters and anion antiporters. The Na^+/Ca^{2+} antiporter is critically important in Ca^{2+} homeostasis and extrudes calcium. The Na^+/Ca^{2+} exchanger of myocytes may operate in both directions. If the Na^+ content of the cell increases, then Ca^{2+} entry and Na^+ extrusion will be favored. It is particularly active in excitable membranes such as those of nervous and heart cells. It exchanges 3 Na^+ per 1 Ca^{2+} and has a large transport capacity, exporting 15–30 nmoles Ca^{2+}/mg membrane protein/sec (Khananshvili, 1990; Tani, 1990; McCormack et al., 1988; Murphy et al., 1988). The Na^+/H^+ antiporter, which is activated when pH_i falls, is a ubiquitous transport system involved in the regulation of intracellular pH and control of cell volume (Tani, 1990; Scharschmidt and Van Dyke, 1987). The Cl^-/HCO_3^- anion antiporter, which is similar to the band 3 protein in the membrane of red blood cells, is activated when pH_i rises and ejects HCO_3^- from the cell in exchange for Cl^-, thereby decreasing the pH_i.

The Na^+ or H^+ ion coupled entry of solutes accomplishes two purposes. First, new solutes are brought into the compartment (cell or vesicle) providing it with nutrients, hormones, transmitters and regulators. Second, the original substrate of the primary group (Na^+ or H^+) is returned to the cell (Johnstone, 1990).

A major physiological role for antiporters is to regulate primary ion gradients created by ion pumps (Na^+/K^+ ATPase, sarcoplasmic and plasma membrane Ca^{2+} ATPases, H^+ ATPase, H^+/K^+ ATPase). These pumps exist in both phosphorylated and dephosphorylated forms and belong to the family of P-type ATPases (Carafoli, 1992; Andersen and Vilsen, 1990).

Na^+/K^+ ATPase is an electrogenic pump (3 $Na^+/2$ K^+) that is required to maintain osmotic balance and stabilize cell volume (Alberts *et al.*, 1989; Yingst, 1988).

Plasma membrane Ca^{2+} ATPase exports calcium from the cell. It is a high-affinity enzyme, but transports calcium with a low total capacity (0.5 nmole of Ca^{2+}/mg of membrane protein/sec). It is a P-type ion pump (aspartylphosphate) whose regulation has been described in hepatocytes and hepatomas (Lotersztajn *et al.*, 1992) and in the brain (Fraser and Sarnacki, 1992). Plasma membrane Ca^{2+} ATPase is stimulated by calmodulin, acidic phospholipids and long-chain polyunsaturated fatty acids, and by controlled proteolysis (Brodin *et al.*, 1992; Lehotsky *et al.*, 1992). Liver plasma Ca^{2+} ATPase has distinctive characteristics as it is not stimulated by calmodulin and has a much higher affinity for calcium (Carafoli, 1992; Grover and Khan, 1992; Wang *et al.*, 1992; Carafoli, 1991; Andersen and Vilsen, 1990; Wier, 1990; Schatzmann, 1989; Carafoli, 1987).

Plasma membrane H^+ ATPase plays a role in regulation of pH_i.

1.2.1. Calcium Homeostasis

Because many cellular functions are highly dependent on $[Ca^{2+}]_i$, the precise value of cytosolic calcium is of critical importance to determine cell function. Cells normally satisfy their calcium demands by using their own calcium pool, either the cytosolic pool or that in the organelles. Cell calcium homeostasis depends on the ionized portion of calcium, which represents only 0.1% of the total intracellular pool, and is controlled by soluble and membrane proteins (Burgoyne, 1991; Tani, 1990; Wier, 1990; Capiod *et al.*, 1989; Carafoli, 1987).

Specific soluble ligands are calmodulin, which is probably the most important since it is present in all eukaryotic cells in considerable amounts, and troponin C, which is found in muscle cells. Their main role is not the buffering of calcium, but the processing of its signal. Conformational changes of these soluble ligands after binding of calcium permit their interaction with proteins whose association with the membrane is regulated by calcium, or with targets inducing the association of actin and myosin (Tani, 1990; Carafoli, 1987).

Calcium transport across plasma membrane depends on Ca^{2+} channels, Na^+/Ca^{2+} exchanger, and Ca^{2+} ATPase.

Calcium transport across intracellular membranes essentially depends on the endoplasmic (and sarcoplasmic) reticulum in which the uptake system is mediated by Ca^{2+} ATPase and the release by the ryanodine-sensitive receptor (Wier, 1990; Berridge, 1987). Sarcoplasmic reticulum Ca^{2+} ATPase is a very abundant protein, well characterized in heart, skeletal, and smooth muscle (Tani, 1990; Schatzmann, 1989; Movsesian *et al.*, 1984). In nonmuscle cells, the release of Ca^{2+} from endoplasmic reticulum appears to be mediated by phosphoinositides (Williamson and Monck, 1989; Berridge, 1987). In muscle cells, the depolarization of the cell membrane is somehow transmitted through the transverse tubules to the interior of the muscle cell, thus causing efflux of calcium from the neighboring cisternae of the reticulum. The nature of the signal has been proposed to be calcium, giving rise to the Ca^{2+}-induced Ca^{2+} release phenomenon. Controversy still exists whether inositoltriphosphate also acts as a second messenger in the release of calcium from the sarcoplasmic reticulum in heart muscle (Mouton et al., 1991).

1.3. The Cytoskeleton and the Extracellular Matrix

The eukaryotic cytoskeleton is presumably responsible for the maintenance of cell shape and plasma membrane integrity, since the plasma membrane is an unstable structure that spontaneously forms small membrane vesicles if isolated or dissociated from the underlying cytoskeleton. An intact cytoskeleton is essential for vital cellular functions such as transport, cell-cell interaction, cell-extracellular matrix interaction, cell motility, and cell cycle (Rogers et al., 1992; Steenbergen et al., 1987).

1.3.1. Microfilaments

Microfilaments are made up of actin monomers. Each actin monomer has one ATP binding site and one high-affinity binding site for divalent cations. Microfilament actin (F-actin) exists in equilibrium with soluble globular actin (G-actin). Filament formation proceeds in two steps, nucleation, and then elongation involving the addition of monomers. The process is regulated by ATP hydrolysis and by monomer binding proteins such as profilin. A large number of actin-binding proteins regulate filament structural processes. Cross-linking and the formation of networks depend on filamin and villin (in stress fibers). When activated by the binding of calcium, gelsolin severs an actin filament and forms a cap on the newly formed plus end of the filament, thus breaking up the network of actin filament. Gelsolin is regulated not only by calcium but also by phosphatidylinositol-4,5-bisphosphate (PIP_2). Microfilaments are essential for cell structure and morphology. In nonmuscle cells they form an actin cortex on the cell periphery made of stress fibers that have connections, via talin, vinculin, and α-actinin, with the extracellular matrix through, for example, the fibronectin receptor, one of the β_1 integrins (see Section 1.4). Microfilaments have been suggested to play a role in cell activation (Rogers et al., 1992; Alberts et al., 1989; Forcher, 1989; Burridge et al., 1988).

1.3.2. Microtubules

Microtubule formation occurs in two steps, nucleation (i.e., oligomer formation) and elongation, when oligomers bind to both ends of the filament but at different rates. Hence, microtubules, like microfilaments, are polar. Tubule formation requires energy. The *in vitro* assembly is strictly dependent on magnesium, whereas micromolecular amounts of calcium, in the presence of calmodulin, are inhibitory. Microtubule-associated proteins speed up the nucleation step and are involved in linking the microtubules to other cell components (Rogers *et al.*, 1992; Alberts *et al.*, 1989; Karr *et al.*, 1980). One membrane protein synapsin appears to interact directly with tubulin (Beaudoin and Grondin, 1991).

1.3.3. Intermediate Filaments

Intermediate filaments are classified into five groups based on the sequence homologies of the 20 or more intermediate filament proteins. When a cell has differentiated it has a characteristic filament protein expression: mesenchymal cells express vimentin; muscle cells express desmin; nerve cells express neurofilaments; and epithelial cells express cytokeratins. Intermediate filaments can assemble *in vitro* without the presence of any other protein; thus the information required for the assembly is present in the primary structure of intermediate filament proteins so that they are mainly formed co-translationally. Intermediate filaments are not polar. They make connections with the nucleus and the plasma membrane and are mechanically integrators of space, cell regulation, and nuclear function (Rogers *et al.*, 1992).

1.3.4. Extracellular Matrix

Extracellular space is filled with an intricate network of macromolecules constituting the extracellular matrix. The two main classes that make up the matrix are glycosaminoglycans and fibrous proteins of two functional types: mainly structural (collagen and elastin) and mainly adhesive (fibronectin and laminin). Fibronectin is a multifunctional molecule, one domain binding to collagen, another to heparin, another to a specific receptor of the integrin family (Ruoslahti, 1991; Alberts *et al.*, 1989).

Basement membrane is a specialized extracellular matrix that forms the supporting structure on which epithelial and endothelial cells grow and surrounds muscle cells, fat cells, and Schwann cells. The basement membrane appears to be a storage compartment for bFGF, which is bound to heparan sulfate proteoglycans in the matrix. It has also been shown to contain an intrinsic proteolytic system (plasminogen, plasminogen activator, and plasminogen activator inhibitor type-1) that may play a role in tissue remodeling. Moreover, complement factor C1q can bind to laminin; C3d has been detected in basement membranes and potentially plays a role in fixation of immune complexes (Paulsson, 1992).

1.4. The Integrins

An important function of plasma membrane integrins is to convey informa-tion to the cell interior and to transmit signals from the cell interior to the cell exterior. They interact either with ligands of the extracellular matrix or with cellular counter-receptors or soluble molecules such as fibrinogen inducing cell-cell adhesion of PMN and aggregation of platelets (Quaranta and Jones, 1991).

1.4.1. Interaction with the Extracellular Matrix

The integrin family involved in cell matrix interactions is known as $\beta1$ integrins (or VLAs) as they are composed of a $\beta1$ subunit and diverse α subunits (Table I).

The recognition site for many of the integrins that bind to extracellular matrix (and platelet adhesion molecules) is the tripeptide RGD (arginine-glycine-aspartic acid). Specific recognition between integrins and fibronectin, vitronectin, col-lagens, fibrinogen, von Willebrand factor, laminin, and other proteins depends on flanking regions that are similar but not identical. The extracellular matrix binding proteins are constitutively activated (Mosher, 1991; Akiyama et al., 1990).

1.4.2. Cell-Cell Adhesion

The β_2 integrins are regulated polymorphonuclear molecules mediating ad-hesion to endothelial cells. They are composed of LFA-1, Mac-1 and p150,90. Leucocyte integrin LFA-1 (or CD11a-CD18 or $\alpha_L \beta_2$) interacts with cell-surface receptors, ICAM-1 and ICAM-2, which are members of the immunoglobulin superfamily. LFA-1 needs to be activated. Mac-1 complex is referred to as CD 11b/CD18 complex (see Table I). Activation of polymorphonuclear leukocytes first results in a qualitative alteration of the heterodimers (or qualitative upregula-tion) (Zimmerman et al., 1992) that are constitutively present in the poly-morphonuclear leukocytes plasma membrane and, second, in a quantitative in-crease of their expression, with neutrophil accumulation at the endothelial surface as a net result (Simpson and Lucchesi, 1987).

1.4.3. Cell Aggregates

In a third mode of aggregation, the major integrin in platelets $\alpha_{IIb}/\beta_{IIIa}$ or $GP_{IIb}/_{IIIa}$ promotes the binding of platelets to one another through soluble multi-valent mediator molecules such as fibrinogen and von Willebrand factor. Such aggregation is a key factor in blood clotting. The main platelet integrin requires activation in order to bind to its ligand (Ruoslahti, 1991).

Table I

The Integrin Family and Adhesion Receptors

Adhesion molecules		Receptors	
Names	Regulation of surface expression	Names	Regulation of surface expression
1. The β_1 integrins involved in cell-extracellular matrix adhesion		Extracellular matrix proteins	
$\alpha_1\beta_1$ (VLA-1)	Constitutively activated	Collagen	
$\alpha_2\beta_1$ (VLA-2)		Collagen types I–IV, VI	
$\alpha_3\beta_1$ (VLA-3)		Fibronectin, laminin, collagen	
$\alpha_4\beta_1$ (VLA-4)		Fibronectin	
$\alpha_5\beta_1$ (VLA-5, fibronectin receptor, GP_{Ia-IIa})		Fibronectin	
$\alpha_6\beta_1$ (VLA-6)		Laminin	
2. The β_2 integrins involved in leukocyte-endothelial cell adhesion		Counter-receptors	
LFA-1 or CD11a/CD18 or α_L/β_2 integrin	C5a, PAF Constitutively present on the plasma membrane and translocated to it by upregulation	ICAM-1 or CD54 ICAM-2	IL1, TNFα, slow response
Mac1 or Mol or CD11b/CD18 or OKM1 or $\alpha_M\beta_2$ integrin	C5a, fMLP, IL8, PAF, LTB$_4$, fast response	ICAM, bound iC3b	IL1, TNFα, fast response

(continued)

Table II
Some Commonly Used Inhibitors of Glycosylation

Inhibitor	Biosynthetic step(s) blocked	Consequences
N-GlcNAc-linked pathway		
Castanospermine	Blocks processing glucosidases	Glycoproteins with $Glc_{2-3}Man_9GlcNAc_2$
N-methyl-deoxynojirimycin		Complex chains decreased.
Deoxymannojirimycin	Blocks initial mannose removal	Glycoproteins with Man_9-$GlcNAc_2$
		Complex chains decreased
Swainsonine	Blocks final mannose removal	Partial processing to hybrid chains
O-GalNAc-linked pathway		
Benzyl-α-GalNAc	Competitively inhibits chain elongation	Free chains made on GalNAc
O-Xyl-linked pathway		
4MU- and pNP-β-xylosides	Competitively inhibits chain elongation	Reduced amounts of O-linked chains
	(can also alter glycolipids)	Free GAG chains made on xylosides
		Reduced amounts of chains on proteoglycans
Glc-ceramide-linked pathway		
PDMP	Inhibits Glc-ceramide synthase	Stops GSL synthesis; builds up ceramide, sphingosine, and sphingomyelin
Multiple pathways		
Fluoro and deoxy-sugar analogues	Metabolized to sugar nucleotides that interfere with various steps; inhibits some processing enzymes	Blocks synthesis early
Tunicamycin	Blocks GlcNAc-P-P-dolichol synthesis and UDP-Gal uptake into Golgi	Decrease in N-linked oligosaccharide synthesis; decrease in galatosylations of many molecules.
Amphomycin	Blocks synthesis of dolichol phosphoryl sugars	No N-linked oligosaccharides of >Man_5 made; incomplete GPI-anchors made
Brefeldin A	Disrupts Golgi trafficking	Many late steps blocked
Monensin	Ionophore, disrupts Golgi integrity	Many late steps blocked

2. ISCHEMIA AND THE ALTERATIONS OF THE PLASMA MEMBRANE

Ischemia involves cessation of circulation to a tissue so that metabolic substrates are not delivered and metabolic products are not removed. Generally, it is accompanied by severe oxygen deprivation, although collateral flow in heart and kidney needs to be considered (Weinberg, 1991). Hypoxia expresses a level of oxygen deprivation below the critical level of oxygen required by the mitochondrial electron transport chain. Anoxia indicates complete absence of oxygen and full reduction of mitochondrial cytochromes.

Adverse effects of a short ischemic period, followed by reperfusion (Forman *et al.*, 1989), are reversible. Time of no return is 20 min to 40 min in myocardium (Entman *et al.*, 1991; Przyklenk, 1988; Simpson and Lucchesi, 1987), 65 min in human liver, and between 60 and 120 min in rat liver (Hasselgren, 1987).

It is generally accepted that the primary cause of irreversible injury after a long period is due to ischemia; reperfusion leads to a hastening of the necrotizing process (Forman *et al.*, 1989; Hasselgren, 1987; Lucchesi and Mullane, 1986; Braunwald and Kloner, 1985).

2.1. Ischaemia: Cause and Site of Injury

The cause of irreversible injury remains to be defined. The major possible causes that have been proposed are (1) oxygen-free radicals (Simpson and Lucchesi, 1987; Braunwald and Kloner, 1985) produced during ischemia and reperfusion; (2) depletion of ATP stores resulting from the failure of mitochondria to produce ATP, leading to mitochondrial and plasma membrane dysfunction and impaired Ca^{2+} and Na^+ homeostasis, with the consequent influx of water and cell swelling; (3) calcium overload, leading to activation of Ca^{2+} phospholipases, proteases, and nucleases; (4) reduced protein synthesis; and (5) influx of inflammatory cells.

Another point of debate is whether a particular subcellular membrane or organelle (or several of them) is the most important site leading to irreversible injury; for example, mitochondria (Mittnacht and Farber, 1981; Mittnacht *et al.*, 1979), lysosomes (Wattiaux and Wattiaux-De Coninck, 1984), or plasma membrane (Carbonnelle, 1987; Carbonelle *et al.*, 1987).

The extent of injury depends on the cells that are under consideration: heart endothelial cells versus myocytes, liver endothelial cells or hepatocytes. Low energy demand and high glycolytic activity may be the cause of the less severe coronary injury caused to endothelium compared to the cardiomyocytes in the ischemic and anoxic heart (Mertens *et al.*, 1990). This susceptibility could also be related to the antioxidant defense power that may differ from cell to cell and from species to species (Vercellotti *et al.*, 1988). For example, endothelial cells possess hardly any catalase activity and contain a large amount of ferritin, which is an important donor of Fe^{2+} initiating the formation of hydroxyl radicals, xanthine oxidase, and cyclooxygenase producing superoxide anions. They are enriched in polyunsaturated fatty acids, which are a target for oxygen-derived active species. The oxygen tension in the vascular region is relatively high, and the endothelial cell layer is under permanent mechanical stress (Koster et al., 1987).

2.2. Reperfusion

Reperfusion in acute myocardial infarction is characterized by arrhythmias, usually ventricular, and by the so-called stunned myocardium—i.e., delay of the recovery of cardiac function and normal ATP content (Billman et al., 1991; Bolli et al., 1989; Kloner, 1988; Przyklenk, 1988; Braunwald and Kloner, 1985). Free radicals produced by xanthine oxidase seem to account for arrhythmias in rats (Downey, 1990; Manning et al., 1988; Watson et al., 1984) as well as for stunning in dogs (Bolli et al., 1989). Protection by superoxide dismutase depends on the model of ischemia and reperfusion. In the isolated rat heart, reperfusion arrhythmias after 10 and 15 min of regional ischemia, induced by occlusion of the left coronary artery, can be prevented by superoxide dismutase. However, superoxide dismutase is without effect after 10 min of global ischemia, obtained by stopping coronary flow completely (Voogd et al., 1991). Myocardial hemorrhage is probably less important as it does not extend the zone where myocytes are irreversibly altered and since it could be caused by reperfusion of already damaged microvasculature at the level of the endothelium.

Reperfusion is associated with endothelial cell swelling, increased vascular permeability, enhanced capillary filtration, and capillary plugging (Kubes et al., 1990). Low or no reflow may be attributed to blebs that may impede microcirculation. These may result from damage to endothelial cells (Forman et al., 1989) and to myocytes (Braunwald and Kloner, 1985), from endothelial cells swelling (Forman et al., 1989), red cell rheological changes, platelet aggregation (Braquet et al., 1989), and plugging by leukocytes (Simpson and Lucchesi, 1987), although leukocytes are not absolutely required for development of the no-reflow phenomenon (Forman et al., 1989).

Reperfusion, and thus reintroduction of oxygen in the ischemic tissue, is accompanied by

1. an increased production of oxygen-free radicals;
2. a normalization of pH;
3. a decrease of ATP content and, as a result, dysfunction of ATP-dependent pumps;
4. a massive influx of sodium and calcium inducing activation of phospho-

lipases, damage of mitochondria and plasma membrane, cell edema and swelling;

5. an alteration of endothelial cells and platelet aggregation;
6. recruitment and adhesion of polymorphonuclear leukocytes (Simpson and Lucchesi, 1987).

2.3. Production of Oxygen-Derived Free Radicals

Oxygen-derived active compounds are the superoxide anion, the hydroxyl radical, and hydrogen peroxide. Their production occurs, to some extent, during ischemia (Bolli *et al.*, 1989; Przyklenk, 1988; Lucchesi and Mullane, 1986) but predominantly at the time of oxygen readmission during reperfusion. They are produced by different sources (Gores *et al.*, 1989). Superoxide anion and hydrogen peroxide may be derived from xanthine oxidase activity, which results from the conversion of xanthine dehydrogenase by a calcium dependent proteolysis and/or from the oxidation of thiol groups of the NADH form of the enzyme (Repine, 1991; McCord, 1988). Although rabbit and human hearts, as well as human kidney, have a very low content of xanthine oxidase (Weinberg, 1991; Lawson *et al.*, 1990; Forman *et al.*, 1989; Grum *et al.*, 1989; Eddy *et al.*, 1987; Muxfeldt and Schaper, 1987), this source seems to account for the production of free radicals in human heart (Werns and Lucchesi, 1990) after ischemia and reperfusion. It is noteworthy that xanthine oxidase is strategically concentrated in endothelial cells (Ward, 1991; Werns and Lucchesi, 1990), causing injury to the endothelial cells or the subendothelial matrix (Sundqvist, 1991). Thus, endothelial cells are not passive partners in the events leading to cytotoxicity (Ward, 1991). Chick embryo heart cells exposed to hypoxic conditions are another source of reactive oxygen species (Murphy *et al.*, 1988). Xanthine oxidase is also active in plasma (Till *et al.*, 1991). Xanthine and hypoxanthine, the substrates of xanthine oxidase, would become available in cells after the breakdown of ATP. Oxygen-free radicals are, on the other hand, produced in damaged mitochondria (Veitch *et al.*, 1992; Vandeplassche, 1989; Malis and Bonventre, 1988). It is to be noted that complexes I and III are the major sources of free radicals produced by mitochondria *in vitro* (Turrens and Boveris, 1980; Cadenas *et al.*, 1977). They can also originate from cytochrome P 450 mediated process, from catecholamine (Noronha-Dutra *et al.*, 1988), and arachidonic acid metabolism (Forman *et al.*, 1989; Simpson and Lucchesi, 1987). Iron catalyzes the production of hydroxyl radicals. During chemical hypoxia induced by KCN and iodoacetate, ·the reductive stress favors formation of toxic oxygen species (Gores *et al.*, 1989). Infiltration of monocytes and neutrophils is another possible source of free radicals, although free-radical-mediated injury has been shown to occur in many leukocyte-free systems (Downey, 1990). In liver, Kupffer cell-derived reactive oxygen species seem to be involved in the initial vascular and parenchymal cell injury and also, indirectly, in the recruitment of neutrophils into the liver (Jaeschke, 1991).

Loss of antioxidant enzymes such as catalase, glutathione peroxidase, and superoxide dismutase, or antioxidant small molecules, such as lipophilic α tocopherol, β carotene and ubiquinol, and hydrophilic ascorbic acid and uric acid, is detrimental to cell defense against free-radical attack (Andreoli et al., 1986; Harlan et al., 1984). Finally, the relative localization of radical-generation antioxidants and protein targets has a major influence on the extent of radical attack on proteins and membranes (Dean et al., 1991).

2.4. Reduction of Intracellular pH

During ischemia, reduced pH has a protecting effect (Weinberg, 1991). Decrease of pH is secondary to lactic acid accumulation, the release of protons during ATP hydrolysis and CO_2 accumulation. (Weinberg, 1991; Hochachka and Mommsen, 1983). The mechanism of protection is unknown: ATP level could be preserved owing to inhibition of H^+ ATPases, Na^+/H^+ exchange, and AMP level by inhibition of 5' nucleotidase. Reduced pH can decrease transmembrane Ca^{2+} flux, the affinity of calmodulin for Ca^{2+}, and the activity of plasma membrane phospholipases A_2 and C, and mitochondrial phospholipase A_2. Correction of pH during reperfusion could exacerbate injury (Currin et al., 1991; Weinberg, 1991).

2.5. Reduction of Energy Metabolism

ATP is essential for myocardial function since hydrolysis of ATP provides the energy for contraction, for the maintenance of ion gradients, and for macromolecular synthesis. Normally, such energy demand can only be met by oxidative phosphorylation. During ischemia, ATP synthesis is markedly diminished at the onset of anaerobic conditions. However, the rate of ATP utilization is also reduced so that enough ATP is present to maintain certain functions for a significant period of time (Jennings and Steenbergen, 1985). ATP is probably not solely responsible for determining cell viability. It is perhaps more likely that other perturbations act synergistically with ATP to cause cell death, namely cellular homeostatic processes dependent on ATP (Hyslop et al., 1988).

Within seconds of the abrupt onset of total ischemia, the myocardium converts from aerobic metabolism to anaerobic glycolysis. However, anaerobic glycolysis does not produce enough ATP to keep pace with cellular utilization of ATP. As a consequence, reserves of ATP, creatine phosphate, and ADP fall (Jones et al., 1982). Concomitantly, there is a transitory increase in AMP followed by its degradation to adenosine, hypoxanthine, and xanthine and other purine catabolites. NAD is also degraded gradually (Schraufstatter *et al.*, 1986). In ischemic rat liver, the ATP level is rapidly restored on recirculation after an ischemic period of less than 2 hr, but remains depressed for 2 hr or more prolonged ischemia (Kamiike *et al.*, 1982). Myocytes can survive reduction of ATP to 30% of control, but ATP levels of less than 10% of control are associated with the death of severely ischemic myocytes *in vivo* (Jones *et al.*, 1982).

During reperfusion of ischemic working heart, 87–96% of total ATP synthesis is due to oxidative phosphorylation, and the maximum glycolytic production does not exceed 13%. The functional importance of the cytosolic phosphorylation, promoted by glucose and supporting the normalization of contractile function, could in fact reflect directly the poise of sarcoplasmic reticulum Ca^{2+} ATPase, with the low phosphorylation potential resulting in diminished Ca^{2+} concentration across the reticulum membrane. Thus, a consistent correlation exists between cytosolic ATP phosphorylation potential and reperfusion contractile function (Mallet *et al.*, 1990; Anundi and De Groot, 1989).

Glycolysis in cells exposed to H_2O_2 was inactivated at the glyceraldehyde-3-phosphate-dehydrogenase step by three different mechanisms: direct inactivation of the enzyme; reduction of the intracellular concentration of NAD^+, which is probably exhausted by poly-ADP-ribose polymerase; and a modest fall in pH, possibly depressing the activity of the enzyme. Moreover, physiological quantities of peroxide inhibit pyruvate dehydrogenase of cultured cardiomyocytes, blocking the aerobic oxidation of glucose and inhibiting the oxidative phosphorylation of ADP (Cochrane, 1991; Vlessis *et al.*, 1991; Hyslop *et al.*, 1988; Schraufstatter *et al.*, 1986).

2.6. Mechanisms of Calcium Overload

Elevated cytosolic-free calcium may uncouple mitochondrial oxidative phosphorylations, activate phospholipases and proteases, and disrupt cytoskeletal structures. Verapamil and nifedipine—slow channel blocking agents—have been found to protect the ischemic myocardium, probably acting by different mechanisms (Orrenius *et al.*, 1989; Cheung *et al.*, 1984).

In the initial phase of altered calcium homeostasis in myocardial ischemic injury, postulated changes include an increase in $[Ca^{2+}]_i$ without necessarily a change in total cell calcium. Only in the later stage of cell injury is there an increase in total cell calcium (Buja *et al.*, 1988).

Thus, calcium overload is not seen in the myocardium, which is not irreversibly injured. On the contrary, both reperfusion and reoxygenation after prolonged ischemia and hypoxia are associated with a large increase in Ca^{2+} content in severely damaged myocardium (Tani, 1990). Increased uptake of calcium correlates the depressed recovery of mechanical function of the myocardium (Billman *et al.*, 1991). Therefore, excessive calcium accumulation has been implicated in irreversible myocardial injury and cell necrosis, as well as in chemical anoxia (Nicotera et al., 1989), oxidative stress cytotoxicity (Nicotera et al., 1988), and toxic cell killing (Orrenius et al., 1989). An important synergism between unpaired energy metabolism and calcium accumulation has been supported by the studies of different laboratories (Buja et al., 1988).

Impairment of calcium homeostasis can result in the release of Ca^{2+} from intracellular stores, enhanced Ca^{2+} influx, and inhibition of Ca^{2+} extrusion at the plasma membrane (Orrenius et al., 1989).

Different sarcolemmal calcium transport systems may be perturbed in ischemic and reperfused myocardium and hence initiate and accelerate the calcium overload (Tani, 1990; Buja et al., 1988).

It has been proposed that the first step of calcium overload is Na^+ overload, which is rapidly converted into Ca^{2+} overload by the Na^+/Ca^{2+} exchanger (Boddeke et al., 1989). Loss of membrane integrity and increase of membrane permeability, resulting from the action of Ca^{2+}-dependent proteases and lipases as well as from the degradative action of oxygen-free radicals, may aggravate calcium overload (Geeraerts et al., 1991).

A recent study supports the hypothesis that global ischemia increases the number of voltage-dependent calcium channels in the canine brain cell membrane, which may allow increased entry of calcium into the cell during ischemia and reperfusion (Hoehner et al., 1992).

Alternatively, it has been proposed that the calcium channel probably does not account for calcium overload during either ischemia, as it is inactivated by decreased pH and ATP level prevailing during myocardial ischemia, or during reperfusion, as it appears to be presumably inactive during reperfusion (Tani, 1990; Murphy et al., 1988). The protective action of pretreatment with calcium antagonists might be explained by the delayed depletion of ATP during ischemia due to reduction of myocardial contractility and protection of mitochondrial oxidative phosphorylation during ischemia and reperfusion (Zernig, 1991; Takeo et al., 1988; Cheung et al., 1984). As a consequence, Na^+/K^+ ATPase and Ca^{2+} ATPase activities should be maintained.

It was suggested that the Na^+ channel, in contrast to the Ca^{2+} channel, was active during ischemia and that a continuous leakage of Na^+ through the Na^+ channel occurs in myocytes. A direct interaction of calcium antagonists with Na^+ channels was demonstrated, thus explaining their protective action.

The Na^+/K^+ ATPase has a primary role in the maintenance of a low and stable $[Na^+]_i$ in cardiac muscle cells. Ischaemia and depletion of ATP reduced cardiac and kidney Na^+/K^+ ATPase activity and led to elevation of $[Na^+]_i$. Reperfusion accelerated the Na^+/K^+ ATPase function which could account for most of the decline of $[Na^+]_i$ during reperfusion as 0.1 mM ouabain administered during reperfusion has been seen to abolish the rapid decline of $[Na^+]_i$ in isolated rat hearts (Tani, 1990, Kim and Akera, 1987; Kako et al., 1988).

The Na^+/Ca^{2+} exchanger directly links the Na^+ ion to the intracellular Ca^{2+} metabolism. The Na^+/Ca^{2+} exchanger is moderately decreased in sarcolemmal preparations of ischemic and/or reperfused myocardium, but is inhibited during severe ischemia when the pH_i falls to approximately pH 6 and when ATP concentration decreases since dephosphorylation inhibits exchanger activity. Flux of Na^+ and Ca^{2+} through this system has an important effect on ion homeostasis because of its high capacity compared to other Ca^{2+} ion transport systems. Reperfusion could be expected to reactivate the exchanger, which can operate in either direction, and thus calcium overload has been proposed to occur via the Na^+/Ca^{2+} exchanger (Tani, 1990; Murphy et al., 1988; McCormack et al., 1988).

Ca^{2+} ATPase is reduced in membrane vesicles from ischemic myocardium, and the reduction is enhanced upon reperfusion. However, the amount of calcium moved by this pump is relatively small as compared to the capacity of the Na^+/Ca^{2+} exchanger. Thus, a complete loss Ca^{2+} pump function would not result in a large increase in net calcium uptake during reperfusion (Tani, 1990; McCormack et al., 1988; Murphy et al., 1988).

Lipid degradation by phospholipases can increase membrane permeability to calcium as lysophospholipids possess detergent properties. However, in ischemia and reperfusion lipid degradation is a relatively late event, whereas calcium overload is observed during reperfusion after 30 min of ischemia or in vitro within the first 30 min after hydrogen peroxide injury to cultured P388D1 cells (Woodley et al., 1991; Tani, 1990; Hyslop et al., 1986).

In heart and skeletal (Quast and Cook, 1989) muscle, K_{ATP} channels are proposed to open following an ischemic injury. The ensuing hyperpolarization would prevent Ca^{2+} entry through voltage-operated calcium channels, thereby reducing calcium overload. Additionally, hyperpolarization in muscle tissue would oppose contractile activity, thereby preserving ATP_i. The opening of K_{ATP} channels could provide an important safety mechanism in the case of ischemic muscle (Quast and Cook, 1989).

2.7. Lipid Peroxidation

Ischemia reperfusion injury is mediated by enhanced calcium levels and production of oxygen-free radicals. Both can directly or indirectly modify membrane fluidity and permeability and hence cell viability (Masaki et al., 1989). However, it has been proposed that lipid peroxidation would not be the primary initiator of pathological processes, but would instead play a secondary role in the development of pathological events associated with myocardial ischemia and oxidative injury (Koster et al., 1987).

It has been reported that H_2O_2 induces peroxidative cleavage of polyunsaturated fatty acids (Block, 1991; Patel and Block, 1988). However, H_2O_2 has a very low reactivity toward unsaturated fatty acids, lipid peroxidation requiring the ferrous ion for the production of hydroxyl radicals in the Fenton reaction (Liu *et al.*, 1990). Because the hydroxyl radical is so reactive it must be generated in the vicinity of unsaturated fatty acids (Radi *et al.*, 1991). Cell reductants, such as superoxide anion, can cause iron mobilization from ferritin, which is needed for the formation of hydroxyl radical. Cytochrome *c* plus H_2O_2 can also generate hydroxyl radicals or hydroxyl radical-like species (Radi *et al.*, 1991). It is to be noted that protection against the superoxide anion-producing systems needs not only superoxide dismutase, dismutating superoxide anion into H_2O_2, but also catalase. This suggests that hydrogen peroxide or the hydroxyl radical may be more important than the superoxide anion (Watanabe *et al.*, 1990).

2.7.1. Membrane Lipid Peroxidation

Peroxidation of membrane lipids leads to the formation of lipid radicals, peroxyradicals, hydroperoxides, and a variety of lipid fragments. Unsaturated fatty acids are particularly involved in the initiation of peroxidation, resulting in alteration of the physical state of the bilayer. Peroxidative cleavage of membrane lipids can lead to alterations in cholesterol/phospholipid ratio, unsaturation index, fatty acyl chain length, and the percentage distribution of fatty acids (Patel and Block, 1988). The overall effects of lipid peroxidation are a decrease in membrane fluidity; aggregation of intramembranous particles with the appearance of particle-free, presumably gel-phase regions observed by freeze-fracture electron microscopy (Florine-Casteel *et al.*, 1991); an increase of the membrane leakiness, in the case of calcium, for example; and modulation of membrane enzymes and carriers (Gleason *et al.*, 1991; Watanabe *et al.*, 1990).

Cross-linking and polymerization of membrane proteins (Freeman *et al.*, 1986), formation of lipid-protein adducts (Parinandi *et al.*, 1990), and a decrease in cholesterol content (Carbonnelle, 1987; Carbonnelle *et al.*, 1987) have been described (Patel and Block, 1988).

Analysis of phospholipid extracts from purified organelle fractions of canine heart subjected to global ischemia and reperfusion revealed that sarcolemmal membranes had the highest content of oxidized phospholipids, compared to sarcoplasmic and mitochondrial membranes. The enhanced sensitivity of sarcolemma could be related to a higher phospholipid to protein ratio and a higher cholesterol to protein ratio (Romaschin *et al.*, 1990). The susceptibility of an organelle is probably a complex function including the contents of polyunsaturated fatty acids and of intramembrane chain-breaking scavengers (α-tocopherol), and the proximity and the nature of the free radical or oxidant source and antioxidant enzymes (Romaschin *et al.*, 1990). In the cell membrane, lipid domains, in which local acyl chain fluidity is decreased, may be regions prone to lipid peroxidation (McLean and Hagaman, 1992).

2.7.2. Activation of Phospholipase A₂

Increased permeability to calcium leads to the fixation of calcium to lipid anionic sites and Ca^{2+}-activation of membrane-associated enzymes such as phospholipases (Rehfeldt *et al.*, 1991; Humes *et al.*, 1989; Otamiri *et al.*, 1986).

The plasma membrane damage after an oxidative stress is not a direct consequence of lipid peroxidation, but could be caused by the membrane disturbances leading to phospholipase A_2 activation, which overwhelms the reacylation capacity, thus producing cytolytic amounts of lysophosphatides. Phospholipase A_2 would be activated in a two-step process: first, activation would be triggered by lipid peroxidation and, second, by increasing amounts of lysophosphatides, leading to increased fluidity and to discontinuities in the lipid phases, thus improving the substrate accessibility by such cracks (Das *et al.*, 1986; Ungemach, 1985).

Phospholipase A_2 has an essential role in the detoxification of phospholipid peroxides. If lipid peroxides are not removed they can break down to free-radical products that initiate chain reactions to produce more lipid peroxides. They can also break down to give aldehydes. Phospholipase A_2 preferentially hydrolyzes peroxidized fatty acids from membranes, and there is an absolute requirement for phospholipase A_2 to allow glutathione peroxidase to reduce fatty acid hydroperoxides in membranes (Van Kuijk *et al.*, 1987).

The pronounced phospholipase A_2 hydrolysis was not necessarily limited to oxidized phospholipids, but rather a host of phospholipids, presumably associated with regions of membrane damage, may become preferred substrates for the enzyme. Evidence has been provided for the phospholipase A_2-dependent release of specific classes of intact fatty acids in addition to removal of oxidized fatty acids in model membranes. The extent of hydrolysis of intact fatty acids correlates approximately with the degree of unsaturation. Besides its possible role in protecting membranes during free-radical induced damage, phospholipase A_2 has a more widely documented role as the rate-determining step in eicosanoid synthesis. It has been suggested that low levels of lipid peroxidation are accompanied by the release of significant amounts of arachidonic acid along with lipid peroxides. It has been hypothesized that peroxides could have a role in prostaglandin synthesis (Sevanian *et al.*, 1988; Das *et al.*, 1986; Sevanian and Kim, 1985; Hemler and Lands, 1980).

2.7.3. Fluidity of the Membrane

Oxidant injury to membrane has been shown both to increase and decrease membrane order. Apparently contradictory results may be related to the time elapsed after oxidative injury and reperfusion, to the variety of analyzed samples, and to the fact that different probes have been used. Such probes include diphenylhexatriene, partitioning in the hydrophobic region of the membrane;

dimethylammonium-diphenylhexatriene anchored at the lipid-water interface; 5-, 12-, 16-doxylstearic acid reporters of membrane fluidity at different depths within the membrane; or 4-Mal TEMPO, a nitroxide analogue of N-ethylmaleimide covalently bound to integral and peripheral membrane proteins.

Exposure of pulmonary artery endothelial cells *in vitro* during 30 min to H_2O_2 and 30 min recovery leads to a rigidification of the glycerol backbone region of the plasma membrane and increased lipid peroxidation but no lactate dehydrogenase leakage. After 6 hr of recovery, an increase in fluidity as measured with dimethylammonium-diphenylhexatriene was associated with an increased release of lactate dehydrogenase and membrane leakiness but no further increase in lipid peroxidation (Block, 1991).

It was demonstrated on single cultured rat hepatocytes that, after ATP depletion, there was an early and sustained increase in lipid order of small plasma membrane blebs, which may predispose to peroxidation injury (McLean and Hagaman, 1992). These numerous plasma membrane blebs fuse and coalesce to form large membrane blebs. Up to this point, injury was reversible. Cell death was found to be initiated by a sudden loss of membrane permeability barrier coincident with the rupture of large blebs (Florine-Casteel *et al.*, 1991; Zernig, 1991).

A decrease in membrane fluidity affects erythrocyte membrane labeled with 5-doxylstearic acid and subjected to xanthine/xanthine oxidase (Watanabe *et al.*, 1990). After global cerebral ischemia (8 min) and reperfusion (15 min), a membrane ordering was shown at the luminal surface of the perfused capillaries in the middle region as reported by 12-doxylstearic acid (Phelan and Lange, 1991).

An increase in fluidity, measured with diphenylhexatriene as a probe, characterizes pulmonary artery endothelial cells exposed to hypoxic conditions (Block *et al.*, 1989) and aorta endothelial cells exposed to xanthine oxidase and labeled with 4-Mal TEMPO (Freeman *et al.*, 1986).

Results in our laboratory have demonstrated a significant decrease of the order parameter in rat liver after 1 and 2 hr *in vivo* ischemia and *in vitro* reperfusion, by fluorescence anisotropy using diphenylhexatriene, and after 2 hr ischemia and reperfusion, when trimethylammonium-diphenylhexatriene was used. This increase in fluidity could be related to the lower cholesterol/phospholipid ratio and possibly to washing out of less fluid blebs by reperfusion (Carbonnelle *et al.*, 1987). However, it cannot explain the aggregation of protein particles observed in freeze-fracture studies of plasma membranes from ischemic liver (Wattiaux-De Coninck *et al.*, 1988). It is possible that that phenomenon originates from alteration of the sustaining cytoskeleton.

2.8. Injury to the Cytoskeleton

Injured and dying cells undergo morphological changes, including swelling of the cytoplasm and blebbing of the plasma membrane. Membrane blebbing has been associated with ATP and NAD$^+$ depletion, with an increase in intracellular calcium and activation of calcium-dependent protease, and with an oxidative stress, evidenced by the glutathione redox state. Apparent redistribution of F-actin and alteration of vinculin in blebbing cells, including endothelial cells, translate the role of cytoskeleton in regulating cell shape in response to injury (Gores et al., 1989; Herman et al., 1988; Hinshaw et al., 1986).

In many cell types, cell-surface blebbing is an early sign of cell injury. Cell-surface bleb formation occurred in three stages over 1–3 hr after chemical anoxia and ATP depletion. Stage I is characterized by the formation of numerous small blebs. Toward the end of stage II, the cells begin to swell. Although swelling is not a prerequisite for blebbing, it ends rapidly with the apparent breakdown of one of the terminal blebs. Breakdown of the bleb membrane initiates stage III of the injury and is coincident with a rapid increase of nonspecific permeability. On reoxygenation, stages I and II are fully reversible. Stage III is not reversible and signals the loss of cell viability. A sustained increase in lipid order in such small blebs has been shown (Florine-Casteel et al., 1991; Gores et al., 1989; Herman et al., 1988; Thor et al., 1988).

Plasma membrane shedding corresponds to a different phenomenon whereby eukaryotic cells release some fragments of the plasma membrane and sometimes some cytoplasmic material into extracellular space. This process, accompanied with the disruption by calpain of actin membrane interactions, is not associated with cell death. It may represent different strategies according to which cells adapt to their extracellular environment, sending messages to other cells, economically replacing some defective parts of the plasma membrane or removing some undesirable components of the cell surface (Beaudoin and Grondin, 1991).

The formation of plasma membrane blebs, leading to cell death, seems mediated by changes in the cortical cytoskeleton. At least two mechanisms may explain the disruption of cytoskeletal elements. Experimental evidence points to the crucial role of thiol oxidation and elevation of calcium concentration in bleb formation (Malorni et al., 1991; Gores et al., 1989; Lemasters et al., 1983).

2.8.1. Oxidative Stress

In chemical anoxia, inducing ATP depletion and the formation of toxic oxygen species (hydrogen peroxide) but involving no calcium increase, blebbing may result from cytoskeletal alteration. An increase of calcium is thus not required for cell death to occur (Gores et al., 1989; Nieminen et al., 1988).

Cytoskeletal structures, and in particular the microfilaments, represent important targets in changes induced by an oxidative stress in cultured cells, leading to the redistribution of several cytoskeletal and membrane proteins as well as the dissociation of the cytoskeletal network from its adhesion domains in the plasma membrane. An oxidative stress on cultured mammalian cells, provoked by menadione, revealed that some surface proteins and adhesion molecules, vinculin, underwent changes in their expression over the bleb surface. Moreover, different behavioral characteristics of actin, microfilaments, vimentin, and keratin intermediate filaments and microtubules were observed. Also, α-actinin, vimentin, and microtubular proteins (tubulin, MAPs, and tau) were detected within the blebs, while actin and keratin filaments appeared to be absent (Malorni et al., 1991).

Oxidative stress induced by menadione may critically oxidize sulfhydryl groups in cytoskeletal proteins, and this, together with other additional factors, may be responsible for increased aggregates of F-actin and bleb formation. Monomeric actin has four thiol groups, three of which are naturally masked in the presence of physiological concentrations of ATP. Since an oxidative stress results in ATP depletion, it seems probable that unmasking of thiol groups in actin due to ATP depletion may increase their susceptibility to oxidation damage and induce cross-linking of actin molecules via disulfide bridges to form large molecular aggregates. (Hinshaw et al., 1991; Mirabelli et al., 1988; Thor et al., 1988; Hinshaw et al., 1986).

2.8.2. Calcium-Induced Proteolysis

Among the constituents of adhesion plaques—vinculin, talin, and α-actinin—talin appears to be the preferred substrate for calpain-II. The colocalization of this calcium-requiring enzyme and talin in adhesion plaques raises the possibility that calcium-dependent proteolytic activity provides a mechanism for regulation of some aspects of adhesion plaque physiology and function via cleavage of talin (Beckerle et al., 1987). In relation to this, calpains can associate with membranes in a calcium-dependent manner, and this association activates the enzyme (Zimmerman and Schlaepfer, 1988; Mellgren, 1987).

Incubation of platelets with menadione induces a direct oxidation of cytosolic polypeptides as well as an increase in cytosolic calcium. The cytosolic calcium ion seems to exert its action either directly, by promoting the dissociation of polypeptides from the cytoskeleton, or indirectly, through the activation of leupeptin-inhibitable proteases that specifically degrade the actin-binding protein (Mirabelli et al., 1989).

Loss of anti-vinculin antibody staining after 120 min of myocardial ischemia demonstrates a direct link between cytoskeletal damage and plasma membrane disruption, and this suggests that the breakdown of the cytoskeletal scaffold underlying the plasma membrane may be responsible for weakening the plasma membrane, allowing it to rupture as a consequence of cell swelling during ischemia. The results are consistent with the hypothesis that the damage occurs because of the activation of calcium-dependent proteases, possibly degrading vinculin and other cytoskeletal proteins (Steenbergen et al., 1987).

There appears to be a strong kinetic association between the loss of $[Ca^{2+}]_i$ regulation and ATP levels necessary for normal excitation-contraction coupling of actin and myosin. A simultaneous rise of Ca^{2+} and fall in ATP may favor sustained contraction and tend to dissociate the cytoskeleton from the plasma membrane, allowing bleb formation (Hinshaw et al., 1986).

Ischemia-induced loss of epithelial cell polarity is associated with the disruption of the cortical actin microfilament network and the opening of cellular-tight junctions. Dissociation of cortical actin from integrins and CAMs as well as an untethering of actin-associated membrane proteins, like Na^+/K^+ ATPase, gives rise to lateral bilayer movement and loss of surface membrane lipid and protein polarity (Molitoris, 1991).

2.8.3. Role of Endothelial Cell Cytoskeleton in the Control of Endothelium Permeability

The maintenance of the endothelial cell integrity is critically related to the structure of its cytoskeleton (Gottlieb et al., 1991; Shasby et al., 1982). Regulation of vascular permeability, flow resistance, and the migration of leukocytes through the endothelium during inflammation could be explained, at least in part, by showing that endothelial cells are contractile. To maintain endothelial integrity when exposed to flowing blood and intraluminal pressure, cells must be able to adhere firmly to the subendothelium and form tight junctions with adjacent cells. The cells probably contract to open intercellular junctions during the inflammatory response, allowing the exit of leukocytes, platelets, and fluid. The endothelial gap formation is ATP dependent, triggered by calcium and mediated by an interaction between actin and myosin (Schnittler et al., 1990; Watanabe et al., 1991).

An oxidative stress mediated by xanthine and xanthine oxidase increased endothelial albumin transfer in a dose-dependent fashion. This was associated with reversible changes in endothelial cell shape and actin filaments and with calcium homeostasis (Shasby et al., 1985). While stress fibers are important in adhesion to extracellular matrix by means of integrins, microtubules are not. Agents that perturb the endothelial cells cytoskeleton at the level of dense peripheral bands—where actin, myosin, tropomyosin, α-actinin, and vinculin colocalize—are hyperoxia, thrombin, and TNF. Endothelial cells incubated with thrombin undergo a reorganization of actin filaments, associated with a loss of vinculin cell-cell plaques. Similarly, alteration by cytochalasin B, which caps actin and induces depolymerization of the endothelial cells cytoskeleton, produces a rapid reversible increase of endothelium permeability (Shasby et al., 1982).

2.9. Injury to the Endothelial Cell Plasma Membrane

Endothelial function is essential for the maintenance of tissue integrity, cell-cell communication, and for maintaining a selective permeability barrier (Hennig and Chow, 1988). During myocardial ischemia, damage to endothelial cells and neutrophil accumulation were not obvious, although in severe ischemia swelling of endothelial cells has been observed. However, after ischemia and reperfusion the endothelium, which is susceptible to injury by free radicals (Bowman et al., 1983), is structurally and functionally altered at a time where neutrophil-endothelial cell interactions are enhanced (Marzi et al., 1992; Forman et al., 1989; Hathaway and March, 1989; Mercandetti et al., 1984; Mullane, 1991). Endothelial dysfunction occurs very soon after reperfusion (Lefer et al., 1991; Viehman et al., 1991).

Different functions of endothelial cells may be perturbed during ischemia. Normally, they synthesize connective tissue components and are endowed with procoagulant, anticoagulant, antiplatelet, and fibrinolytic properties. They interact with polymorphonuclear leukocytes and release molecules that modulate vessel wall tone and control cell proliferation (Davies, 1989; Mantovani and Dejana, 1989; Ryan, 1989; Jaffe, 1988).

2.9.1. Endothelial Cells, Hemostasis, and Fibrinolysis

The subendothelium is endowed with procoagulant properties but is normally covered with endothelial cells. Intact endothelial cells are nonthrombogenic, and unstimulated platelets do not adhere to the surface of intact confluent monolayers of endothelial cells both in vivo and in vitro. This property seems to be intrinsic to the endothelial cells plasma membrane and is unrelated to prostacyclin production. Antithrombogenic properties depend on the luminal plasma lining of the endothelial cells, supporting a negatively charged glycocalyx, which limits adherence, and also containing heparin, antithrombin III, and thrombomodulin.

Injury to the endothelial cells results in the exposure of the subendothelium, collagen type IV and V, and microfibrils that causes platelet aggregation and the release of thromboxane. Adhesion of platelets is dependent on the von Willebrand factor synthesized by endothelial cells and stored in the Weibel-Palade body (Newby and Henderson, 1990). Blood coagulation via the extrinsic clotting cascade and adhesion of platelets also depends on tissue factor, a procoagulant cell surface cofactor (Stern et al., 1988) synthesized by injured endothelial cells (Scarpati and Sadler, 1989), and factors V, IX, X, and XII (Jaffe, 1988). Factor XII, or Hageman factor, is activated inside the contact system after binding on a negatively charged surface such as vascular basement membrane (Kozin and Cochrane, 1988).

Inhibition of adhesion of stimulated platelets to endothelial cells is highly prostacyclin (PGI_2)-dependent. PGI_2 is a potent vasodilator and inhibitor of platelet function, which is synthesized by endothelial cells from arachidonic acid (Newby and Henderson, 1990). Moreover, endothelial cells possess ectoenzymes that rapidly metabolize ADP, which recruits nearby platelets into the developing platelet plug (proaggregatory), and ATP, a vasodilator, to AMP and adenosine, which is a vasodilator and a strong inhibitor of platelet function (antiaggregatory). The major source of adenosine is, however, the cardiac myocyte (Ely and Berne, 1992; Ignarro, 1990; Selwyn et al., 1990; Jaffe, 1988). Thus, alterations of endothelial cell membrane function during ischemia and reperfusion stimulate hemostasis.

The fibrinolytic properties of endothelial cells reside in the synthesis and secretion of tissue plasminogen activator and plasminogen activator inhibitor. Plasmin, plasminogen, and tissue plasminogen activator all bind to endothelial cells, and the activation of bound plasminogen is more efficient than that of fluid-phase plasminogen (Dudani et al., 1991; Sanzo et al., 1990; Jaffe, 1988). After anoxia and reoxygenation, a significant decrease of both plasminogen activator inhibitor 1 and tissue plasminogen activator secretion was observed (Shatos et al., 1990).

2.9.2. Regulation of Vascular Smooth Muscle Tone

Endothelial dysfunction may occur in the absence of significant morphologic or ultrastructural damage (Lefer et al., 1991). Brief ischemia and reperfusion not only induce an increase in microvascular permeability but also provoke profound endothelium-dependent vasodilator dysfunction, loss of vasodilator response, and development of abnormal constriction (Dauber et al., 1990; Selwyn et al., 1990; Furchgott and Vanhoutte, 1989; Haselton et al., 1989). Endothelial dysfunction does not occur during myocardial ischemia in vivo, at least in ischemic periods up to 90 min (Lefer et al., 1991; Viehman et al., 1991), but every early after reperfusion and is prevented by recombinant human superoxide dismutase (Lefer et al., 1991). Even in the absence of blood cells in isolated perfused rat hearts (Lefer et al., 1991), a markedly reduced endothelium-dependent relaxation occurs at 2.5 min after reperfusion, concomitantly with the production of superoxide anions (the same time as that observed in ischemia reperfusion in vivo). This demonstrates that the early burst of superoxide anion does not originate from neutrophils. Endothelial cells are the most likely source of superoxide anions (Lefer et al., 1991; Tsao and Lefer, 1990; Tsao et al., 1990.

Endothelium is involved in the regulation of vascular smooth muscle tone by secreting the relaxing factors endothelium-derived relaxing factor (EDRF), prostacyclin (PGI_2), and endothelium-derived contraction factors, endothelin, thromboxane, leukotrienes, and superoxide anions (Martin et al., 1992; Mehta et al., 1992; Mehta et al., 1989; Vanhoutte and Katusic, 1988; Simionescu and Simionescu, 1986). The endothelium plasmalemma contains receptors for many vasoactive compounds, including acetylcholine, histamine, serotonin, catecholamines, bradykinin, and adenine nucleotides, reflecting the integral involvement of the endothelium in regulating vascular smooth muscle tone.

The intact endothelium releases EDRF in response to endogenous circulating compounds such as acetylcholine and serotonin (Ignarro, 1990; Newby and Henderson, 1990; Furchgott and Vanhoutte, 1989; Ignarro, 1989). EDRF is indistinguishable from nitric oxide (NO) and produces smooth muscle relaxation, inhibits platelet aggregation (Radomski et al., 1991), and adherence of polymorphonuclear leukocytes (McCall et al., 1988). Synthesis of EDRF in endothelial cells results from the conversion of arginine to NO plus citrulline by a plasma membrane-bound NADPH-dependent monooxygenase, which is activated by calcium. NO diffuses out of the cell and binds to cytosolic guanylate cyclase in the adjacent smooth muscle cells; the elevated cGMP leads to smooth muscle relaxation by stimulation of intracellular binding of free calcium (McCall and Vallance, 1992).

On the other hand, it has been shown that EDRF is very unstable and is destroyed by superoxide anions (Lawson et al., 1990; Gryglewski et al., 1986; Vanhoutte et al., 1986), although contradictory results have been obtained concerning the protective action of superoxide dismutase (Horwitz et al., 1990; Mehta et al., 1989). As a result, NO disappearance may lead to decreased endothelial-dependent relaxation, increased platelet aggregation, and adherence of polymorphonuclear leukocytes.

Endothelin, a 21-residue peptide, is the most potent vasoconstrictor peptide known. It is synthesized by endothelial cells, acts directly on arterial smooth muscle, and provokes strong and sustained contraction (Furchgott and Vanhoutte, 1989; Hiley, 1989; Kanse et al., 1989; Yanagisawa and Masaki, 1989).

The possible pathological roles of endothelin-1 in myocardial infarction, cerebral ischemia, and reperfusion-induced calcium overload need to be considered (Nayler, 1990). The levels of endothelin are increased in acute myocardial ischemia (Mehta et al., 1992). Twenty-minute global ischemia causes externalization of endothelin-1 receptors from internal sites to the cell surface (Godfraind and Govoni, 1989; Liu et al., 1989). Binding of endothelin to receptors activates inositoltriphosphate metabolism and raises cytosolic calcium (Godfraind and Govoni, 1989). Since Ca^{2+} overloading signals cell death, this increase of endothelin binding sites may be of pathophysiological importance. However, to answer the question of whether endothelin-1 was responsible for the anoxia-induced contraction of various blood vessels (Hiley, 1989), it was argued that the time of response to endothelin-1 (which is slow in the onset) and the endothelin-dependent response to decreased pO_2 (which is rapid) are too different to be accounted for by the same substance (Hiley, 1989).

2.10. Recruitment and Adhesion of Polymorphonuclear Leukocytes and Inflammaters

Recruitment and adhesion of polymorphonuclear leukocytes to endothelial cells is an early and requisite event in acute inflammation (Zimmerman et al., 1992). Polymorphonuclear leukocytes are also important determinants for the extension of myocardial damage beyond the initial injury.

Migration of leukocytes into the ischemic myocardium occurs already during ischemia (Forman et al., 1989; Braunwald and Kloner, 1985) and is greatly enhanced during reperfusion (Forman et al., 1989). Neutrophil depletion or suppression of neutrophil adherence or function results in a significant salvage of the ischemic myocardium (Jaeschke et al., 1990; Hernandez et al., 1987; Simpson and Lucchesi, 1987), although sometimes no beneficial effect has been found (Werns and Lucchesi, 1988). It is to be noted that free-radical-mediated injury has been shown to occur in many leukocyte-free systems (Downey, 1990; Braunwald and Koner, 1985). Factors contributing to the accumulation of leukocytes are:

1. microvascular trapping of neutrophils;
2. production of cytokines IL-1 and TNF-α;
3. release of chemotactic factors and stimulation of intercellular adhesion between polymorphonuclear leukocytes and endothelial cells (Entman et al., 1991).

2.10.1. Microvascular Trapping of Neutrophils

Reduction in the vessel lumen appears to be due to the release of vasoconstrictors, such as thromboxane A_2 (Otani et al., 1986), and eicosanoids, like peptidoleukotriene C_4, which also activates platelet adherence and aggregation and would contribute to neutrophil accumulation (Entman et al., 1991). Peptidoleukotrienes LTC_4 and LTD_4, produced by endothelial and vascular smooth muscle cells from leukotrienes A_4 supplied by neutrophils, are potent vasoconstrictors. However, it is not known whether they play an important role in ischemia (Entman et al., 1991; Welbourn et al., 1991; Serhan et al., 1982).

2.10.2. Production of Cytokines

TNF-α and IL-1, which are secreted mainly by macrophages but also by endothelial cells, have an enormous impact on inflammation and a wide spectrum of action (Voltarelli and Garovoy, 1988). TNF-α enhances the ability of mature macrophages and polymorphonuclear leukocytes to secrete reactive oxygen intermediates and intracellular granule contents, and to adhere to the endothelium. TNF-α production is increased during ischemia and reperfusion (Colletti et al., 1990; Beutler and Cerami, 1989). IL-1 elicits immediate inflammatory reactions, characterized by binding of blood polymorphonuclear leukocytes to the vessel walls, and induces the production of secondary mediators including platelet-activating factor (Tse and Rosenthal, 1988). IL6, contained in postischemic cardiac lymph, induces binding of activated polymorphonuclear leukocytes to cardiac myocytes (Youker et al., 1990).

2.10.3. Release of Chemotactic Factors and Stimulation of Intracellular Adhesion between Endothelial Cells and Polymorphonuclear Leukocytes

There is a continuous physiological interaction of neutrophils and endothelial cells with about one-tenth of total blood polymorphonuclear leukocytes emigrating from the blood stream each hour (Forman et al., 1989; Harlan, 1985). The hallmark of acute inflammation is the recruitment of neutrophils and their increased adherence to endothelium. Adhesion of granulocytes to endothelial cells requires regulated expression of molecules on both the endothelial cells and the granulocytes. These proadhesion molecules act either to tether the two cells together or as signals that induce activation-dependent adhesion events (Zimmerman et al., 1992).

The precise steps linking xanthine oxidase and reactive oxygen metabolites to chemotactic activity are not well understood. Moreover, the exact nature of the chemotactic agents operative in ischemia are unknown, although complement fragments appear to be of key importance (Welbourne et al., 1991).

Damaged mitochondria release f-Meth-peptides, which are chemotactic in vitro, whereas nonformylated mitochondrial proteins are not (Carp, 1982).

Participation of the complement system in myocardial ischemic injury is obtained from studies in which complement depletion resulted in a reduction of the infarcted zone (Lucchesi, 1990). Subcellular fractions from heart muscle have been suggested to activate both the classic and alternate complement pathways (Entman et al., 1991; Kagiyama et al., 1989; Rossen et al., 1988; Rossen et al., 1985; McManus et al., 1983; Pinckard et al., 1975). It has been established that myocardial infarction can activate the third component of complement (Lucchesi, 1990; Simpson and Lucchesi, 1987; Rossen et al., 1985; McManus et al., 1983; Pinckard e al., 1983; Hill and Ward, 1971). This could be related to the fact that ischemic myocardial tissue gives rise to a tissue protease that activates the third component of complement (Giclas et al., 1979) and that lysosomal proteases are also potent activators of complement (Entman et al.,

1991). Extracellular fluid (Smith *et al.*, 1991) in the reperfused cardiac tissue contains complement-activating proteins and chemotactic activity blocked by anti-C5a antibodies (Smith *et al.*, 1991).

LTB_4 is a potent chemotactic agent produced during ischemia and reperfusion, which may also activate polymorphonuclear leukocytes (Entman *et al.*, 1991; Welbourn *et al.*, 1991; Evers *et al.*, 1985; Serhan *et al.*, 1982; Gimbrone *et al.*, 1984). A direct LTB4 receptor antagonist, L4223982n, reduces myocardial infarct size in rabbit (Entman *et al.*, 1991).

Platelet activating factor (PAF) antagonists in combination with other agents may prove to have extensive therapeutic potential in the treatment of ischemic disease (Abbott and Page, 1989; Lemasters *et al.*, 1989). PAF, a mediator of oxygen-free-radical production in polymorphonuclear leukocytes, is released in the first minutes during reperfusion of the ischemic heart (Kubes *et al.*, 1990; Montrucchio *et al.*, 1989; Lewis *et al.*, 1988). Its synthesis by endothelial cells is induced by hydrogen peroxide (Hathaway and March, 1989; Lewis *et al.*, 1988). PAF is expressed on the cell surface of endothelial cells (and not released as in other cells). PAF associated with endothelial cells activates polymorphonuclear leukocytes by binding to a cell-surface receptor, inducing upregulation of CD11a-CD18 and CD11b-CD18 (Zimmerman *et al.*, 1992).

IL8, a potent chemotactic factor for polymorphonuclear leukocytes, synthesized by activated endothelial cells, is released into the fluid phase but may also associate with the endothelial cell surface. It activates polymorphonuclear leukocytes by binding to a specific receptor, stimulating upregulation of β_2 integrins (Zimmerman *et al.*, 1992).

Treatment with allopurinol or administration of superoxide dismutase dramatically decreases the extent of granulocyte infiltration, suggesting that xanthine oxidase-derived superoxide anions may play a role in the recruitment of neutrophils during the ischemic period (Komatsu *et al.*, 1992; Grisham *et al.*, 1986). McCord and co-workers (McCord, 1986; Petrone *et al.*, 1980) suggested that superoxide anions induced formation of a neutrophil activating factor in serum, consisting of an as yet unidentified lipid component, noncovalently bound to serum albumin (McCord, 1987). Superoxide dismutase should not only prevent tissue injury from direct attack (McCord, 1986), but also indirectly by inhibiting polymorphonuclear leukocyte accumulation by this chemoattractant.

Binding of chemotactic factors to the endothelial cell surface is a possible explanation for the localization and augmentation of neutrophil adherence (Harlan, 1985). High-affinity receptors for f-Methionyl-Leucyl-Phenylalanine (f-MLP) have been reported on the surface of endothelial cells in culture.

Such adherence is necessary for leukocyte accumulation at sites of inflammation, as well as in ischemic and reperfused myocardium (Smith *et al.*, 1991). It is possible to inhibit the process with monoclonal antibodies against leukocyte integrins, CD11a/CD18 (LFA-1) and CD11b/CD18 (Mac-1). Such antibodies reduce infarct size and neutrophil accumulation by inhibiting adhesion of neutrophils to the endothelium (Dean *et al.*, 1991; Smith *et al.*, 1991; Simpson *et al.*, 1988; Veder *et al.*, 1988; Arfors *et al.*, 1987).

Polymorphonuclear leukocyte adhesion to activated endothelial cells requires, in a first step, rapidly coordinated expression of PAF and GMP-140 onto endothelial cell surface, both of which bind to receptors on polymorphonuclear leukocytes (Table I). Binding of PAF of the endothelial cell membrane with its receptor on polymorphonuclear leukocytes serves as a signal for activation-dependent alteration of LFA-1 and Mac-1. This activation makes them able to bind, in a second step. to counter-receptors on the endothelial cells or myocytes. ICAM-1 and ICAM-2 on endothelial cells, and ICAM-1 on myocytes, are the counter-receptors for LFA-1; ICAM-1 (and bound C3bi) is the counter-receptor for Mac-1 (Carveth *et al.*, 1992; Dejana *et al.*, 1992; Zimmerman *et al.*, 1992; Entman *et al.*, 1991; Smith *et al.*, 1991; Kubes *et al.*, 1990; Zimmerman *et al.*, 1990; Hogg, 1989; Hynes, 1987). The counter-receptor for GMP140 is unknown (Zimmerman *et al.*, 1992). A monoclonal antibody to ICAM-1 intercellular adhesion molecule, which is expressed on multiple cell types including endothelial cells, reduces central nervous system ischemic injury (Clark *et al.*, 1991). GMP140, localized in Weibel-Palade bodies of endothelial cells, is rapidly shifted to plasma membrane in response to such stimuli as oxygen radicals (Patel *et al.*, 1991). GMP140 appears to be a natural anti-inflammatory molecule that may prevent inappropriate activation of neutrophils in the circulation (Wong *et al.*, 1991). At least two other adhesion molecules, ELAM-1 and LAM-1, influence neutrophils (Harlan, 1985) but with different kinetics (Dejana *et al.*, 1992) ELAM-1 synthesized by endothelial cells is co-expressed with ICAM-1 after stimulation by IL1 and TNFα.

Activated neutrophils mediate vascular damage by releasing reactive oxygen species, products of membrane phospholipases, and granule enzymes (Harlan, 1985). Lysosomal elastase and collagenase increase vascular permeability, and azurophilic acid and neutral proteases digest vascular basement membrane (Harlan, 1985; Harlan *et al.*, 1981). Moreover, α-1-antiprotease is inactivated by neutrophil-derived oxidants, thereby permitting an unchecked activity of proteolytic enzymes.

3. CONCLUSIONS

Ischemia, whether or not it is followed by reperfusion, induces deterioration of the plasma membrane as well as of other subcellular membranes. A question that still remains unanswered concerns the lesion that is the determining factor in causing cell death. Despite the large amount of experimental observations, one cannot unequivocally attribute the preponderant role in structural and functional modifications that will kill the cell to a well-defined membrane alteration. The problem is important to solve not only to obtain a better understanding of the mechanisms leading to cell death but also to find the most appropriate therapeutic procedures to cure ischemic diseases.

4. REFERENCES

Abbott, A., and Page, C., 1989, PAF: New antagonists, new roles in disease and a major role in reproductive biology, *Trends Pharmacol. Sci.* **10**:256–257.

Akiyama, S. K., Nagata, K., and Yamada, K. M., 1990, Cell surface receptors for extracellular matrix components, *Biochim. Biophys. Acta* **1031**:91–110.

Alberts, B., Bray, D., Lewis, J., Raff, M., Roberts, K., and Watson, J. D., 1989, *The Cell*, 2nd ed., M. Robertson, New York.

Andersen, J. P., and Vilsen, B., 1990, Primary ion pumps, *Curr. Opin. Cell Biol.* **2**:722–730.

Andreoli, S. P., Mallett, C. P., and Bergstein, J. M., 1986, Role of glutathione in protecting endothelial cells against hydrogen peroxide oxidant injury, *J. Lab. Clin. Med.* **108**:190–198.

Anundi, I., and De Groot, H., 1989, Hypoxic liver cell death: Critical pO_2 and dependence of viability on glycolysis, *Am. J. Physiol.* **257**:G58–G64.

Arfors, K-E., Lundberg, C., Lindbom, L., Lundberg, K., Beatty, P. G., and Harlan, J. M., 1987, A monoclonal antibody to the membrane glycoprotein complex CD 18 inhibits polymorphonuclear leukocytes accumulation and plasma leakage *in vivo*, *Blood* **68**:338–340.

Bean, B. P., 1989, Classes of calcium channels in vertebrate cells, *Annu. Rev. Physiol.* **51**:367–384.

Beaudoin, A. R., and Grondin, G., 1991, Shedding of vesicular material from the cell surface of eukaryotic cells: Different cellular phenomena, *Biochim. Biophys. Acta* **1071**:203–219.

Beckerle, M. C., Burridge, K., DeMartino, G. N., and Croall, D. E., 1987, Colocalization of calcium-dependent protease II and one of its substrates at sites of cell adhesion, *Cell* **51**:569–577.

Berridge, M. J., 1987, Inositol triphosphate and diacylglycerol: Two interacting second messengers, *Annu. Rev. Biochem.* **56**:159–193.

Beutler, B., and Cerami, A., 1989, The biology of cachectin/TNF—a primary mediator of the host response, *Annu. Rev. Immunol.* **7**:625–655.

Billman, G. E., McIlroy, B., and Johnson, J. D., 1991, Elevated myocardial calcium and its role in sudden cardiac death, *FASEB J.* **5**:2586–2592.

Block, E. R., 1991, Hydrogen peroxide alters the physical state and function of the plasma membrane of pulmonary artery endothelial cells, *J. Cell. Physiol.* **146**:362–369.

Block, E. R., Patel, J. M., and Edwards, D., 1989, Mechanism of hypoxic injury to pulmonary artery endothelial cell plasma membrane, *Am. J. Physiol.* **257**:C223–C231.

Boddeke, E., Hugtenburg, J., Jap, W., Heynis, J., and Van Zwieten, P., 1989, New anti-ischaemic drugs: Cytoprotective action with no primary haemodynamic effects, *Trends Pharmacol. Sci.* **10**:397–400.

Bolli, R., Jeroudi, M. O., Patel, B. S., DuBose, C. M., Lai, E. K., Roberts, R., and McCay, P. B., 1989, Direct evidence that oxygen-derived free radicals contribute to postischaemic myocardial dysfunction in the intact dog, *Proc. Natl. Acad. Sci. USA* **86**:4695–4699.

Bowman, C. M., Butler, E. N., and Repine, J. E., 1983, Hyperoxia damages cultured endothelial cells causing increased neutrophil adherence, *Am. Rev. Respir. Dis.* **128**:469–472.

Braquet, P., Paubert-Braquet, M., Koltai, M., Bourgain, R., Bussolino, F., and Hosford, D., 1989, Is there a case for PAF antagonists in the treatment of ischemic states?, *Trends Pharmacol. Sci.* **10**:23–30.

Braunwald, E., and Kloner, R. A., 1985, Myocardial reperfusion: A double-edged sword?, *J. Clin. Invest.* **76**:1713–1719.

Brodin, P., Falchetto, R., Vorherr, T., and Carafoli, E., 1992, Identification of two domains which mediate the binding of activating phospholipids to the plasma-membrane Ca^{2+} pump, *Eur. J. Biochem.* **204**:939–946.

Brown, A. M., 1991, A cellular logic for G protein-coupled ion channel pathways, *FASEB J.* **5**:2175–2179.

Buja, L. M., Hagler, H. K., and Willerson, J. T., 1988, Altered calcium homeostasis in the pathogenesis of myocardial ischemic and hypoxic injury, *Cell. Calcium* **9**:205–217.

Burgoyne, R. D., 1991, Locating intracellular calcium stores, *Trends Biochem. Sci.* **16**:319–320.

Burridge, K., Fath, K., Kelly, T., Nuckolls, G., and Turner, C., 1988, Focal adhesions: Transmembrane junctions between the extracellular matrix and the cytoskeleton, *Annu. Rev. Cell. Biol.* **4**:487–525.

Cadenas, E., Boveris, A., Ragan, C. I., and Stoppani, A. O. M., 1977, Production of superoxide radicals and hydrogen peroxide by NADH-ubiquinone reductase and ubiquinol-cytochrome C reductase beef-heart mitochondria, *Arch. Biochem. Biophys.* **180**:248–257.

Capiod, T., Mauger, J-P., Binet, A., and Claret, M., 1989, Regulation of calcium in non-excitable cells, *Curr. Opin. Cell. Biol.* **1**:211–214.

Carafoli, E., 1987, Intracellular calcium homeostasis, *Annu. Rev. Biochem.* **56**:395–433.

Carafoli, E., 1991, The calcium pumping ATPase of the plasma membrane, *Annu. Rev. Physiol.* **53**:531–547.

Carafoli, E., 1992, The Ca^{2+} pump of the plasma membrane, *J. Biol. Chem.* **5**:2115–2118.

Carbonnelle, V., 1987, Etude de la membrane plasmique du foie de rat au cours de l'ischémie, Thèse doctorale, Facultés des Sciences, Facultés Universitaires N.D. de la Paix, Namur.

Carbonnelle, V., Wattiaux, R., and Wattiaux-De Coninck, S., 1987, Plasma membrane of normal and ischaemic rat liver, *Biochem. Soc. Trans.* **16**:537–538.

Carp, H., 1982, Mitochondrial *N*-formylmethionyl proteins as chemoattractants for neutrophils, *J. Exp. Med.* **155**:264–275.

Carveth, H. J., Shaddy, R. E., Whatley, R. E., McIntyre, T. M., Prescott, S. M., and Zimmerman, G. A., 1992, Regulation of platelet-activating factor (PAF) synthesis and PAF-mediated neutrophil adhesion to endothelial cells activated by thrombin, *Semin. Thromb. Hemost.* **18**:126–134.

Cheng, K-H., Lepock, J. R., Wen Hui, S., and Yeagle, P. L., 1986, The role of cholesterol in the activity of reconstituted Ca-ATPase vesicles containing unsaturated phosphatidylethanolamine, *J. Biol. Chem.* **261**:5081–5087.

Cheung, J. Y., Leaf, A., and Bonventre, J. V., 1984, Mechanism of protection by verapamil and nifedipine from anoxic injury in isolated cardiac myocytes, *Am. J. Physiol.* **246**:C323–C329.

Clark, W. M., Madden, K. P., Rothlein, R., and Zivin, J. A., 1991, Reduction of central nervous system ischemic injury by monoclonal antibody to intercellular adhesion molecule, *J. Neurosurg.* **75**:623–627.

Cochrane, C. G., 1991, Cellular injury by oxidants, *Am. J. Med.* **91**:23S–30S.

Colletti, L. M., Remick, D. G., Burtch, G. D., Kunkel, S. L., Strieter, R. M., and Campbell, D. A., 1990, Role of tumor necrosis factor-α in the pathophysiologic alterations after hepatic ischemia/reperfusion injury in the rat, J. Clin. Invest. **85**:1936–1943.

Currin, R. T., Gores, G. J., Thurman, R. G., and Lemasters, J. J., 1991, Protection by acidotic pH against anoxic cell killing in perfused rat liver: Evidence for a pH paradox, *FASEB J.* **5**:207–210.

Das, D. K., Engelman, R. M., Rousou, J. A., Breyer, R. H., Otani, H., and Lemeshow, S., 1986, Role of membrane phospholipids in myocardial injury induced by ischemia and reperfusion, *Am. J. Physiol.* **251**:H71–H79.

Dauber, M., VanBenthmysen, K. M., McMurtry, I. F., Wheeler, G. S., Lesnefsky, E. J., Horwitz, L. D., and Weil, J. V., 1990, Functional coronary microvascular injury evident as increased permeability due to brief ischemia and reperfusion, *Circ. Res.* **66**:986–998.

Davies, P. E., 1989, How do vascular endothelial cells respond to flow?, *News Physiolog Sci* **4**:22–25.

Dean, R. T., Hunt, J. V., Grant, A. J., Yamamoto, Y., and Niki, E., 1991, Free radical damage to proteins: The influence of the relative localization of radical generation, antioxidants, and target proteins, *Free Radic. Biol. Med.* **11**:161–168.

Dejana, E., Needham, L., and Gordon, J., 1992, Endothelial cell adhesive interactions, in *Endothelial Cell Dysfunctions* (N. Simionescu and M. Simionescu, eds.), pp. 153–167, Plenum Press, New York.

Downey, J. M., 1990, Free radicals and their involvement during long-term myocardial ischemia and reperfusion, *Annu. Rev. Physiol.* **52**:487–504.

Dudani, A. K., Hashemi, S., Aye, M. T., and Ganz, P. R., 1991, Identification of an endothelial cell surface protein that binds plasminogen, *Mol. Cell. Biochem.* **108**:133–140.

Eddy, L. J., Stewart, J. R., Jones, H. P., Engerson, T. D., McCord, J. M., and Downey, J. M., 1987, Free radical-producing enzyme, xanthine oxidase, is undetectable in human hearts, *Am. J. Physiol.* **253**:H709–H711.

Ely, S. W., and Berne, R. M., 1992, Protective effects of adenosine in myocardial ischemia, *Circulation* **85**:893–904.

Entman, M. L., Michael, L., Rossen, R. D., Dreyer, W. J., Anderson, D. C., Taylor, A., and Smith, W., 1991, Inflammation in the course of early myocardial ischemia, *FASEB J.* **5**:2529–2537.

Evers, A. S., Murphree, S., Saffitz, J., Jakschick, B. A., and Needleman, P., 1985, Effects of endogenously produced leukotrienes, thromboxane, and prostaglandins on coronary vascular resistance in rabbit myocardial infarction, *J. Clin. Invest.* **75**:992–999.

Florine-Casteel, K., Lemasters, J. J., and Herman, B., 1991, Lipid order in hepatocyte plasma membrane blebs during ATP depletion measured by digitized video fluorescence polarization microscopy, *FASEB J.* **5**:2078–2084.

Forcher, P., 1989, Calcium and polyphosphoinositide control of cytoskeletal dynamics, *Trends Neurol. Sci.* **12**:468–474.

Forman, M. B., Puett, D. W., and Virmani, R., 1989, Endothelial and Myocardial injury during ischemia and reperfusion: Pathogenesis and therapeutic implications, *J. Am. Coll. Cardiol.* **13**:450–459.

Fortunati, E., and Bianchi, V., 1989, Plasma membrane damage detected by nucleic acid leakage, *Mol. Toxicol.* **2**:27–38.

Franciolini, F., and Petris, A., 1990, Chloride channels of biological membranes, *Biochim. Biophys. Acta* **1031**:247–259.

Fraser, C. L., and Sarnacki, P., 1992, Regulation of plasma membrane-bound Ca^{2+}-ATPase pump by inositol phosphates in rat brain, *Am. J. Physiol.* **262**(31):F411–F416.

Freeman, B. A., Rosen, G. M., and Barber, M. J., 1986, Superoxide perturbation of the organization of vascular endothelial cell membranes, *J. Biol. Chem.* **261**:6590–6593.

Furchgott, R. F., and Vanhoutte, P. M., 1989, Endothelium-derived relaxing and contracting factors, *FASEB J.* **3**:2007–2018.

Geeraerts, M. D., Ronveaux-Dupal, M-F., Lemasters, J. J., and Herman, B., 1991, Cytosolic free Ca^{2+} and proteolysis in lethal oxidative injury in endothelial cells, *Am. J. Physiol.* **261**:C889–C895.

Giclas, P. C., Pinckard, R. N., and Olson, M. S., 1979, *In vitro* activation of complement by isolated heart subcellular membrane, *J. Immunol.* **122**:146–151.

Gimbrone, M. A., Brock, A. F., and Schafer, A. I., 1984, Leukotriene B4 stimulates polymorphonuclear leukocytes adhesion to cultured vascular endothelial cells, *J. Clin. Invest.* **74**:1552-1555.

Gleason, M. M., Medow, M. S., and Tulenko, T. N., 1991, Excess membrane cholesterol alters calcium movements, cytosolic calcium levels, and membrane fluidity in arterial smooth muscle cells, *Circ. Res.* **69**:216-226.

Godfraind, T., and Govoni, S., 1989, Increasing complexity revealed in regulation of Ca^{2+} antagonist receptor, *Trends Pharmacol. Sci.* **10**:297-301.

Gordon, D., 1990, Ion channels in nerve and muscle cells, *Curr. Opin. Cell Biol.* **2**:695-707.

Gores, G. J., Flarsheim, C. E., Dawson, T. L., Nieminen, A-L., Herman, B., and Lemasters, J. J., 1989, Swelling, reductive stress, and cell death during chemical hypoxia in hepatocytes, *Am. J. Physiol.* **257**:C347-C354.

Gottlieb, A. I., Langille, B. L., Wong, M. K. K., and Kim, D. W., 1991, Biology of disease: Structure and function of the endothelial cytoskeleton, *Lab. Invest.* **65**:123-137.

Grisham, M. B., Hernandez, L. A., and Granger, D. N., 1986, Xanthine oxidase and neutrophil infiltration in intestinal ischemia, *Am. J. Physiol.* **251**:G567-G574.

Grover, A. K., and Khan, I., 1992, Calcium pump isoforms: Diversity, selectivity and plasticity, *Cell Calcium* **13**:9-17.

Grum, C. M., Gallagher, K. P., Kirsh, M. M., and Shlafer, M., 1989, Absence of detectable xanthine oxidase in human myocardium, *J. Mol. Cell. Cardiol.* **21**:263-267.

Gryglewski, R. J., Palmer, R. M. J., and Moncada, S., 1986, Superoxide anion is involved in the breakdown of endothelium-derived vascular relaxing factor, *Nature* **320**:454-456.

Harlan, J. M., 1985, Leukocyte-endothelial interactions, *Blood* **65**:513-525.

Harlan, J. M., Killen, P. D., Harker, L. A., and Striker, G. E., 1981, Neutrophil-mediated endothelial injury *in vitro*, *J. Clin. Invest.* **68**:1394-1403.

Harlan, J. M., Levine, J. D., Callahan, K. S., Schwartz, B., and Harke, L. A., 1984, Glutathione redox cycle protects cultured endothelial cells against lysis by extracellularly generated hydrogen peroxide, *J. Clin. Invest.* **73**:706-713.

Haselton, F. R., Mueller, S. N., Howell, R. E., Levine, E. M., and Fishman, A. P., 1989, Chromatographic demonstration of reversible changes in endothelial permeability, *J. Appl. Physiol.* **67**:2032-2048.

Hasselgren, P-O., 1987, Prevention and treatment of ischemia of the liver, *Surg. Gynecol. Obstet.* **164**:187-196.

Hathaway, D. R., and March, K. L., 1989, Molecular cardiology: New avenues for diagnosis and treatment of cardiovascular disease, *J. Am. Coll. Cardiol.* **13**:265-282.

Hemler, M. E., and Lands, W. E. M., 1980, Evidence for a peroxide-initiated free radical mechanism of prostaglandin biosynthesis, *J. Biol. Chem.* **255**:6253-6261.

Hennig, B., and Chow, C. K., 1988, Lipid peroxidation and endothelial cell injury: Implications in atherosclerosis, *Free Radic. Biol. Med.* **4**:99-106.

Herman, B., Nieminen, A-L., Gores, G. J., and Lemasters, J. J., 1988, Irreversible injury in anoxic hepatocytes precipitated by an abrupt increase in plasma membrane permeability, *FASEB J.* **2**:146-151.

Hernandez, L. A., Grisham, M. B., Twohig, B., Arfors, K. E., Harlan, J. M., and Granger, D. N., 1987, Role of neutrophils in ischemia-reperfusion-induced microvascular injury, *Am. J. Physiol.* **253**:H699-H703.

Hiley, C. R., 1989, Functional studies on endothelin catch up with molecular biology, *Trends Pharmacol. Sci.* **10**:47-50.

Hill, J. H., and Ward, P. A., 1971, The phlogistic role of C3 leukotactic fragments in myocardial infarcts in rats, *J. Exp. Med.* **133**:885-900.

Hinshaw, D. B., Burger, J. M., Beals, T. F., Armstrong, B. C., and Hyslop, P. A., 1991, Actin polymerization in cellular oxidant injury, *Arch. Biochem. Biophys.* **288**:311-316.

Hinshaw, D. B., Sklar, L. A., Bohl, B., Shraufstatter, I. U., Hyslop, P. A., Rossi, M. W., Spragg, R. G., and Cochrane, C. G., 1986, Cytoskeletal and morphologic impact of cellular oxidant injury, *Am. J. Physiol.* **123**:454–464.

Hochachka, P. W., and Mommsen, T. P., 1983, Proton and anaerobiosis, *Science* **219**:1391–1397.

Hoehner, P. J., Blanck, T. J. J., Roy, R., Rosenthal, R. E., and Fiskum, G., 1992, Alteration of voltage-dependent calcium channels in canine brain during global ischemia and reperfusion, *J. Cereb. Blood Flow Metab.* **12**:418–424.

Hogg, N., 1989, The leukocyte integrins, *Immunol. Today* **10**:111–114.

Horwitz, L. D., van Benthuysen, K. M., Sheridan, F. M., Lesnefsky, E. J., Dauber, I. M., and McMurtry, I. F., 1990, Coronary endothelial dysfunction from ischemia and reperfusion: Effect of reactive oxygen metabolite scavengers, *Free Radic. Biol. Med.* **8**:381–386.

Humes, H. D., Nguyen, V. D., Cieslinski, D. A., and Messana, J. M., 1989, The role of free fatty acids in hypoxia-induced injury to renal proximal tubule cells, *Am. J. Physiol.* **256**:F688–F696.

Hynes R. O., 1987, Integrins: A family of cell surface receptors, *Cell* **48**:549–554.

Hyslop, P. A., Hinshaw, D. B., Halsey, W. A., Schraufstatter, I. U., Sauerheber, R. D., Spragg, R. G., Jackson, J. H., and Cochrane, C. G., 1988, Mechanisms of oxidant-mediated cell injury: The glycolytic and mitochondrial pathways of ADP phosphorylation are major intracellular targets inactivated by hydrogen peroxide, *J. Biol. Chem.* **263**:1665–1675.

Hyslop, P. A., Hinshaw, D. B., Schraufstatter, I. U., Sklar, L. A., Spragg, R. G., and Cochrane, C. G., 1986, Intracellular calcium homeostasis during hydrogen peroxide injury to cultured $P388D_1$ cells, *J. Cell. Physiol.* **129**:356–366.

Ignarro, L. J., 1989, Endothelium-derived nitric oxide: Actions and properties, *FASEB J.* **3**:31–36.

Ignarro, L. J., 1990, Biosynthesis and metabolism of endothelium-derived nitric oxide, *Annu. Rev. Pharmacol. Toxicol.* **30**:535–560.

Jaeschke, H., 1991, Reactive oxygen and ischemia/reperfusion injury of the liver, *Chem. Biol. Interact.* **79**:115–136.

Jaeschke, H., Farhood, A., and Smith, C. W., 1990, Neutrophils contribute to ischemia/reperfusion injury in rat liver in vivo, *FASEB J.* **4**:3355–3359.

Jaffe, E. A., 1988, Endothelial cells, in *Inflammation: Basic Principles and Clinical Correlates* (J. L. Gallin, I. M. Goldstein, and R. Snyderman, eds.), pp. 559–576, Raven Press, New York.

Jain, S. K., 1985, *In vivo* externalization of phosphatidylserine and phosphatidylethanolamine in the membrane bilayer and hypercoagulability by the lipid peroxidation of erythrocytes in rats, *J. Clin. Invest.* **76**:281–286.

Jennings, R. B., and Steenbergen, C., 1985, Nucleotide metabolism and cellular damage in myocardial ischemia, *Annu. Rev. Physiol.* **47**:727–749.

Johnstone, R. M., 1990, Ion-coupled cotransport, *Curr. Opin. Cell Biol.* **2**:735–741.

Jones, R. N., Reimer, K. A., Hill, M. L., and Jennings, R. B., 1982, Effect of hypothermia on changes in high-energy phosphate production and utilization in total ischemia, *J. Mol. Cell. Cardiol.* **14**:123–130.

Kagiyama, A., Savage, H. E., Michael, L. H., Hanson, G., Entman, M. L., and Rossen, R. D., 1989, Molecular basis of complement activation in ischemic myocardium: Identification of specific molecules of mitochondrial origin that bind human c_1q and fix complement, *Circ. Res.* **64**:607–615.

Kako, K., Kako, M., Matsuoka, T., and Mustapha, A., 1988, Depression of membrane bound Na^+–K^+-ATPase activity induced by free radicals and by ischemia of kidney, *Am. J. Physiol.* **254**:C330–C337.

Kamiike, W., Watanabe, F., Hashimoto, T., Tagawa, K., Ikeda, Y., Nakao, K., and Kawashima, Y., 1982, Changes in cellular levels of ATP and its catabolites in ischemic rat liver, *Biochem. J.* **91**:1349–1356.

Kanse, S. M., Ghatei, M. A., and Bloom, S. R., 1989, Endothelium binding sites in porcine aortic and rat lung membranes, *Eur. J. Biochem.* **182**:175–179.

Karr, T. L., Kristofferson, D., and Purich, D. L., 1980, Calcium ion induces endwise depolymerization of bovine brain microtubules, *J. Biol. Chem.* **255**:11853–11856.

Khanashvili, D., 1990, Cation antiporters, *Curr. Opin. Cell Biol.* **2**:731–734.

Kim, M-S., and Akera, T., 1987, O_2 free radicals: Cause of ischemia-reperfusion injury to cardiac Na^+/K^+-ATPase, *Am. J. Physiol.* **252**:H252–H257.

Kloner, R. A., 1988, Introduction to the role of oxygen radicals in myocardial ischemia and infarction, *Free Radic. Biol. Med.* **4**:5–7.

Komatsu, H., Koo, A., Ghadishah, E., Zeng, H., Kuhlenkamp, J. F., Inoue, M., Guth, P. H., and Kaplowitz, N., 1992, Neutrophil accumulation in ischemic reperfused rat liver: Evidence for a role for superoxide free radicals, *Am. J. Physiol.* **262**:G669–G676.

Koster, J. F., Biemond, P., and Stam, H., 1987, Lipid peroxidation and myocardial ischaemic damage: Cause or consequence?, in *Lipid Metabolism in the Normoxic and Ischaemic Heart* (H. Stam and G. J. Van der Vusse, eds.), pp. 253–260, Springer-Verlag, New York.

Kozin, F., and Cochrane, C. G., 1988, The contact activation system of plasma: Biochemistry and pathophysiology, in *Inflammation: Basic Principles and Clinical Correlates* (J. L. Gallin, I. M. Goldstein, and R. Snyderman, eds.), pp. 101–120, Raven Press, New York.

Kubes, P., Ibbotson, G., Russel, J., Wallace, J. L., and Granger, D. N., 1990, Role of platelet-activating factor in ischemia/reperfusion-induced leukocyte adherence, *Am. J. Physiol.* **259**:G300–G305.

Lawson, D. L., Mehta, J. L., and Nichols, W. W., 1990, Coronary reperfusion in dogs inhibits endothelium-dependent relaxation: Role of superoxide radicals, *Free Radic. Biol. Med.* **8**:373–380.

Lefer, A. M., Tsao, P. S., Lefer, D. J., and Ma, X-L., 1991, Role of endothelial dysfunction in the pathogenesis of reperfusion injury after myocardial ischemia, *FASEB J.* **5**:2029–2034.

Lehotsky, J., Raeymaekers, L., Missiaen, L., Wuytack, F., De Smedt, H., and Casteels, R., 1992, Stimulation of the catalytic cycle of the Ca^{2+} pump of porcine plasma-membranes by negatively charged phospholipids, *Biochem. Biophys, Acta* **1105**:118–124.

Lemasters, J. J., Stemkowski, C. J., Ji, S., and Thurman, R. J., 1983, Cell surface changes and enzyme release during hypoxia and reoxygenation in the isolated, perfused rat liver, *J. Cell, Biol.* **97**:778–786.

Lemasters, J. L., Caldwell-Kenkel, J. C., Currin, R. T., Tanaka, Y., Marzi, I., and Thurman, R. G., 1989, Endothelial cell killing and activation of Kupffer cells following reperfusion of rat liver stored in Euro-Collins solution, in *Cells of the Hepatic Sinusoid* (E. Wisse, D. L. Knook, K. Decker, eds.) Kupffer Cell Foundation, the Netherlands, pp. 277–280.

Lewis, M. S., Whatley, R. E., Cain, P., McIntyre, T. M., Prescott, S. M., and Zimmerman, G. A., 1988, Hydrogen peroxide stimulates the synthesis of platelet-activating factor by endothelium and induces endothelial cell-dependent neutrophil adhesion, *J. Clin. Invest.* **82**:2045–2055.

Liu, J., Casley, D. J., and Nayler, W. G., 1989, Ischaemia causes externalization of endothelin-1 binding sites in rate cardiac membranes, *Biochem. Biophys, Res. Commun.* **164**:1220–1225.

Liu, X., Prasad, M. R., Engelman, R. M., Jones, R. M., and Das, D. K., 1990, Role of iron on membrane phospholipid breakdown in ischemic-reperfused rat heart, *Am. J. Physiol.* **259**:H1101–H1107.

Lotersztajn, S., Pavoine, C., Deterre, P., Capeau, J., Mallat, A., LeNguyen, D., Dufour, M., Rouot, B., Bataille, D., and Pecker, F., 1992, Role of G protein β/γ subunits in the regulation of the plasma membrane Ca^{2+} pump, *J. Biol. Chem.* **267**:2375–2379.

Low, M. G., 1989, Glycosyl-phosphatidylinositol: A versatile anchor for cell surface proteins, *FASEB J.* **3**:1600–1604.

Lucchesi, B. R., 1990, Modulation of leukocyte-mediated myocardial reperfusion injury, *Annu. Rev. Physiol.* **52**:561–576.

Lucchesi, B. R., and Mullane, K. M., 1986, Leukocytes and ischemia-induced myocardial injury, *Annu. Rev. Pharmacol. Toxicol.* **26**:201–224.

Malis, C. D., and Bonventre, J. V., 1988, Susceptibility of mitochondrial membranes to calcium and reactive oxygen species: Implications for ischemic and toxic tissue damage, in *Biological Membranes: Aberrations in Membrane Structure and Function* (pp. 235–259), Alan R. Liss, New York.

Mallet, R. T., Hartman, D. A., and Bünger, R., 1990, Glucose requirement for postischemic recovery of perfused working heart, *Eur. J. Biochem.* **188**:481–493.

Malorni, W., Iosi, F., Mirabelli, F., and Bellomo, G., 1991, Cytoskeleton as a target in menadione-induced oxidative stress in cultured mammalian cells: Alterations underlying surface bleb formation, *Chem. Biol. Interact.* **80**:217–236.

Manning, A., Bernier, M., Crome, R., Little, S., and Hearse, D., 1988, Reperfusion-induced arrhythmias: A study of the role of xanthine oxidase-derived free radicals in the rat heart, *J. Mol. Cell. Cardiol.* **20**:35–45.

Mantovani, A., and Dejana, E., 1989, Cytokines as communication signals between leukocytes and endothelial cells, *Immunol. Today* **10**:370–375.

Martin, L. D., Barnes, S. D., and Wetzel, R. C., 1992, Acute hypoxia alters eicosanoid production of perfused pulmonary artery endothelial cells in culture, *Prostaglandins* **43**:371–382.

Marzi, I., Knee, J., Bühren, V., Menger, M., and Trentz, O., 1992, Reduction by superoxide dismutase of leukocyte-endothelial adherence after live transplantation, *Surgery* **111**:90–97.

Masaki, N., Kyle, M. E., and Farber, J. L., 1989, *tert*-Butylhydroperoxide kills cultured hepatocytes by peroxidizing membrane lipids, *Arch. Biochem. Biophys.* **269**:390–399.

McCall, T., and Vallance, P., 1992, Nitric oxide takes centre-stage with newly defined roles, *Trends Pharmacol. Sci.* **13**:1–5.

McCall, T., Whittle, B. J. R., Broughton-Smith, N. K., and Moncada, S., 1988, Inhibition of FMLP-induced aggregation of rabbit neutrophils by nitric oxide, *Br. J. Pharmacol.* **95**:517.

McCord, J. M., 1986, Superoxide radical: A likely link between reperfusion injury and inflammation, *Free Radic. Biol. Med.* **2**:325–345.

McCord, J. M., 1987, Oxygen-derived radicals: A link between reperfusion injury and inflammation, *Fed. Proc.* **48**:2402–2406.

McCord, J. M., 1988, Free radicals and myocardial ischemia: overview and outlook, *Free Rad. Biol. Med.* **4**:9–14.

McCormack, J. G., Boyett, M. R., Jewell, B. R., and Orchard, C. H., 1988, Ion movement and contractility in heart cells, *Trends Pharmacol. Sci.* **9**:343–345.

McLean, L. R., and Hagaman, K. A., 1992, Effect of lipid physical state on the rate of peroxidation of liposomes, *Free Radic. Biol. Med.* **12**:113–119.

McManus, L. M., Kolb, W. P., Crawford, M. H., O'Rourke, R. A., Grover, F. L., and Pinckard, R. N., 1983, Complement localization in ischemic baboon myocardium, *Lab. Invest.* **48**:436–477.

Mehta, J. L., Lawson, D. L., and Nichols, W. W., 1989, Attenuated coronary relaxation after reperfusion: Effects of superoxide dismutase and TxA2 inhibitor U 63557A, *Am. J. Physiol.* **26**:H1240–H1246.

Mehta, J. L., Nicolini, F. A., Donnelly, W. H., and Nichols, W. W., 1992, Platelet-leukocyte-endothelial interactions in coronary artery disease, *Am. J. Cardiol.* **69**:8B–13B.

Mellgren, R. L., 1987, Calcium-dependent proteases: An enzyme system active at cellular membranes?, *FASEB J.* **1**:110–115.

Mercandetti, A. J., Lane, T. A., and Colmerauer, M.E.M., 1984, Cultured human endothelial cells elaborate neutrophil chemoattractants, *J. Lab. Clin. Med.* **104**:370–380.

Mertens, S., Noll, T., Spahr, R., Krützfeldt, A., and Piper, H. M., 1990, Energetic response of coronary endothelial cells to hypoxia, *Am. J. Physiol.* **258**:H689–H694.

Mirabelli, F., Salis, A., Marinoni, V., Finardi, G., Bellomo, G., Thor, H., and Orrenius, S., 1988, Menadione-induced bleb formation in hepatocytes is associated with the oxidation of thiol groups in actin, *Arch. Biochem. Biophys.* **264**:261–269.

Mirabelli, F., Salis, A., Vairetti, M., Bellomo, G., Thor, H., and Orrenius, A., 1989, Cytoskeletal alterations in human platelets exposed to oxidative stress are mediated by oxidative and Ca^{2+}-dependent mechanisms, *Arch. Biochem. Biophys.* **270**:478–488.

Mittnacht, S., and Farber, J. L., 1981, Reversal of ischemic mitochondrial dysfunction, *J. Biol. Chem.* **256**:3199–3206.

Mittnacht, S., Sherman, S. C., and Farber, J. L., 1979, Reversal of ischemic mitochondrial dysfunction, *J. Biol. Chem.* **254**:9871–9878.

Molitoris, B. A., 1991, Ischemia-induced loss of epithelial polarity: Potential role of the actin cytoskeleton, *Am. J. Physiol.* **260**:F769–F778.

Montrucchio, G., Alloatti, G., Tetta, C., De Luca, R., Saunders, R. N., Emmanuelli, G., and Camussi, G., 1989, Release of platelet-activating factor from ischemic-reperfused rabbit heart, *Am. J. Physiol.* **256**:H1236–H1246.

Mosher, D. F., 1991, How do integrins integrate?, *Curr. Opin. Cell Biol.* **1**:394–396.

Mouton, R., Huisamen, B., and Lochner, A., 1991, The effect of ischaemia and reperfusion on sarcolemmal inositol phospholipid and cytosolic inositol phosphate metabolism in the isolated perfused rat heart, *Mol. Cell. Biochem.* **105**:127–135.

Movsesian, M. A., Nishikawa, M., and Adelstein, R. S., 1984, Phosphorylation of phospholamban by calcium-activated, phospholipid-dependent protein kinase, *J. Biol. Chem.* **259**:8029–8032.

Mullane, K., 1991, Neutrophil and endothelial changes in reperfusion injury, Trends in Cardiovascular Medicine, **1**:282–289.

Murphy, J. G., Smith, T. W., and Marsh, J. D., 1988, Mechanisms of reoxygenation-induced calcium overload in cultured chick embryo heart cells, *Am. J. Physiol.* **254**:H1133–H1141.

Muxfeldt, M., and Schaper, W., 1987, The activity of xanthine oxidase in heart and liver of rats, guinea pigs, pigs, rabbits and human beings, *Circulation* **76**:IV-198.

Nargeot, J., 1991, Mixing and matching calcium channels, *Curr. Opin. Cell Biol.* **1**:350–352.

Nayler, W., 1990, Endothelin: Isoforms, binding sites and possible implications in pathology, *Trends Pharmacol. Sci.* **11**:96–99.

Newby, A. C., and Henderson, A. H., 1990, Stimulus-secretion coupling in vascular endothelial cells, *Annu. Rev. Physiol.* **52**:661–674.

Nicotera, P., McConkey, D., Svensson, S-A., Bellomo, G., and Orrenius, S., 1988, Correlation between cytosolic Ca^{2+} concentration and cytotoxicity in hepatocytes exposed to oxidative stress, *Toxicology* **52**:55–63.

Nicotera, P., Thor, H., and Orrenius, S., 1989, Cytosolic-free Ca^{2+} and cell killing in hepatoma 1c1c7 cells exposed to chemical anoxia, *FASEB J.* **3**:59–64.

Nieminen, A-L., Gores, G. J., Wray, B. E., Tanaka, Y., Herman, B., and Lemasters, J. J., 1988, Calcium dependence of bleb formation and cell death in hepatocytes, *Cell Calcium* **9**:237–248.

Noronha-Dutra, A., Steen-Dutra, E., and Woolf, N., 1988, Epinephrine-induced cytotoxicity of rat plasma: Its effects on isolated cardiac myocytes, *Lab. Invest.* **59**:813–817.

Orrenius, S., McConkey, D. J., Bellomo, G., and Nicotera, P., 1989, Role of Ca^{2+} in toxic cell killing, *Trends Pharmacol. Sci.* **10**:281–285.

Otamiri, T., Sjödahl, R., and Tagesson, C., 1986, Lysophosphatidylcholine potentiates the increase in mucosal permeability after small-intestinal ischaemia, *Scand. J. Gastroenterol.* **21**:1131–1136.

Otani, H., Engelman, R. M., Rousou, J. A., Breyer, R. H., and Das, D. K., 1986, Enhanced prostaglandin synthesis due to phospholipid breakdown in ischemic-reperfused myocardium, *J. Mol. Cell. Cardiol.* **18**:953–961.

Parinandi, N. L., Weis, B. K., Natarajan, V., and Schmid, H. H. O., 1990, Peroxidative modification of phospholipids in myocardial membranes, *Arch. Biochem. Biophys.* **280**:45–52.

Patel, J. M., and Block, E. R., 1988, The effect of oxidant gases on membrane fluidity and function in pulmonary endothelial cells, *Free Radic, Biol. Med.* **4**:121–134.

Patel, K. D., Zimmerman, G. A., Prescott, S. M., McEver, R. P., and McIntyre, T. M., 1991, Oxygen radicals induce human endothelial cells to express GMP-140 and bind neutrophils, *J. Cell Biol.* **112**:749–759.

Paulsson, M., 1992, Basement membrane proteins: Structure, assembly and cellular interactions, *Crit. Rev. Biochem. Mol. Biol.* **27**:93–127.

Petrone, W. F., English, D. K., Wong, K., and McCord, J. M., 1980, Free radicals and inflammation: Superoxide-dependent activation of a neutrophil chemotactic factor in plasma, *Proc. Natl. Acad. Sci. USA* **77**:1159–1163.

Phelan, A. M., and Lange, D. G., 1991, Ischemia/reperfusion-induced changes in membrane fluidity characteristics of brain capillary endothelial cells and its prevention by liposomal-incorporated surperoxide dismutase, *Biochem. Biophys. Acta* **1067**:97–102.

Pinckard, R. N., O'Rourke, R. A., and Crawford, M. H., 1983, Complement localization and mediation of ischemic injury in baboon myocardium, *J. Clin. Invest.* **66**:1050–1056.

Pinckard, R. N., Olson, M. S., Giclas, P. C., Terry, R., Boyer, J. T., and O'Rourke, R., 1975, Consumption of classical complement components by heart subcellular membrane *in vitro* and in patients after acute myocardial infarction, *J. Clin. Invest.* **56**:740–750.

Przyklenk, K., 1988, Oxygen-derived free radicals and "stunned myocardium," *Free Radic. Biol. Med.* **4**:39–44.

Quaranta, V., and Jones, J. C. R., 1991, The internal affairs of an integrin, *Trends Cell. Biol.* **1**: 2–4.

Quast, U., and Cook, N. S., 1989, Moving together: K^+ channel openers and ATP-sensitive K^+ channels, *Trends Pharmacol. Sci.* **10**:431–434.

Radi, R., Turrens, J. F., and Freeman, B. A., 1991, Cytochrome *c*-catalyzed membrane lipid peroxidation by hydrogen peroxide, *Arch. Biochem. Biophys.* **288**:118–125.

Radomski, M. W., Palmer, R. M., and Moncada, S., 1991, Modulation of platelet aggregation by an L-arginine-nitric oxide pathway, *Trends Pharmacol. Sci.* **12**:87–88.

Rampe, D., and Triggle, D. J., 1990, New ligands for the L-type Ca^{2+} channels, *Trends Pharmacol. Sci.* **11**:112–115.

Rehfeldt, W., Hass, R., and Goppelt-Struebe, M., 1991, Characterization of phospholipase A_2 in monocytic cell lines, *Biochem. J.* **276**:631–636.

Repine, J. E., 1991, Oxidant-antioxidant balance: Some observations from studies of ischemia-reperfusion in isolated perfused rat hearts, *Am. J. Med.* **91**:45S–53S.

Rogers, K. R., Morris, C. J., and Blake, D. R., 1992, The cytoskeleton and its importance as a mediator of inflammation, *Ann. Rheum. Dis.* **51**:565–571.

Romaschin, A. D., Wilson, G. J., Thomas, U., Feitler, D. A., Tumiati, L., and Mickle, D.A.G., 1990, Subcellular distribution of peroxidized lipids in myocardial reperfusion injury, *Am. J. Physiol.* **259**:H116–H123.

Rossen, R. D., Michael, L. H., Kagiyama, A., Savage, H. E., Hanson, G., Reisburg, M. A., Moake, J. N., Kim, S. H., Self, D., Weakley, S., Giannini, E., and Entman, M. L., 1988, Mechanism of complement activation after coronary artery occlusion: Evidence that myocardial ischemia in dogs causes release of constituents of myocardial subcellular origin that complex with human C1q *in vivo*, *Circ. Res.* **62**:572–584.

Rossen, R. D., Swain, J. L., Michael, L. H., Weakley, S., Giannini, E., and Entman, M. L., 1985, Selective accumulation of the first component of complement and leukocytes in ischemic canine heart muscle: A possible initiator of an extra myocardial mechanism of ischemic injury, *Circ. Res.* **57:**119–130.

Ruoslahti, E., 1991, Integrins as receptors for extracellular matrix, in *Cell Biology of Extracellular Matrix*, 2nd ed. (E. D. Hay, ed.), pp. 343–349, Plenum Press, New York.

Ryan, U. S., 1989, Endothelium as a transducing surface, *J. Mol. Cell. Cardiol.* **21:**85–90.

Sanzo, M. A., Howard, S. C., Wittwer, A. J., and Cochrane, H. M., 1990, Binding of tissue plasminogen activator to human aortic endothelial cells, *Biochem. J.* **269:**475–482.

Scarpati, E. M., and Sadler, J. E., 1989, Regulation of endothelial cell coagulant properties: Modulation of tissue factor, plasminogen activator inhibitors and thrombomodulin by phorbol 12-myristate 13-acetate and tumor necrosis factor, *J. Biol. Chem.* **264:**20705–20713.

Scharschmidt, B. F., and Van Dyke, R. W., 1987, Proton transport by hepatocyte organelles and isolated membrane vesicles, *Annu. Rev. Physiol.* **49:**69–85.

Schatzmann, H. J., 1989, The calcium pump of the surface membrane and of the sarcoplasmic reticulum, *Annu. Rev. Physiol.* **51:**473–485.

Schnittler, H. J., Wilke, A., Gress, T., Suttorp, N., and Drenckhahn, D., 1990, Role of actin and myosin in the control of paracellular permeability in pig, rat and human vascular endothelium, *J. Physiol.* **431:**379–401.

Schraufstatter, I. U., Hinshaw, D. B., Hyslop, P. A., Spragg, R. G., and Cochrane, C. G., 1986, Oxidant injury of cells: DNA strand-breaks activate polyadenosine diphosphate-ribose polymerase and lead to depletion of nicotinamide adenine dinucleotide, *J. Clin. Invest.* **77:**1312–1320.

Schultz, G., Rosenthal, W., and Hescheler, J., 1990, Role of G proteins in calcium channel modulation, *Annu. Rev. Physiol.* **52:**275–292.

Selwyn, A. P., Vita, J. A., Vekstein, V. I., Yeung, A., Ryan, T., and Ganz, P., 1990, Myocardial ischemia: Pathogenic role of disturbed vasomotion and endothelial dysfunction in coronary atherosclerosis. Silent myocardial ischemia: A critical appraisal, *Adv. Cardiol.* **37:**42–52.

Serhan, C. N., Radin, A., Smolen, J. E., Korchak, H., Samuelson, B., and Weissmann, G., 1982, Leukotriene B4 is a complete secretagogue in human neutrophils: A kinetic analysis, *Biochem. Biophys. Res. Commun.* **107:**1006–1012.

Sevanian, A., and Kim, E., 1985, Phospholipase A_2-dependent release of fatty acids from peroxidized membranes, *Free Radic. Biol. Med.* **1:**263–271.

Sevanian, A., Wratten, M. L., McLeod, L. L., and Kim, E., 1988, Lipid peroxidation and phospholipase A_2 activity in liposomes composed of unsaturated phospholipids: A structural basis for enzyme activation, *Biochim. Biophys. Acta* **961:**316–327.

Shasby, D. M., Lind, S. E., Shasby, S. S., Goldsmith, J. C., and Hunninghake, G. W., 1985, Reversible oxidant-induced increases in albumin transfer across cultured endothelium: Alterations in cell shape and calcium homeostasis, *Blood* **65:**605–614.

Shasby, D. M., Shasby, S. S., Sullivan, J. M., and Peach, M. J., 1982, Role of endothelial cell cytoskeleton in control of endothelial permeability, *Circ. Res.* **51:**657–661.

Shatos, M. A., Doherty, J. M., Stump, D. C., Thompson, E. A., and Collen, D., 1990, Oxygen radicals generated during anoxia followed by reoxygenation reduce the synthesis of tissue-type plasminogen activator and plasminogen activator inhibitor-1 in human endothelial cell culture, *J. Biol. Chem.* **265:**20443–20448.

Sher, E., Biancardi, E., Passafaro, M., and Clementi, F., 1991, Physiopathology of neuronal voltage-operated calcium channels, *FASEB J.* **5:**2677–2683.

Simionescu, M., and Simionescu, N., 1986, Functions of the endothelial cell surface, *Annu. Rev. Physiol.* **48:**279–293.

Simpson, P. J., and Lucchesi, B. R., 1987, Free radicals and myocardial ischemia and reperfusion injury, *J. Lab. Clin. Med.* **110:**13–30.

Simpson, P. J., Todd, R. F., Fantone, J. C., Mickelson, J. K., Griffin, J. D., and Lucchesi, B. R.,

1988, Reduction of experimental canine myocardial reperfusion injury by a monoclonal anti-body (Anti-Mol, Anti-CD11b) that inhibits leukocyte adhesion, *J. Clin. Invest.* **81**:624–629.

Smith, C. W., Anderson, D. C., Taylor, A. A., Rossen, R. D., and Entman, M. L., 1991, Leukocyte adhesion molecules and myocardial ischemia, *Trends Cardiovasc. Med.* **1**:167–170.

Steenbergen, C., Hill, M. L., and Jennings, R. B., 1987, Cytoskeletal damage during myocardial ischemia: Changes in vinculin immunofluorescence staining during total *in vitro* ischemia in canine heart, *Circ. Res.* **60**:478–486.

Stern, D. M., Handley, D. A., and Nawroth, P. P., 1988, Endothelium and the regulation of coagulation, in "Endothelial Cell Biology in Health and disease" (N. Simionescu and M. Simionescu eds.), pp. 275–306, Plenum Press, New York, N.Y.

Sundqvist, T., 1991, Bovine aortic endothelial cells release hydrogen peroxide, *J. Cell. Physiol.* **148**:152–156.

Takeo, S., Tanonaka, K., Tazuma, Y., Fukao, N., Yoshikawa, C., Fukumoto, T., and Tanaka, T., 1988, Diltiazem and verapamil reduce the loss of adenine nucleotide metabolites from hypoxic hearts, *J. Mol. Cell. Cardiol.* **20**:443–456.

Tani, M., 1990, Mechanisms of Ca^{2+} overload in reperfused ischemic myocardium, *Annu. Rev. Physiol.* **52**:543–559.

Thor, H., Mirabelli, F., Salis, A., Cohen, G. M., Bellomo, G., and Orrenius, S., 1988, Alterations in hepatocyte cytoskeleton caused by redox cycling and alkylating quinones, *Arch. Biochem. Biophys.* **266**:397–407.

Till, G. O., Friedl, H. P., and Ward, P. A., 1991, Lung injury and complement activation: Role of neutrophils and xanthine oxidase, *Free Radic. Biol. Med.* **10**:379–386.

Tsao, P., Aoki, N., Lefer, D. J., Johnson, G., and Lefer, A. M., 1990, Time course of endothelial dysfunction and myocardial injury during myocardial ischemia and reperfusion in the cat, *Circulation* **82**:1402–1412.

Tsao, P. S., and Lefer, A. M., 1990, Time course and mechanism of endothelial dysfunction in isolated ischemic and hypoxic-perfused rat hearts, *Am. J. Physiol.* **259**:H1660–H1663.

Tse, H. Y., and Rosenthal, A. S., 1988, Lymphocytes: Interaction with macrophages, in *Inflammation: Basic Principles and Clinical Correlates* (J. L. Gallin, I. M. Goldstein, and R. Snyderman, eds.), pp. 631–649, Raven Press, New York.

Tsien, R. W., Ellinor, P. T., and Horne, W. A., 1991, Molecular diversity of voltage-dependent Ca^{2+} channels, *Trends Pharmacol. Sci.* **12**:349–355.

Turrens, J. F., and Boveris, A., 1980, Generation of superoxide anion by the NADH dehydrogenase of bovine heart mitochondria, *Biochem. J.* **191**:421–427.

Ungemach, F. R., 1985, Plasma membrane damage of hepatocytes following lipid peroxidation: Involvement of phospholipase A_2, in *Free Radicals in Liver Injury* (G. Poli, K. H. Cheeseman, M. U. Dianzani, and T. F. Slater, eds.), pp. 127–134, IRL Press, Oxford, UK.

Vandeplassche, G., Hermans, C., Thone, F., and Borgers, M., 1989, Mitochondrial hydrogen peroxide generation by NADH-oxidase activity following myocardial ischemia in the dog, *J. Mol. Cell. Cardiol.* **21**:383–392.

Vanhoutte, P. M., and Katusic, Z. S., 1988, Endothelium-derived contracting factor: Endothelin and/or superoxide anion?, *Trends Pharmacol. Sci.* **9**:229–230.

Vanhoutte, P. M., Rubanyi, G. M., Miller, V. M., and Houston, D. S., 1986, Modulation of vascular smooth muscle contraction by the endothelium, *Annu. Rev. Physiol.* **48**:307–320.

van Kuijk, F. J. G. M., Sevanian, A., Handelman, G. J., and Dratz, E. A., 1987, A new role for phospholipase A_2: Protection of membranes from lipid peroxidation damage, *Trends Biochem. Sci.* **12**:31–34.

Vedder, N. B., Winn, R. K., Rice, C. L., Chi, E. Y., Arfors, K.-E., and Harlan, J. M., 1988, A monoclonal antibody to the adherence-promoting leukocyte glycoprotein CD18 reduces organ injury and improves survival from hemorrhagic shock and resuscitation in rabbits, *J. Clin. Invest.* **81**:933–944.

Veitch, K., Hombroeckx, A., Caucheteux, D., Pouleur, H., and Hue, L., 1992, Global ischaemia induces a biphasic response of the mitochondrial respiratory chain, *Biochem. J.* **281**:709–715.

Vercellotti, G. M., Dobson, M., Schorer, A. E., and Moldow, C. F., 1988, Endothelial cell heterogeneity: Antioxidant profiles determine vulnerability to oxidant injury (42652), *Proc. Soc. Exp. Biol. Med.* **187**:181–189.

Viehman, G. E., Ma, X-L., Lefer, D. J., and Lefer, A. M., 1991, Time course of endothelial dysfunction and myocardial injury during coronary arterial occlusion, *Am. J. Physiol.* **261**:H874–H881.

Vlessis, A. A., Muller, P., Bartos, D., and Trunkey, D., 1991, Mechanism of peroxide-induced cellular injury in cultured adult cardiac myocytes, *FASEB J.* **5**:2600–2605.

Voltarelli, J. C., and Garovoy, M. R., 1988, Graft-versus-host disease, in *Inflammation: Basic Principles and Clinical Correlates* (J. L. Gallin, I. M. Goldstein, and R. Snyderman, eds.), pp. 719–731, Raven Press, New York.

Voogd, A., Sluiter, W., and Koster, J. F., 1991, Contradictory effects of superoxide dismutase after global or regional ischemia in the isolated rat heart, *Free Radic. Biol. Med.* **11**:71–75.

Wang, K.K.W., Villalobo, A., and Roufogalis, B. D., 1992, The plasma membrane calcium pump: A multiregulated transporter, *Trends Cell. Biol.* **2**:46–52.

Ward, P. A., 1991, Mechanism of endothelial cell killing by H_2O_2 or products of activated neutrophils, *Am. J. Med.* **91**:3C-89S–94S.

Watanabe, H., Kobayashi, A., Yamamoto, T., Suzuki, S., Hayashi, H., and Yamazaki, N., 1990, Alterations of human erythrocyte membrane fluidity by oxygen-derived free radicals and calcium, *Free Radic. Biol. Med.* **9**:507–514.

Watanabe, H., Kuhne, W., Spahr, R., Schwartz, P., and Piper, H. M., 1991, Macromolecule permeability of coronary and aortic endothelial monolayers under energy depletion, *Am. J. Physiol.* **260**:H1344–H1352.

Watson, B. D., Busto, R., Goldberg, W. J., Santiso, M., Yoshida, M., and Grinberg, M., 1984, Lipid peroxidation *in vivo* induced by reversible global ischemia in rat brain, *J. Neurochem.* **42**:268–274.

Wattiaux, R., and Wattiaux-De Coninck, S., 1984, Effects of ischemia on lysosomes, *Int. Rev. Exp. Pathol.* **26**:85–106.

Wattiaux-De Coninck, S., Carbonnelle, V., and Wattiaux, R., 1988, Plasma membrane of normal and ischaemic rat liver: Morphological aspects, Proceedings of the IVth Asia-Pacific Conference and Workshop on Electron Microscopy, Bangkok, pp. 787–788.

Weinberg, J. M., 1991, The cell biology of ischemic renal injury, *Kidney Int.* **39**:476–500.

Welbourn, C. R. B., Goldman, G., Paterson, I. S., Valeri, C. R., Shepro, D., and Hechtman, H. B., 1991, Pathophysiology of ischaemia reperfusion injury: Central role of the neutrophil, *Br. J. Surg.* **78**:651–655.

Werns, S. W., and Lucchesi, B. R., 1988, Leukocytes, oxygen radicals, and myocardial injury due to ischemia and reperfusion, *Free Radic. Biol. Med.* **4**:31–37.

Werns, S. W., and Lucchesi, B. R., 1990, Free radicals and ischemic tissue injury, *Trends Pharmacol. Sci.* **11**:161–166.

Wier, W. G., 1990, Cytoplasmic $[Ca^{2+}]$ in mammalian ventricle: Dynamic control by cellular processes, *Annu. Rev. Physiol.* **52**:467–485.

Williamson, J. R., and Monck, J. R., 1989, Hormone effects on cellular Ca^{2+} fluxes, *Annu. Rev. Physiol.* **51**:107–124.

Wong, C. S., Gamble, J. R., Skinner, M. P., Lucas, C. M., Berndt, M. C., and Vadas, M. A., 1991, Adhesion protein GMP140 inhibits superoxide anion release by human neutrophils, *Proc. Natl. Acad. Sci. USA* **88**:2397–2401.

Woodley, S. L., Ikenouchi, H., and Barry, W. H., 1991, Lysophosphatidylcholine increases cyto-solic calcium in ventricular myocytes by direct action on the sarcolemma, *J. Mol. Cell. Cardiol.* **23**:671–680.

Yanagisawa, M., and Masaki, T., 1989, Molecular biology and biochemistry of the endothelins, *Trends Pharmacol. Sci.* **10**:374–378.

Yeagle, P. L., 1989, Lipid regulation of cell membrane structure and function, *FASEB J.* **3**:1833–1842.

Yingst, D. R., 1988, Modulation of the Na^+/K^+ ATPase by Ca and intracellular proteins, *Annu. Rev. Physiol.* **50**:291–303.

Youker, K., Smith, C. W., Shappell, S. B., Michael, L. H., Rossen, R. D., Anderson, D. C., and Entman, M. L., 1990, Interleukin-6 contained in post-ischemic cardiac lymph induces CD18-dependent binding of activated neutrophils to isolated cardiac myocytes, *Clin. Res.* **38**:250A.

Zernig, G., 1990, Widening potential for Ca^{2+} antagonists: Non-L-type Ca^{2+} channel interaction. *Trends Pharmacol. Sci.* **11**:38–44.

Zernig, G., 1991, Clinical future for Ca^{2+} antagonists looks more promising, *Trends Pharmacol. Sci.* **12**:439–443.

Zimmerman, G. A., McIntyre, T. M., Mehra, M., and Prescott, S. M., 1990, Endothelial cell-associated platelet-activated factor: A novel mechanism for signaling intercellular adhesion, *J. Cell Biol.* **110**:529–540.

Zimmerman, G. A., Prescott, S. M., and McIntyre, T. M., 1992, Endothelial cell interactions with granulocytes: Tethering and signaling molecules, *Immunol. Today* **13**:93–99.

Zimmerman, U-J. P., and Schlaepfer, W. W., 1988, Calcium-activated neutral proteases (calpains) are carbohydrate binding proteins, *J. Biol. Chem.* **263**:11609–11612.

Index

The manufacturer's authorised representative in the EU is Springer
Nature Customer Service Centre GmbH, Europaplatz 3, 69115 Heidelberg,
Germany. If you have any concerns regarding our products, please
contact ProductSafety@springernature.com

Printed and bound by CPI Group (UK) Ltd, Croydon, CR0 4YY
23/04/2026
02095623-0002